Seismic Design of Steel Structures

Seismic Design of Steel Structures

Victor Gioncu and
Federico Mazzolani

CRC Press
Taylor & Francis Group
Boca Raton London New York

CRC Press is an imprint of the
Taylor & Francis Group, an **informa** business

CRC Press
Taylor & Francis Group
6000 Broken Sound Parkway NW, Suite 300
Boca Raton, FL 33487-2742

First issued in paperback 2017

Version Date: 20130808

ISBN 13: 978-0-415-24263-9 (hbk)
ISBN 13: 978-1-138-07537-5 (pbk)

Library of Congress Cataloging-in-Publication Data

Gioncu, Victor.
Seismic design of steel structures / Victor Gioncu, Federico Mazzolani.
pages cm
Includes bibliographical references and index.
ISBN 978-0-415-24263-9 (hbk. : alk. paper)
1. Building, Iron and steel--Earthquake effects. 2. Earthquake resistant design. 3. Steel, Structural--Earthquake effects. 4. Steel, Structural--Ductility. I. Mazzolani, Federico M. II. Title.

TA684.G57.5 2013
624.1'762--dc23 2013030664

Visit the Taylor & Francis Web site at
http://www.taylorandfrancis.com

and the CRC Press Web site at
http://www.crcpress.com

Contents

Preface

This volume completes the triad of books prepared by Victor Gioncu and myself. Their titles are

- *Ductility of Seismic Resistant Steel Structures* (2002)
- *Earthquake and Structural Engineering* (2011)
- *Seismic Design of Steel Structures* (2013)

Facing the task to write the Preface to this last volume, I am full of sorrow, crying for the recent absence of Victor. In fact, he passed away on 30 March 2013 in his hometown, Timisoara. First, we have to recognize that our academic community will miss an outstanding scientist, a refined researcher, an appreciated teacher, a clever designer, in one word, a very complete engineer. I am personally missing my best friend after many years of sincere friendship and fruitful scientific cooperation in writing papers and books as well as in participating together on research projects, conferences, seminars, and many other events.

I met Victor in Naples when he visited Italy for the first time in 1985, overcoming the difficulties to go abroad from Romania during the communist regime. Despite this, he was able to come again to Naples in December 1990, a few days before the commencement of the Romanian revolution. Since 1991, our frequentation was continuous, without any interruption, giving rise to our very effective cooperation. The Polytechnic University of Timisoara, after many years of isolation, was at last allowed to be involved in many European projects, and I was there for many scientific purposes at least once per year, with a great pleasure to join my new friends.

In 1994, Victor had a very brilliant idea: the problem of the behavior of seismic-resistant steel structures could become an important subject to be proposed at the international level by organizing an ad hoc forum for collecting the contributions of outstanding specialists working in this field. Based on this idea, we started the chain of specialty conferences that today are the well-known STESSA Conferences on the Behaviour of Steel Structures in Seismic Areas. The first conference was held in Timisoara in 1994, followed by a series of successful conferences every three years in important towns of different continents: Kyoto, 1997; Montreal, 2000; Naples, 2003; Yokohama, 2006; Philadelphia, 2009; and Santiago de Chile, 2012. With the exception of the last two conferences, Victor was always actively present.

I remember Victor's enthusiasm when we cochaired the one-week course on Design of Seismic Resistant Steel Structures held in Udine at the International Center for Mechanical Sciences in 2000 and we coedited the corresponding lecture volume.

In the same period, we were completing our first book, *Ductility of Seismic Resistant Steel Structures,* which represented a comprehensive activity of "four hands writing." We observed that steel is generally considered to be an excellent material for seismic-resistant structures. Its strength and ductility enable it to withstand large elastic and plastic deformations when compared to other constructional materials. However, it was also noticed that unexpected failures during the 1994 Northridge and 1995 Kobe earthquakes showed the need for more research into the performance of steel structures under exceptional loading conditions that are not predicted and codified in current regulations. There was an urgent need to improve practical guidance. This was, indeed, the scope of our first book, providing a critical review of recent progress in the conception, design, and construction of seismic-resistant steel structures and improving their performance during any kind of ground motion. Among our comprehensive review of the analytical techniques, particular emphasis was placed on the assessment of structural ductility of steel members as the most efficient way to prepare and protect structures against unexpected strong catastrophic events. That book was equipped with a free Windows-based computer program, DUCTROCT M, for evaluating the ductility of members and connections. This computer program was set up by Victor's daughter, Dana Pectu, who transformed our physical ideas into software. In particular, the first book concentrated on the local ductility aspects of steel members.

We intended to extend this approach to global ductility at the level of whole steel structures by preparing a second volume. Recognizing the paramount importance of the type of earthquake on the global behavior of buildings, we decided to start developing the aspects related to this issue, which is strictly related to seismology. Going more and more in detail into this general subject, it is easy to see that structural design has grown within the new multidisciplinary development fields of engineering seismology and earthquake engineering, the former as a branch of seismology with the purpose of transferring new seismological knowledge to structural engineers for practical use, the latter with the task of providing an analysis of structures under seismic actions. Both must be integrated in order to produce important progress in structural seismic design. Proceeding in the book preparation, we discovered that one book was not enough to treat the relationship between engineering seismology and earthquake engineering. To mix it with the design of seismic-resistant steel structures was neither appropriate nor homogeneous.

So we devoted the second book to seismic design in general, bringing together engineering seismology and earthquake engineering under the title *Earthquake and Structural Engineering,* which was valid for all kinds of structural materials. It focused on the seismological aspects of design by analyzing various types of earthquakes and how they differently affect structures. Understanding the distinction between these earthquake types and their different impact on buildings can show the difference between whether the building stands or fails, or at least it allows engineers to know how much it costs to be repaired.

This book, titled *Seismic Design of Steel Structures,* represents the natural conclusion of twenty years of common activity between Victor and myself, which was finalized to give a substantial contribution to seismic design in general and in particular to seismic-resistant steel structures. The predicted passage from local to global behavior was developed by taking into account the output of the first and second volumes. The whole material has been divided into six chapters.

Chapter 1, "Failure of a Myth," starts from the consideration that steel is usually considered an ideal material for seismic-resistant applications, but some accidents in the last decades of the twentieth century undermined this reputation. These cases are listed in the first chapter, and the main reasons for the anomalous behavior of steel structures, due to exceptional situations connected both to a lack of knowledge and/or design errors, are clearly identified and explained case by case. The main recent investigations come from the lessons learned from these tragic events.

Chapter 2, "Steel against Earthquakes," shows how to use steel to resist seismic actions together with the reasons for the excellent behavior of seismic-resistant steel structures in many applications around the world, demonstrating that good design principles generally lead to successful results.

Chapter 3, "Challenges in Seismic Design," is devoted to identifying the gap in knowledge on the effects of different types of earthquakes on structural behavior and also considers some important lessons to be learned. It provides a suitable relationship with the main output of the second volume, *Earthquake and Structural Engineering.* A distinction is given between strong and low-to-moderate seismic regions also from the point of view of the methodological approach.

Chapter 4, "New Generation of Steel Structures," represents the heart of the book, illustrating the most advanced seismic-resistant structural systems based on the use of steel. Starting from the systems that improve existing solutions, that is, EBF (eccentric braces), RBS (dog-bone), and BRB (buckling restrained braces), new bracing systems have been analyzed based on the use of steel and aluminum panels in the form of full-bay, partial-bay, and bracing types. These innovative systems are also applied to the seismic upgrading of existing RC buildings. Finally, new connecting systems for beam-to-column nodes, based on posttensioned energy-dissipating (PTED) connections, are also illustrated.

Chapter 5, "Advances in Steel Beam Ductility," shows the results of the extensive use of the DUCTROCT M computer program, whose features were presented in our first volume, *Ductility of Seismic Resistant Steel Structures*. It is used to evaluate the seismic available ductility, both monotonic and cyclic, for different types of earthquakes (near-field and far-field). The effects of earthquake type on the behavior of steel beams are presented in detail, giving possible explanations for the bad performance of beam-to-column connections during both the 1994 Northridge and the 1995 Kobe earthquakes.

Chapter 6, "Fire after Earthquake," closes the volume with an analysis of a particular loading condition, that is, when a structure damaged, maybe seriously, by an important earthquake is subjected to a sudden fire. This situation has been very common in many earthquakes (Lisbon, 1755; San Francisco, 1906; Kobe, 1995; Tohoku, 2011), sometimes producing more damage and fatalities than the earthquake alone. It is a very new subject, and it opens the way to a new research field.

As mentioned at the beginning, this book completes the series of three books, which together represent an essential and comprehensive guide for structural engineers concerned with the design of more economical but safer steel structures in seismic zones. The three books serve as an authoritative reference for academics and postgraduate students in this important area of structural design.

At the end of this effort, I am very sad that Victor cannot participate in this final conclusion, which basically consists of the simple satisfaction for an author to just hold his new book in his hands. However, I can testify that Victor actively participated in the preparation of this volume until the end, giving a valuable contribution as usual, despite his difficult health situation in a continuous home–hospital alternation during these last two painful years.

Good-bye, Victor, we will continue to be in touch ... in spirit.

Federico M. Mazzolani
April 2013

Chapter 1

Failure of a myth

1.1 THE MYTH OF STEEL AS A PERFECT MATERIAL FOR SEISMIC-RESISTANT STRUCTURES

1.1.1 Why steel is considered a perfect material

Before starting modern engineering, the world of architecture was dominated by traditional materials, such as timber, clay brick, and stone. All buildings were made by using these materials, which are characterized by high vulnerability in earthquake-prone areas. Therefore, the beginning of using steel as a new structural material gave one hope that it would be the ideal material for buildings in seismic areas. The good behavior of steel structures during some earthquakes consolidated the conviction that a perfect material for solving these problems had been discovered.

Indeed, in modern practice, it is generally accepted that steel is an excellent material for structural applications. This myth is based on the following issues (Gioncu, 2006):

1. Its performance in terms of native properties: very good strength and ductility. In normal conditions, it is capable of withstanding substantial inelastic deformation without losing strength, in contrast to other construction materials, such as concrete or masonry.
2. The good performance of steel structures during many earthquakes during the last century when the principles of seismic design were developed. According to the modern approach, it is also possible to reduce the seismic design forces and control the structural damage in the case of strong earthquakes. This good performance of steel structures gives confidence in the design methods developed, starting from the middle of the last century.

1.1.2 Seismic events justifying this myth

1.1.2.1 1906 San Francisco earthquake

The first very important seismic event produced at the beginning of the 20th century was the 1906 San Francisco earthquake (M 7.8) (Figure 1.1) (Gioncu and Mazzolani, 2011). The earthquake damaged large portions of the city, especially the oldest buildings made of masonry and timber. The importance of this earthquake for seismic design was the establishment of the first relation between a fault and an earthquake, being the starting of the scientific conception about the nature of earthquakes. Despite the wide damage in many zones of the city, the steel buildings behaved very well, this good behavior being attributed to the steel structure characteristics (Figure 1.2). The USGS 1907 report describes some of the minor damage to

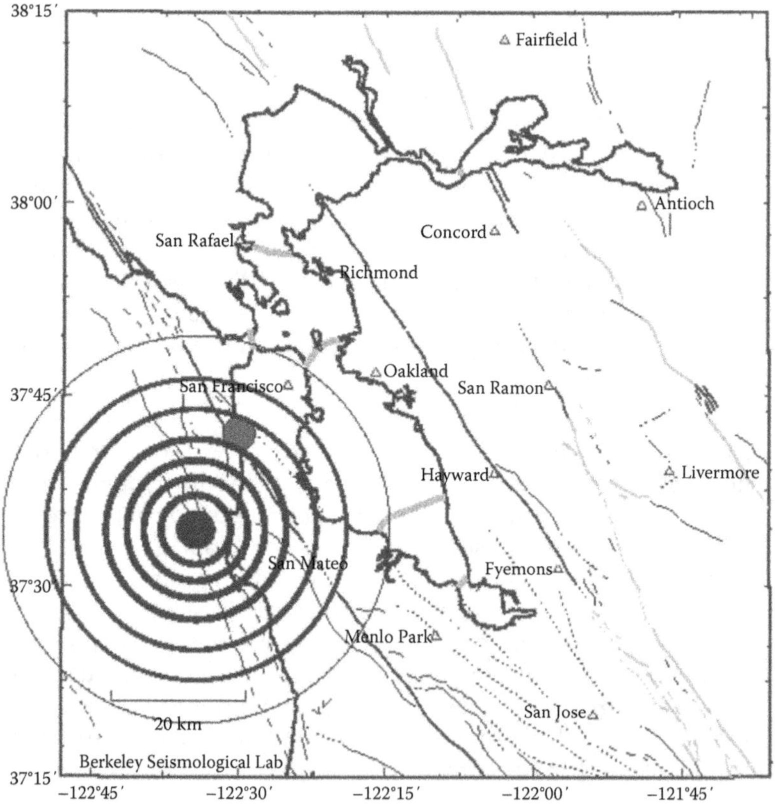

Figure 1.1 Location of the April 18, 1906 San Francisco earthquake. (After Gioncu, V., Mazzolani, F.M. 2011: *Earthquake Engineering for Structural Design*. Spon Press, London.)

Figure 1.2 Good behavior of steel structures during the 1906 San Francisco earthquake: (a) Old City Hall; (b) library building of the Stanford University; (c),(d) general views. (USGS (nd): 1906 San Francisco Earthquake Photos, http://earthquake.usgs.gov/regional/nca/1906/18april/photos.php.)

steel structures even if, in some cases, the collapse of the exterior masonry claddings made the steel structures vulnerable to fire. Today, based on the actual knowledge, this satisfactory behavior can be explained by the fact that, at that period, the connections of steel structures were made by using rivets, which functioned well during earthquakes due to the absence of important stress concentrations. Some damage to riveted connections was attributed to constructional defects.

The destruction resulting from this earthquake opened the eyes of people to the hazards of living in an earthquake-prone zone. At the same time, the behavior of steel buildings gave one the conviction that steel was a reliable material for structures in seismic areas. This seismic event can be also considered as the place where this myth was born.

1.1.2.2 1923 Kanto earthquake

The second event contributing to the trust in steel structures was the 1923 Japan Kanto earthquake in the area around Sagami Bay (M 8.3) (Figure 1.3) (Gioncu and Mazzolani, 2011). The origin of this earthquake was the

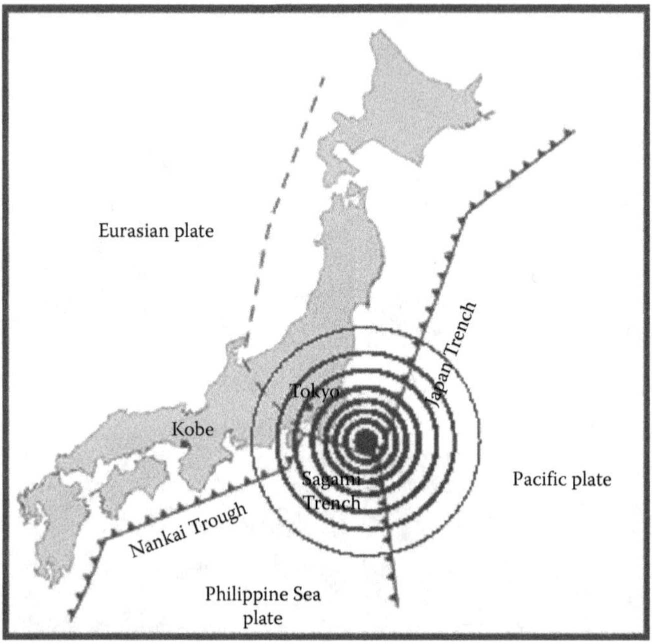

Figure 1.3 The 1923 Kanto earthquakes. (USGS (nd): Historic earthquakes. Kanto, Japan, 1923 September 01, M 7.9, http://www.earthquake.usgs.gov/earthquakes/world/events/1923_09_01.php.)

presence of a fault system associated with the four tectonic plates near the Japan isles. The large movements of the ground in the Tokyo–Yokohama area had catastrophic consequences; more than half of the building stock in these two cities was severely damaged or ruined completely.

The use of steel in Japan at the time of the Kanto earthquake was relatively new, the first steel structures being erected only 5 years before the earthquake: four large buildings were completed and two almost. The structures of these buildings suffered little to no damage, but serious damage occurred to the infill masonry of facades, leaving the structures unprotected from fire. Even if the information at that time was not detailed, the quite good performance of these steel structures is visible in some historical photos among the overall destruction of the city (Figure 1.4).

1.1.2.3 1957 Mexico City earthquake

For structural engineering, the supreme demonstration of the good behavior of steel structures in the seismic areas was the excellent performance of the *Torre Latino Americana* building in Mexico City (Figure 1.5a) during

(a)

(b)

Figure 1.4 Steel structure surrounded by ruins in Tokyo. (USGS (nd).)

the 1957 earthquake (M 7.9), when a lot of surrounding buildings collapsed due to the bad soil conditions (see Section 1.2.1). This tower was built in 1956 and was the first Mexican skyscraper and, until 1984, the tallest building in Mexico City. The notoriety of this building was recognized by the American Institute for Steel Construction, conferring it the merit for *the tallest building ever exposed to huge seismic force*. Indeed, it was really the first skyscraper built in a very active seismic zone (Wikipedia, nd).

The structural scheme is a symmetrical steel moment-resisting frame (44 floors) with heavy built-up H columns, wide flange composite beams, and rigid riveted connections (Figure 1.5b). A 13 m deep basement was constructed to reduce the net bearing pressure on the piles raft. The foundation

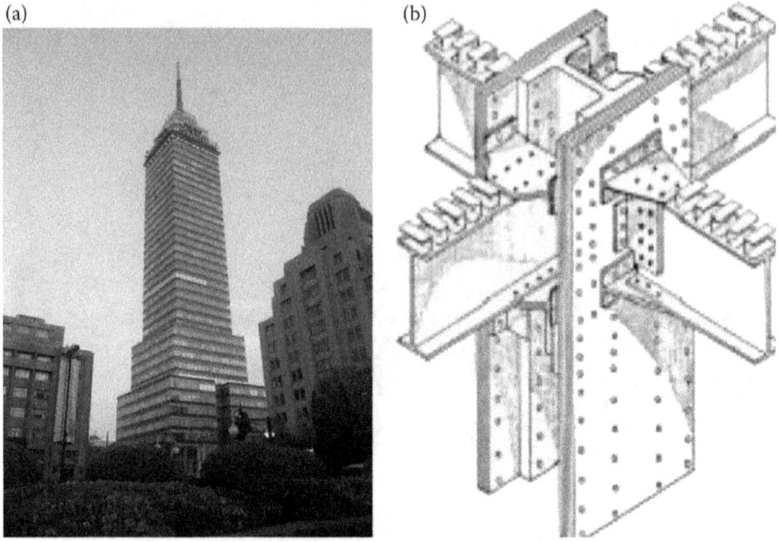

Figure 1.5 Latino Americana Tower, Mexico City: (a) view of building; (b) structural system.

is a rigid mat supported by 361 pylons, which were driven to a depth of 34 m below the street level to reach a 5 m thick stratum of sand followed by firm to stiff clays and sands (Figure 1.6) (Tomlison and Boorman, 2001).

The good behavior of this building is mainly because it was well designed and well erected. The foundation system increases the fundamental period of the building, in such a way to be larger than the site ground motion periods.

1.1.2.4 Is this myth justified?

Despite the not accurate information about the performance of steel structures during these earthquakes, the myth of their very good behavior was born. Why? The good or bad behavior of a structure during a strong earthquake is judged in function of its *ductility*, considered as the ability of the structure to undergo large plastic deformations without losing the strength. Before the 1960s, the notion of ductility was used only for characterizing the material behavior. After Baker's studies in plastic design and Housner's research works in earthquake problems, this concept was extended at the level of structure.

Therefore, considering the good material ductility and the little information about the steel structure behavior under earthquakes, the myth of steel as an excellent material for seismic-resistant structures arose.

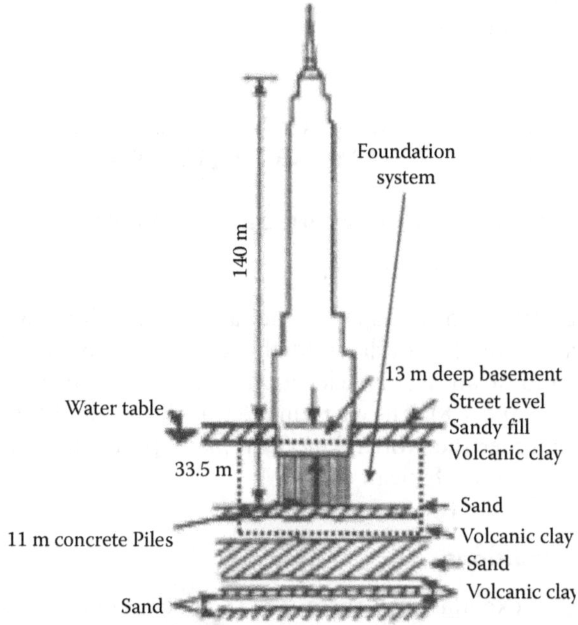

Figure 1.6 Foundation arrangement of the Latino Americana Tower. (After Tomlison, M.J., Boorman, R. 2001: *Foundation Design and Construction*. Longman Scientific Technical.)

1.1.3 Seismic decade 1985–1995: Failure of a myth

Contrary to the above examples of good behavior of steel structures under earthquakes, the seismic decade 1985–1995, with the severe earthquakes of Mexico City (1985), Northridge (1994), and Kobe (1995), has seriously compromised this myth of steel as being the most suitable material for seismic-resistant structures. It was a providential signal that, in the same place, Mexico City, where the case of the Torre Latino Americana building was assumed as an example of good performance of steel structures, the first overall collapse (and fortunately a single case) of a steel structure occurred in a condition of very bad soil: the Pino Suarez building.

The bad performance of the joints in steel structures during the Northridge and Kobe earthquakes, having the same characteristics of damage, showed that there were some general mistakes in the design concepts. The brittle fracture due to pulse earthquake type, in the case of the Ashiyahama buildings in Kobe, was a clear example of this. The general remark that the damage resulted even when both the design and detailing were done in perfect accordance with the design philosophy and the code provisions amplifies the challenge addressed to structural engineers for future activities (Gioncu and Mazzolani, 2002, 2011).

1.2 BEHAVIOR OF STEEL STRUCTURES DURING AMERICAN AND ASIAN EARTHQUAKES

1.2.1 1985 Mexico City earthquake (Mexico)

1.2.1.1 Earthquake characteristics

On September 19, 1985, a major earthquake of magnitude 8.1 occurred, with an epicenter in the south of Michoacan state, about 400 km from Mexico City and at an average elevation of 2236 m above the mean sea level (Figures 1.7 and 1.8). The earthquake was the result of the subduction of Cocos plate under the North American plate, producing a lot of severe earthquakes along the Mexican coast.

1.2.1.2 Soil conditions

This earthquake explains why there was so much damage at such a long distance from the epicenter, when the normal attenuation law suggests that a low level of ground motion would be expected at such a distance. The

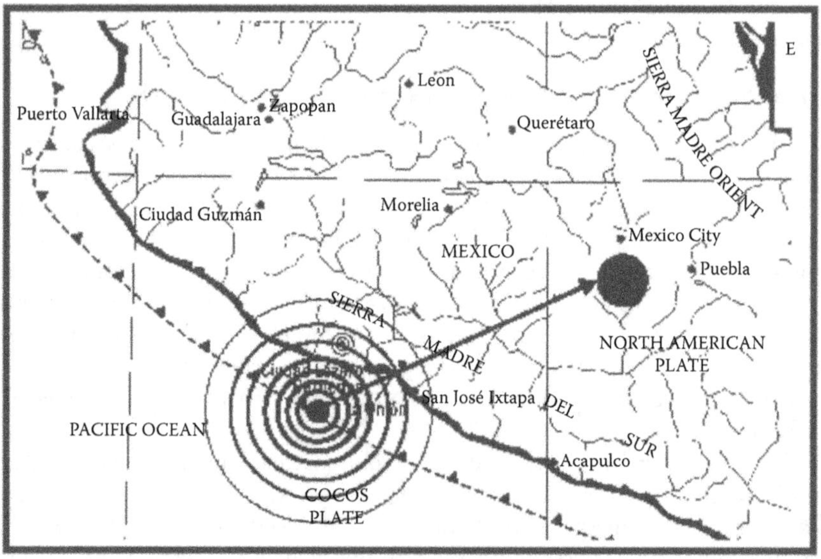

Figure 1.7 Mexico City earthquake of 1985. (After Encyclopedia Britannica (nd): Mexico City earthquake of 1985 http://www.britannica.com/EBchecked/topic/1421132/Mexico-city-earthquake-of-1985.)

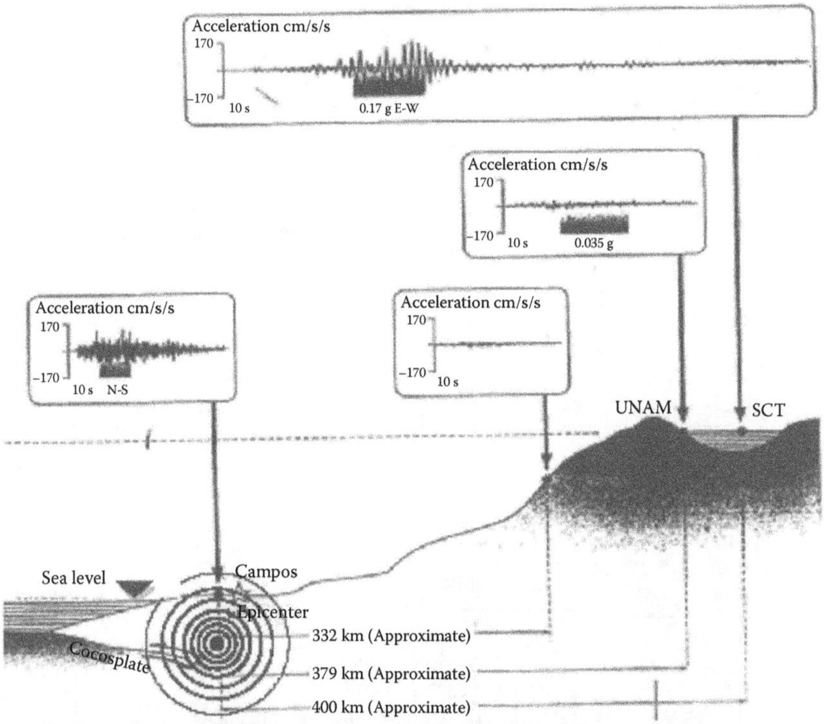

Figure 1.8 Recorded accelerations in the epicenter and in Mexico City. (After Davidovici, V. 1999: *La Construction en Zone Seismique*. Le Moniteur, Paris.)

first answer results from the fact that the earthquake source is located at a great depth, and therefore, the area of influence was very large. But the main explanation lies in the bad soil conditions in a part of Mexico City. In fact, the main part of the Mexico City Metropolitan area was developed on Lake Texcoco, used by the pre-Columbian civilizations until the 16th century. The valley of Mexico City is a closed basin, which was filled by water and wind-laden transported materials during the ancient period. Owing to the disintegration of rocks, the surrounding hills were gradually eroded and the finest elements were transported by water into the basin (Gioncu and Mazzolani, 2002). The lake area consists of a layer of alluvial soil of 20–30 m in depth.

Figure 1.8 shows the increasing duration due to these soil conditions; it was about 30 s in the epicenter and hard zones, while over 180 s in the lake area, producing a lot of important cycles.

Concerning the accelerations, one can see that the attenuation is normally till the base of the mountains surrounding the Mexico Valley. By contrast, in the center of the valley, the amplification was very high (Figure 1.9a) with the increasing of natural time (s) periods of ground motions. The

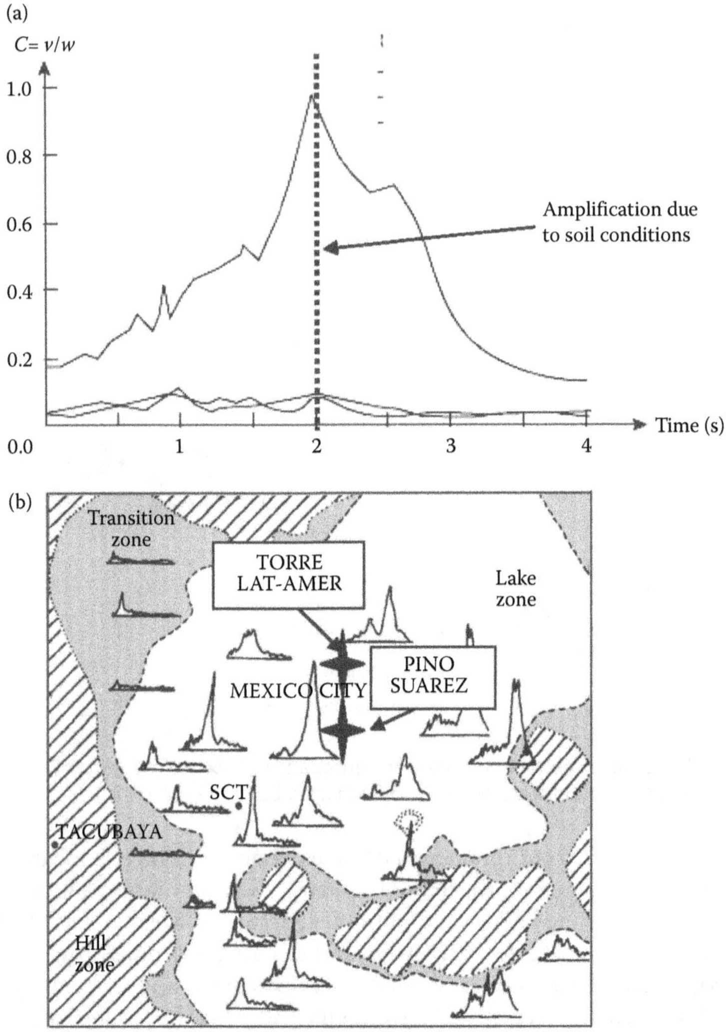

Figure 1.9 Accelerations amplification: (a) response spectra. (After Scawthorn, C. 2003: *Earthquake Engineering Handbook* (eds. W.F. Chen and C. Scawthorn), CRC Press.) (b) Distribution of spectra for Mexico City Valley. (From Gioncu, V., Mazzolani, F.M. 2002: *Ductility of Seismic Resistant Steel Structures*. Spon Press, London.)

acceleration amplification and the increase in dominant periods are presented in Figure 1.9b for the tree zones: hill, transition, and lake. The periods vary between 0.5 s for the hill zone up to 5.2 s for lake areas and the acceleration amplification reaches 12.7 times in some zones. Hence, this earthquake was one of the most devastating events in the context of structures with long fundamental periods, such as the multistoried steel moment-resisting frames.

1.2.1.3 Damage of some steel structures

This earthquake was surely the first event in which a significant number of steel buildings, including the modern ones, were subjected to the severe test of an exceptional earthquake. More than 100 steel structures were tested and the results were unfavorable, mainly for buildings having 6–20 stories (Osteraas and Krawinkler, 1989; Gioncu and Mazzolani, 2002). The main cause of this unexpected behavior was the double resonance phenomena, seismic wave-soil, and soil structure, giving rise to a required ductility exceeding the normal demand. The influence of higher vibration modes, which were more active than the first one, caused damage on the upper stories and also collisions between adjacent buildings.

The most frequently used steel structures (41%) were the moment-resistant frames. This system consisted of box columns (two channels and cover plates or four welded plates), H-section column and beams (either hot-rolled or welded), or truss girders built up with angle sections. Frames having less than 10 stories generally behaved well. For multistoried structures having a long fundamental period, the damage was concentrated in the welded columns or in truss girders, producing a buckling of the compression members. The behavior of buildings with truss girders for floor structures indicated that this was not a good solution when the site conditions require high ductility for beams.

The second type was the steel dual system, consisting of bracings located in some bays. One can observe that, in many cases, the position of bracing systems was eccentric (see the Pino Suarez buildings), introducing important torsional effects in the structure.

The third type was the mixed dual system with steel frames and concrete shear walls. In this system, truss girders were also used for floor beams and the damage was primarily concentrated in these elements.

The one-bay moment-resisting frame of the 11-storied 77 Amsterdam Street building was a special case, where a serious damage in welded connections was observed, for the first time, in the first four stories of the transversal frames (Gioncu and Mazzolani, 2002).

As already mentioned, the Torre Latino Americana building behaved very well during this earthquake.

1.2.1.4 Collapse of Pino Suarez buildings

The Pino Suarez complex, shown in Figure 1.10, is composed of five high-rise steel buildings: two identical 15-story structures (A and E buildings), and three identical 22-story structures (B, C, and D buildings). The

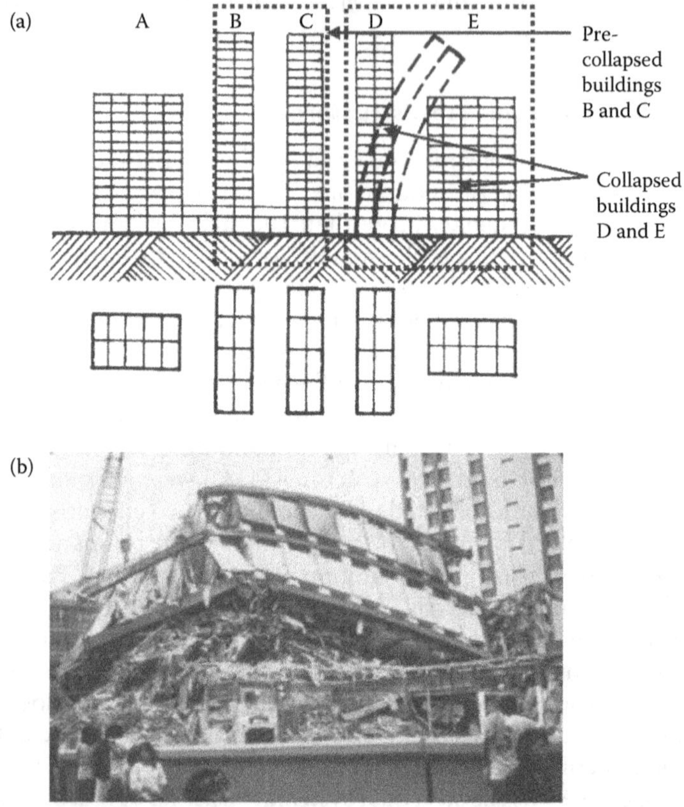

Figure 1.10 Pino Suarez complex: (a) collapsed and precollapsed buildings. (Gioncu, V., Mazzolani, F.M. 2002: *Ductility of Seismic Resistant Steel Structures*. Spon Press, London; after Ger, J.F., Cheng, F.Y., Lu, L.W. 1993: *Journal of Structural Engineering*, 119(3), 852–870. (b) Collapsed buildings D and E. (FEMA 451B 2007: NEHRP recommended provisions for new buildings and other structures. Training and instructional materials, June 2007.)

buildings are standing on a two-level reinforced concrete subway station, which acts as a rigid foundation common for all the five buildings.

During the earthquake, building D collapsed on building E and buildings B and C were seriously damaged, being in the stage of precollapse. This situation represents a very rare occasion giving one the opportunity to study a building just before the collapse, at the ultimate state level.

The structural system is of dual type, composed by moment-resisting frames and braced frames around the service core with two transversal X-braced bays and one V-braced frame in the exterior longitudinal frame (Figure 1.11a). The braces were placed in eccentric positions, introducing important torsional effects. The main conceptual defect consisted in the lack of any torsional

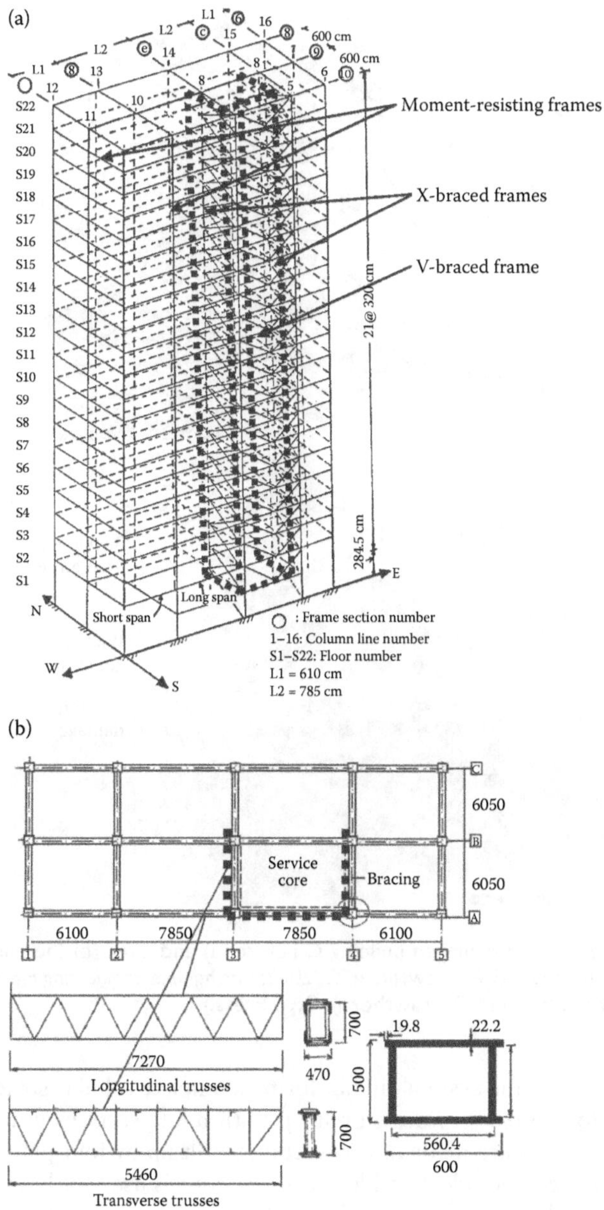

Figure 1.11 Structural system of Pino Suarez building: (a) Structure configuration. (After Ger, J.F., Cheng, F.Y., Lu, L.W. 1993: *Journal of Structural Engineering*, 119(3), 852–870.) (b) Structural details. (Gioncu, V., Mazzolani, F.M. 2002: *Ductility of Seismic Resistant Steel Structures*. Spon Press, London; after Ger, J.F., Cheng, F.Y., Lu, L.W. 1993: *Journal of Structural Engineering*, 119(3), 852–870.)

(a)

(b)

Figure 1.12 View of the damaged building C before (a) and after (b) the elimination of claddings. (After Scawthorn, C. 2003: *Earthquake Engineering Handbook* (eds. W.F. Chen and C. Scawthorn), CRC Press.)

rigidity, as the bracing system did not form a complete tubular structure. The beams are double truss girders, built up with angle sections and plate elements. All columns are made of box sections built up by four plates continuously fillet welded. Double T-welded cross-sections is used for braces.

Figure 1.12 shows the damaged building C standing in the precollapse stage (Figure 1.12a) and when all claddings and partition walls were removed, as well as some top stories that were demolished (Figure 1.12b). Examining the large number of localized failures, it can be observed that they are mainly concentrated around the bracing system at the fourth story. Local buckling of the truss girder elements (Figure 1.13a), as well as local

(a)

(b)

(c)

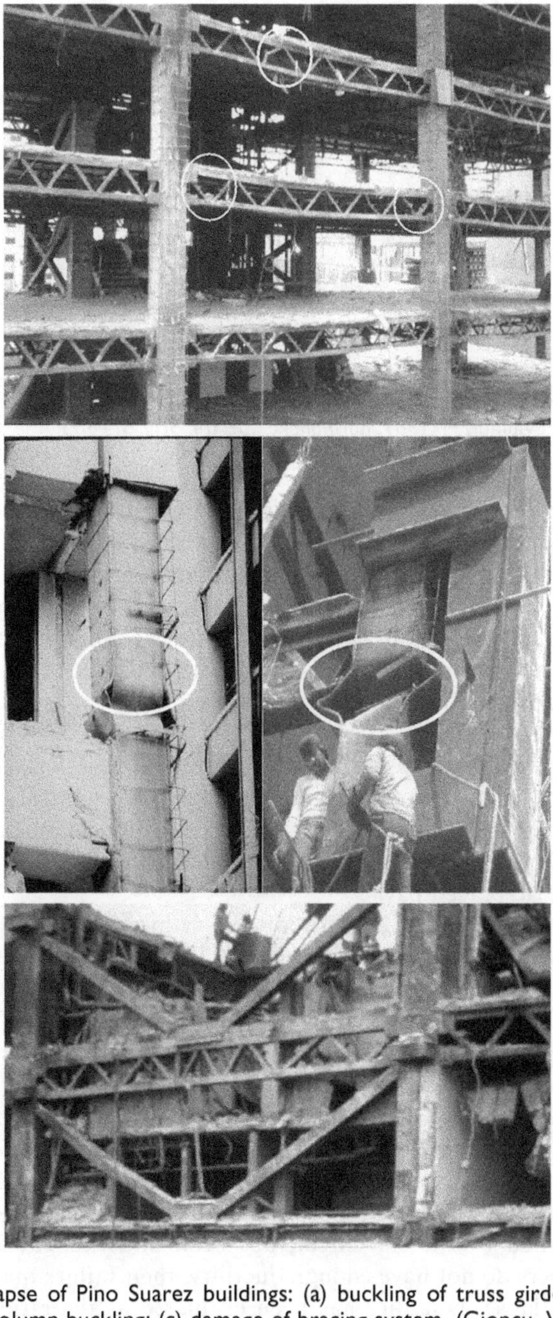

Figure 1.13 Collapse of Pino Suarez buildings: (a) buckling of truss girder members; (b) column buckling; (c) damage of bracing system. (Gioncu, V., Mazzolani, F.M. 2002: *Ductility of Seismic Resistant Steel Structures*. Spon Press, London; USGS, nd.)

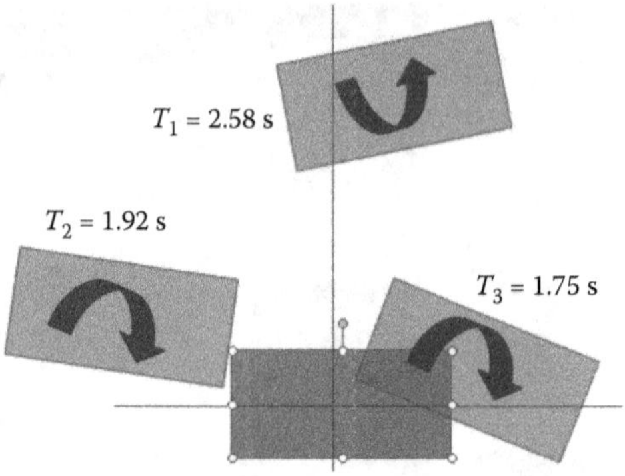

Figure 1.14 Fundamental periods and modal displacements of roof. (After Osteraas, J., Krawinkler, H. 1989: *Earthquake Spectra*, 5(1), 51–88.)

buckling produced by a fracture of welding (Figure 1.13b) and damage of bracing elements (Figure 1.13c), was observed.

The most likely collapse scenario for building D is that the exterior box columns at the braced zone buckled locally and lost most of their load-carrying capacity. The X bracings and the story columns were overloaded, producing the development of a story mechanism. So, building D collapsed in an overturning mode over building E (Osteraas and Krawinkler, 1989).

The elastic analysis performed by Osteraas and Krawinkler (1989) (Figure 1.14) has demonstrated that the first three natural periods of the structure coincide with the maximum acceleration amplification (Figure 1.9a). The modal displacements at the top level have also shown the effects of the eccentric-braced bay.

A very well-conducted analysis of this failure has been performed by Cheng et al. (1992), Ger and Cheng (1992), and Ger et al. (1993). These research works contain both theoretical and experimental tests. The inelastic analysis has shown some very important aspects.

First of all, the location of local buckling in columns is presented in Figure 1.15, showing that column failures are concentrated near the braced bays. The first column to buckle locally is situated at the 4th floor, in good accordance with the observed failure. Owing to the redistribution effect, local buckling also occurred in the adjacent columns.

Second, the required ductility, considering the actual on-site accelerations, amplified by the soil effect, has been undermined, showing that, when members do not have enough ductility, their failure may produce the structure collapse. Ductility has been evaluated as the ratio between the rotation at the maximum end and the one at the first yielding moment. The required ductility for columns is presented in Figure 1.16a, the maximum

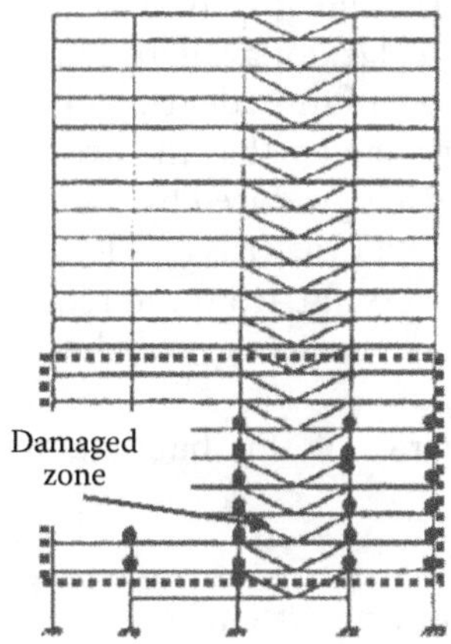

Figure 1.15 Column local buckling locations. (After Ger, J.F., Cheng, F.Y., Lu, L.W. 1993: *Journal of Structural Engineering*, 119(3), 852–870.)

being situated at the 9th story with a value of about 7. For longitudinal truss girders (Figure 1.16b), the maximum required ductility is situated at the 7th story, being about 7.5. Compared to the value 4 of the design ductility for both columns and girders, the determined required ductility is two times the design one. It means that the columns and girders situated in the

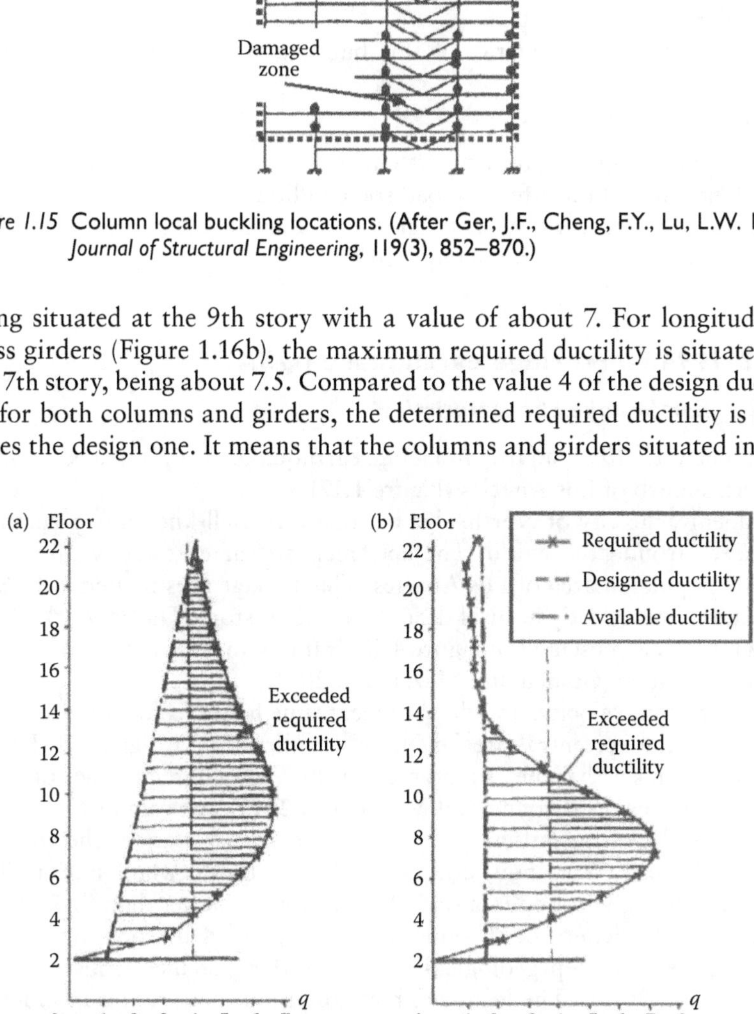

Figure 1.16 Required, designed, and available ductilities: (a) for columns; (b) for long direction truss girders. (Gioncu, V., Mazzolani, F.M. 2002: *Ductility of Seismic Resistant Steel Structures*. Spon Press, London.)

weak zone, between the 4th and 10th floors, may fail and lose their load-carrying capacity due to the lack of available ductility.

The experimental tests on a full-scale truss girder and box column specimens of this building have shown that the available ductility is 1.715 for truss girders, the failure being produced by the local buckling of top chord members, clearly showing that the truss girders have insufficient ductility. The available ductility is very far from the required and designed ductility. The column failures are due to the very high level of axial forces and reduced bending moments.

In conclusion, the collapse of this building was due to the following reasons:

1. Asymmetry of the bracing system
2. Ignorance of the effects of bad soil conditions
3. Use of a structural system that does not assure the available ductility in concordance with the required ductility

1.2.2 1994 Northridge earthquake (USA)

1.2.2.1 Earthquake characteristics

On January 17, 1994, a fairly moderate earthquake (M 6.7) struck a northwestern suburb of Los Angeles (Figure 1.17).

Suddenly, the city of Northridge became very well known to structural engineers around the world. The epicenter is situated directly under the densely populated area of Los Angeles. The damage is estimated at US$30 billion, the most costly natural disaster in U.S. history. The network of the known faults is presented in Figure 1.17, framing in the category of crustal blind trust faults (Gioncu and Mazzolani, 2011).

Acceleration, velocity, and displacement time histories for some stations situated in the epicentral area in the north direction are given in Figure 1.18. One can see that the ground motions have the characteristics of near-source earthquakes (Gioncu and Mazzolani, 2011), with important velocity and displacement pulses: one pulse in the Sylmar station (the shortest distance from the epicenter) and two pulses in the Newhall station. The anomalously high ground motion in Santa Monica, larger than those experienced in the vicinity of the epicenter, is explained by focusing seismic waves caused by deep geological structures that act like acoustic lenses (Davis et al., 2000). The localized high amplitudes of ground motions at the Tarzana station (acceleration of 1.82 g and velocity about 400 cm/s, Figure 1.19) can be explained by the relief effects, the station being situated on the top of a hill. The vertical ground motion components were very large; the ratios of vertical to horizontal components are shown in Figure

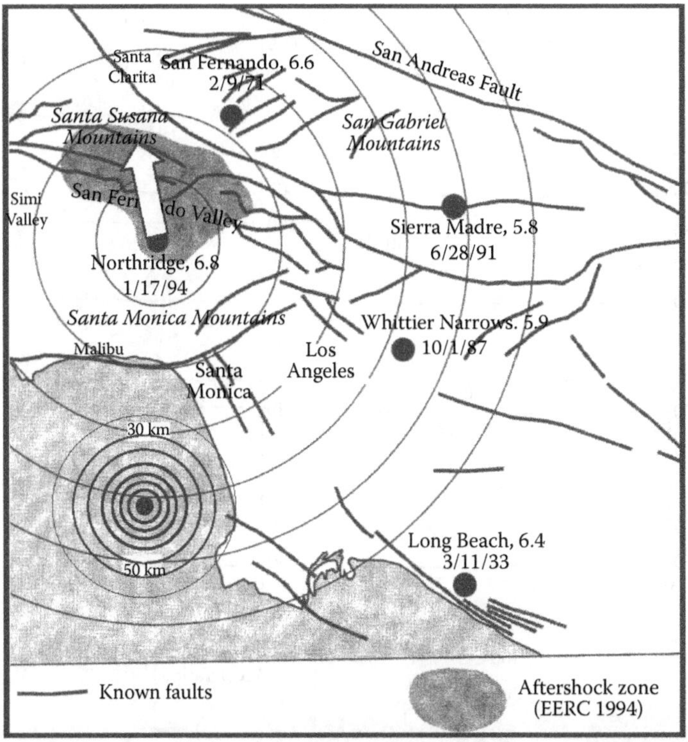

Figure 1.17 Known faults, location of the epicenter of the Northridge earthquake, and directivity. (After USGS 1996: Surface expression of coactive slip on a blind fault during Northridge earthquake sequences, California. Open report 96-698.)

1.20. The increase of this ratio at small distances from the epicenter is dramatic (Gioncu and Mazzolani, 2002).

Therefore, the characteristics of the Northridge earthquake possess the typical properties of near-source ground motions: strong pulse velocity, reduced number of cycles, and important vertical components. The phenomena, like the focus of seismic waves or the effects of relief, are always present in these areas, giving great uncertainties in determining the design values. Owing to the large damage produced during the Northridge earthquake, many specialists consider that it was anomalous, in comparison with the ordinary earthquakes. However, in the frame of the new knowledge about the effects of diverse earthquake types (Gioncu and Mazzolani, 2011), the Northridge earthquake is a normal one, having all the near-source characteristics. The reason it was considered as anomalous was because the specialists were surprised that an important earthquake, occurring under an important city, produced new unexpected effects on structures.

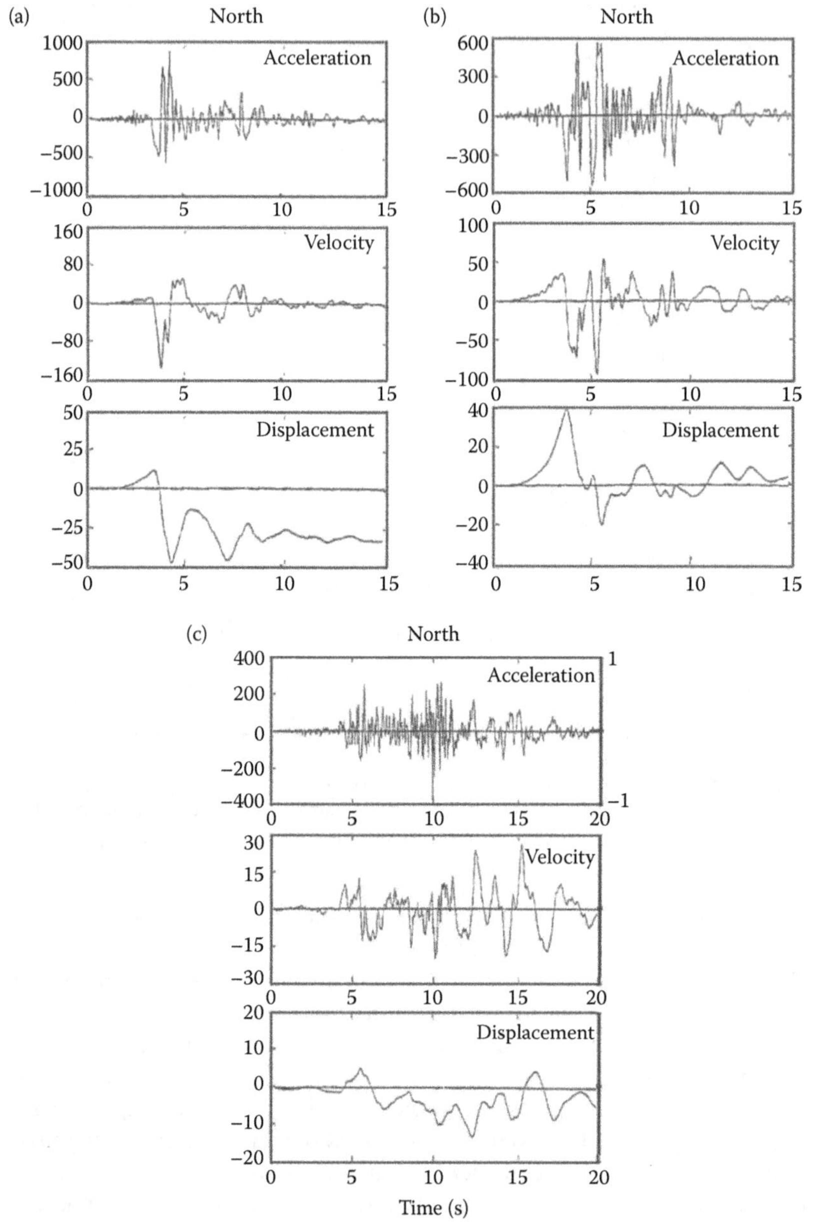

Figure 1.18 Northridge motion time–history and distance from epicenter: (a) Sylmar, 16 km north; (b) Newhall, 20 km north; (c) Santa Monica, 30 km south. (Gioncu, V., Mazzolani, F.M. 2002: *Ductility of Seismic Resistant Steel Structures*. Spon Press, London; after Iwan, W.D. 1995: Drift demand spectra for selected Northridge sites. SAC Report 95-05.)

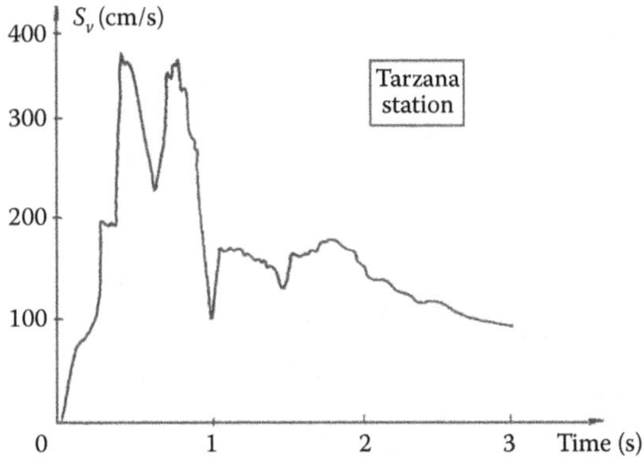

Figure 1.19 Velocity spectrum for Tarzana station. (From Gioncu, V., Mazzolani, F.M. 2002: *Ductility of Seismic Resistant Steel Structures.* Spon Press, London.)

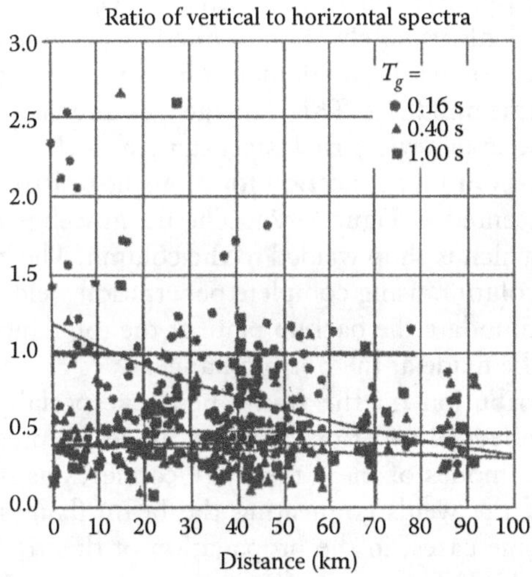

Figure 1.20 Ratio of vertical to horizontal spectra. (Gioncu, V., Mazzolani, F.M. 2002: *Ductility of Seismic Resistant Steel Structures.* Spon Press, London; after Hudson, R.L., Skyers, B.D., Lew, M. 1996: Vertical strong motion characteristics of the Northridge earthquake. In *11th World Conference on Earthquake Engineering*, Acapulco, June 23–28, No. 728.)

1.2.2.2 Damage of connections

The biggest issue emerging from the Northridge earthquake is the surprisingly poor performance of many steel buildings. Since it was reported that none of the steel structures collapsed, in comparison with many reinforced concrete buildings that collapsed, the first very optimist announcement

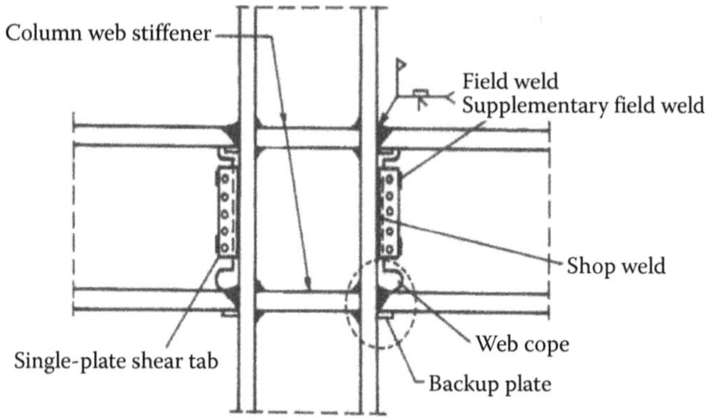

Figure 1.21 Typical beam–column connection. (From Gioncu, V., Mazzolani, F.M. 2002: *Ductility of Seismic Resistant Steel Structures*. Spon Press, London.)

Northridge earthquake confirms the steel superiority was published (Vannacci, 1994). Unfortunately, more refined inspections, after removing the finishing, have shown relevant damage on beam–column connections in up to 140 steel frame buildings. This damage was unprecedented, because a near-source earthquake never struck steel structures (Bertero et al., 1994).

A typical beam–column connection for moment-resisting frames, extensively used, is presented in Figure 1.21. The beam web is field bolted to a single shear tab, which is shop welded to the column. The beam flanges are field welded to a column, using complete penetration welds. Web copes are required to accommodate the backup plate at the top flange and to permit the bevel weld to be made at the bottom flange.

A fracture distribution on the frame height, especially in the middle zone, due to the effect of superior vibration modes, is presented in Figure 1.22a. The failure modes of these typology connections mainly consisted in the fracture of the welds connecting the beam flanges to the column flanges and, in some cases, in the propagation of this fracture within the column (Figure 1.22b). The views of some connection fractures are shown in Figure 1.23.

From the examination of the frame damage, it must be mentioned that the evidence of plastic hinges that actually formed in the beams was rare. It seemed that the seismic energy passed directly to the connections, overloading them and causing fractures there. Therefore, the global plastic mechanism with plastic hinges at the beam extremities, which was able to dissipate seismic energy, was not formed. The seismic design philosophy, based on the seismic energy dissipation by plastic rotations in plastic hinges, failed.

The investigations faced to find the causes of damage in steel buildings during the Northridge earthquake gave rise to wide discussions within the scientific community (Gioncu and Mazzolani, 2002).

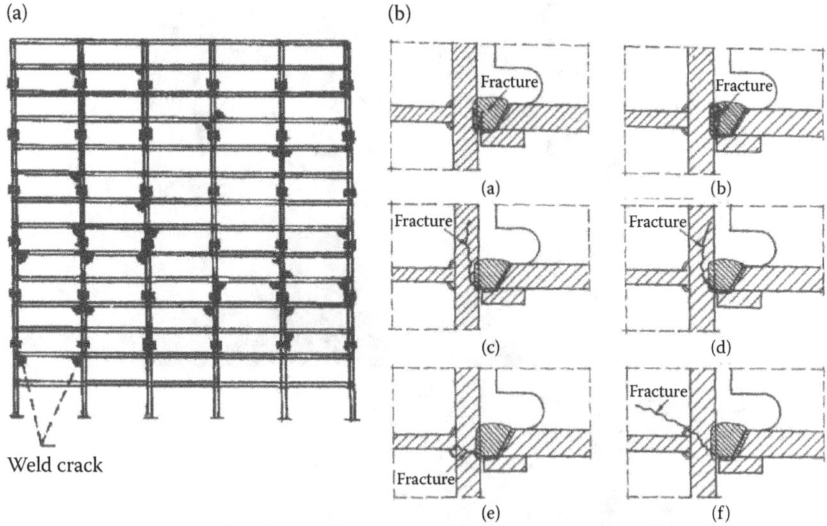

Figure 1.22 Damage of moment-resisting frames. (a) Location of connection fractures. (From Gioncu, V., Mazzolani, F.M. 2002: *Ductility of Seismic Resistant Steel Structures*. Spon Press, London, after Astaneh-Asl, A., Shen, J.H., D'Amore, E. 1995: *Giornale Italiane della Costruzione in Acciaio*, Riva del Garda, October 15–18, vol. 1, 69–79.) (b) Typical fracture of column–beam connections. (Gioncu, V., Mazzolani, F.M. 2002: *Ductility of Seismic Resistant Steel Structures*. Spon Press, London; after Engelhardt, M.D., Sabol, T.A. 1997: Seismic-resistant steel moment connections: Development since the 1994 Northridge earthquake. Progress in Structural Engineering and Materials. Construction Research Communication Limited, 68–77.)

On the one hand, the responsibility can be ascribed to the poor workmanship in field welding and to the bad welding details and procedures. For instance, the presence of a backing bar produces an artificial crack, which initiates the rupture (Figure 1.24). But fractures in the column base plate, made without any backing bar, were also observed (Figure 1.25).

On the other hand, these connection fractures can be ascribed to the ground motion characteristics of the Northridge earthquake, having typical near-source properties (Gioncu and Mazzolani, 2011): velocity pulse, with reduced number of cycles and important vertical components. These factors have induced a high strain rate in the structure, increasing the fragility of steel elements and reducing the capacity to dissipate seismic energy, leading to an actual level of seismic actions that is larger than the design one. So, the building response searches for the weakest parts of the structure and finds that they are the welded connections. The argument that the recorded velocities are not so high to produce the important strain rate in structures can be dismissed, because the high variability of ground motions

(a) (b) (c)

Figure 1.23 Cracked connections. (FEMA 355A 2000: State of the art report on base metals and fracture. SAC Joint Venture.)

in the epicentral area cannot be covered by the present seismic stations. It is only by chance that the maximum value of ground motion velocity was detected (see Tanzana station).

1.2.3 1995 Kobe earthquake (Japan)

1.2.3.1 Earthquake characteristics

The earthquake that shook Kobe on January 17, 1995, measuring 7.2 magnitude, was the most devastating earthquake to strike Japan since the Kanto event in 1923. The epicenter of the main shock was located below the Awaji Island, 20 km southwest from Kobe downtown (Figure 1.26a). The fault directivity was extended from the southwest toward the northeast, directly under the Kobe town. One must mention that this earthquake occurred far from the known faults and was produced by an unknown fault, which was a strike-slip one. The strong shaken area stood on the alluvial deposits of

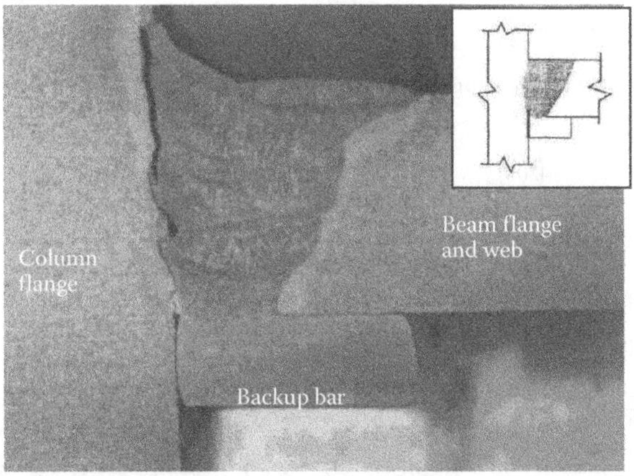

Figure 1.24 Fracture of the welds due to backup bar. (After Chi, W.M., Deierlein, G., Ingraffea, A. 2000: *Journal of Structural Engineering*, 126(1), 88–97.)

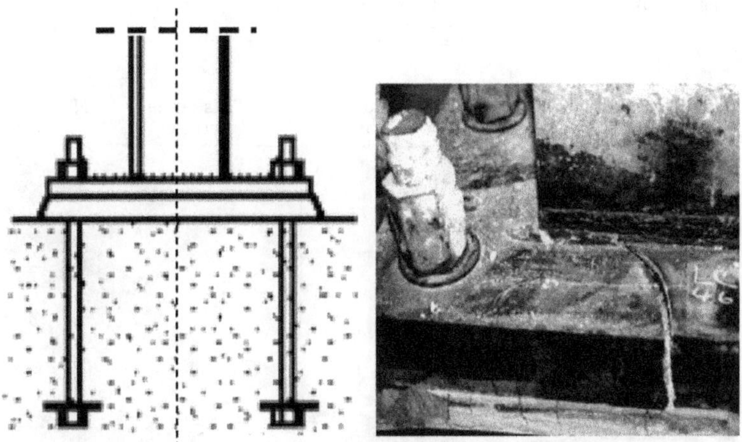

Figure 1.25 Fracture of column base plate. (CSUN (nd): Oviatt Library. Earthquake gallery, http://C\Users\user\Desktop\Earthquake Gallery Oviatt Library.mht.)

the Rokko mountains in the north and the coastline in the south (Figure 1.26b). The examination of the ground motions suggests that the earthquake consisted of three subevents (Figure 1.27). The maximum recorded acceleration was equal to 0.835*g*. But unfortunately, in spite of a very dense network of stations, it seems that no ground motions have been recorded in the most severely shacked areas. The ground motions were amplified by the presence of a thick layer of alluvial soil, which caused an extensive liquefaction over the area of the harbor district of Kobe. The vertical components

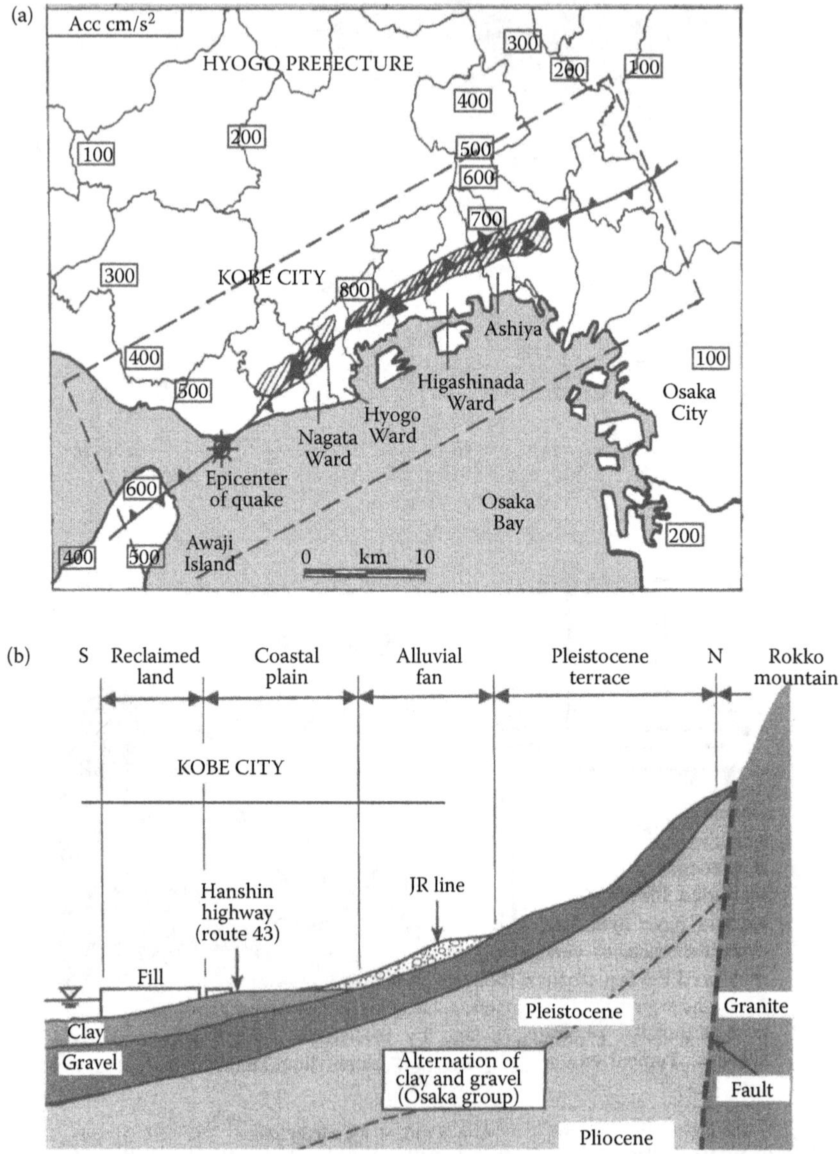

Figure 1.26 The Kobe earthquake features: (a) Epicenter location and fault rupture. (Gioncu, V., Mazzolani, F.M. 2002: *Ductility of Seismic Resistant Steel Structures*. Spon Press, London.) (b) Alluvial deposits under town. (After Hamada, M., Wakamatsu, K. 1999: Liquefaction, ground deformation and their related damage to structure. Report on 1995 Kobe earthquake, INCEDE Report, 1999-2003, 57–112.)

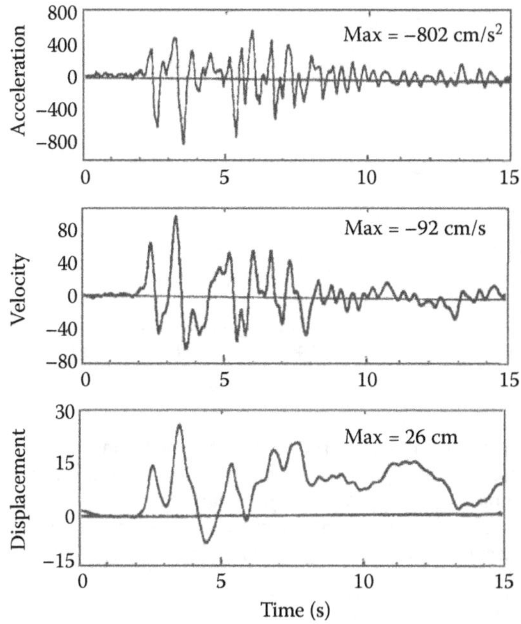

Figure 1.27 Ground motions of Kobe earthquake—JMA Station. (Gioncu, V., Mazzolani, F.M. 2002: *Ductility of Seismic Resistant Steel Structures*. Spon Press, London; after Iwan, W.D. 1996: The drift demand spectrum and its application to structural design and analysis. In *11th World Conference on Earthquake Engineering*, Acapulco, June 22–28, No. 1116.)

were vital in the epicentral area (Figure 1.28). The recorded velocities made it possible to appreciate that the maximum values were about 500 cm/s.

The characteristics of the Kobe earthquake were the typical near-source ones: pulse velocities, very high velocities, short duration, and important vertical components. After the Northridge event, the Kobe earthquake was the second case when important ground motions occurred under an important city and where all the main damage happened to the near-source effects.

1.2.3.2 Damage in some steel structures

Member ductility is conditioned by the global buckling phenomena, as can be seen in Figure 1.29. During the Kobe earthquake, the overall buckling of bracing members produced large permanent deformations, leading to a reduction in the energy absorption capacity. In some cases, the damage consists in the fracture of the bracing connections, due to the presence of the bolt holes, which generates a section with reduced resistance (Figure

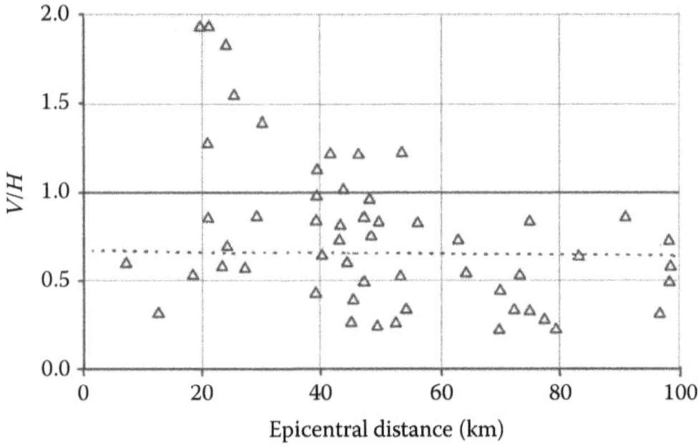

Figure 1.28 Vertical components related to the horizontal ones. (Gioncu, V., Mazzolani, F.M. 2002: *Ductility of Seismic Resistant Steel Structures*. Spon Press, London; after Elnashai, A.S., Bommer, J.J., Martinez-Pereira, A. 1998: Engineering implication of strong motion records from recent earthquakes. In *11th European Conference on Earthquake Engineering*, Paris, September 6–11, No. 59.)

1.30). It is very important to underline that even if some structural members collapsed, the structure did not suffer a total failure (Figure 1.31). Only some partial structural collapse of the steel structures was observed (Figure 1.32).

1.2.3.3 Connection damage

The fracture of welded beam-to-column connections in moment-resisting frames is undoubtedly the most widespread type of failure that occurred in steel structures during both the Northridge and Kobe earthquakes (AIJ, 1995). However, it has been observed (Akiyama and Yamada, 1995) that the fracture mechanisms were different between the two earthquakes, mainly because different trends regarding the detailing of beam-to-column connections are developed in the United States and Japan.

The U.S. structural system is based on H-shaped beams and columns, the beams being connected by welding to continuous columns (Figure 1.33a); the only part resisting seismic actions is the beam-to-column-welded connection. In contrast, the current Japanese practice uses H-shaped beams, which are field welded to box columns (Figure 1.33b). The column is not continuous in the node, due to the presence of continuity plates at the position of the beam flanges and the short segment of the column. The stresses of the beam flanges are safely transmitted by means of continuity plates welded through the box column. The transmission of the beam web stresses

Figure 1.29 Buckling of bracings. (Gioncu, V., Mazzolani, F.M. 2002: *Ductility of Seismic Resistant Steel Structures*. Spon Press, London; after Fischinger, M. 1997: EASY, Earthquake Engineering Slide Information System. FGG-IKPIP Institute Slovenia.)

involves the out-of-plane bending resistance of the column wall plates, and therefore it is not perfect. This causes high stress concentration at the ends of the beam flanges, which is further increased by the web copes, which are introduced for an easier welding (Mazzolani, 1998). The fractures of welded joints are located in the columns, in the continuity plates; in beam-to-column connections, the amount of damage is in general more dangerous than in the case of the U.S.-welded connections.

There are basically two types of joints: the first is shop-welded and the second is site-welded (Figure 1.34). From the comparison between fractures or cracks in the shop- or site-welding types, it is very clear that the influence of welding conditions plays a very important role: the number of damaged

Figure 1.30 Collapse of bolted connections in bracings. (Gioncu, V., Mazzolani, F.M. 2002: *Ductility of Seismic Resistant Steel Structures*. Spon Press, London; after Fischinger, M. 1997: EASY, Earthquake Engineering Slide Information System. FGG-IKPIP Institute Slovenia.)

Figure 1.31 Damage of a steel structure. (Gioncu, V., Mazzolani, F.M. 2002: *Ductility of Seismic Resistant Steel Structures*. Spon Press, London; after Fischinger, M. 1997: EASY, Earthquake Engineering Slide Information System. FGG-IKPIP Institute Slovenia.)

Figure 1.32 Partial collapse of a building corner. (Courtesy of Fischinger, M. 1997.)

connections of site-welding type being 2.6 times greater than that of the shop-welding type.

The fracture types in the beam-to-column connections are shown in Figure 1.35. The initiation of fractures is caused by the same backup bar, as in the case of the Northridge earthquake, even if there are technological differences in execution (Nakashima, 2000a; Nakashima et al., 2000). The similitude of these fractures with the Northridge ones can be observed by comparing Figure 1.34 with Figure 1.22. Two different locations of fractures in the beam-to-column joints are shown in Figure 1.36.

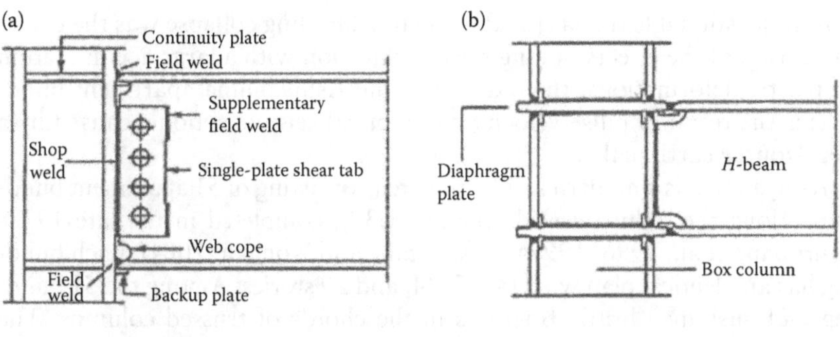

Figure 1.33 Comparison of different beam-to-column connections: (a) U.S. system; (b) Japanese system. (Mazzolani, F.M. 1998: *Control of the Semi-Rigid Behaviour of Steel Engineering Structural Connections* (ed. R. Maquoi), COST C1 Conference, Liége, September 17–18, 371–384.)

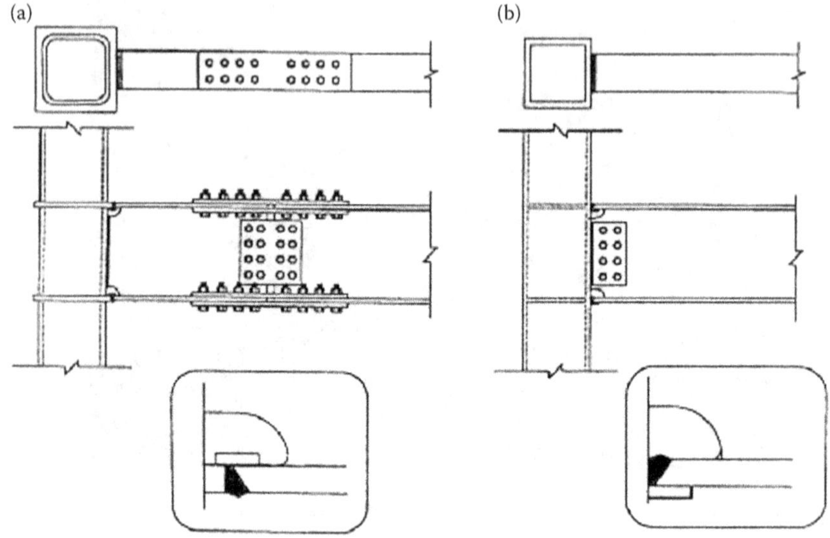

Figure 1.34 Typical beam-to-column connections: (a) shop-welding type; (b) site-welding type. (Gioncu, V., Mazzolani, F.M. 2002: *Ductility of Seismic Resistant Steel Structures*. Spon Press, London; after Nakashima, M. 2000b: Overview of damage of steel building structures observed in the 1995 Hyogoken-Nanbu (Kobe) earthquake. Disaster Prevention Research Institute Report.)

There is no doubt that the main reason for these connection fractures in Kobe is the same as in Northridge, which is characteristic of the effect of near-source seismic actions.

1.2.3.3.1 Damage in the Ashiyahama buildings in Kobe

It is unquestionable that as the Pino Suarez building collapse was the exemplification of the effects of long seismic duration with a large accumulation of plastic deformations, the damage in the Ashiyahama apartment buildings is the result of pulse velocity and vertical seismic actions, caused by a near-source earthquake.

Ashiyahama is a modern residential area consisting of 51 apartment buildings, along the Ashiya seaside (Figure 1.37), completed in the late 1970s (Kurobane et al., 1996, 1997a; Takabatake and Nonaka, 2001). Each building has an identical plan with 14, 19, 24, and 29 stories. Among the 51 buildings, 21 sustained brittle fractures in the chords of trussed columns. The most remarkable damage occurred in the 19th and 24th storied buildings.

The structures are shown in Figure 1.38. The front elevation is a so-called mega moment frame with trussed girders and columns, while the

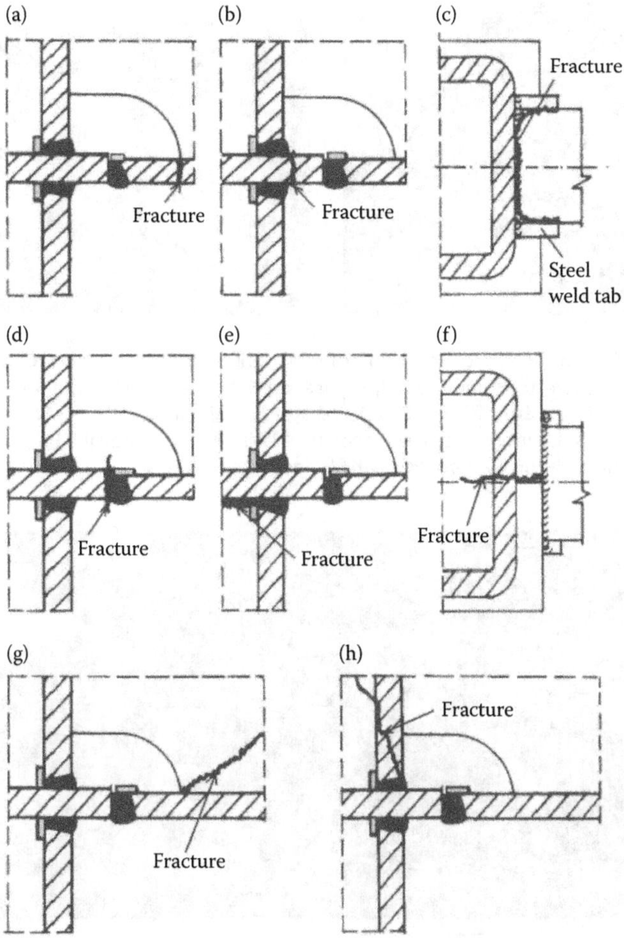

Figure 1.35 Typical connection fractures. (Gioncu, V., Mazzolani, F.M. 2002: *Ductility of Seismic Resistant Steel Structures.* Spon Press, London; after JSSS, 1997: Kobe earthquake damage to steel moment connections and suggested improvement. Technical report, No. 39.)

transversal structure is a braced frame. All buildings are supported by 38 m long pipe pile foundations.

Each house unit is made of precast concrete panels, mounted in place on the beams, sized for gravity loads only. The trussed columns have welded box-section chords, the braces are wide-flange elements, and the trussed girders have wide-flange sections. The box sections are made of two longitudinal welded channel sections, with the exception of the first story columns, which are made of four plates welded together.

(a) (b)

Figure 1.36 Failure of some beam-to-column connections: (a) fracture of continuity plate-to-column weld; (b) fracture of beam-to-column weld. (Gioncu, V., Mazzolani, F.M. 2002: *Ductility of Seismic Resistant Steel Structures.* Spon Press, London; after Fischinger, M. 1997: EASY, Earthquake Engineering Slide Information System. FGG-IKPIP Institute Slovenia.)

Figure 1.37 Ashiyahama apartment ensemble. (Gioncu, V., Mazzolani, F.M. 2002: *Ductility of Seismic Resistant Steel Structures.* Spon Press, London; after Fischinger, M. 1997: EASY, Earthquake Engineering Slide Information System. FGG-IKPIP Institute Slovenia.)

The brittle fracture occurred in the chord of the trussed columns, but never in the girders. Fracture patterns are classified into two main types:

- Fractures out of welded sections, across the full section of the chord (especially at the ground floor), or starting from the brace connection (Figure 1.39)
- Fracture in the welds at the chord splice (especially at the middle levels due to the influence of high modes of vibrations) (Figure 1.40)

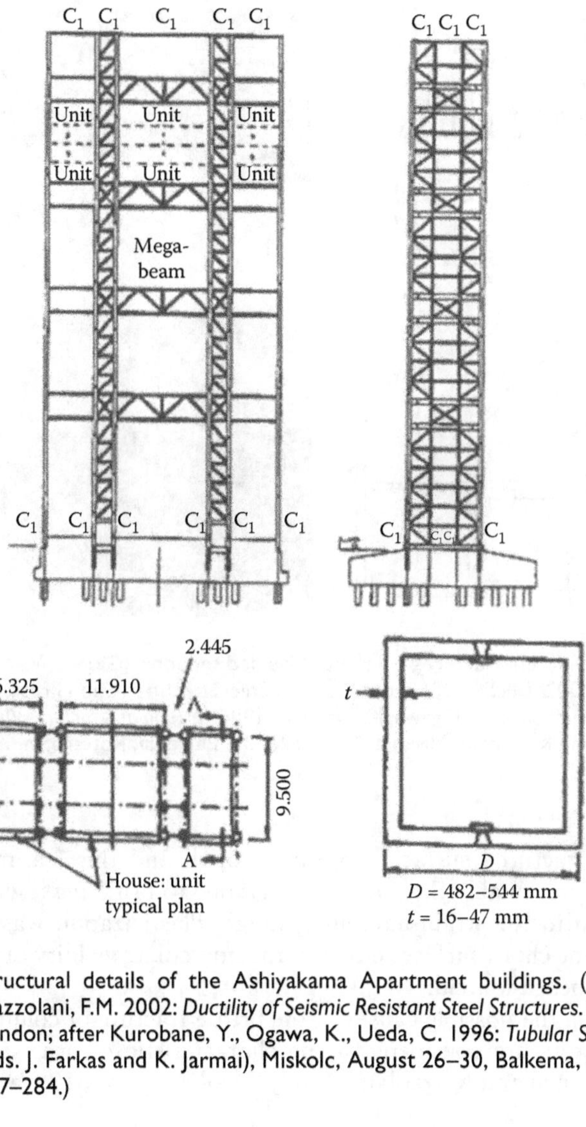

Figure 1.38 Structural details of the Ashiyakama Apartment buildings. (Gioncu, V., Mazzolani, F.M. 2002: *Ductility of Seismic Resistant Steel Structures*. Spon Press, London; after Kurobane, Y., Ogawa, K., Ueda, C. 1996: *Tubular Structures VII* (eds. J. Farkas and K. Jarmai), Miskolc, August 26–30, Balkema, Rotterdam, 277–284.)

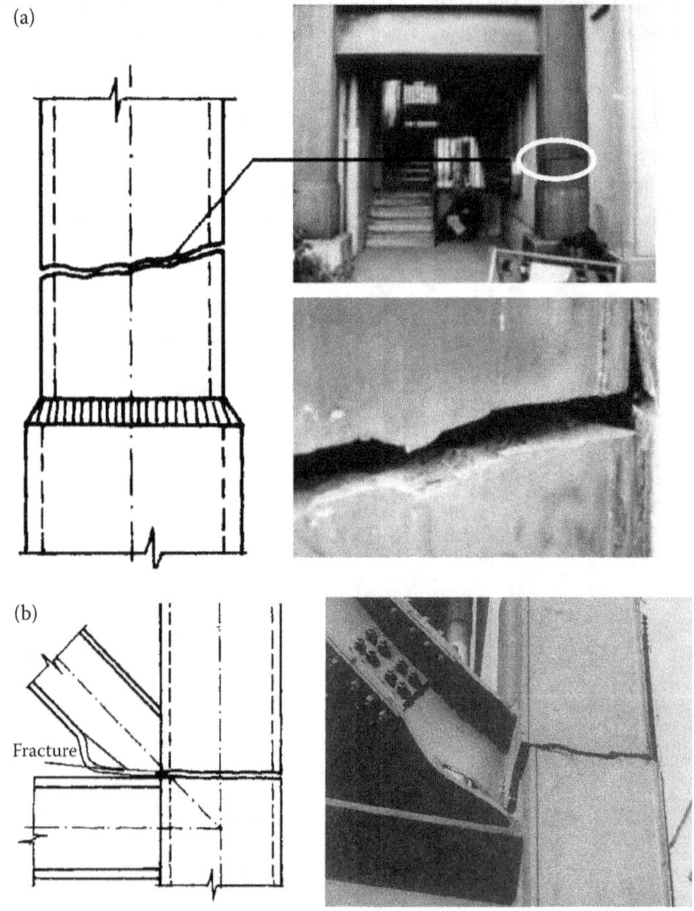

Figure 1.39 Different fractures type out of welded sections. (Gioncu, V., Mazzolani, F.M. 2002: *Ductility of Seismic Resistant Steel Structures*. Spon Press, London; after Kurobane, Y., Ogawa, K., Ueda, C. 1996: In *Tubular Structures VII* (eds. J. Farkas and K. Jarmai), Miskolc, August 26–30, Balkema, Rotterdam, 277–284.)

All the fractured surfaces appeared rough and this feature led us to think that presumably they occurred just due to brittleness, without being strained up to the strain-hardening range. Plasticization was recognized only on some chord surfaces either in the immediate vicinity of cracks or in the region near the weld.

Analyzing the possible causes of the brittle fracture of columns, the following factors are considered to be characteristic of near-source earthquakes (Gioncu and Mazzolani, 2002; Kurobane et al., 1997a, b):

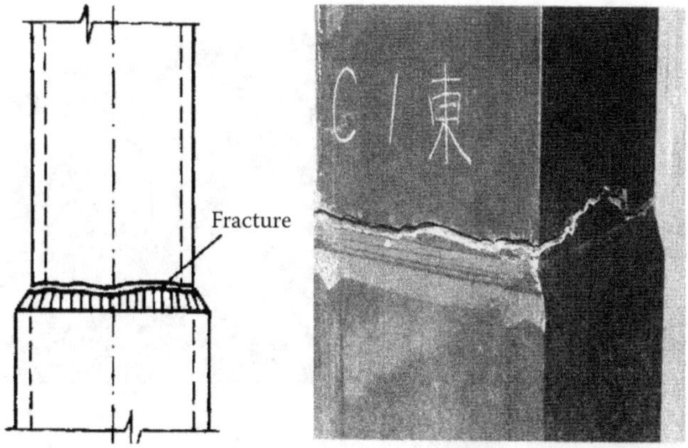

Figure 1.40 Fracture in welds. (Gioncu, V., Mazzolani, F.M. 2002: *Ductility of Seismic Resistant Steel Structures*. Spon Press, London; after Kurobane, Y., Azuma, K., Ogawa, K. 1997a: Brittle fracture in steel building frames. Comparative study of Northridge ND Kobe earthquake damage. *International Institute of Welding, Annual Assembly*, 1997, 1–30.)

- Influence of high strain rate due to pulse loading, both in the horizontal and vertical directions
- Increase in the influence of strain rate due to low temperature, which during the earthquake was estimated to be about 0°C in the chords
- Influence of important vertical components
- Influence of welding on the ductility due to residual stresses

A common problem associated with earthquakes in urban areas is fire (see Chapter 6). Shaking and ground displacement often cause the severing of electrical and gas lines, leading to the development of many localized fires. The response to this event is usually not effective, because shock waves also produce the rupture of pipes carrying water. After the earthquake of Kobe, there were hundreds of fires, many ignited by toppled gas cookers and kerosene stoves (Figure 1.41).

From the very high cost of losses produced by the earthquake, an important part can be attributed to the fire. The fire raged unchecked in several places. Firemen were unable to do anything to stop them, because there was no water and the fires spread rapidly due to the strong wind.

1.2.4 1999 Kocaeli earthquake (Turkey)

1.2.4.1 Earthquake characteristics

The northern part of Anatolia was struck by the *1999 Kocaeli earthquake* (Figure 1.42) (also called the Izmit earthquake), which can be considered

Figure 1.41 Fire after earthquake in Kobe: (a) area interested by fire; (b) aerial view of fire; (c) a building destroyed by earthquake followed by fire.

among the largest seismic events that occurred in the Eastern Mediterranean Basin during the last century, due to the number of fatalities and damage. The rupture of the Anatolian strike-slip fault (with similarity to the San Andreas U.S. fault) had a length of 145 km in a zone with a population of 20 million inhabitants (one-third of Turkey's total population), which includes nearly half of Turkey's industry.

Owing to the similarity between the North Anatolian and San Andreas faults, it was expected that the damage that occurred during the Kocaeli earthquake was at a similar level to the one that occurred in the United States. This was true only in respect of reinforced concrete (RC) structures. The predominant structural system used for buildings in Turkey consists of reinforced concrete frames, with masonry claddings. Unfortunately, poor structural conception (soft stories, strong beams, and weak columns) and execution (low material qualities and poor detailing) caused the collapse

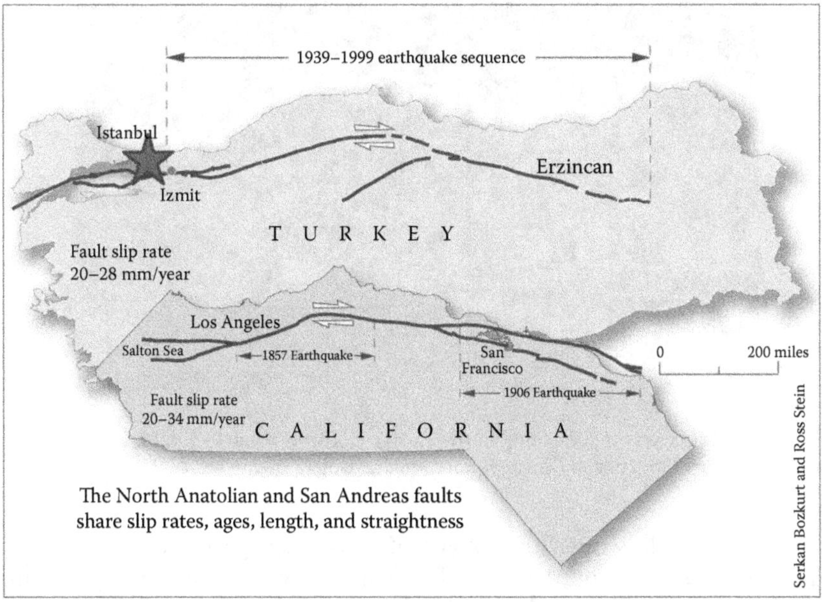

Figure 1.42 The North Anatolian strike-slip fault in comparison with the San Andreas fault.

of a large number of buildings. This earthquake clearly demonstrated that improperly conceived and constructed buildings kill people.

1.2.4.2 Steel structure damage

Steel, being the most expensive construction material in Turkey, has been used mainly in industrial structures, not in buildings. Some of them were damaged by this earthquake. Typical causes for collapse include failure of anchor bolts at the column bases and structural instability under overturning forces. Other evidence of damage includes fracture of brace connections, buckling of braces, and local buckling in concrete-filled steel hollow pipes. One example is the quite old steel structure of the railcar factory (Figure 1.43). Typical causes of failure in steel structures were the inadequacy of anchor bolts at the column bases and failure of brace connections (Erdik, 2000).

Many industrial facilities suffered damage and interruption of operations, as a result of the fire after the earthquake. The most significant fire damage occurred at the Tüpras refinery (Izmit) (Figure 1.44). Three separate fires started after the earthquake, one of which took several days to be brought under control, with substantial fire damage.

Figure 1.43 Damaged steel building of the railcar factory. (After Erdik, M. 2000: Report on 1999 Kocaeli (Turkey) Earthquake. Bogazici University Report 2000.)

Figure 1.44 illustrates the extent of tank damage of the four tanks, which were affected by the fire. The main fire was caused by a pipe rupture during the earthquake, exposing oil products at extreme temperatures to oxygen.

1.2.5 2003 Bam earthquake (Iran)

1.2.5.1 Earthquake characteristics

The *2003 Iran Bam earthquake*, which was produced at the fault between the Arabian and Eurasian tectonic plates (Figure 1.45), destroyed the ancient citadel of Bam, dating back around 2500 years (during the Persian period) and considered to be the biggest adobe construction of the world. The city benefited from tourism, with an increasing number of people visiting the ancient citadel. The earthquake destroyed 70% of the city of Bam and its citadel was leveled to the ground, producing a very large number of fatalities due to the collapse of old buildings. This earthquake was responsible for the irreparable destruction of one of the main important monumental cities in the world.

1.2.5.2 Steel structure damage

It is worth noting that the use of steel is not always to be considered as "a blessing," which helps buildings to survive during destructive earthquakes. It is the case of some steel buildings erected in Iran, which in principle were conceived as moment-resisting frames (Figure 1.46a), but their connections, the so-called Korjini type, were so poor as to be the weakest

Figure 1.44 Fire after earthquake: (a) fire in a petrochemical complex; (b) damage of tanks produced by fire after earthquake. (After Erdik, M. 2000: Report on 1999 Kocaeli (Turkey) Earthquake. Bogazici University Report 2000; Saatcioglu M. et al. 2001: *Canadian Journal of Civil Engineering*, 28, 715–737.)

point of the structure, being completely devoid of strength and ductility. In many cases, traditional hollow brick-reinforced concrete floors and heavy brick walls completed the steel skeleton. The different rigidity of the two parts was responsible for the collapse of the whole structure (Figure 1.46b). This kind of buildings is common not only in Bam but also in many other cities. Another reason for the damage to steel buildings is the combination of moment steel frames in one direction and braced frames in

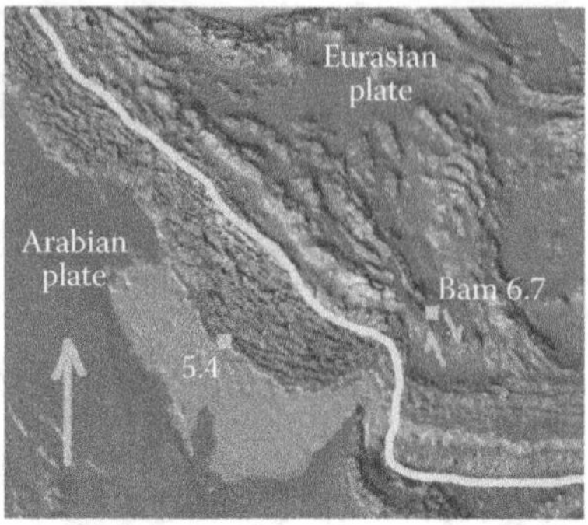

Figure 1.45 The Bam earthquake in Iran.

the perpendicular direction. Their behavior was generally satisfactory, but some exceptions rose when the bracings were eccentrically or incorrectly located (Figure 1.46c).

The 2003 Iranian Bam earthquake has shown that the use of steel beams and columns without observing proper provisions for earthquake resistance did not provide any improvement with respect to nonengineered buildings (Zahrai and Heidarzadeh, 2004).

1.2.6 2010 Maule earthquake (Chile)

1.2.6.1 Earthquake characteristics

On February 27, 2010, a magnitude M 8.8 earthquake struck the central part of Chile. The rupture occurred in the contact between the Nazca and the South American plates, with an approximate extension of 450 km in the north-south direction (Figure 1.47a). This subduction-type earthquake affected an area of approximately 160,000 km^2, which houses approximately 75% of the population of Chile. Figure 1.47b shows a map of the affected area and the approximated rupture zone.

About 55–65% of the damage is believed to have occurred to residential buildings, 20–30% to commercial buildings, and 15–20% to industrial facilities. The majority of buildings in Chile are made of masonry or reinforced concrete. A small percentage of high-rise commercial buildings in the major cities, such as Santiago, are constructed using steel framing. The industrial facilities are normally built using steel structures. Owing to code provisions (large overstrength) and a very well-organized supervision, the damage was reduced in comparison with a very large earthquake magnitude (Herrera et al., 2012).

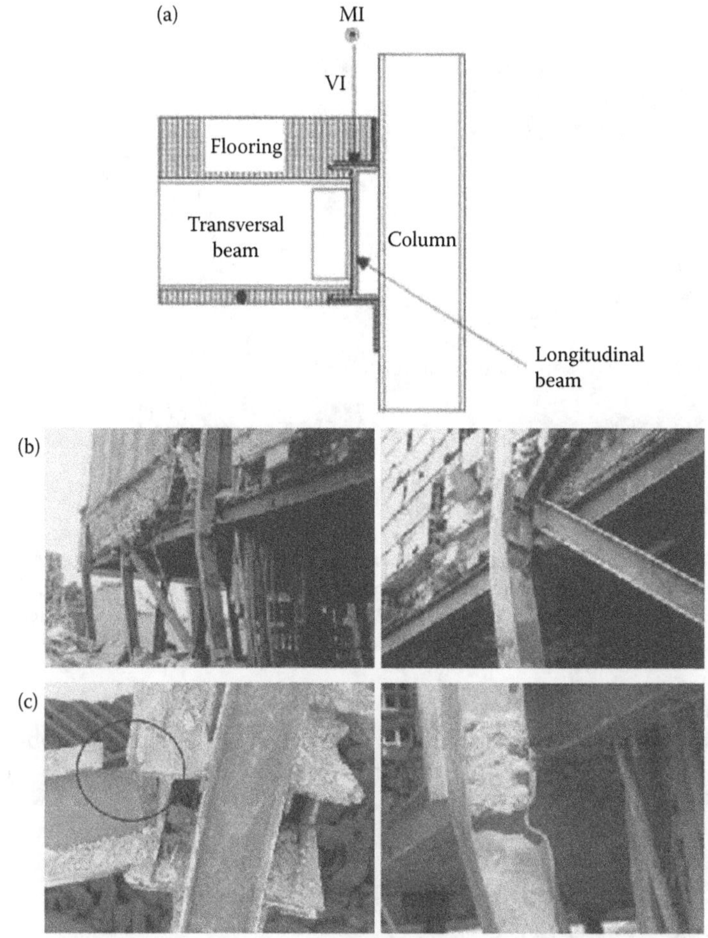

Figure 1.46 Steel structure erected in Iran: (a) beam-to-column end-span connection of Khorjini type; (b) collapse of braced frame; (c) connection failure. (After Zahrai, S.M., Heidarzadeh, M. 2004: *13th World Conference on Earthquake Engineering*, Vancouver, August 1–6, Paper No. 1715.)

1.2.6.2 Steel structure damage

Industrial structures, which were mainly made of steel, performed adequately. Damage was generally limited to the oldest structures not designed according to the latest seismic code or facilities affected by the tsunami caused by the earthquake. The majority of industrial steel structures were either concentrically braced frames or moment-resisting

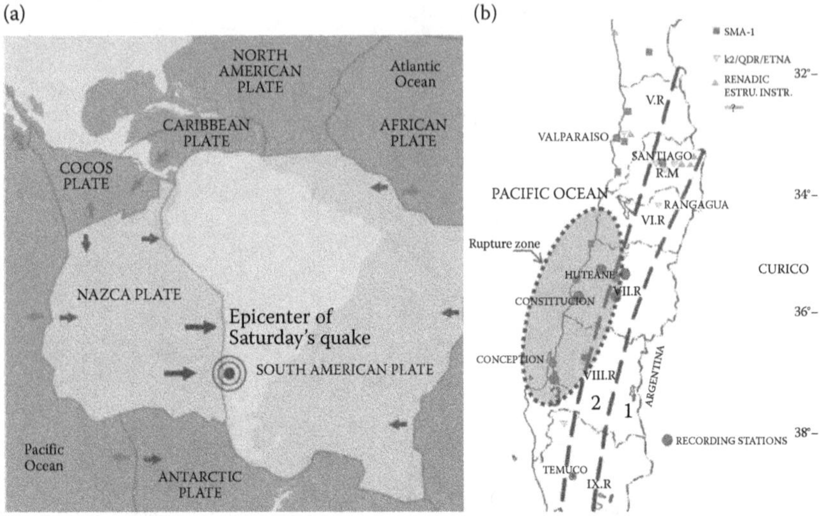

Figure 1.47 The Maule earthquake in Chile: (a) contact between Nazca and South American plates; (b) affected area. (After Herrera R.A., Beltran J.F., Aguirre C., Verdugo A. 2012: *Behavior of Steel Structures in Seismic Areas, STESSA 2012* (eds. F.M. Mazzolani, R.A. Herrera), Santiago, January 9–11, CRC Press, 37–43.)

frames, or a combination of both systems. The most commonly observed type of damage was the fracture of bracing members and of the anchor bolt (Figure 1.48).

1.2.7 2011 Christchurch earthquake (New Zealand)

1.2.7.1 Earthquake characteristics

The February 2011 Christchurch earthquake was a powerful natural event, which severely damaged New Zealand's second largest city. This was produced by a strike-slip fault (Alpide fault) in the South Island of New Zealand (Figure 1.49). It is situated on the Ring of Fire, where the Pacific and Australian plates push each other sideways. The Alpide fault is the largest active fault in New Zealand and extends over 650 km. The 2011 Christchurch earthquake (magnitude M 6.3) followed nearly 6 months after the magnitude M 7.1 earthquake of September 4, 2010, which caused significant damage to Christchurch and the central Canterbury region, but there were no direct fatalities. These two damaging earthquakes were the strongest from a series of 2010–2011 earthquakes. This series comprised six earthquakes: one each on September 4 and December 26, 2010,

Figure 1.48 Fracture of braces and bolt anchors. (Herrera R.A., Beltran J.F., Aguirre C., Verdugo A. 2012: *Behavior of Steel Structures in Seismic Areas, STESSA 2012* (eds. F.M. Mazzolani, R.A. Herrera), Santiago, January 9–11, CRC Press, 37–43.)

February 22, June 6, and two on June 13, 2011 (Clifton et al., 2011). The earthquake that occurred in February caused a widespread damage across Christchurch, especially in the central part of the city and eastern suburbs, which was exacerbated by the buildings and infrastructure already being weakened by the September 4, 2010 earthquake and its aftershocks. This was the first time in modern history that a damaging earthquake occurred in an urban area twice in the same year, followed by a second important event, which produced significantly more damage than the first one. So, it is anticipated that new lessons will be learned as a result (Bruneau, 2011). At the same time, owing to the fault strike-slip special characteristics, this earthquake belongs to a series of very damaging ground motions, without being very strong in magnitude with respect to the previous famous earthquakes: 1989 Loma Prieta, M 7.1; 1994 Northridge (SUA), M 6.7; 1995 Kobe, M 6.9; 1999 Kocaeli, M 7.4.

Taking into account the damage produced by these moderate earthquakes, it was not surprising to find a similar scenario during the Christchurch earthquakes.

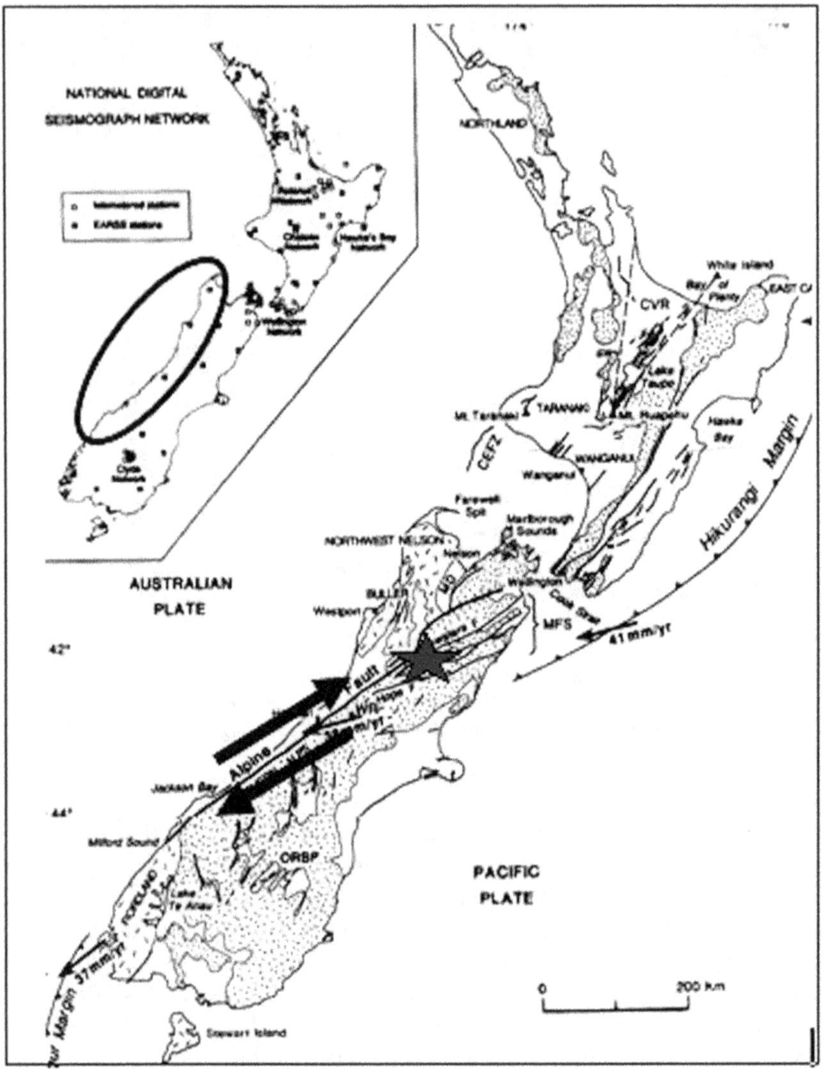

Figure 1.49 Christchurch earthquake produced by strike-slip Alpide fault. (Gioncu, V., Mazzolani, F.M. 2011: *Earthquake Engineering for Structural Design*. Spon Press, London.)

1.2.7.2 Steel structure damage

The number of modern steel structures is relatively low in the Christchurch area; constructions of steel multistoried buildings or parking garages started after 1990, being designed with the latest seismic provisions. This is the reason why, with few notable exceptions, the steel structures performed

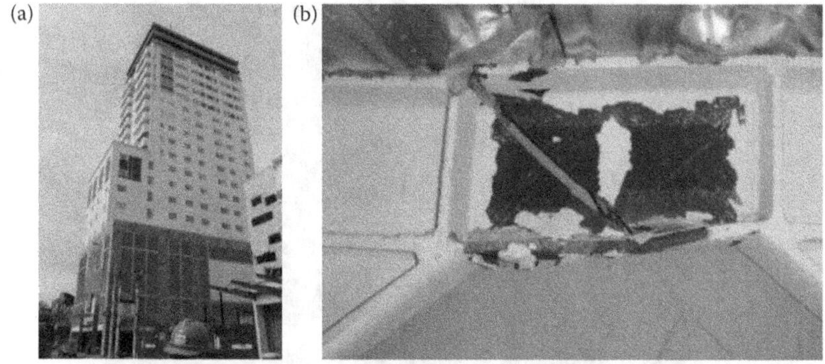

Figure 1.50 Pacific Tower: (a) general view; (b) damage of link beam in the eccentrically braced frame. (Clifton Ch. et al. 2011: *Bulletin of the New Zealand Society for Earthquake Engineering*, 44(4):297–318.)

satisfactorily well during this earthquake series (Clifton et al., 2011). The exceptions refer to some fractures of eccentrically braced frames and the buckling of some braces in centrically bracing systems.

The 22-storied Pacific Tower (Figure 1.50a), which was completed in 2010, consisted of perimetral and eccentrically braced frames around the elevator. Some yielding in the link beam webs were observed after the earthquakes (Figure 1.50b) (Clifton et al., 2011).

But the most significant link beam damage in two eccentrically braced frames was observed in a hospital parking garage (Figure 1.51a). Damage consists of a fracture in the welding zone or web yielding of link beams (Figure 1.51b). A preliminary report shows that the level of excitation of the earthquake was at least as large as or has exceeded the design basis of the NZ seismic code in an area, where steel eccentrically braced frames were used. This is the first time that this type of damage has been observed after an earthquake. The earthquake provides the first opportunity to observe how these types of structures perform and/or fail during large earthquakes (Clifton et al., 2011).

For concentrically braced frames used in a parking garage, the buckling of some braces or the fracture of brace-to-column connections was observed (Figure 1.52).

1.2.8 2011 Tohoku earthquake (Japan)

1.2.8.1 Earthquake characteristics

On March 11, 2011, an earthquake with a magnitude of M 9.0 occurred off the Pacific coast of Tohoku, often mentioned in Japan as the Tohoku earthquake or the Great East Japan earthquake. It was an undersea megathrust earthquake off the coast of Japan, produced by the fault between

Figure 1.51 Parking garage: (a) eccentrically braced system; (b) damage of the link beam in the welding zone and yielding in the web. (Clifton Ch. et al. 2011: *Bulletin of the New Zealand Society for Earthquake Engineering*, 44(4):297–318.)

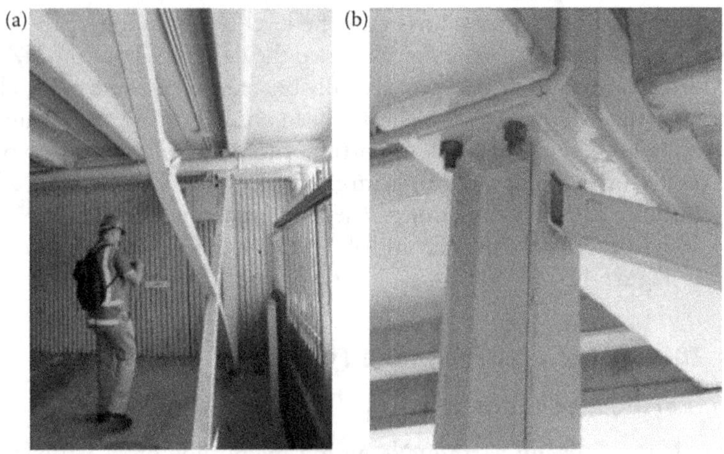

Figure 1.52 Damage in concentrically braced frames: (a) buckling of braces; (b) fracture of brace-to-column connection. (Clifton Ch. et al. 2011: *Bulletin of the New Zealand Society for Earthquake Engineering*, 44(4):297–318.)

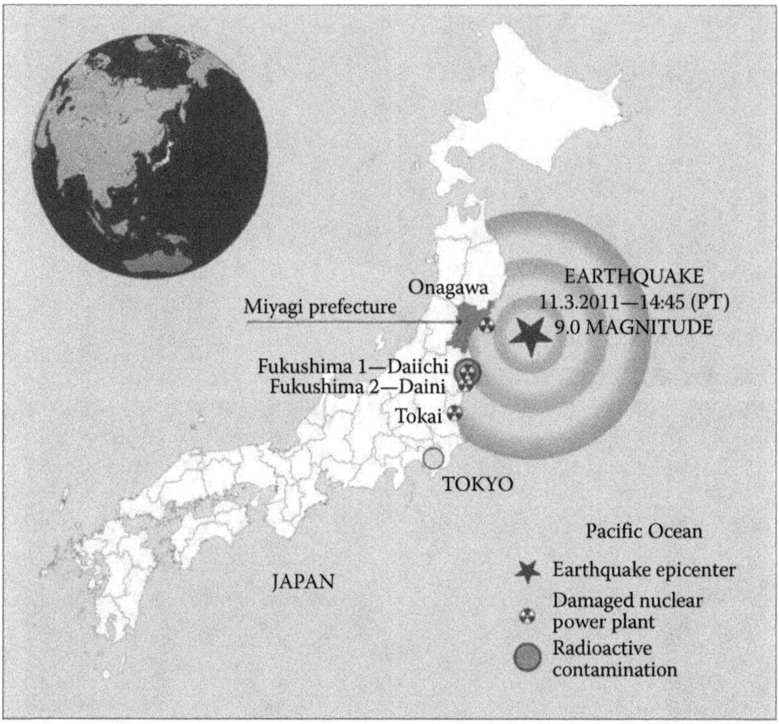

Figure 1.53 Tohoku earthquake and tsunami in Japan.

the North American and Pacific tectonic plates (Figure 1.53) (NILIM-BRI, 2011). It was the most powerful earthquake known, which hit Japan and is one of the five most powerful earthquakes in the world. This earthquake produced powerful tsunami waves, which reached heights up to 40.5 m in some zones. As a consequence, the structural damage was produced either by an earthquake or by a tsunami.

1.2.8.2 Steel structure damage

Both the earthquake and the tsunami caused extensive damage to steel structures. The observed damage can be classified into the one caused by ground motions and the one caused by the tsunami. Severe ground motions caused damage to the beam-to-column connections, buckling or fracture of the diagonal braces, yielding or fracture of the anchor bolts (Figure 1.54). A distinctive feature of this earthquake was the damage caused by the tsunami. The steel buildings were damaged, experiencing great lateral deformations (Figure 1.55) (Midorikawa et al., 2012). More than 300 places on fire after the earthquake were identified in the affected area (Figure 1.56).

Figure 1.54 Damage or fractures of connections, braces, and anchor bolts due to ground motions. (After Midorikawa M. et al. 2012: Earthquake and tsunami damage on steel buildings caused by the 2011 Tohoku Japan earthquake. *International Symposium on Engineering Lessons Learned from the 2011 Great East Japan Earthquake*, Tokyo, March 1–4, 1061–1076.)

Figure 1.55 Structure damage and collapse due to tsunami. (After Midorikawa M. et al. 2012: Earthquake and tsunami damage on steel buildings caused by the 2011 Tohoku Japan earthquake. *International Symposium on Engineering Lessons Learned from the 2011 Great East Japan Earthquake*, Tokyo, March 1–4, 1061–1076.)

Figure 1.56 Structure damage in the Fukushima Nuclear Plant Station. (BBC Asia News 2012: Japan panel. Fukushima nuclear disaster "man-made". Comments 248.)

The tsunami caused a nuclear accident in three reactors in the Fukushima nuclear power plant. This accident caused nuclear contamination in the surrounding environment (Figure 1.53).

1.3 BEHAVIOR OF STEEL STRUCTURES DURING EUROPEAN EARTHQUAKES

1.3.1 General

By contrast with to the cases presented in the previous sections, the damage to steel structures in Europe is very rare, due to two main reasons:

- Different earthquake types, prevalently being intraplate low and moderate earthquakes
- Reduced number of steel structures, reinforced concrete being the most popular structural material in the earthquake-prone countries

Four typical earthquake scenarios, focusing on the damage to steel structures, are briefly illustrated: the historic 1977 earthquake in Bucharest (Romania), the 1999 earthquake in Athens (Greece), and the two more recent Italian earthquakes in Abruzzo (2009) and in Emilia (2012).

In general, it can be observed that the main damage was recorded in industrial buildings or industrial facilities. However, many cases showed that the level of damage to steel structures was quite limited. Also, there were some interesting cases in which the few existing steel structures performed very well.

1.3.2 1977 Vrancea earthquake (Romania)

1.3.2.1 Earthquake characteristics

The *1977 Romania Vrancea earthquake* (M 7.2) (Figure 1.57) (Balan et al., 1982) is considered as one of the most damaging earthquakes in Europe, because a very densely populated area was affected.

Most of the damage was concentrated in the Romanian capital of Bucharest, at a distance of 160 km from the epicenter, where about 33 large RC buildings collapsed. Owing to the depth of the source (94 km) (intermediate intraslab earthquake type), the affected area was very large.

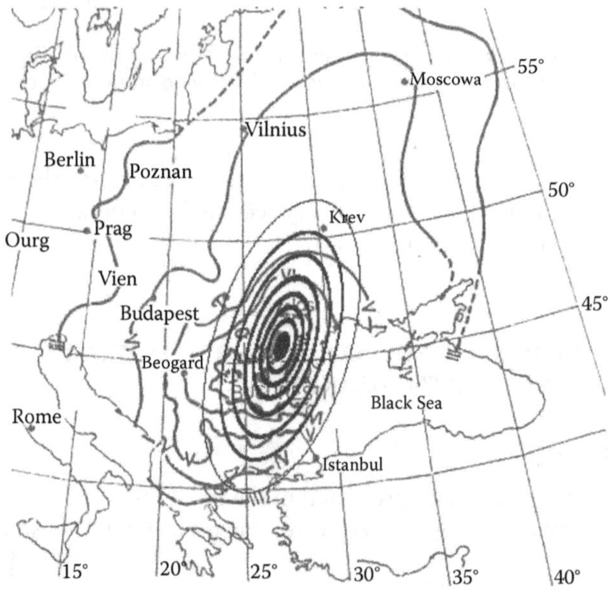

Figure 1.57 Location of the Vrancea earthquake and the affected area. (Gioncu, V., Mazzolani, F.M. 2011: *Earthquake Engineering for Structural Design*. Spon Press, London.)

1.3.2.2 Damage to one-story steel buildings

Damage to steel structures is mainly the result of bad structural conception and bad solutions for details. During the strong 1977 earthquake in Vrancea, there were some collapses of single-storied buildings, produced by the difference in rigidity between two load-bearing structures supporting the roof, such as elastic columns and rigid frames (Figure 1.58) or masonry walls with different rigidity, and the inadequate solution for anchorage of

(a)

(b)

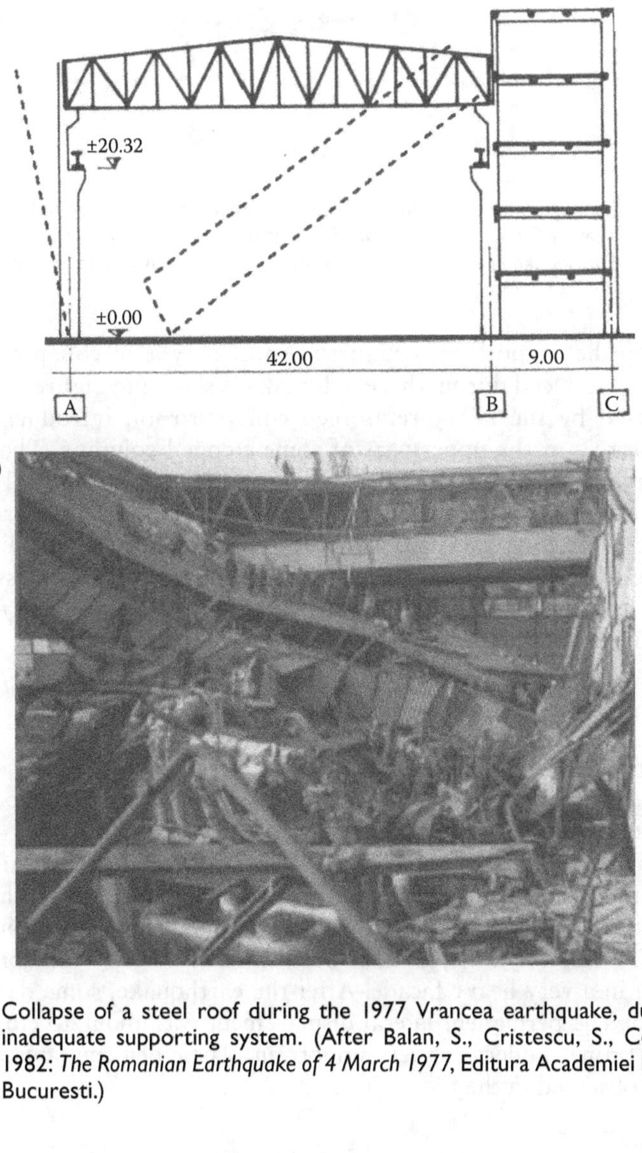

Figure 1.58 Collapse of a steel roof during the 1977 Vrancea earthquake, due to an inadequate supporting system. (After Balan, S., Cristescu, S., Cornea, I. 1982: *The Romanian Earthquake of 4 March 1977*, Editura Academiei Romane, Bucuresti.)

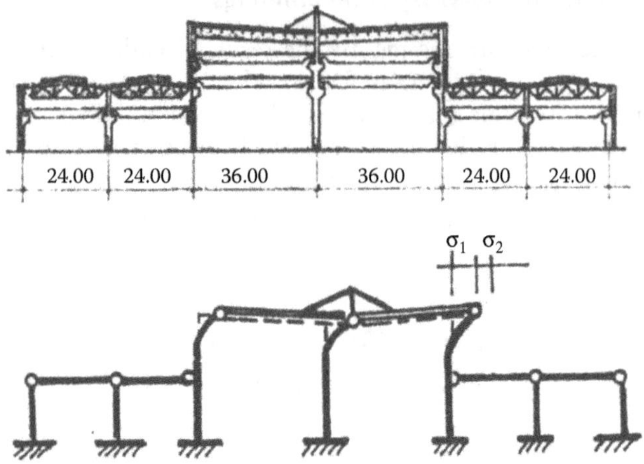

Figure 1.59 Damage of a steel industrial building during the Vrancea earthquake, due to heavy RC roof. (After Balan, S., Cristescu, S., Cornea, I. 1982: *The Romanian Earthquake of 4 March 1977*, Editura Academiei Romane, Bucuresti.)

trusses to the supporting structures. Another type of collapse of a steel structure produced during this earthquake is shown in Figure 1.59, which was caused by the heavy reinforced concrete roof, introducing lateral seismic forces at the upper part of some stepped columns. The result of these forces produced lateral plastic deformations of 8–85 cm. The difference in rigidity of the central part of the structure, in comparison with the general rigidity, was also a reason for this bad behavior. In another case, the horizontal displacements of the supports of steel trusses, made as roller bearing, were so large that the trusses fell down from these supports (Figure 1.60).

1.3.2.3 Damage to multistoried steel buildings

Owing to the tradition in Romania, reinforced concrete was used for the structure of multistoried buildings. So, there were a small number of steel structures involved in the 1977 earthquake.

The *Victoria telephone exchange office* (Figure 1.61a), 12 stories, was erected in 1931–1933, conforming to an American project. The scheme of the steel structure is an MRF with columns and beams jointed with rivets. The exterior walls, made of bricks, were covered with stone blocks, resulting in a very heavy facade. After the earthquake, some damage was observed. The permanent lateral displacement was about 15 cm, showing that the frames underwent very important plastic deformations. No damage was observed in the rivet joints.

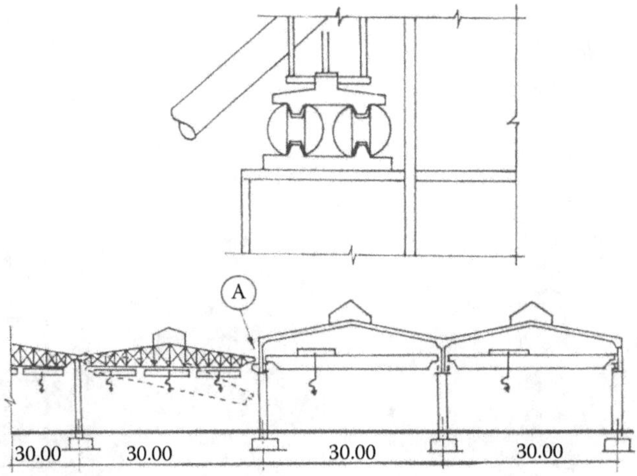

Figure 1.60 Collapse of the steel roof of an industrial building during the Vrancea (Romania), due to fall down of the roller support. (After Balan, S., Cristescu, S., Cornea, I. 1982: *The Romanian Earthquake of 4 March 1977,* Editura Academiei Romane, Bucuresti.)

The *Adriatica building* (Figure 1.61b) was erected just before the World War II, the structural system being MRF with riveted joints. The floors are made of heavy reinforced concrete. During the earthquake, very important lateral displacements were produced and, owing to plastic deformations, permanent lateral displacements of about 20–25 cm were measured at the first level. No damage to joints was reported.

The *Ministry of Transport building* (Figure 1.61c), 12 stories, erected in 1940, was designed in composite steel–concrete MRFs. No structural damage was observed, but due to the great deformability of the structure, the brick panels were damaged by very important cracks.

One must mention that these three buildings, before the 1977 earthquake, were subjected to the 1940 Vrancea earthquake, which already produced some damage.

1.3.3 1999 Athens earthquake

1.3.3.1 Earthquake characteristics

The 1999 Athens earthquake, registering a moment magnitude of 6.0, occurred on September 7, 1999. The tremor was epicentered approximately at 17 km to the northwest of the Athens center, in a sparsely populated area. More than 100 buildings (including three major factories) in these areas collapsed, being Greece's deadliest natural disaster in almost half a century. This event surprised the Greek seismologists, as it came from a

Figure 1.61 Multistoried buildings damaged during the 1977 Vrancea earthquake: (a) Victoria telephone exchange office; (b) Adriatica building; (c) Ministry of Transport building.

previously unknown fault, originating in an area that was for a long time considered to be of a particularly low seismicity.

1.3.3.2 Steel structure damage

There were very few steel structures in the affected area, so the related damage was poor. Figure 1.62 shows some roof bracings fractures during the 1999 Athens earthquake, produced by the bolts fracture (Altay, 2005).

Figure 1.62 Bracing fractures during the 1999 Athens earthquake. (Altay, G. 2005: Post-earthquake damage assessment in the Mediterranean regions. Earthquake protection of historical buildings by reversible mixed technologies, *PROHITECH*, WP 2.)

1.3.4 2009 Abruzzo earthquake (Italy)

1.3.4.1 Earthquake characteristics

On April 6, 2009, the Abruzzo region was rocked by an M 5.8 earthquake, whose epicenter was located at a depth of about 8 km, just 4 km away from the city center of the town of L'Aquila (Figure 1.63). The affected area was limited to 26 × 50 km. Many buildings made of RC and masonry were seriously damaged. Some small villages with masonry houses were totally destroyed.

The economical losses for the damage to the cultural heritage (churches, historical palaces) reached an amount of 1.5 billions Euros, about 40% of the total.

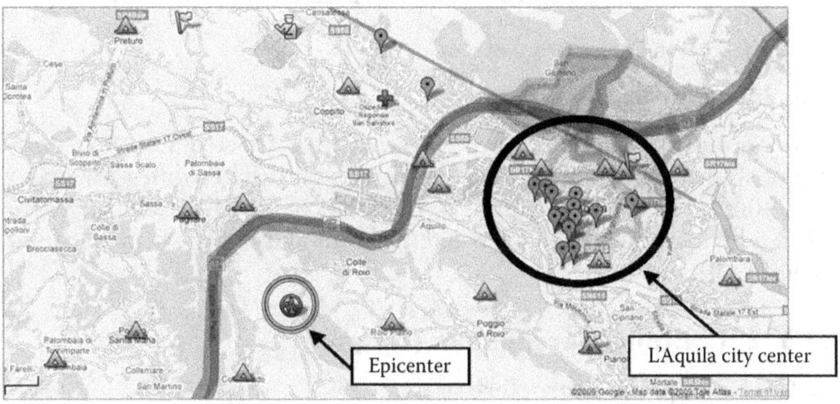

Figure 1.63 Location of the epicenter of the 2009 Abruzzo earthquake respect to the city center of L'Aquila.

Figure 1.64 Local and global failure of silos.

1.3.4.2 Steel structure damage

No important steel buildings were present in the affected area. The presence of steel structures was limited to some industrial facilities (Faggiano et al., 2009). The metallic cylindrical reservoirs of the Milk Company were affected by local buckling in the external wall and few of them collapsed (Figure 1.64).

1.3.5 2012 Emilia earthquake (Italy)

1.3.5.1 Earthquake characteristics

During 2012 springtime, a series of earthquakes shook the Emilia region in the north of Italy, the most important being the one M 5.9 on 20 May

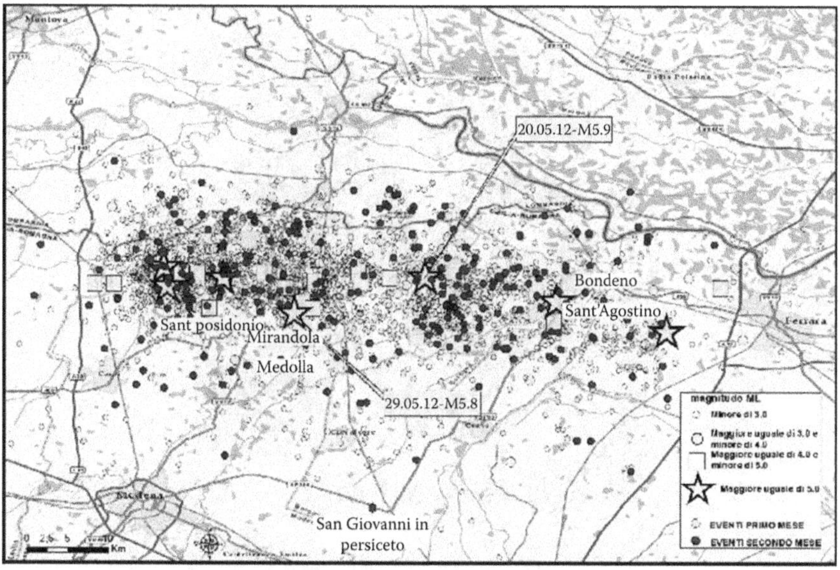

Figure 1.65 The distribution of the seismic events during the series of earthquakes in Emilia (Italy). (http:ingvterremoti.wordpress.com.)

with an epicenter at a depth of 6.3 km and the next one M 5.8 on 29 May. After the second earthquake, three other earthquakes with a magnitude greater than M 5 followed. The locations of these five big earthquakes were identified with stars on the map of Figure 1.65. The minor earthquakes of the seismic series are shown by black and white circles.

Important damage manifested in RC and masonry buildings in many towns and villages located in a wide area belonging to seven provinces. Serious damage manifested in the cultural heritage (historical buildings, churches). A total of 80% of the industrial buildings, generally made of precast RC structures, collapsed (Savoia et al., 2012). It is worth noting that this zone was included in a seismic area just a few years before this earthquake, and therefore the majority of buildings were not designed according to the seismic code.

1.3.5.2 Steel structure damage

The cases of damage are limited to industrial constructions, which can be subdivided into buildings, silos, and shelves (Faggiano et al., 2012). The storage building of a ceramic factory collapsed (Figure 1.66). The structure was composed of two multilevel MRFs, longitudinally and transversally connected at the top by trusses. The collapse was caused by constructional defects in the connection details of column-to-column, beam-to-column,

Figure 1.66 The collapse of an industrial building during the Emilia earthquake. (Faggiano, B. et al. 2012: Steel buildings within the Emilia earthquake (in Italian). Progettazione Sismica, n.3, IUSS Press, ISSN 1973–7432.)

and column-to-foundation. A steel silo of an industrial plan collapsed because of the formation of a plastic hinge at the base (Figure 1.67).

The most vulnerable steel structures were the industrial shelves used for storing medium and large volumes of material. They are usually designed just for gravity loads and composed by modular cold-formed, thin-walled elements bolted together. The most frequent failure mode of big shelves was the overturning, mainly due to the lack of anchorage at the bases, which was facilitated by insufficient lateral stiffness and excessive slenderness of vertical elements, affected also by the out-of-straightness and eccentricities (Figure 1.68). Error in design, defect of erection, and a lack of maintenance were mainly responsible for such failures (Faggiano et al., 2012). These events demonstrated how this special typology of industrial facilities was unfairly underestimated from the structural point of view.

Besides these accidents to industrial steel structures, it is worth stressing that many important steel buildings for special purposes, also characterized

Figure 1.67 The collapse of a steel silo during the Emilia earthquake. (Faggiano, B. et al. 2012: Steel buildings within the Emilia earthquake (in Italian). Progettazione Sismica, n.3, IUSS Press, ISSN 1973–7432.)

Figure 1.68 The collapse of industrial shelves during the Emilia earthquake. (Faggiano, B. et al. 2012: Steel buildings within the Emilia earthquake (in Italian). Progettazione Sismica, n.3, IUSS Press, ISSN 1973–7432.)

by valuable architectural features, such as the Ferrari Museum, behaved very well under this unexpected series of earthquakes (Faggiano et al., 2012).

1.4 ENGINEERING LESSONS LEARNED FROM THE LAST STRONG EARTHQUAKES

1.4.1 Advances in structural design

Recent investigations have enabled some causes of damage to steel buildings to be identified. There are some causes belonging to the design principles (Mexico City, Northridge, and Kobe earthquakes) and other to detailing design (Kocaeli, Bam, Chile, New Zealand, and Japan earthquakes). The conclusions resulting especially after the Northridge and Kobe earthquakes improved the new design codes (Gioncu and Mazzolani, 2011).

The *Northridge earthquake* pointed out two important aspects. On the one hand, the damage can be ascribed to the use of field welding, so poor workmanship is solely to blame, and therefore it is necessary to tighten site supervision and to improve welding details and procedures. On the other hand, the causes of damage can be attributed to an excess of seismic loading, together with defective design guidance, leading to an available rotation ductility or fracture rotation in the connections lower than the earthquake-imposed demand. Some other factors influencing the fracture modalities are related to material properties, temperature, strain rate, joint geometry, plate thickness, and so forth.

More extensive causes can be drawn from the *Kobe earthquake* (Mazzolani, 1995):

- At the material level, damage can be imputed to the concurrence of high velocity of load application and very low temperatures. These conditions are not reproducible by means of the usual laboratory tests and until now they have been ignored by the codes.
- Local buckling of cross-sections in Kobe can be attributed to the important vertical component of the ground motions, which are normally underestimated or sometimes neglected by seismic codes.
- A distinction must be drawn between *far-field* earthquakes, which are characterized by the cyclic alternation of actions, and *near-field* earthquakes, where the impulsive feature of the ground motion is predominant, as it occurred in Northridge and Kobe.

These two earthquakes have changed the direction of earthquake engineering research throughout the world. The reason lies not in the number of dead, but in their economic costs. Each earthquake sets a record loss for natural disasters for both the United States and Japan. Following these earthquakes, it was immediately apparent that the old principles for seismic design had to be changed. While the previous principles had been primarily oriented to safeguard buildings against collapse, the new and more refined rules are devoted to reducing the damage costs, by keeping the nonstructural elements and the structures at an acceptable damage level. So, in structural design, two limit states were introduced: ultimate and service limit states (Gioncu and Mazzolani, 2011).

Regarding the strength requirements, the American code UBC 97 provides for special moment-resisting frames in a reduction factor R_w equal to 12, which is equivalent to a value of the European q-factor equal to 8. By contrast, Eurocode 8 provides a q-factor value equal to 6 and the Japanese code a structural coefficient D_s equal to 0.25, which corresponds to $q = 4$. Therefore, concerning the damage experienced by steel buildings during the Northridge earthquake, it could be attributed to an excess of local ductility demand due to the high earthquake intensity, as compared to the adopted design level. The code provisions, mainly based on constructive measures, do not assure these high levels of local demands.

Furthermore, in the U.S. practice, the beam flange is welded to the column flange by provisionally using a bolted connection in the beam web during the erection. In such a connection, the bending moment at the beam web is hardly transmitted to the column, resulting in a stress concentration in the flanges at the beam ends. This normally occurs in a beam-to-column connection because, while the stress transmission between the beam flange and the column is completely made through the diaphragm plate, the bending moment in the web of the beam cannot be transmitted completely to the column, since the stress transmission is made through the out-of-plane bending of the column flange. The diaphragm plate is usually thicker and wider than the beam flange, and therefore the fracture develops on the side of the beam (Mazzolani, 1998).

These effects are increased by the fact that the weak-beam type is preferred both in Japan and the United States as a yield mechanism. In addition, the use of compact beam sections, with a small width-to-thickness ratio, in order to avoid local buckling, gives rise to stress concentration in the beam flanges and a brittle mode of failure in the connection. Besides, defects in material of the heavy column must be an incentive for the propagation of brittle cracks on the side of the column.

Both in the United States and in Japan, extensive programs of experimental tests have been carried out on beam-to-column subassemblages, in bidimensional and tridimensional frames. It has been evidenced (Bertero et al., 1994) that the types of failure that occurred in welded beam-to-column connections during the earthquakes have already been observed in laboratory experiments. Numerical analyses of the seismic response of steel-framed buildings damaged during the Northridge earthquake have pointed out that there were several ground motions, recorded during this earthquake, able to significantly lead the structure into the inelastic range.

In many cases, the plastic rotation demand at the beam-ends exceeded 0.02 rad; therefore, on the basis of the available experimental data and codes provisions (UBC 97 and AIJ 90), it is clear that the cracking occurring in the connections cannot be considered as unusual.

In the experimental tests carried out in the United States, a plastic rotation supply equal to 0.02 rad has been used as a benchmark to judge the seismic performance of beam-to-column connections, because it was believed to be sufficient to withstand severe earthquakes. As the recent experience demonstrated that this limit value could be exceeded, it is clear that the attention should be focused on the design value of the q-factor, which could be reduced to limit the plastic rotation demands occurring during severe earthquakes or, as an alternative, on the improvement of the seismic performance of the dissipative zones.

The SAC steel project has been developed to derive new design procedures accounting for the lessons learned from the Northridge earthquake. In particular, the structural design philosophy—as well as the characteristics of welding and of the structural details—the velocity of load application, and the influence of the earthquake vertical component are the main issues under investigation within the SAC project. A special session has been devoted to the SAC official presentations during the STESSA 2000 Conference in Montreal (Mazzolani and Tremblay, 2000).

However, on examining the theoretical and experimental SAC results from an observation of the damage in steel structure connections after the earthquakes of Northridge and Kobe, it seems that there is an urgent need to investigate new topics, such as the influence of the strain rate on the cyclic behavior of beam-to-column joints for near-field earthquakes.

In the last period, a wide activity has been developed at the world level to review the whole background of modern seismic codes to grasp the design rules that failed (Mazzolani, 1998).

From a comparison of the European activity to the American and Japanese ones, the following question can rise: *Can the results of the American and Japanese "on field" experience be applied to the European practice?* First, one must mention that the main important damage to steel structures refers to zones in the American and Asian continents. This is mainly because the seismic conditions in Europe are very different from those in those zones; in fact, any important seismic faults, comparable to the ones producing the above analyzed damage, do not cut Europe (with the exception of Vrancea and South Italy), the majority of this area being framed into the so-called stable continental areas with low-to-moderate earthquakes (see Chapter 3).

However, specialists have considered that it is appropriate to analyze if the lesson from the last few important earthquakes could help in the improvement of the European codes. The first analysis has shown that there are important differences in the steel grade, the chemical composition, and the mechanical characteristics of steel. Also, the welding techniques are different. In addition, the different strength requirements and different levels of seismic input lead to different plastic rotation demands. In this context, a group of eight European countries (Italy, Romania, Greece, Portugal, France, Belgium, Bulgaria, and Slovenia) developed a joint research project (INCO-COPERNICUS), the so-called *RECOS (Reliability of Connections of Steel Frame Buildings in Seismic Areas)*, sponsored by the European Commission (Mazzolani, 1999a). The program of the joint project has been established to provide an answer to the above questions, by accomplishing the following objectives:

- Analysis and synthesis of research results, including code provisions, in relation with the evidence of the Northridge and Kobe earthquakes. Particular attention has also been devoted to those research results, which were not accounted for in preparing the provisions of the modern seismic codes.
- Assessment of new criteria for selecting the behavior factor for different structural typologies and definition of the corresponding range of validity. In particular, in this field, the need to provide simplified methods for evaluating the q-factor is felt to be extremely urgent. The aim is to provide the designer with an operative tool that allows to be aware of the inelastic performances of the designed structure.
- Identification and evaluation of the structural characteristics of connections influencing the seismic response of steel buildings. Therefore, the research has been devoted to the strength and stiffness evaluation of moment-resistant connections and to the prediction of their degradation under cyclic loads by also considering the effect of the strain rate and temperature. Low cycle fatigue has to be also investigated.
- Definition of criteria for designing and detailing beam-to-column connections for seismic-resistant structures, also considering the seismicity of the site (far-source and near-source effects).

The RECOS project was completed in 1999 (Mazzolani, 1999b) and the main results have been compiled in a single volume (Mazzolani, 2000).

Nowadays, the code revision activity in Europe is aimed at the improvement of Eurocode 8 *Design of structures for earthquake resistance*, whose application is going to be widespread within the earthquake-prone Mediterranean countries. Since the beginning, a fundamental help to this activity was given by the Technical Committee 13 *Seismic Design* (TC 13) of the European Convention for Constructional Steelwork (ECCS), which was one of the main contributors to the present version of EN 1998-1:2004 "Eurocode 8: Design of structures for earthquake resistance—Part 1: General rules, seismic actions and rules for buildings." In particular, the *European Recommendations for steel structures in seismic zones*, published in 1988 under the chairmanship of F. M. Mazzolani, represented an important background document for the steel section of Eurocode 8, part 1.

However, the design codes are now in a continuous development and TC13, under the chairmanship of R. Landolfo, following its mission, is tracing the guidelines for updating the European seismic codes, according to the worldwide research trends. The TC13 action plan is based on short-term objectives, which start from the analysis of the current status of European and worldwide codification and the identification of further research priorities and critical issues in the technical specifications. In this perspective, the committee focused on the issues in the current version of EN 1998-1, that needed clarification and/or development, aiming at contributing to a new generation of European codes. The outcome of the activity of this committee with respect to the aforementioned objectives in the field of codification and technical specifications can be summarized through the following topics (Landolfo, 2012):

> Material overstrength, selection of steel toughness, local ductility, design rules for connections in dissipative zones, new links in eccentrically braced frames, behavior factors, capacity-design rules, design of concentrically braced frames, dual structures, drift limitations and second-order effects, new structural types, and low-dissipative structures.

1.4.2 Challenges in seismic design

In spite of the important advances in seismic design that resulted due to the impact of the Northridge and Kobe earthquakes, there are many aspects insufficiently reflected in code provisions. Owing to the progress in model-based computer simulation, the structural response can be predicted fairly confidently, but these achievements remain without real effects if the accurate determination of the seismic actions is doubtful (Gioncu and Mazzolani, 2006). The *elimination of the gap* between the advances in engineering seismology and earthquake engineering is a major challenge of the new approach in seismic design (Gioncu and Mazzolani, 2011).

One of these aspects lies in the fact that both earthquakes struck very densely populated areas. Only looking to these experiences, it is sufficient to appreciate the catastrophic potential that could be produced by even a moderate earthquake, when its epicenter is situated around an urban area. The difference in damage costs can be justified by the fact that the epicenter of the Kobe earthquake was located beneath a highly urbanized region, while the one of the Northridge earthquake was positioned beneath the northern edge of Los Angeles. But this enormous damage can be considered as a minor one in comparison with the potential losses in big cities situated in seismic areas, such as San Francisco, Tokyo, and so on. The growing world urban population increases the number of cities, which are located in seismic areas over some earthquake sources. The Northridge and Kobe earthquakes generate a new research direction, the effects of *near-source earthquakes* on the structures situated in epicentral areas (Gioncu and Mazzolani, 2011). Therefore, the code provisions must consider two different situations for the structure design: far-field and near-field earthquakes. The last research works show that, while for far-field earthquakes, the cyclic approach and accumulation of plastic deformation are the prevailing issues, for the near-field earthquakes, the seismic wave propagation with great velocity and high strain-rate effects are very important, producing the connection fracture.

The second very important challenge for design provisions is the *consideration of source type*. The interplate and intraslab sources produce very strong earthquakes, while the intraplate sources provoke low-to-moderate earthquakes. The interplate sources have different characteristics, function of source type: subduction, strike-slip, and collision. The intraslab sources can be considered as intermediate or deep ones. The intraplate sources result in the fracture of some weak crustal zones. There are very important differences in the structural response, so it is a mistake to neglect this aspect (Gioncu and Mazzolani, 2011). Generally, the code provisions consider strong earthquakes only, assuming that using these provisions also for low-to-moderate earthquakes is to be on the safety side. But, in the last period, a new direction of research presented itself that was very interesting: behavior of structures under low-to-moderate earthquakes. For the United States, the *Annals of the New York Academy of Science* published a volume devoted to the differences in earthquake hazards in the eastern and western seismic zones of the United States (Jakob and Turkstra, 1985). The *Electronic Journal of Structural Engineering* (Lam, 2008) published a special issue devoted to low-to-moderate seismic regions of Southeast Asia and Australia. For Europe, the symposium organized by Liege University (Degee, 2008) considers the effects of low and moderate earthquakes produced in north-western Europe.

The third challenge consists of the introduction, besides the verification of ultimate and service limit states, of a new verification, named the *survivability limit state*, especially for severe earthquakes, which concerns the check of the structural ductility by comparing the required ductility with the available ductility. The actual codes are based on the constructional provisions, which are inadequate for strong earthquakes, when the whole structure must form a preselected plastic mechanism and the plastic hinges must have an adequate rotation capacity. Special comprehensive methodology must be elaborated for these situations (Gioncu and Mazzolani, 2002) to ensure that the structure, after reaching the ultimate limit state, has a controlled postlimit behavior.

However, the most important challenge in structural design is to introduce in practice the *new generation of steel structures* that are able to protect the buildings against earthquakes. This new generation aims to improve the existing typologies (moment-resisting frames with protection of connections against fracture, braced frames with braces protected against buckling) or to propose new ones (frames braced with steel or aluminum panels) (see Chapter 5).

REFERENCES

AIJ 1995: Reconnaissance report on damage to steel building structures observed from the 1995 Hyogoken-Nanbu (Hanshin-Awaji) Earthquake. AIJ, Steel Committee of Kinki Branch.

Akiyama, H., Yamada, S. 1995: Damage of steel buildings in the Hyogoken-Nanbu earthquake. In *Proceedings of EASEC '95. Gold Coast*, Australia.

Altay, G. 2005: Post-earthquake damage assessment in the Mediterranean regions. Earthquake protection of historical buildings by reversible mixed technologies, *PROHITECH* project, WG2 document.

Astaneh-Asl, A., Shen, J.H., D'Amore, E. 1995: Seismic vulnerability of welded WMRF damages during 1994 Northridge earthquake. *Giornale Italiane della Costruzione in Acciaio*, Riva del Garda, October 15–18, vol. 1, 69–79.

Balan, S., Cristescu, S., Cornea, I. 1982: *The Romanian Earthquake of 4 March 1977*. Editura Academiei Romane, Bucuresti.

BBC Asia News 2012: Japan panel. Fukushima nuclear disaster "man-made". Comments 248.

Bertero, V.V., Anderson, J.C., Krawinkler, H. 1994: Performance of steel building structures during the Northridge Earthquake. Earthquake Engineering Research Center, Report No. UBC/EERC-94/09, University of California, Berkeley.

Bruneau, M. 2011: Observations of the performance of eccentrically braced frames during the 2011 Christchurch New Zealand earthquake. MCEER report.

Cheng, F.Y., Lu, L.W., Ger, J.F. 1992: Observations on behaviour of tall steel building under earthquake excitations. *SSRC 1992 Annual Technical Session Earthquake Stability Problems in Eastern North America*, Pittsburg, April 6–7, 15–26.

Chi, W.M., Deierlein, G., Ingraffea, A. 2000: Fracture toughness demands in welded beam-column moment connections. *Journal of Structural Engineering*, 126(1), 88–97.

Clifton, Ch., Bruneau, M., MacRae, G., Leon, R., Fussell, A. 2011: Steel structures damage from the Christchurch earthquake series of 2010 and 2011. *Bulletin of the New Zealand Society for Earthquake Engineering*, 44(4), 297–318.

CSUN (nd): Oviatt Library. Earthquake gallery, http://C\Users\user\Desktop\Earthquake Gallery Oviatt Library.mht.

Davidovici, V. 1999: *La Construction en Zone Seismique*. Le Moniteur, Paris.

Davis, P.M., Rubinstein, J.L., Liu, K.H., Gao, S.S., Knopoff, L. 2000: Northridge earthquake damage caused by geologic focusing of seismic waves. *Science*, 289(8 September), 1746–1750.

Degee H. (ed) 2008: *Seismik Risk 2008*. Earthquakes in North-Western Europe. Liege, 11–12 September.

Elnashai, A.S., Bommer, J.J., Martinez-Pereira, A. 1998: Engineering implication of strong motion records from recent earthquakes. In *11th European Conference on Earthquake Engineering*, Paris, September 6–11, No. 59.

Encyclopedia Britannica (nd): Mexico City earthquake of 1985 http://www.britannica.com/EBchecked/topic/1421132/Mexico-city-earthquake-of-1985.

Engelhardt, M.D., Sabol, T.A. 1997: Seismic-resistant steel moment connections: Development since the 1994 Northridge earthquake. Progress in Structural Engineering and Materials. Construction Research Communication Limited, 68–77.

Erdik, M. 2000: Report on 1999 Kocaeli (Turkey) earthquake. Bogazici University Report 2000.

Faggiano, B., Formisano, A., D'Aniello, M., Landolfo, R. 2012: Steel buildings within the Emilia earthquake (in Italian). Progettazione Sismica, n. 3, IUSS Press, ISSN 1973–7432.

Faggiano, B., Iervolino, I., Magliulo, G., Manfredi, G. Vanzi, I. 2009: il comportamento delle strutture industriali nell'evento de l'Aquila. Progettazione Sismica, n.3, IUSS Press, ISSN 1973–7432.

FEMA 355A 2000: State of the art report on base metals and fracture. SAC Joint Venture.

FEMA 355E 2000: State of the art report on past performance of steel moment-frame buildings in earthquakes. SAC Joint Venture.

FEMA 451B 2007: NEHRP recommended provisions for new buildings and other structures. Training and instructional materials, June 2007.

Fischinger, M. 1997: EASY, Earthquake Engineering Slide Information System. FGG-IKPIP Institute Slovenia.

Ger, J.F., Cheng, F.Y. 1992: Collapse assessment of a tall building damaged by 1985 Mexico earthquake. *The 10th World Conference on Earthquake Engineering*, Madrid, July 19–24, Balkema, Rotterdam, 51–59.

Ger, J.F., Cheng, F.Y., Lu, L.W. 1993: Collapse behaviour of Pino Suarez building during 1985 Mexico City earthquake. *Journal of Structural Engineering*, 119(3), 852–870.

Gioncu, V. 2006: Advances in seismic codification for steel structures. *Costruzioni Metalliche*, 6, 69–87.

Gioncu, V., Mazzolani, F.M. 2002: *Ductility of Seismic Resistant Steel Structures*. Spon Press, London.

Gioncu, V., Mazzolani, F.M. 2006: Influence of earthquake type on the design of seismic-resistant steel structures. Part 1: Challenges of new design approaches, Part 2: Structural responses for different earthquake types, *Proceedings of the STESSA Conference on Behaviour of Steel Structures in Seismic Areas* (eds. F.M. Mazzolani, A. Wada), Yokohama, Japan, August 14–17, 2006.

Gioncu, V., Mazzolani, F.M. 2011: *Earthquake Engineering for Structural Design*. Spon Press, London.

Hamada, M., Wakamatsu, K. 1999: Liquefaction, ground deformation and their related damage to structure. Report on 1995 Kobe earthquake, INCEDE Report, 1999-03, 57–112.

Herrera, R.A., Beltran, J.F., Aguirre, C., Verdugo, A. 2012: Seismic performance of steel structures during the 2010 Maule earthquake. In *Behavior of Steel*

Structures in Seismic Areas, STESSA 2012 (eds. F.M. Mazzolani, R.A. Herrera), Santiago of Chile, January 9–11, CRC Press, 37–43.

Hudson, R.L., Skyers, B.D., Lew, M. 1996: Vertical strong motion characteristics of the Northridge earthquake. In *11th World Conference on Earthquake Engineering*, Acapulco, June 23–28, No. 728.

Iwan, W.D. 1995: Drift demand spectra for selected Northridge sites. SAC Report 95-05.

Iwan, W.D. 1996: The drift demand spectrum and its application to structural design and analysis. In *11th World Conference on Earthquake Engineering*, Acapulco, June 22–28, No. 1116.

Jakob, H.K., Turkstra, C.L. (eds) 1985: Earthquake hazard and design of constructional facilities in the Eastern United States. *Annals of the New York Academy of Sciences*, 558, XI–XIV, 457–588.

JSSS, 1997: Kobe earthquake damage to steel moment connections and suggested improvement. Technical report, No. 39.

Kurobane, Y., Azuma, K., Ogawa, K. 1997a: Brittle fracture in steel building frames. Comparative study of Northridge ND Kobe earthquake damage. *International Institute of Welding, Annual Assembly*, 1–30.

Kurobane, Y., Ogawa, K., Ueda, C. 1996: Kobe earthquake damage to high-rise Ashiyahama apartment buildings: Brittle tensile failure of box section columns. In *Tubular Structures VII* (eds. J. Farkas and K. Jarmai), Miskolc, August 26–30, Balkema, Rotterdam, 277–284.

Kurobane, Y., Wang, B., Azuma, K., Ogawa, K. 1997b: Brittle fracture in steel frames. In *Behaviour of Steel Structures in Seismic Areas* (eds. F.M. Mazzolani and H. Akiyama), Kyoto, August 2–8, 10/17 Salerno, 833–844.

Lam N.T.K. (ed) 2008: Special issue: Earthquake engineering in the low and moderate seismic regions of Southeast Asia and Australia. *Electronic Journal of Structural Engineering*, 1, 1–141. ISSN: 1443-9255.

Landolfo, R. (ed) 2012: *Assessment of EC8 Provisions for Seismic Design of Steel Structures (First Edition)*. ECCS Publication, Brussels.

Mazzolani, F.M. 1995: Some simple considerations arising from Japanese presentation on the damage caused by the Hanshin earthquake. In *Stability of Steel Structures* (ed. M. Ivanyi), SSRC Colloquium September 21–23, Budapest, Akademiai Kiado, vol. 2, 1007–1010.

Mazzolani, F.M. 1998: Design of steel structures in seismic regions: The paramount influence of connections. In *Control of the Semi-Rigid Behaviour of Steel Engineering Structural Connections* (ed. R. Maquoi), COST C1 Conference, Liége, 17–18 September, 371–384.

Mazzolani, F.M. 1999a: Reliability of moment resistant connections of steel building frames in seismic areas: The first year of activity of the RECOS project. In *2nd European Conference on Steel Structures EUROSTEEL*. Prague, May 26–29.

Mazzolani, F.M. 1999b: Reliability of moment resistant connections of steel building frames in seismic areas. In *Seismic Engineering for Tomorrow, International Seminar in Honor of Professor Hiroshi Akiyama*. Tokyo, Japan, November 26.

Mazzolani, F.M. (ed) 2000: *Moment Resisting Connections of Steel Frames in Seismic Areas: Design and Reliability*. E & FN SPON, New York.

Mazzolani, F.M., Tremblay, R. (eds) 2000: Behaviour of Steel Structures in SeismicAreas, STESSA 2000, Montreal, August 21–24.

Midorikawa, M., Nishiyama, I., Tada, M., Terada, T. 2012: Earthquake and tsunami damage on steel buildings caused by the 2011 Tohoku Japan earthquake. In *International Symposium on Engineering Lessons Learned from the 2011 Great East Japan Earthquake*, Tokyo, 1–4 March 2012, 1061–1076.

Nakashima, M. 2000a: Quality assurance for welding of Japanese welded beam-to-column. In *Behaviour of Steel Structures in Seismic Areas, STESSA 2000* (eds. F. M. Mazzolani and R. Tremblay), Montreal, August 21–24, Balkema, Rotterdam, 223–230.

Nakashima, M. 2000b: Overview of damage of steel building structures observed in the 1995 Hyogoken-Nanbu (Kobe) earthquake. Disaster Prevention Research Institute Report.

Nakashima, M., Roeder, C., Maruoka, Y. 2000: Steel moment frames for earthquakes in United States and Japan. *Journal of Structural Engineering*, 126(8), 861–868.

NILIM-BRI, 2011: The off the Pacific coast of Tohoku earthquake. BRI Report.

Osteraas, J., Krawinkler, H. 1989: The Mexico City earthquake of September 19, 1985. Behavior of steel buildings. *Earthquake Spectra*, 5(1), 51–88.

Saatcioglu, M. et al. 2001: The August 17 1999 Kocaeli (Turkey) earthquake-damage to structures. *Canadian Journal of Civil Engineering*, 28, 715–737.

Savoia, M., Mazzotti, C., Buratti, N., Ferracuti, B., Bova, M., Ligabue, V., Vincenzi, L. 2012: Damages and collapses in industrial precast buildings after the Emilia earthquake. Special issue of "Earthquake Engineering" dedicated to the Emilia Earthquake, May 2012, n. 2–3, April–September, 120–131.

Scawthorn, C. 2003: Earthquake: A historical perspective, Chapter 1. In *Earthquake Engineering Handbook* (eds. W.F. Chen and C. Scawthorn), CRC Press, London.

Takabatake, H., Nonaka, T. 2001: Numerical study of Ashiyahama residential building damage in the Kobe earthquake. *Earthquake Engineering and Structural Dynamics*, 30, 879–897.

Tomlison, M.J., Boorman, R. 2001: *Foundation Design and Construction*. Longman Scientific Technical, UK.

USGS (nd): 1906 San Francisco Earthquake Photos, http://earthquake.usgs.gov/regional/nca/1906/18april/photos.php.

USGS (nd): The behaviour of column, http://www.smate.wwu.edu/teched/geology/eq-column.html.

USGS (nd): Historic earthquakes. Kanto, Japan, 1923 September 01, M 7.9, http://www.earthquake.usgs.gov/earthquakes/world/events/1923_09_01.php.

USGS 1996: Surface expression of coactive slip on a blind fault during Northridge earthquake sequences, California. Open report 96-698.

Vannacci, G. 1994: 17 Gennaio, 94: Il terremoto di Northridge conferma le superiorita dell'acciaio. *Costructioni Metalliche*, 5, 9–12.

Wikipedia (nd): Torre Latinoamericana. http://en.wikipedia.org/wiki/Torre_Latinoamericana.

Zahrai, S.M., Heidarzadeh, M. 2004: Seismic performance of existing buildings during the Bam earthquake. In *13th World Conference on Earthquake Engineering*, Vancouver, 1–6 August 2004, Paper No. 1715.

Chapter 2

Steel against earthquakes

2.1 STEEL AS THE MATERIAL OF CHOICE FOR SEISMIC AREAS

Steel structures have always been considered as a suitable solution for constructions in high seismicity areas, due to the very good strength and ductility exhibited by the structural material, the high quality assurance guaranteed by the industrial production of steel shapes and plates, and the reliability of connections built up both in the workshop and in the field (Mazzolani and Piluso, 1996; Mazzolani and Gioncu, 2000). To exemplify this good behavior, in many papers, the excellent performance, during the 1957 earthquake, of the Torre Latino Americana building in Mexico City (Figure 2.1a) was mentioned as an example, in contrast with the reinforced concrete buildings, which frequently underwent many collapses. Based on these considerations, for a long time, no provisions for steel structures were included in the codes, considering that no problems occurred in seismic area.

But during the seismic decade 1985–1995, the specialists recognized that the good behavior of steel structures, under particular conditions, may be a dogma, which is denied by reality. In fact, the Mexico City (1985), Northridge (1994), and Kobe (1995) earthquakes have seriously compromised the ideal image of steel as being the perfect material for seismic areas. It was a providential sign that in the same place, Mexico City, where the Torre Latino Americana building was assumed to be an example of good performance of steel structures, the first overall collapse of a steel structure, the Pino Suarez building (Figure 2.1b), occurred. The bad performance of joints in steel structures, both in the Northridge and Kobe earthquakes, having the same characteristics of damage, shows that there are some general mistakes in design concept. And the fact that in both cases damage also arose when the design and detailing were performed in perfect accordance with the design philosophy and code provisions amplifies the challenge addressed to structural engineers (Gioncu and Mazzolani, 2002, 2003). The main conclusion, from the damage analysis during both the

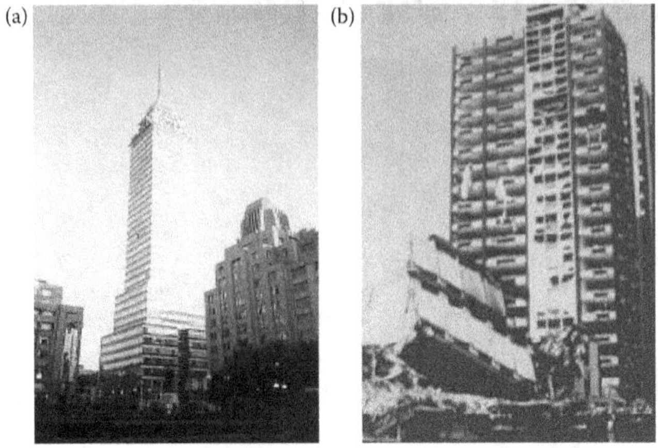

Figure 2.1 Good and bad steel structure behaviors: (a) Torre Latino Americana; (b) Pino Suarez building.

abovementioned earthquakes, is that these events belong to the category of *exceptional actions* not included in current design codes (Mazzolani, 2002).

At the same time, one must mention that, in contrast to reinforced concrete structures, the case of the overall collapse of steel structures during these exceptional actions has been extremely rare. Contrarily, steel structures during strong earthquakes suffered many local brittle failures at the beam-to-column connections, which do not produce an overall collapse, but it shows that the knowledge about the behavior of the steel structures is not complete.

To improve the constructional details and to propose new design solutions for achieving a correspondence between the design requirements and the actual structural response, the scientific community began to deepen the reasons for this poor behavior: does it depend on the material quality, on the design concept, on the structural scheme, on the constructional detail, on the code provision or on the seismic input occurring (Mazzolani, 1998, 2000b)? Most of these questions are still being analyzed, but much more has been understood about the seismic behavior of steel structures. Consequently, during the last few years, research works made some important advances in the seismic design of steel structures, which will be presented in the following chapters. One must mention the great influence of the following research activities:

1. In the United States, by the SAC Joint Venture (Structural Engineers Association of California, Applied Technology Council, California Universities for Research in Earthquake Engineering) (Roeder, 2000), founded by the Federal Emergency Management Agency (FEMA).

2. In Europe, by RECOS (Reliability of Moment Resistant Connections of Steel Building Frame in Seismic Areas), sponsored by the European Community within the INCO-COPERNICUS joint research projects, developed by partners from eight European countries (Belgium, Bulgaria, France, Italy, Portugal, Romania, and Slovenia) (Mazzolani, 1999c, 2000a); the series of the STESSA (Behaviour of Steel Structures in Seismic Areas) Conferences (1994, Timisoara; 1997, Kyoto; 2000, Montreal; 2003, Napoli; 2006, Yokohama; 2009, Philadelphia; 2012, Santiago de Chile; the next in 2015 is planned in Shanghai).

These results have already been or are going to be introduced into the structural design provisions for the seismic-resistant design in many earthquake-prone countries, giving rise to a new generation of seismic codes. In spite of the above-isolated bad behaviors, which can be eliminated by a proper design, steel structures certainly continue to be the best solution for structures situated in seismic area.

One question arises from the above-presented damage produced during the North American and Japan earthquakes (Mazzolani, 1999a,b,c):

Why did these important earthquakes damaging steel structures not affect the European countries?

The first answer is related to the fact that the majority of earthquakes that took place in Europe are very different from the ones occurring in other parts of the world (see Chapter 3). The second explanation, which is even more important, is connected to the fact that the most important users of steel structures in Europe are situated in low seismicity areas. The annual Statistical Report of the ECCS (see Figure 2.2) presents the total European production of steel used for constructions (kilotonnes/million inhabitants) for each country belonging to this organization. The countries are classified in terms of their seismicity, namely, low seismicity (magnitude less than 4.5), moderate seismicity (magnitude between 4.5 and 6.5), and high seismicity (higher than 6.5). To these data, one can add the low consumption of other countries not included in the ECCS organization, such as Spain, Greece, and Balkan countries, but situated in high seismic areas. This is a confirmation that the steel consumption is very high in countries with low seismicity and very low for countries with high seismicity.

Considering the recognized good behavior of steel structures, this situation seems to be strange, but this is related to the economical reasons and consolidated traditions. The Northern countries use steel and composite structures more than the Southern countries do, mainly for economical reasons. In contrast, the Southern countries are less rich than the Northern countries and they traditionally use both masonry and reinforced concrete as the most common constructional materials and also sometimes less expensive materials (Mazzolani, 2006). Therefore, the best solution for the

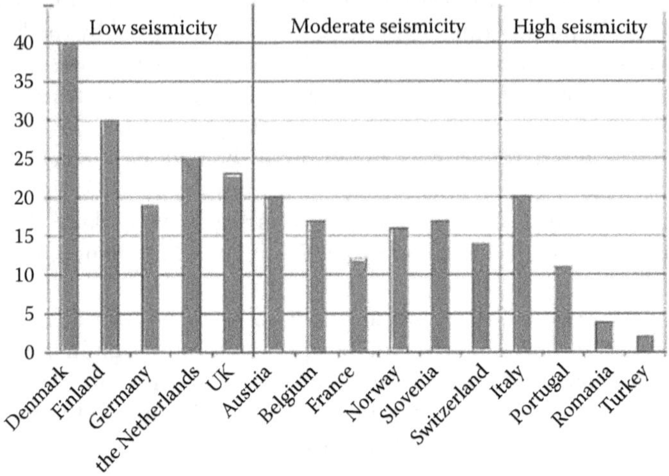

Figure 2.2 Steel consumption in the ECCS countries.

countries situated in seismic area should be to renounce their traditions and use steel structures on a large scale for assuring their seismic protection.

2.2 DEVELOPMENT OF STEEL STRUCTURAL SYSTEMS

2.2.1 Early development

The use of steelwork in seismic-resistant structures is quite new, being experienced for the first time since the beginning of the 20th century. This practice was slowly inherited from the wood technology, which was used in the past to reinforce the masonry buildings in some earthquake-prone countries of the Mediterranean basin.

Many regions in the Balkan Peninsula, Greece, Italy, and Turkey were devastated by catastrophic earthquakes during the past centuries, giving rise to spontaneous, but not engineered, building systems to protect the population from the recurring seismic effects.

From a historical overview of the development of seismic-resistant solutions, it seems that one of the first ideas was raised after the 1627 earthquake of Benevento in Campania (Italy), when a new constructional system, the so-called *baraccato*, was imposed by the authorities. It consisted of a wooden skeleton with regular mesh whose columns were fixed in the masonry foundation or directly in the ground. The wooden frames were covered on both sides by thin canes or wicker panels reinforced by clay or mortar followed by plastering. The lightness of these walls, together with the wood strength, optimized the seismic behavior by reducing the inertia forces.

A century later, the ravaging effects of the 1755 earthquake of Lisbon (Portugal) compelled the Marquis of Pombal to issue technical provisions imposing the use of wooden frames similar to the ones of Benevento.

A similar system, called *case baraccate*, was introduced in Calabria (Italy) after a devastating earthquake at the end of the 18th century (1783), and it was imposed by law as a compulsory system by King Ferdinand IV in the Reign of the Two Sicilies, hoping to reduce the damage due to further earthquakes. This special type of housing represented a new composite system, in which the masonry walls are reinforced by wooden frames. The wooden frames are inserted and completely hidden inside the masonry (Figure 2.3a), according to the structural scheme of a frame with vertical, horizontal, and diagonal bars, which anticipated the scheme of the modern-braced structures (Figure 2.3c), where wood is substituted by steel.

Similar systems, using the structural cooperation between masonry and wood as the *case baraccate*, can be also found in other Mediterranean countries for seismic protection purposes. The use of the combination of wood-masonry was also very popular in northern European countries, such as France, the United Kingdom, Germany, Austria, and the Netherlands, where the wooden elements have the scope to reinforce the building, but not for seismic protection purposes, contemporarily giving an esthetic feature to the facade (Figure 2.3b). The *case baraccate* can be surely considered as the ancestors of the modern seismic-resistant steel structures with megadiagonals on the facade (Figure 2.3c).

2.2.2 Development in the United States

The steel frame buildings mainly started to be constructed in the United States at the beginning of the 20th century. Some early examples are the tall buildings of San Francisco at the end of the 19th century, by using steel frames to support gravity loads, surrounded by masonry (brick or stone) perimeter walls to provide lateral load resistance.

This type of buildings performed well during the San Francisco earthquake (1906), so it was also used later on (Elsesser, 2004).

The good performance of steel encouraged the development of its use in new structural systems during the period 1906–1940, particularly in California, where steel frames were designed to also carry lateral loads by using different new systems: knee bracing, belt trusses on the floors to limit drift, rigid-frame moment connections using top and bottom girder flange connections to columns. Since the 1950s in the United States, the riveted connections were substituted by high-strength bolted joints and, starting from 1960, the moment connections to columns were made by partially welded girders, making the welded steel moment frame the primary structural system for resisting lateral loads.

(a)

(b)

(c)

Figure 2.3 Evolution of bracing systems: (a) *case baraccate*; (b) north European houses; (c) modern-braced buildings.

After the 1989 Loma Prieta and 1994 Northridge earthquakes, which marked many fractures in welded connections (Gioncu and Mazzolani, 2002), some new conceptions for steel structure design were developed, in particular the new solutions to improve the connection behavior. Another improvement considers the possibility of providing the structural scheme with seismic energy dissipation, leading the engineers to design the so-called dual systems: one for providing strength and the other for adding both damping and energy dissipation.

To sum up, the chronological development of steel structures, mainly in the United States, can be listed as follows:

1. 1890 to 1930, the system consisted of steel framework with brick and after 1910 with reinforced concrete as in-filled panels. In this early period, steel buildings were composed by members that are built up from plates, angles, channels, tie-plates and lacing with haunch, knee brace and large gusset plate connections, as illustrated in Figure 2.4 (Roeder, 2002). During this period, many structures of this type were erected, not only in the strong seismic areas of California but also in New York and Chicago with low seismicity. Figure 2.5 shows the buildings of the Royal Globe Insurance in San Francisco and New York.
2. 1930 to 1960, steel moment-resisting frames (MRFs) were created by using riveted (or bolted) top and bottom girder connections to the column (Figure 2.6) (Roeder, 2002). The characteristic buildings of

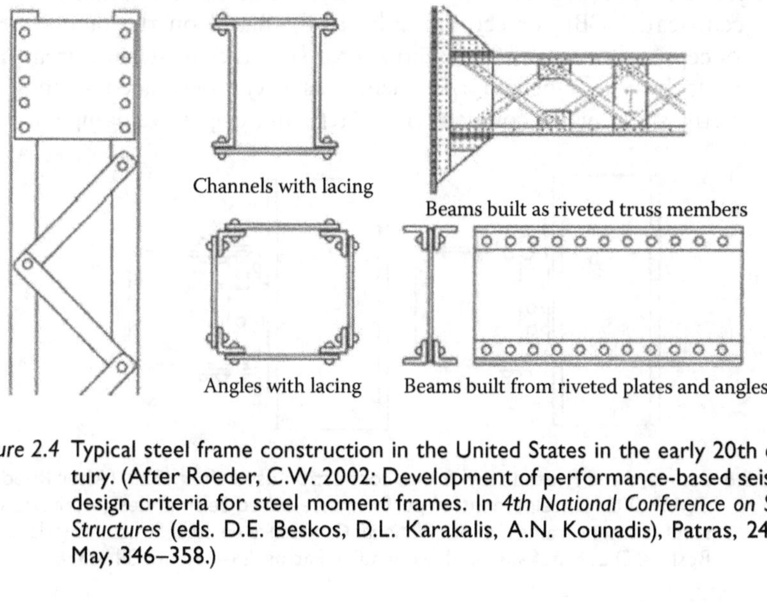

Figure 2.4 Typical steel frame construction in the United States in the early 20th century. (After Roeder, C.W. 2002: Development of performance-based seismic design criteria for steel moment frames. In *4th National Conference on Steel Structures* (eds. D.E. Beskos, D.L. Karakalis, A.N. Kounadis), Patras, 24–25 May, 346–358.)

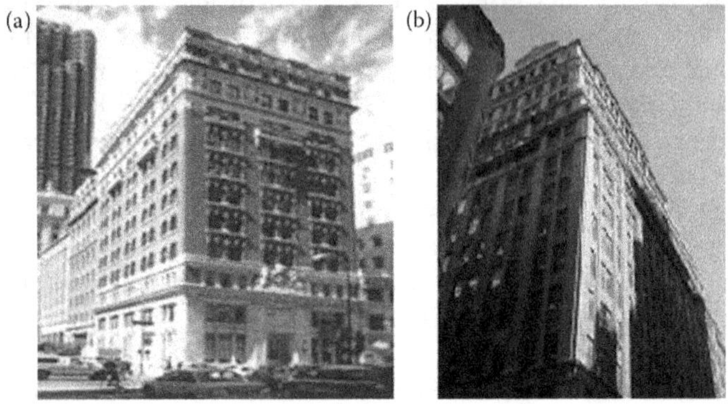

Figure 2.5 Royal globe insurance buildings: (a) 1907 San Francisco; (b) 1927 New York.

this period are the San Francisco Shell Oil Building and the New York Empire State Building (Figure 2.7).

3. 1960 to 1995, steel moment frames with all welded beam–column connections were used (Figure 2.8). During this period, very important buildings were erected (Figure 2.9), having some special megabracing systems on the facade, pyramidal form, or mixed steel–concrete structures (Figure 2.10).
4. After the 1995 Northridge earthquake, when the conventionally welded frames were generally considered as vulnerable schemes, a great number of very innovative solutions were studied. Many satisfactory seismic-resistant solutions were proposed by using new connection types (based mainly on bolted connections) and braced frames (centrically, CBF, or eccentrically, EBF). Based on the dual-system concept, other significant solutions for structure in seismic areas use stable hysteretic-behavior elements (as eccentric-braced frames or ductile shear walls) coupled with MRFs and/or passive dampers.

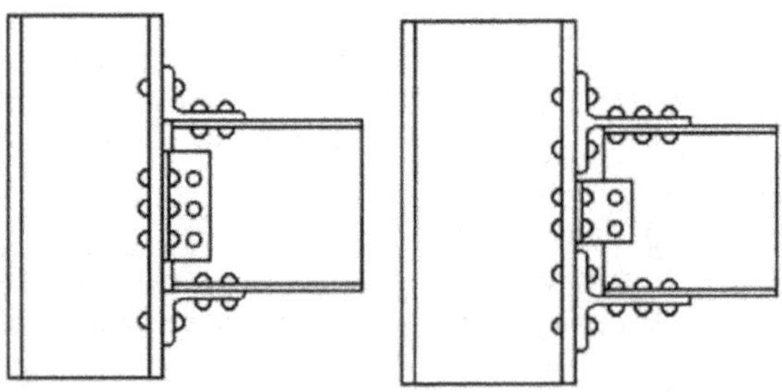

Figure 2.6 Standardized riveted steel frame connection used after 1920. (After Roeder, C.W. 2002: Development of performance-based seismic design criteria for steel moment frames. In *4th National Conference on Steel Structures* (eds. D.E. Beskos, D.L. Karakalis, A.N. Kounadis), Patras, 24–25 May, 346–358.)

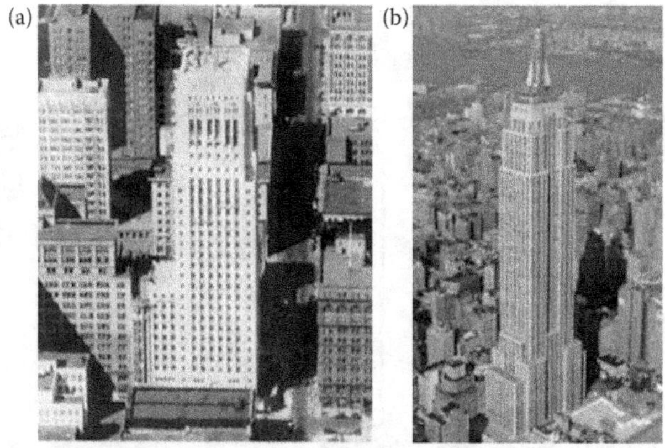

Figure 2.7 First tall buildings in the United States: (a) 1930 San Francisco Shell Oil Building; (b) 1931 New York Empire State Building.

These solutions, in addition to providing strength, limit the frame joint rotation, add damping to the system, and reduce seismic loads. Reliable systems, completely different from the classical ones, are designed using the seismic isolation concept. Thus, the building structure is supported on isolation bearings, which effectively separate the superstructure from the ground. From the early 1980s, this concept has been thoroughly studied, researched, and analyzed, and now it is used in significant projects in regions of high seismicity.

The main advanced solutions for steel structures in seismic areas will be presented in Chapter 4. It can be certainly recognized that almost all new solutions for steel structures in seismic areas come from the innovative ideas proposed in the United States.

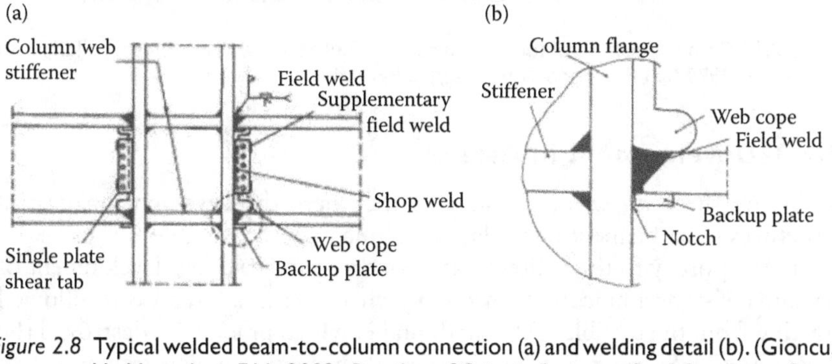

Figure 2.8 Typical welded beam-to-column connection (a) and welding detail (b). (Gioncu, V., Mazzolani, F.M. 2002: *Ductility of Seismic Resistant Steel Structures*. Spon Press, London.)

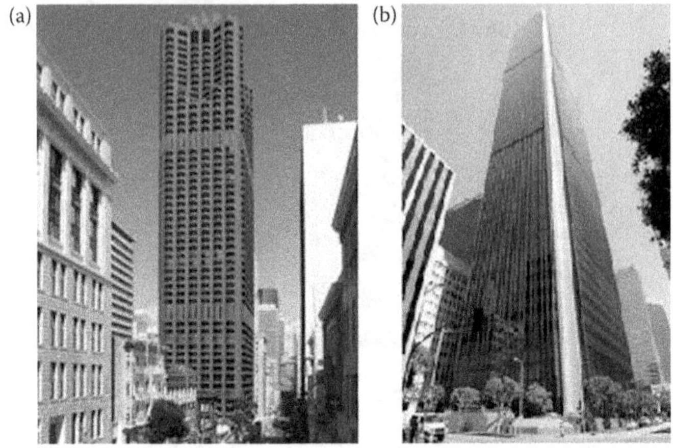

Figure 2.9 Building erected in high seismicity areas: (a) San Francisco Bank of America; (b) 1973 Los Angeles Aon Center.

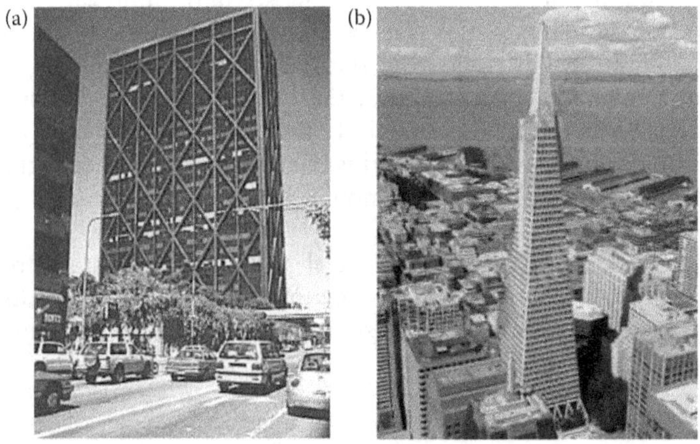

Figure 2.10 Some special structural systems: (a) 1964 San Francisco Alcoa Building; (b) 1972 San Francisco Transamerica Pyramid.

2.2.3 Development in Asia

There are three important seismic zones where the development of steel structures must be mentioned: Japan, China, and Taiwan.

Contemporary to the United States, one assisted to a big development of seismic-resistant applications in *Japan*, where a similar trend was followed starting from the terrible 1923 earthquake, which practically destroyed the city of Tokyo. Since that event, the seismic protection policy grew in parallel with the growing of the steel culture. Today, the majority of building

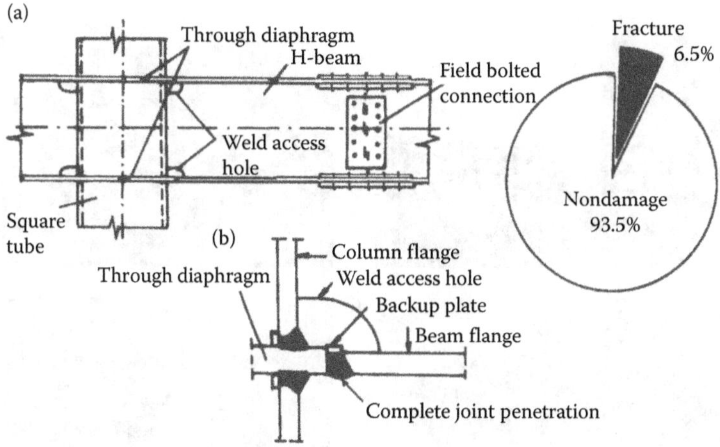

Figure 2.11 "Column tree" system used in Japan: (a) prefabricated beam-to-column joint; (b) beam-to-column welded connection.

structures in Japan are made of steel, according to the main structural typologies and by using very high-industrialized technology. Steel columns are shop-welded and transported to the site, where they are assembled. The so-called *column tree* system is often used for normal building, in order to facilitate the erection on site of the prefabricated shop-welded elements (Figure 2.11). In this way, the rigid node is completely welded in a shop under quality control.

All the main structural types are used (Figure 2.12):

1. Moment resisting frames (MRFs)
2. Steel frames with reinforced concrete (RC) cores
3. Dual moment frames with bracings

The MRF's structural scheme is very popular in many important tall buildings that have been erected in Tokyo (Figure 2.13).

After the tragic experience of the Kobe's earthquake, also in Japan the problem of the use of energy dissipation devices has been greatly felt. Now, a majority of multistoried buildings are equipped with special protection systems.

The introduction of dissipative stiffened panels within the framed structure is today very common in constructional practices (Figure 2.14).

The tuned active damper (TAD) system placed at the top of a building is used to minimize the wind and earthquake vibration of the Yokohama Landmark Tower (Figure 2.15).

China is in the process of rapid demographic and social change involving the nationwide engineering and building construction on a massive scale. Devastating earthquakes have been recorded through the Chinese history.

Figure 2.12 Steel seismic-resistant structures in Japan: (a) MRF; (b) MRF + RC cores; (c) MRF + CBF; (d) MRF + EBF.

Figure 2.13 Tokyo Metropolitan Building: (a) during the erection; (b) completed.

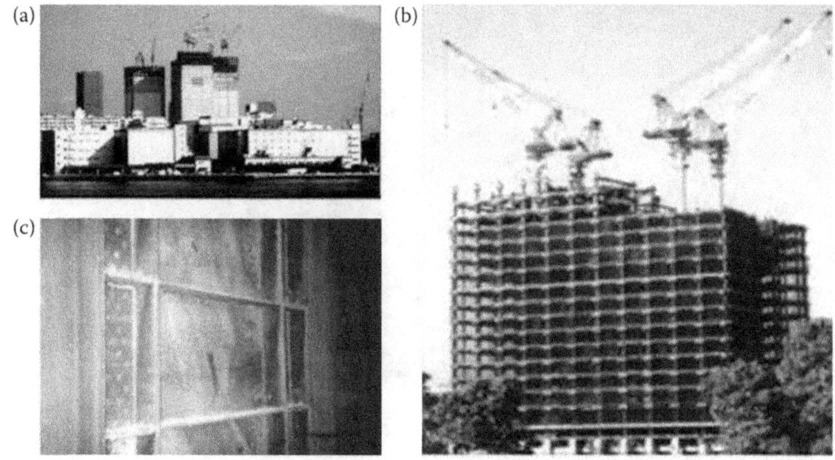

Figure 2.14 New steel buildings in Tokyo Bay: (a) general view; (b) during erection; (c) dissipative panels.

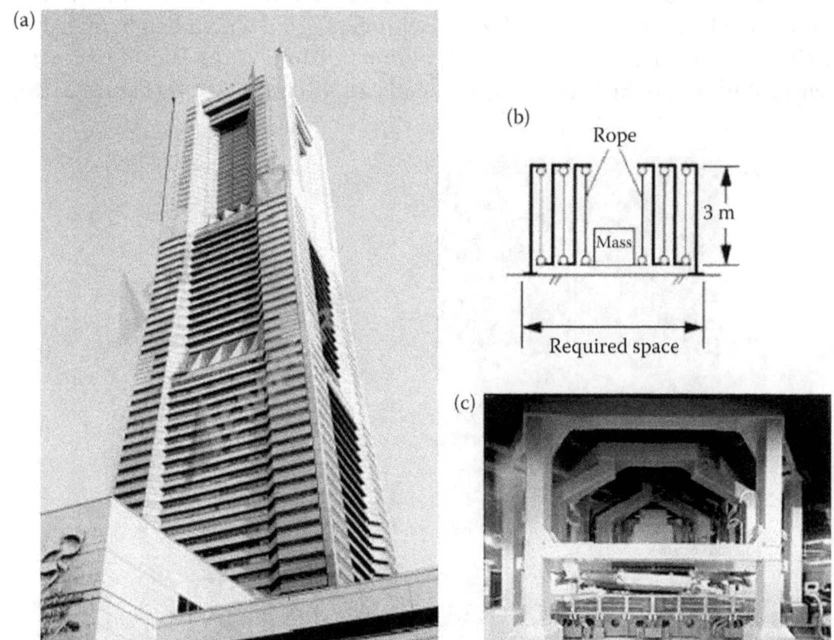

Figure 2.15 Yokohama Landmark Tower: (a) view; (b) and (c) TAD system.

Figure 2.16 High-rise buildings in Shanghai.

So, owing to the existence of a lot of vulnerable building stock, there is a large process to replace them with modern high-rise buildings (Figure 2.16), the steel structure being placed in a very good position, considering the large production of steel in China (Bolton and Cole, 2006). Figure 2.17 shows, as an example, two well-known buildings: the Bank of China Tower from Hong Kong and the Jin Mao Tower in Shanghai with a height of 421 m in 1999. At the moment, the tallest building in China is the World Financial Centre Building in Shanghai with a height of 480 m in 2008 (Figure 2.18).

Figure 2.17 Well-known tall buildings from China: (a) 1990 Hong Kong Bank of China Tower; (b) 1998 Shanghai Jin Mao Tower.

(a) (b)

Figure 2.18 The World Financial Centre in Shanghai (a), with the *window to the sky* (b).

Taiwan sits on an extremely active tectonic region, with seismicity and rate of crustal motion among the highest in the world. The country is also home to a dense population and numerous industrial facilities. Thus, it took immense courage to build over a very active fault the world's tallest building (until 2004), the Taipei 101 Tower with a height of 508 m, the so-called *a giant glass pagoda* (Figure 2.19). For reducing the lateral accelerations due to wind and earthquake, the building was equipped with a tuned mass damper (TMD) system, designed and installed at the building top. This system consisted of a steel ball weighing 606 tonnes, made of a stack of steel plates of varying dimensions, suspended from the 92nd floor. This passive TMD system is more reliable and economical than the active TAD system.

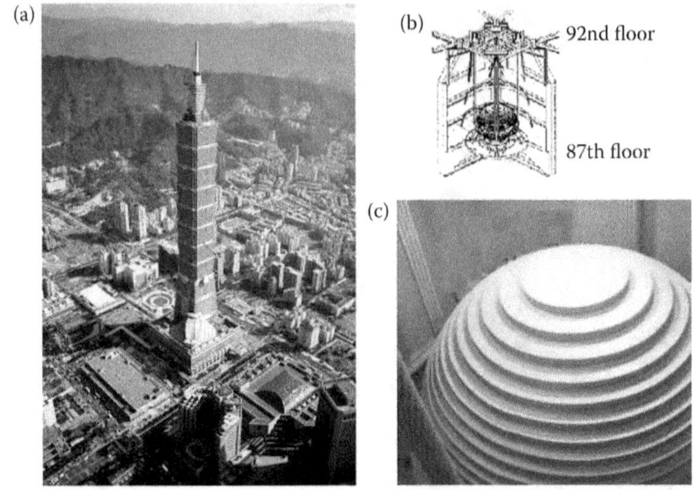

Figure 2.19 2004 Taipei 101 Tower: (a) View of the Tower; (b) tuned mass damper system; (c) detail of the mass.

With these challenging examples, Asia has proved to be the most dynamic region in the world, where new advanced antiseismic solutions are applied with great courage.

2.2.4 Development in Europe

The same development as in the United States and Asian countries was not so evident in the earthquake-prone countries of *Europe*, where reinforced concrete structures continue to be the most popular technology, and also for seismic-resistant applications.

A slow but continuous improvement of the building quality, also in the European earthquake-prone countries, started to be apparent in the last decades of the 20th century. The lessons learned in the last decades from the more severe earthquakes in south Europe (Friuli, Italy, 1976; Bucharest, Vrancea, Romania, 1977; Campania and Basilicata, Italy, 1980; Banat, Romania, 1991; Erzincan, Turkey, 1992; Dinar, Turkey, 1995; Umbria, Italy, 1997; Adana, Turkey, 1998; Izmit and Duzce, Turkey, 1999; Athens, Greece, 1999; Abruzzo, Italy, 2009; Emilia, Italy, 2010) have shown that the behavior of reinforced concrete structures, which are the majority there, has generally been very poor for many unquestionable reasons, that is, nonseismically designed structures, or formally seismically designed, but characterised by wrong detailing and bad execution with poor materials, or seismically designed, but with loads lower than the actual ones.

In contrast, the performance of steel and composite structures, even if not very numerous, has been satisfactorily good everywhere, with a very limited amount of damage and total absence of collapsed buildings. This evidence is going to produce a slow, but continuous, increase in the use of steelwork in seismic-resistant structures also in South Europe (Mazzolani 1999a,b, 2003, 2006).

Since 1980, this tendency was clearly evident in Italy, mainly in the construction of buildings that play a fundamental role during the emergency, such as hospitals, fire stations, electrical facilities, town halls, and so on. In the last decades, also in other southern European countries (Bulgaria, Greece, Romania, and Portugal), this process started and some steel and composite buildings have been erected for important applications.

In *Italy*, the Naples Management Centre (CDN) is an example of this new trend (Figure 2.20a). It is the largest urban development initiative currently underway in Italy and one of the largest in Europe. It consists of the creation of a new city quarter, which provides residential and business functions in such as a way to meet the increasingly growing demand in the Naples metropolitan area (Mazzolani, 1995d). After the Campania-Basilicata earthquake of 1980, the city of Naples was included in the third seismicity zone, the lowest one, which corresponds to a seismic intensity coefficient equal to 0.04, according to the national seismic code. At that time for buildings already designed, but not yet under construction or at their early beginning, it was necessary to provide the upgrading of the structural scheme by keeping the same typology and introducing suitable strengthening changes to fulfil the new seismic requirements. It was the case with many buildings of the Management Centre. They range from 25 to 120 m high and most of them belong to the relevant category of primary importance for civil protection. Many different solutions have been conceived for the seismic-resistant structures, both classical and innovative. The most used classical solution is the mixed one, in which reinforced concrete cores provide to stabilize the simply pinned steel skeleton against the effect of horizontal loads (Figure 2.20b). Many of the following applications shown here belong to this typology.

The Law Court building comprises three towers, which are equal in plan but have different height varying from 78 to 117 m (Figure 2.21a). They

(a) (b)

Figure 2.20 Naples Management Centre, Italy: (a) general view; (b) during erection of MRF + RC cores.

are connected by a horizontal building in the front part, containing the courtrooms.

For each tower, the structural system is made of reinforced concrete curved walls and cores, which provide strength and stability under horizontal loads (wind, earthquakes) and a steel skeleton resisting vertical loads only. The steel columns with a square box section are located along the perimeter. The main floor beams have a radial direction and they are pinned both to the steel columns and to the reinforced concrete walls.

Two twin towers make up the electrical department headquarters of Naples, 120 m high (Figure 2.20b). Their plan has a lozenge shape with dimensions 58 × 14. Two reinforced concrete cores, containing stairs and elevators, connected at the top by a box-section girder to which the 29 stories are suspended, represent the seismic-resistant structure of each tower. The suspended structure is made of steel ties and steel–concrete composite floor beams. The horizontal connection between the cores and composite floor structure is provided by means of elastic–plastic dissipative devices,

(a)

(b)

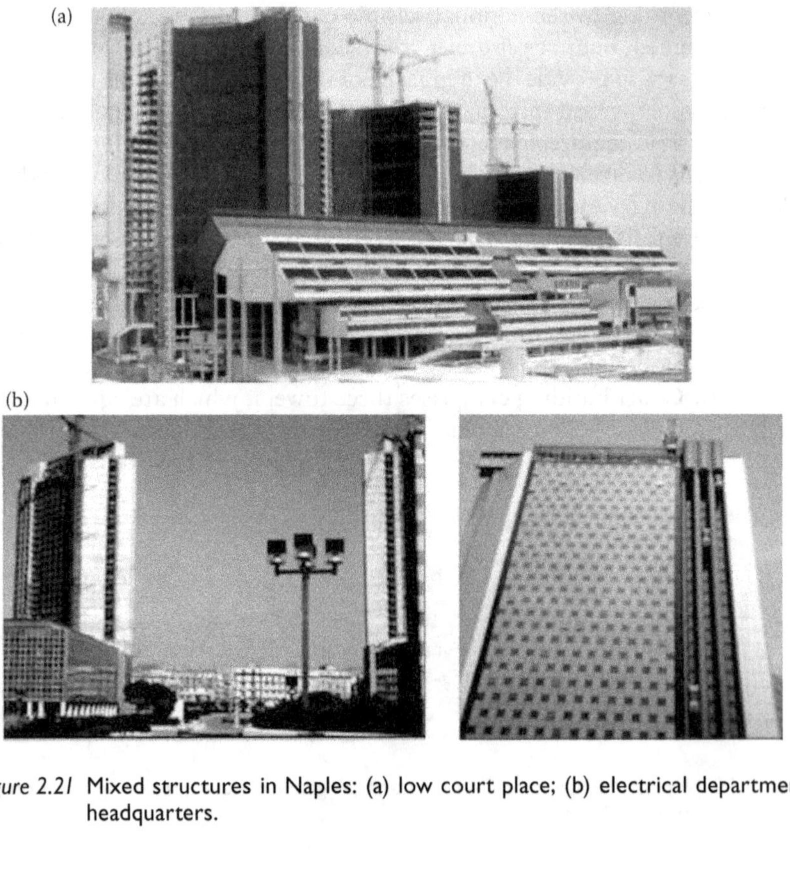

Figure 2.21 Mixed structures in Naples: (a) low court place; (b) electrical department headquarters.

which allow a significant reduction of the seismic effects, mainly at the base of the cores, where the bending moment and shear are reduced to 70%. Numerical and physical models have been used in order to validate this structural solution. The model in scale 1:20 has been tested on a shaking table with six controlled degrees of freedom. The experimental results confirmed the advantages of the use of such dissipative devices in a suspended structure (Belli and Gurracino, 1997).

The new fire station in Naples comprises eight buildings in total (Figure 2.22a). They have been designed by using steel and composite structures with both traditional and innovative solutions for seismic protection.

Building A, containing the garage, headquarters, lodgings, and canteen for the fire brigade, is the most important at least for two reasons: first, because this building, initiated in 1981 and completed in 1985, was the first example in Italy of a base-isolated structure; second, because it received the award of the ECCS in 1987. The structural scheme is based on a composite structure, in which the concrete cores, containing the stairs, elevators, and poles, are spaced about 18 × 18 m.

They provide the resistance against horizontal forces. The steel skeleton is suspended to a top grid by means of vertical ties. The top grid is connected to the upper part of the concrete cores by means of special devices, which isolate the steel skeleton from the vertical and horizontal ground motions transmitted to the towers.

The bearing devices (Figure 2.22b), made of a combination of neoprene and teflon, play a double role by allowing free movements under serviceability conditions and by providing damping and energy absorption during an earthquake (Mazzolani and Serino, 1997a).

Building B, containing the garage, pole, storehouse, guesthouse, and training, has a similar plan distribution with the same structural mesh as the previous building, but the structural solution is different (Figure 2.22c). The only vertical elements are four couples of steel towers, which have a double function of resisting seismic actions and supporting the floor slabs by means of neoprene bearings. The connection between the main floor beams and the adjacent steel towers is made by means of shock-block devices made of oleo-dynamic cylinders. They act as provisional restraints, which allow the free movements of the structure under service loads (live loads, wind, temperature, etc.), but become rigid under seismic actions. They produce the effect of changing the structural scheme in a more favorable way under low-intensity earthquakes and reducing the degree of damageability of the structure under severe earthquakes (Mazzolani and Serino, 1997b).

Building E, containing the headquarters, offices, and lodgings, in contrast to the first two, has a classical scheme. The steel structure is braced by means of reticular concentric bracings (CBF), which stabilize the structure under horizontal loads. Primary and secondary beams supported by columns, the beam-to-column joints being simply pinned, compose the floor structure.

Figure 2.22 Fire brigade center of Naples: (a) general view; (b) Building A; (c) Building B; (d) Building E.

Different typologies are used in Italy in higher seismicity areas, like in Messina (Sicily), where eccentric bracing frames (EBFs) were used for the structure of the seven-storied building built in 1996 for the library of the new Department of Literature and Philosophy of the university (Figure 2.23). The shape of the building is an octagonal prism. The eccentrically braced system of "K"-shaped bracings has been calibrated to achieve the simultaneous engagement into plastic range of the links located at the different stories. All connections in the MRFs are rigid and of full-strength type (D'Amore and D'Amore, 1997).

A seismic-resistant steel structure with eccentric bracings (EBFs) has been designed for the first time in Romania (Georgescu, 1996). The Mara Hotel (Figure 2.24a), located in the picturesque mountain town of Sinaia, has a structure composed of a gravitational system for vertical loads and an EBF system resisting seismic actions and providing the required stiffness. EBFs are of K-braced (Figure 2.24b) and D-braced (Figure 2.24c) types for to architectural reasons. The links are short and the link-to-column connections are made of a rigid segment shop-welded to the column and field-bolted to the link. Also, the link-to-brace connection is field-bolted. The "dog-bone" solution has been adopted at the base of the columns to allow the formation of a plastic hinge there and to consequently guarantee a global failure mechanism (Figure 2.24d).

Again, EBFs were used more recently in the Banc Post Building in Timisoara (Figure 2.25a). The steel skeleton of five levels is a space MRF composed by cruciform columns and integrated by K-bracings (Figure 2.25b). The bolted end-plate beam-to-column connections (Figure 2.25c) have been preliminarily tested under cyclic loading at the laboratory of the University of Timisoara (Dubina et al. 2000). This building in Timisoara

Figure 2.23 Steel EBF structure of the Messina University library.

Figure 2.24 Mara Hotel in Sinaia, Romania: (a) during erection; (b) K-braced EBF; (c) D-braced EBF; (d) dog-bone at the column base.

Figure 2.25 Banc Post Building in Timisoara, Romania: (a) facade; (b) steel structure under erection; (c) K-braced EBF.

followed the first steel building erected there in the early 1990s for the City Bank office (Figure 2.26).

The International Centre Tower of Bucharest (Figure 2.27a) was until 2006 the tallest building in Romania, erected in a very seismic area, Bucharest being the European capital with the highest seismic risk (Dubina et al. 2006). The building has 26 storys, 3 basements, and is 120 m in height. The steel building structure consists of a dual structural configuration, formed by MRFs and central braced frames (CBFs) (Figure 2.27b). Belt trusses are placed at the 14th story. The cross-sections of the columns are in cruciform shape made of hot-rolled profiles (Figure 2.27c). To increase the strength, rigidity, and fire resistance, columns are partially encased with reinforced concrete to obtain composite sections. The beams are made of IPE sections and the braces of HEB sections.

Mixed solutions are very popular in Portugal. In the campus of the University of Lisbon, two 10-storied buildings, shaped as mushrooms, have been erected (1994–1999) for hosting the Department of Electronic and Chemistry. Both buildings consist of a central reinforced concrete core for the stairs and elevators, which supports large cantilever beams at every three levels.

The apartment and office buildings are erected in the EXPO area of Lisbon (Figure 2.28). Also, in these buildings, the bracing function is fulfilled by a reinforced concrete core and the floor structures are composite.

Steel structures are also starting to appear in Greece, where the reinforced concrete tradition is very strong. One of the first steel and composite buildings (Figure 2.29a) is the Faculty of Physics and Mathematics at the Aristotle University in Tessaloniki (1997). The structural scheme is a CBF with inverted V bracings, which are visible in some part of the facade, giving an architectural feature to the building. Until now, EBF structures have been used in Greece just for power plant structures (Figure 2.29b and c) and for upgrading a damaged building (Figure 2.29d).

(a)

(b)

Figure 2.26 City Bank Building in Timisoara, Romania: (a) facade; (b) steel structure under erection.

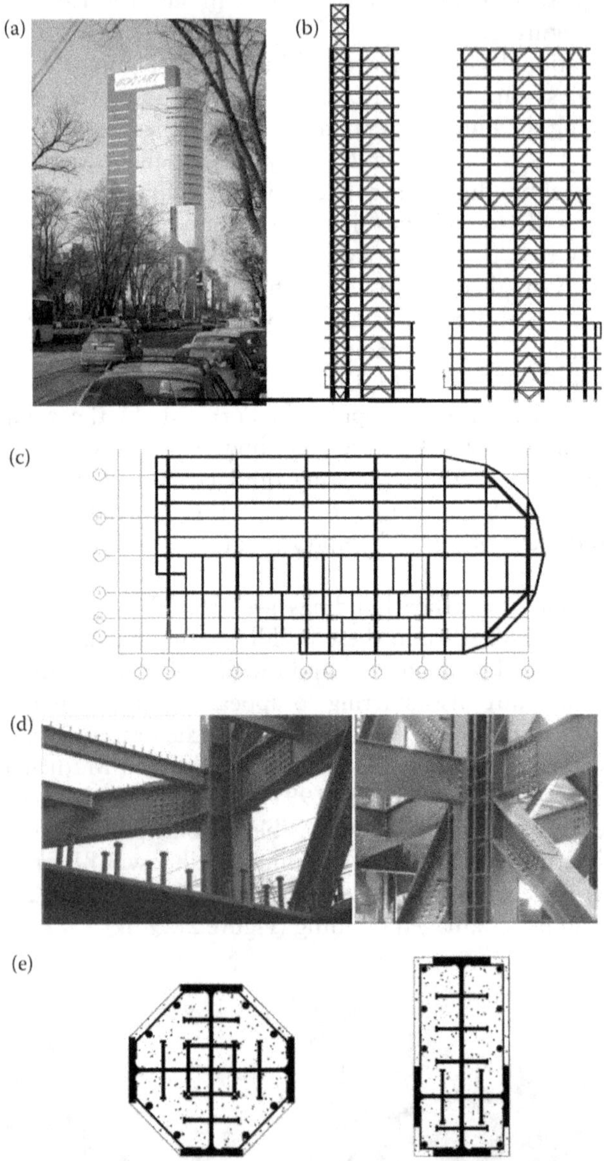

Figure 2.27 International Centre Tower, Bucharest: (a) view; (b) structural system; (c) current floor plan; (d) column sections; (e) view during erection of steel structure. (From Dubina, D. et al. 2006: Analysis and design considerations regarding the project of Bucharest Tower Centre steel structure. In *Proceedings of the International Conference in Metal Structures*, Poiana Brasov, Romania, 20–22 September, Steel—A New and Traditional Material for Building, (eds. Dubina, Ungureanu), Taylor & Francis Group, London, 601–608.)

Figure 2.28 Steel structures in Lisbon, Portugal, for new buildings in the Expo area.

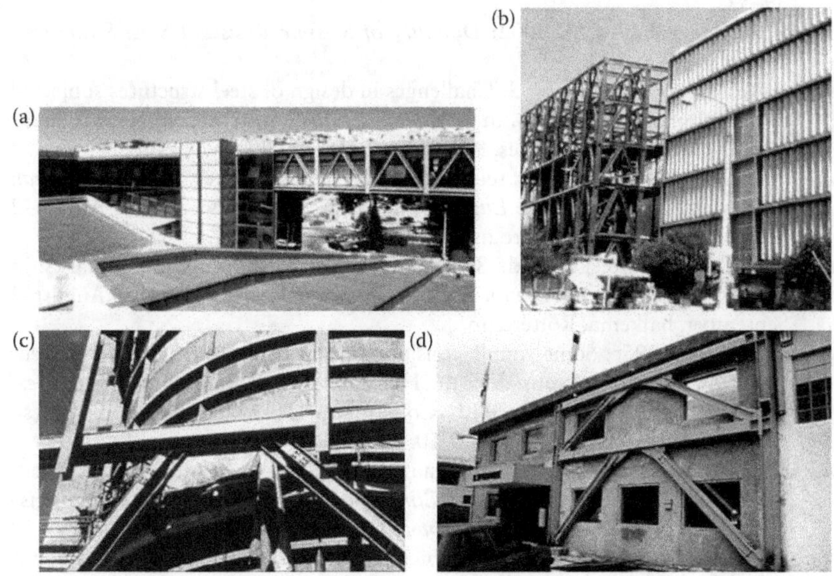

Figure 2.29 Steel structures in Greece: (a and b) Tessaloniki University; (c) EBF for power plant; (d) EBF for upgrading of damaged building.

REFERENCES

Belli, P., Gurracino, F. 1997: Experimental testing and damage evaluation for a typology of damping devices. In *Proceedings of the STESSA Conference on Behaviour of Steel Structures in Seismic Areas* (eds. F.M. Mazzolani, H. Akiyama), Kyoto, 3–8 August 1997, Eolezioni 10/17, Salerno (Italy), 700–707.

Bolton, P., Cole, S. 2006: Earthquakes and a brave new China. Benefied Hazard Research Centre Report. http://www.benefieldhrc.org/activities/issues6/pages/eq_china.htm

D'Amore, A., D'Amore, E. 1997: The library in the new Department of Literature and Philosophy, University of Messina. *Costruzioni Metalliche*, 6, 25–32.

Dubina, D., Dinu, F., Ciutina, A., Stratan, A. 2000: The multi-storey structure of Banc Post Timisoara building: Essential design problems. In *9th International Conference on Metal Structures, ICMS 2000*, Timisoara, 19–22 October.

Dubina, D., Dinu, F., Stratan, A., Ciutina, A. 2006: Analysis and design considerations regarding the project of Bucharest Tower Centre steel structure. In *Proceedings of the International Conference in Metal Structures*, Poiana Brasov, Romania, 20–22 September, *Steel—A New and Traditional Material for Building*, (eds. D. Dubina, V. Ungureanu), Taylor & Francis Group, London, 601–608.

Elsesser, E. 2004: Seismically resistant design-past present, future. In *13th World Conference on Earthquake Engineering*, Vancouver, 1–6 August, Paper No. 2034.

Georgescu, D. 1996: Recent developments in theoretical and experimental results on steel structures: Seismic resistant braced frames. *Costruzioni Metalliche*, 1, 39–52.

Gioncu, V., Mazzolani, F.M. 2002: *Ductility of Seismic Resistant Steel Structures*. Spon Press, London.

Gioncu, V., Mazzolani, F.M. 2003: Challenges in design of steel structures subjected to exceptional earthquakes. In *Behaviour of Steel Structures in Seismic Areas* (ed. F.M. Mazzolani), Naples, 9–12 June, Balkema, 89–95.

Mazzolani, F.M. 1995a: Design of seismic resistant steel structures. In *10th European Conference on Earthquake Engineering* (ed. G. Duma). Vienna, 28 August–2 September, Balkema, Rotterdam.

Mazzolani, F.M. 1995b: Eurocode 8—Chapter Steel: Background and remarks. In *10th European Conference on Earthquake Engineering*, Vienna, 28 August–2 September, Balkema, Rotterdam.

Mazzolani, F.M. 1995c: Some simple considerations arising from Japanese presentation on the damage caused by the Hanshin earthquake. In *Stability of Steel Structures* (ed. M. Ivanyi), SSRC Colloquium, 21–23 September, Budapest, Akademiai Kiado, Vol. 2, 1007–1010.

Mazzolani, F.M. 1995d: Seismic—Resistant solutions in the new Management Center of Naples. In *Fifth World Conference on Habitat and the High-Rise Council on Tall Buildings and Hurban Habitat*, Amsterdam.

Mazzolani, F.M. 1998: Design of steel structures in seismic regions: The paramount influence of connections. In *Control of the Semi-Rigid Behaviour of Steel Engineering Structural Connections* (ed. R. Maquoi), COST C1 Conference, Liége, 17–18 September, 371–384.

Mazzolani, F.M. 1999a: Principles of design of seismic resistant steel structures. In *4th National Conference on Metal Structures*. Ljiubljana (Slovenia), 20 May, 27–42.
Mazzolani, F.M. 1999b: Design and construction of steelworks in seismic zones. In *XVII C.T.A. Congress*, Napoli, 3–6 October.
Mazzolani, F.M. 1999c: Reliability of moment resistant connections of steel building frames in seismic areas: The first year of activity of the RECOS project. In *2nd European Conference on Steel Structures Eurosteel*. Prague, 26–29 May.
Mazzolani, F.M. 1999d: Reliability of moment resistant connections of steel building frames in seismic areas. In *Seismic Engineering for Tomorrow, International Seminar in Honor of Professor Hiroshi Akiyama*. Tokyo, Japan, 26 November.
Mazzolani, F.M. (ed). 2000a: *Moment Resisting Connections of Steel Frames in Seismic Areas: Design and Reliability*. E & FN SPON.
Mazzolani, F.M. 2000b: Steel structures in seismic areas. In *Seismic Resistant Steel Structures* (eds. F.M. Mazzolani, V. Gioncu), CIMS Courses, Udine, Springer, Wien.
Mazzolani, F.M. 2002: Structural integrity under exceptional actions: Basic definitions and field of activity. In *Improving Building's Quality by New Technologies, COST C12 Seminar*, Lisbon, 19–20 April, 67–80.
Mazzolani, F.M. 2003: Steel and composite structures in European seismic areas. *Earthquake Spectra*, 19(2), 415–452.
Mazzolani, F.M. 2006: Steel against earthquake. In *Steel, A New and Traditional Material for Buildings* (eds. D. Dubina, V. Ungureanu), Poiana Brasov, 20–22 September, Taylor & Francis group, London, 31–43.
Mazzolani, F.M., Gioncu, V. (eds). 2000: *Seismic Resistant Steel Structures*. CIMS Courses, Udine, Springer, Wien.
Mazzolani, F.M., Piluso, V. 1996: *Theory and Design of Seismic Resistant Steel Frames*. E & FN SPON, an imprint of Chapman & Hall, London.
Mazzolani, F.M., Serino, G. 1997a: Top isolation of suspended steel structures: Modelling, analysis and application. In *Behaviour of Steel Structures in Seismic Areas STESSA 97* (eds. F.M. Mazzolani, H. Akiyama), Kyoto, 3–8 August, 10/17 Salerno, 734–743.
Mazzolani, F.M., Serino, G. 1997b: Viscous energy dissipation devices for steel structures: Modelling, analysis and application. In *Behaviour of Steel Structures in Seismic Areas STESSA 97* (eds. F.M. Mazzolani, H. Akiyama), Kyoto, 3–8 August, 10/17 Salerno, 724–733.
Roeder, C.W. 2000: SAC program to assure ductile connection performance. In *Proceedings of the STESSA Conference on Behaviour of Steel Structures in Seismic Areas* (eds. F.M. Mazzolani, R. Tremblay), Montreal 21–24 August, Balkema, Rotterdam, 659–666.
Roeder, C.W. 2002: Development of performance-based seismic design criteria for steel moment frames. In *4th National Conference on Steel Structures* (eds. D.E. Beskos, D.L. Karakalis, A.N. Kounadis), Patras, 24–25 May, 346–358.
STESSA '94. 1995: In *Proceedings of 1st International Conference on Behaviour of Steel Structures in Seismic Areas* (eds. F.M. Mazzolani, V. Gioncu), E & FN SPON, London.
STESSA '97. 1997: In *Proceedings of 2nd International Conference on Behaviour of Steel Structures in Seismic Areas* (eds. F.M. Mazzolani, H. Akiyama) Kyoto, 3–8 August, 10/17, Salerno.

STESSA 2000. 2000: In *Proceedings of 3rd International Conference on Behaviour of Steel Structures in Seismic Areas* (eds. F.M. Mazzolani, R. Tremblay), Montreal 21–24 August, Balkema, Rotterdam.

STESSA 2003. 2003: In *Proceedings of 4th International Conference on Behaviour of Steel Structures in Seismic Areas* (ed. F.M. Mazzolani), Naples, 9–12 June, Balkema, Rotterdam.

STESSA 2006. 2006: In *Proceedings of 5th International Conference on Behaviour of Steel Structures in Seismic Areas* (eds. F.M. Mazzolani, A. Wada), Yokohama, 14–17 August, Taylor & Francis, London.

STESSA 2009. 2009: In *Proceedings of 6th International Conference on Behaviour of Steel Structures in Seismic Areas* (eds. F.M. Mazzolani, J.M. Ricles, R. Sausa), Philadelphia, CRC Press, Boca Raton, London.

STESSA 2012. 2012: In *Proceedings of 7th International Conference on Behaviour of Steel Structures in Seismic Areas* (eds. F.M. Mazzolani, R. Herrera), Santiago de Chile, CRC Press, Boca Raton, London.

Chapter 3

Challenges in seismic design

3.1 GAP IN SEISMIC DESIGN METHODOLOGIES

3.1.1 Seismic loading versus structural response

The basic concepts of today's earthquake engineering codes were born almost 70 years ago, when knowledge about seismic actions and structural response was rather poor. Today, the earthquake-resistant design is grown within the new multidisciplinary fields of engineering seismology and earthquake engineering, wherein many exciting developments are predictable in the near future. The challenge for a proper seismic structural design is to solve the balance between seismic demand and structure capacity. Seismic demand corresponds to the effect of earthquakes on the structure and depends on ground motion modeling. Structural capacity is the structural ability to resist these effects without failure.

"Looking to the developments of the Engineering Seismology and Earthquake Engineering, it is very clear that the major efforts of researchers were directed toward the structural response analysis. Therefore, the structural response can be predicted fairly confidently, but these achievements remain without real effects if the accurate determining of the seismic actions is doubtful. In fact, the prediction of the ground motions is still far from a satisfactory level, due to both the complexity of the seismic phenomena and the communication lacks between seismologists and engineers... Therefore, the structural designer must be fully aware of the fact that it useless to determine with great accuracy the structure response for a given seismic action if this one is established with a great level of incertitude" (Gioncu and Mazzolani, 2006a, 2011).

Indeed, all research works in earthquake engineering are involved in the improvement of structural response analysis, the researchers being very happy if the results of their studies demonstrate an improvement of 5% of this response. But they forget that the actual seismic action can exceed by more than 100% the considered seismic load given by the present codes;

therefore, this entire effort does not produce any practical effect if the seismic action is not correctly evaluated.

So, the reduction of uncertainties in ground motion modeling is now the main challenge in structural seismic design. This target is possible only if the impressive progresses in seismology are transferred into structural design (Gioncu and Mazzolani, 2006b). Three important developments during the second half of the last century contributed to the rapid advances in seismology:

1. The development of the tectonic plate theory, which now offers a clear and conceivable framework for the generation of the majority of earthquakes.
2. The establishment of fault mechanisms producing different earthquake types having very different ground motion characteristics.
3. The computer technology, which opens up new possibilities to analyze a large amount of data and to model the fault rupture process, in order to simulate the ground motions.

Any progress in earthquake engineering is impossible without considering this new amount of knowledge, recently cumulated in seismology. Today, after these improvements in knowledge, it is very clear that earthquakes can be classified in function of the site–source distance in *far-field* and *near-field earthquakes* and in function of source characteristics: *interplate, intraslab,* and *intraplate earthquakes.* The characteristics of these different earthquake types should be considered in the structural design. The differences between ground motions generated by these earthquakes are so large that the ignorance of these aspects can be considered as a shortcoming of the code provisions. In this context, engineering seismologists are now expressing great concern to establish the differences in the main characteristics among these earthquakes. At the same time, the task of earthquake engineers is to take greater care about the structural response to these earthquake types.

3.1.2 Critics of current design methodologies

Since the introduction of earthquake engineering by Housner in the 1950s, the response spectrum has become an essential tool in structural analysis and design. But today, after the impressive development of the earthquake engineering science, some shortcomings in the current practice have been identified (Gioncu and Mazzolani, 2002, 2011):

- The codes were historically developed based on the experience of a few recorded ground motions not sufficiently close to the causative faults, due to the absence of a dense network of recording stations.

In the last period, this situation has changed in some very urbanized seismic zones and the recent earthquakes have yielded more records. But unfortunately, there are still a large number of seismic zones without adequate information. Now, each earthquake-prone country must create a data bank, which represents the medium for transferring the record and the digitized accelerograms for storage. There are already some very well-organized data banks. In Italy, the Enea-Enel data bank collected more than one thousand records of earthquakes that occurred in Italy during the last few years. In England, the ICSTM (Imperial College of Science, Technology and Medicine) organized a data bank for Europe and the surrounding areas. Very important data banks also exist in the United States and Japan, which have been enriched after the last great events with a lot of new records.

- In spite of very well-organized network stations, owing to the existence of some unknowledgeable fault network (see the Kobe earthquake, produced along an unknown fault) or some active faults different from the known ones (e.g., the Northridge earthquake produced by the Elysian Park fault far from the very well-known San Andreas fault), it is practically impossible to design a recording network station in a such a way as to have records on the sites with maximum ground motions. So, the recorded values have a great incertitude, and it is only by chance that some of them are situated near the fault (see the Northridge and Chi-Chi earthquakes, where, in spite of a large amount of recorded data, a much-reduced number of records are available for near-source ground motions). In addition, these records can be influenced by some local site stratifications, so that they are available only for the recording station (see the 1977 Vrancea ground motion, recorded in Bucharest on a very poor soil condition). Therefore, the design spectra developed based on these recorded motions are not generally available for other seismic sites.
- The last few important earthquakes recorded in the past decades (1994 Northridge, USA; 1994 Kobe, Japan; 1999 Izmit, Turkey; 1999 Chi-Chi, Taiwan; 2003 Bam, Iran; 2008 Sichuan, China; 2009 L'Aquila, Italy; 2009 Port-au-Prince, Haiti; 2010 Chile; 2012 Emilia, Italy) occurred in very densely populated zones, underlining the importance of introducing the structural design rules for near-field earthquakes in the codes. This imperious demand is supported by the observation that the tremendous growth during the past decades of urban areas in seismically active regions increases the risk that an earthquake could occur where there is a large concentration of population (Figure 3.1a and b). The damage produced in these cases is well illustrated by the abovementioned earthquakes. So, the so-called near-field earthquake

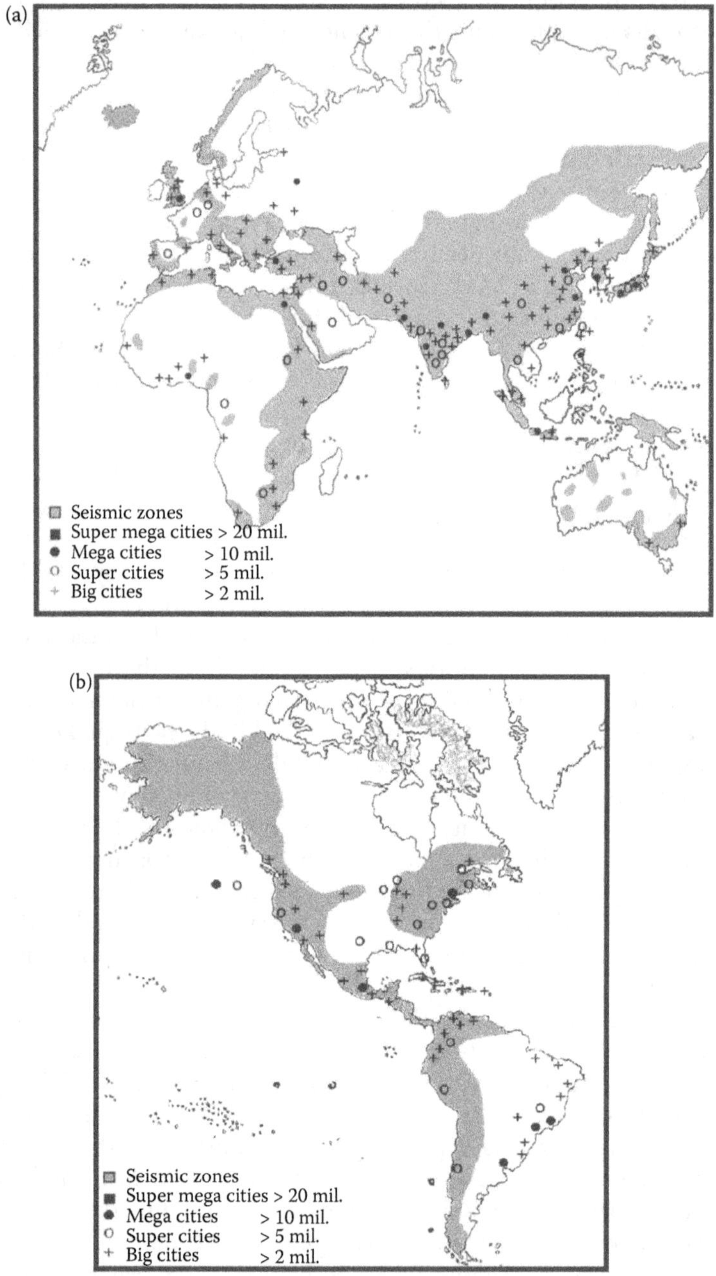

Figure 3.1 Global urbanization and seismic areas: (a) Africa, Eurasia, and Australia; (b) North, Central, and South America. (Gioncu, V., Mazzolani, F.M. 2011: *Earthquake Engineering and Structural Design*. Taylor & Francis, London.)

must be taken into account in structural design, but these aspects are not considered by the response spectra used in practice. Only the USA-UBC 97 and the Chinese GBJ11-2001 consider some aspects of near-field earthquakes. UBC proposes an increase in the acceleration values in the Californian near-source zones, in function of the distance from the fault (but who knows this distance exactly?). The GBJ11-2001 suggests different corner periods for near- and far-field earthquakes. However, both proposals do not represent a correct solution to this problem.

- Peak ground acceleration, in the frame of the used design spectra, is the parameter associated to the severity of an earthquake. However, it has been generally recognized that this is a poor parameter for evaluating the real earthquake damaging potential. For instance, one of the more significant shortcomings of the current design spectra is the fact that they do not account for the duration of the input ground motions. The dissipation of seismic energy, as it is considered in the design philosophy today, is strongly affected by this duration. The near-fault impulse type of ground motion results in a sudden burst of energy into the structure, which must be dissipated immediately by very few large yield excursions. In contrast, for long-duration earthquakes (in case of far-field or deep earthquakes), the cyclic-type ground motion with numerous yield reversals requires a more steady dissipation of energy over a long period of time. The actual design spectra do not provide this differentiation. So, the question is whether the structures are able to face these different situations.
- Knowledge about the fault type characteristics for each zone plays a crucial role in case of regions with low-to-moderate levels of seismicity, due to the lack of earthquake records, the required data to establish reliable spectra being very scarce. It is a real mistake to use data from other strong seismic zones, without considering if there are some available data for the respective zones.

3.1.3 Needs and challenges for the next design practice

The improvement of these aspects gives rise to new approaches in the seismic design of structures, for which a deep knowledge of each type of earthquake characteristics is required. In the last period, a very intensive activity in analytical and numerical modeling for generating earthquakes has been developed, contributing greatly to the process of understanding the intricate phenomena produced during an earthquake and allowing the identification of the different earthquake types. The main task of the specialists is to transfer this new knowledge into structural practice. But this is a very complex activity due to the following reasons (Gioncu and Mazzolani, 2011):

- The definition of a major earthquake action inevitably calls for a series of seismological knowledge, structural engineering judgments, and safety policy. *The reduction (not the elimination) of uncertainties in ground motion modeling is now the main challenge in structural seismic design.* For this reason, the proper seismic design should be elaborated in close collaboration between seismologists, experts in faults characteristics, and structural engineers, who know very well the structural response to a given ground motion. In many cases, this collaboration is very difficult, due to the different education, actually consisting in the differences between science and engineering. In order to fill this gap, engineers must especially work in the engineering seismology branch to transform the qualitative knowledge of seismology into quantitative values to be used by earthquake engineering and to partially eliminate the incertitude in determining the seismic actions as a function of the fault type.
- A significant gap is recognizable between the existing level of the research results and the provisions of the design codes. In many cases, research works are performed by professors and researchers, who are more interested in publishing their results for promotion among their colleagues, than in the transmission of new knowledge to those who will apply. In contrast, structural engineers are conservative professionals and not researchers in new directions of structural design. Their activities are driven by the need to deliver the design in a timely and cost-effective manner. They may also resist new concepts, unless these concepts are set in the context of their present mode of operation. Therefore, during the elaboration of new codes, it must be kept in mind that the design engineering community tends to be conservative. So, the new code provisions must be a compromise between new and old knowledge and procedures. Otherwise, the structural designers will reject the new methodologies.
- The implementation of new concepts in codes is constrained by the need to keep the design process simple and verifiable. Today, the progress of computer software has made it possible to predict the actual behavior of structures subjected to seismic loads using nonlinear analysis. But designers are always pressed by the deadline to deliver the project; therefore, they can accept the use of advanced design methodologies only for very special structures. For the majority of designed buildings, they require to use simple procedures only. This is the reason why, for instance, despite the progress in design methodologies for evaluating the structural ductility, the codes contain

only simple constructional provisions, which do not always assure the required plastic rotation capacity in the members.
- Recognizing the need for code development based on transparent methodology, one must also admit that it is necessary to underline some dangers of this operation; in particular, they are due to the oversimplification, overgeneralization, and immediate application in practice of the latest research results, without an adequate period of time during which these results can be verified.

After these critical points of view concerning the improvement of design methodology, it may be concluded that the process to set up a reliable design philosophy is very complex because so many factors, very difficult to be accurately accounted for, are involved. In this situation, a justified question arises: *What must be the solution of this matter of fact, considering that in the next future the use of a methodology based on design spectra will remain available for the current design?* The answer is not simple and satisfactory, due to the complexity of the structural response during an earthquake. However, considering the development today and in the near future, the proposed solution is based on two points:

- Diversification of design spectra in function of earthquake types.
- Introduction of explicit ductility demand in function of earthquake type, duration and number of large yield cycles, ground motion velocities, effects of strain rate, and so on (Gioncu and Mazzolani, 2002).
- Only the USA-UBC 97 and the Chinese GBJ11-2001 consider some aspects of near-field earthquakes. UBC proposes an increase in acceleration values in the Californian near-source zones, in function of the distance from the fault (but who knows this distance exactly?). The GBJ11-2001 suggests different corner periods for near- and far-field earthquakes. However, the increase of acceleration or the reduced corner periods are not sufficient measures to consider all the very damaging effects of near-field earthquakes.

The first step to have *different spectra* in function of source distance and earthquake type is already given in some codes, but the application rules are not always clear. As already mentioned, the American UBC97 (1997) and Chinese code GB-5011 (2001) consider some aspects of the near-field earthquakes. The UBC97 code introduced, for the first time, the possibility of considering the effects of near-field ground motions, by increasing the accelerations in function of the distance from faults. But the new elaborated American codes NEHRP (2008) and the international code IBC (2006) renounced these provisions concerning the effects of near-field earthquakes, considering that it is very difficult to establish the site–source distance. In contrast, the values of design accelerations are increased. GB5011 (2001) proposes different corner periods for near- and far-field earthquakes. EUROCODE 8 (2004) proposes two types of design spectra, in function of the magnitude range, which can be considered as corresponding to two earthquake types: crustal interplate earthquakes (type 1, for high seismic zones) and crustal intraplate earthquakes (type 2, for region with low seismicity). The provisions of the Japanese code "Seismic Design Standard for Railway Facilities" (Sato et al., 1999) give three different spectra for near-field, interplate, and intraplate earthquakes.

Unfortunately, the code provisions are very poor for *ductility demand*, even if there are some constructional requirements to obtain good ductility. Based on the SAC connection tests (1996), performed after the Northridge joint failures, it is accepted that members and connections should be capable of developing a minimum plastic rotation capacity of the order of 0.025–0.030 radian (FEMA 267, 1997). But these values were obtained by using quasi-static procedures, with increasing displacement history and a considerable number of cycles until connection failure. Can these tests interpret the dynamic action of ground motions with the same characteristics as the ones developed in near-field zones? It is well known that pulse velocities can be so high that they cannot be reproduced in laboratory tests.

Another very important question that must be answered refers to the steel moment frame buildings situated in low or moderate seismic zones, or in zones with a very long duration of ground motions. Are they vulnerable to the same type of damage that occurred during the Northridge and Kobe earthquakes? These questions can have answers only if the code provisions contain different spectra and ductility requirements in function of earthquake type.

3.2 EARTHQUAKE TYPES

3.2.1 Plate tectonics

In the early 1960s, the plate tectonic theory produced a revolution in earth science (Gioncu and Mazzolani, 2006b). In the last period, scientists have verified and refined this theory by producing a more clear understanding on how our planet has been shaped by plate tectonic processes.

Today, it is accepted that the earth is covered by some rigid plates that move across its surface, over and on a partially molten internal layer. So, the surface of the earth is broken up into the so-called *tectonic plates*. *Plate tectonics* is the theory of geology developed to explain the phenomenon of continental drift. This theory defines tectonic plates and their boundaries. In function of their surface, *tectonic plates* are divided into major and minor plates. Tectonic plates can include continental crust, oceanic crust, or both. The distinction between continental crust and oceanic crust is based on the density of constituent materials. Continental crust is composed primarily of granite, so it is relatively light. Oceanic crust is denser than continental crust, being composed of basalt. As a result, oceanic crust generally lies below the sea level, while continental crust is situated above the sea level. There are 52 important tectonic plates (USGS, nd). The 14 *major tectonic plates* are shown in Figure 3.2. Among them, the seven most important plates are

- African plate, covering Africa (continental plate)
- Antarctic plate, covering Antarctica (continental plate)
- Australian plate (also known as the Indo-Australian plate), covering Australia (continental plate)
- Eurasian plate, covering Europe and Asia (continental plate)
- North American plate, covering North America and North-East Siberia (continental plate)
- South American plate, covering South America (continental plate)
- Pacific plate (the largest), covering the Pacific Ocean (oceanic plate)

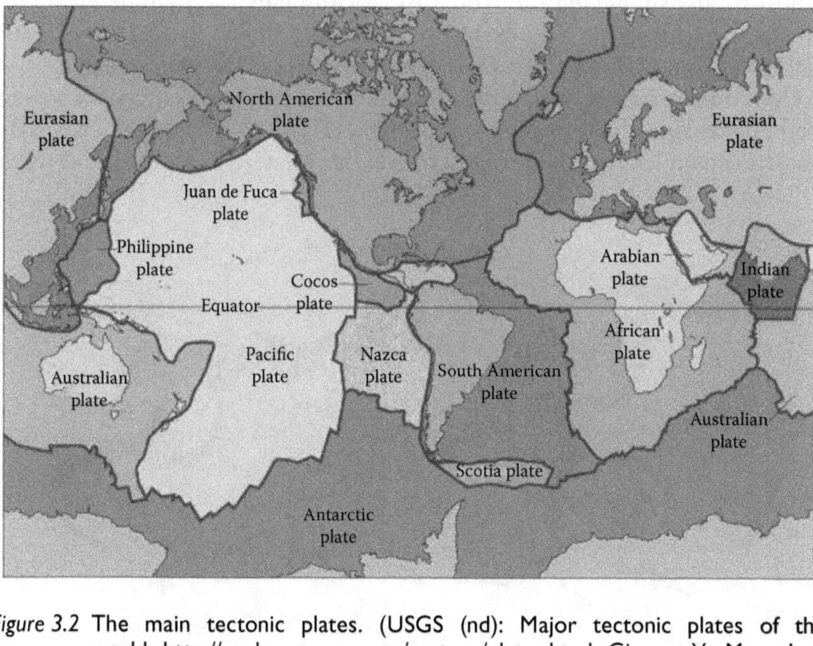

Figure 3.2 The main tectonic plates. (USGS (nd): Major tectonic plates of the world, http://geology.er.usgs.gov/eastern/plates.html; Gioncu, V., Mazzolani, F.M. 2011: *Earthquake Engineering and Structural Design.* Taylor & Francis, London.)

In addition to the above great tectonic plates, the following can be also considered as major plates: Arabian plate (continental plate), Caribbean plate, Cocos plate, Juan de Fuca plate, Nazca plate, and Philippine plate (all being oceanic plates). It is very important to underline that tectonic plates do not coincide with the continent forms.

There are 38 *minor tectonic plates* (Bird, 2003): Aegean Sea, Altiplano, Amurian, Anatolian, Burma, Banda Sea, Balmoral Reef, Birds Head, Caroline, Conway Reef, Easter, Futuna, Galapagos, Juan Fernandez, Kermadec, Manus, Mariana, Maoke, Molucca Sea, New Hebrides, Niuafo'ou, North Andes, North Bismarck, Okhotsk, Okinawa, Panama, Rivera, Sandwich, Scotia, Shetland, Somali, Solomon Sea, South Bismarck, Sunda, Timor, Tonga, Woodlark, and Yangtze plates.

To these major and minor plates, one must add some hundreds of *micro tectonic plates*, formed during the continental drift, such as the Iberian, Adria, Turkish, Black Sea, and other microplates.

The above configuration corresponds to today's situation, ignoring the history of the formation of each plate. There are also some *ancient plates*, which disappeared during the genesis of the present configuration. For instance, the major Eurasian plate is divided into two parts, European and Asian plates, in contact along the Ural mountains, because, about 400 million years ago, they were different continents. The Farallon and Kula plates almost totally subducted the North American plate, the Juan de Fuca and Gorda plates being the remains of this plate. The China plateau is the result of the collision of an ancient minor plate with the major Asian plate.

The comparison between the distribution of the epicenters of strong earthquakes and the tectonic plate borders (Figure 3.3) shows very clearly that the main cause of earthquakes is the relative movement between

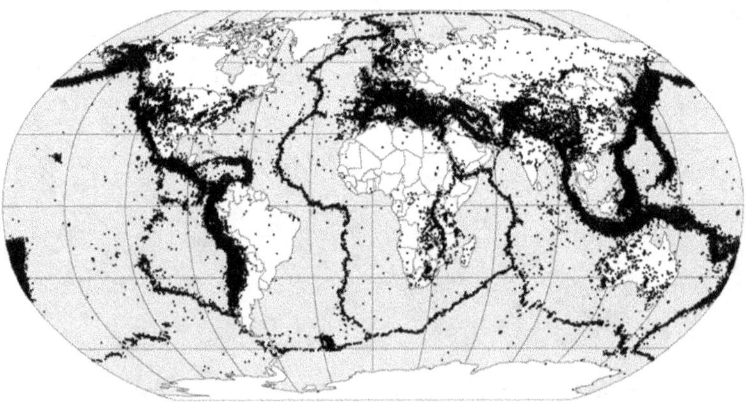

Figure 3.3 Earthquakes and tectonic borders. (Gioncu, V., Mazzolani, F.M. 2011: *Earthquake Engineering and Structural Design*. Taylor & Francis, London.)

tectonic plates. The majority of very strong earthquakes in the world occur along these borders, named *interplate earthquakes,* but there are also low-to-moderate earthquakes shaking the zones within the plate itself, far from the plate borders, the so-called *stable continental area*, producing earthquakes, named *intraplate earthquakes.*

3.2.2 Factors influencing earthquakes

3.2.2.1 Source depth

The depth of a source being a very important factor, there are two earthquake types (Figure 3.4):

- Crustal earthquakes are produced in the earth's crust having the hypocenter situated at depths from 0 to 40–50 km, whose effects are limited only to short distances around the epicenter in a small area.
- Subcrustal earthquakes are produced under the earth's crust, at depths up to 300 km for intermediate earthquakes and over 300 km for deep earthquakes, which may produce great events in the region far from the epicenter, affecting a large area.

3.2.2.2 Epicentral distance

The site–source distance plays a leading role in the design of structures and a classification is absolutely necessary. In function of this distance, the following classification may be considered (Figure 3.5):

- *Epicentral site,* including the area around the epicenter, generally with a radius equal to the source depth
- *Near-field site,* including the area within a distance of 25–40 km around the epicenter

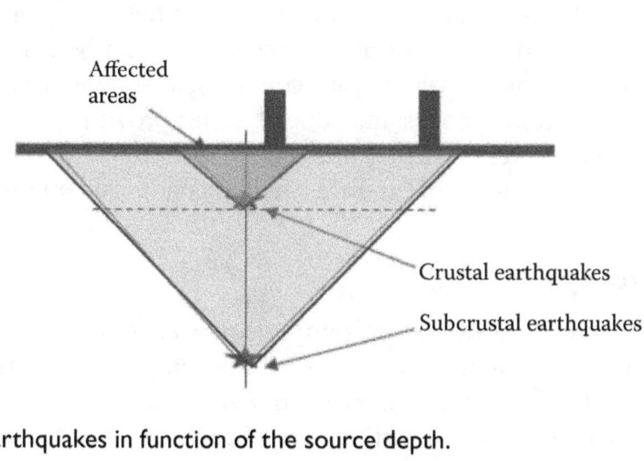

Figure 3.4 Earthquakes in function of the source depth.

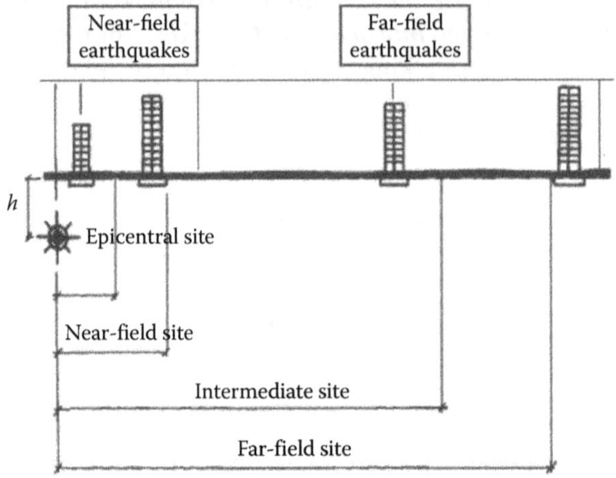

Figure 3.5 Earthquake types in function of epicentral distance. (Gioncu, V., Mazzolani, F.M. 2002: *Ductility of Seismic Resistant Steel Structures*. Spon Press, London; Gioncu, V., Mazzolani, F.M. 2011: *Earthquake Engineering and Structural Design*. Taylor & Francis, London.)

- *Far-field site*, with a distance of about four to five times the source depth
- *Intermediate-field site*, including the area between near- and far-field sites

Considering the properties of the earthquakes produced in these areas, another classification can be proposed:

- *Near-field earthquakes*, produced in epicentral and near-field sites
- *Far-field earthquakes*, produced in intermediate and far-field sites

This is the main important classification of earthquake types, because the differences in earthquake characteristics and their influence on the structural response are so large that ignorance of these aspects can be considered as a shortcoming of the code provisions. Unfortunately, as shown in Section 3.1.2, only a very few national codes introduced some provisions for near-field earthquakes, but they were insufficient to characterize the differences in comparison to the characteristics of far-field earthquakes considered in the code.

3.2.2.3 Source types

A better understanding of the *seismological aspects* of the earthquake phenomena results from the cooperation between geophysicists and seismologists. In spite of the complexity of an earthquake manifestation, real progress was marked in the classification of earthquakes. Nowadays, it is very clear that there are some basic earthquake types with very different characteristics in function of fault types (Gioncu and Mazzolani, 2011) (Figure 3.6). The fault movements can be characterized by the tendency of tectonic plates to move away (creating faults, valleys, and gulfs), to move toward each other (creating mountains due to subduction or crust folding), or to slip. Each of these fault movements produces earthquakes with very different characteristics. Fault types can be classified as follows:

Interplate crustal faults, produced in the crust at the contact edges of two major tectonic plates; the Californian, Anatolian, or Iranian faults are the most representative for this earthquake type. One must emphasize that these contact edges represent only the present configuration of tectonic plates, ignoring the ancient steps in the earth's history. The positions of active faults are very well known because the tectonic plate boundaries are well defined. There are some main faults along the Circum-Pacific Ring and Alpide-Himalayan belt, which regularly produce great earthquakes. Generally, the boundary plate contacts are characterized by the presence of many parallel or diffuse faults, which can produce large earthquakes at the same level as the principal fault. The well-known San Andreas fault has some parallel faults, such as Hayward, Rodgers, Calaveras, Greenville, and San Gregory. One can see that today the probability of the occurrence of a strong earthquake (the predicted *Big One*) is higher for the Hayward fault than for the San Andreas fault.

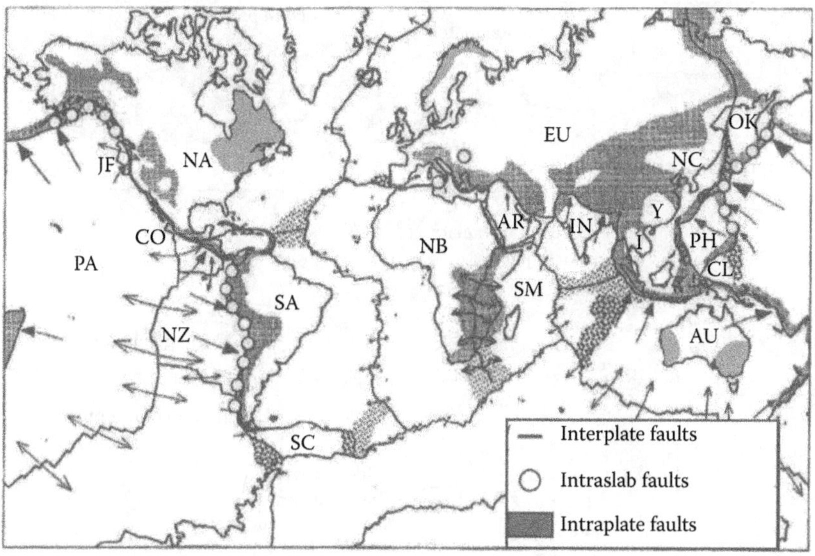

Figure 3.6 Different fault types. (Gioncu, V., Mazzolani, F.M. 2011: *Earthquake Engineering and Structural Design*. Taylor & Francis, London.)

The interplate faults are characterized by some very long fault zones, which appear to be segmented. So, the sources move from one segment to another, the case of the San Andreas, Mexican, and Anatolian faults being significant for these migratory earthquakes.

Intraplate crustal faults, produced in the same plate along some existing inland faults, the most representative for this type being the Central European, Chinese, Japanese, Indian, and Australian earthquakes. These faults were produced during the formation of the actual earth's crust, due to the ancient collisions of some tectonic plates or by the rupture of the existing plates due to internal forces. The position of blind faults is more difficult to be detected, being inactive for many years, but potentially they occur in near stable continental regions. The 1976 Tangshan China earthquake (250,000–660,000 fatalities) is an example of the damaging potential of this fault type. Unfortunately, the geological basis for intraplate seismogenesis is still poorly understood. A truth is established concerning these earthquake types: the presence of a fault potentially indicates a possible earthquake, but there is no certainty that this will occur in the future.

Intraslab faults, produced in the same subducting slab, under the crust at large depths, the most representative for this type being the Canadian, Japanese, and South American earthquakes, as well as the Sicily, Vrancea (Romania), and Southern Greece European earthquakes.

In function of the earthquake position, there are the following main earthquake types (Figure 3.7):

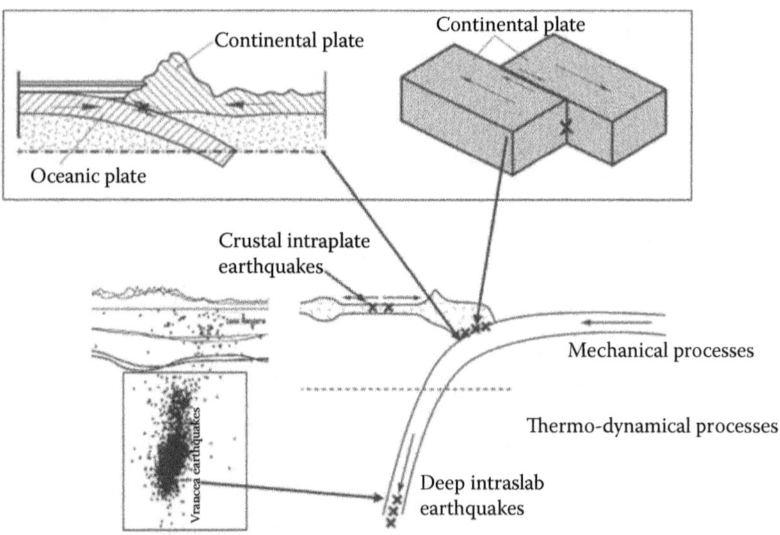

Figure 3.7 Earthquakes types: crustal interplate and intraplate earthquake and deep intraslab earthquakes. (Anastasiadis, A., Mosoarca, M., Gioncu, V. 2012: *Journal of Constructional Steel Research*, 69(2012), 176–191.)

Crustal interplate earthquakes, which are produced along the tectonic convergent plate boundaries, where a fault exists due to previous plate movements. Along the fault, shear friction forces between the two plates equilibrate the gigantic tectonic plate forces and the elastic strain energy is stored. When the shear force capacity is reached, a sudden slip releases the large elastic strain energy stored in the interface rocks. So, in the case of interplate earthquakes, the slipping occurs along these tectonic plate boundaries and their positions are well defined. Only in the case of existing parallel faults, the activated one is an unknown. The depth of these earthquakes does not exceed 40–50 km. The most important type is the interplate earthquake, which is produced by subduction (subduction of an oceanic plate under a continental one), collision (converging of two continental plates), or strike-slip (two tectonic plates slide and crash against each other). These earthquakes are dominated by mechanical processes (shear forces between the two tectonic plates).

Crustal intraplate earthquakes, which are produced in the interior crust of a tectonic plate, far from the boundaries, in the diffuse zones where a network of crustal faults are produced due to the weakness of the earth's crust. In this case, the ground movements are caused by the previous rupture or by a new one, due to the reach of the rock-bearing capacity. Thus, generally, an intraplate earthquake occurs in the same seismic area, but not in the same place, as in the case of interplate earthquakes. So, the position of an intraplate earthquake is generally very difficult to be determined in advance.

Intraslab earthquakes, which are affected by thermal phenomena, and the earthquake occurs in a slab (subducted oceanic plate), at a shallow or great depth, the source depth being situated in a range of 50–300 km. Very special features, different from the cases presented above, characterize this source type. Subcrustal earthquakes are situated under the crust, where the solid rock begins to be transformed into molten lava, due to the high temperature and pressure. Therefore, deep earthquakes are dominated by thermal phenomena, while the intermediate ones, up to 150 km, are dominated by a combination between mechanical and thermal phenomena.

3.2.3 World seismic zones

Generally, the maps showing the world's epicenters do not a make a distinction between the earthquake types, showing only the epicenter locations. The advances in knowing the nature of faults allow us to establish the expected earthquake type in each seismic area of the world and to choose, consequently, the most adequate design methodology, specific for the respective type of earthquake. In what follows, these seismic areas will be described by underlining their main characteristics. The corresponding zones are numbered in function of the geographical position and the

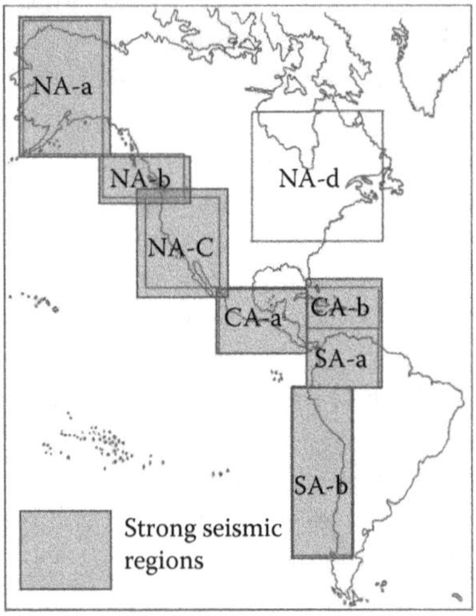

Figure 3.8 Seismic zones in the American continents.

earthquake type. The regions affected by strong earthquakes are marked in Figures 3.8 and 3.9.

The seismic zones in the American continent are presented in Figure 3.8.

North American seismic zones (NA). The western coast is the most seismically active area, being influenced by the Circum-Pacific Ring. The crustal interplate subduction earthquakes, resulting from the subduction of the ancient Kula plate under the North American plate, characterize the Alaska zone (NA-a). In this seismic zone, the Anchorage, the second largest

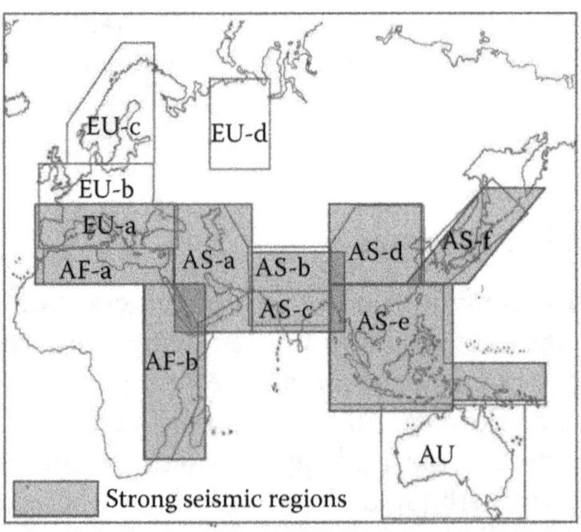

Figure 3.9 Seismic zones in Africa, Europe, Asia, and Australia.

ever-recorded earthquake, occurred in 1964, producing very important avalanches, landslides, and tsunamis. The character of this diffuse seismic zone is marked by the existence, in the continental part, of some intraplate earthquakes, dominated by the Dinali strike-slip fault (like the San Andreas fault, but situated in a scarcely populated region). The same interplate and intraslab earthquake types characterize the western Canadian coast, being produced by the subduction of the Kula and Juan de Fuca plates beneath the North American plate (NA-b). But the most well-known zone of North America is the western coast of the United States (NA-c), dominated by the San Andreas fault and the corresponding diffuse zone, where interplate crustal strike-slip earthquakes are the main ground motion types, due to the oblique movements between the Pacific and the North American plates. The eastern part of North America (NA-d) is dominated by diffuse intraplate earthquakes, resulting from continental movements during a previous cycle of the Pangaea formation. Especially Missouri, the eastern coast of the United States and Canada are subjected to normal, reverse, and thrust earthquake types.

Central American seismic zones (CA). The territory of Mexico is dominated by crustal subduction interplate and intraslab earthquakes, produced by the subduction of the Cocos plate beneath the North American plate (CA-a). In the zone of the Caribbean Isles (CA-b), some subduction interplates occur due to the interaction between the Caribbean and North American plates. The intraplate earthquakes, produced in the seismic areas without very clear fault systems, dominate the eastern zones. The central zone is practically immune from earthquakes.

South American seismic zones (SA). Generally, this seismic area is regular, without important changes along the fault. In the north part of South America, in Venezuela (SA-a), some strike-slip earthquakes occurred, due to the interaction between the Caribbean and the South American plates. Subduction crustal interplate and intraslab earthquakes dominate the western seismic area (SA-b) as a result of the subduction of the Nazca plate beneath the South American plate. The eastern South American zone is immune from earthquakes.

The seismic areas in the African, European, Asian, and Australian continents are presented in Figure 3.9.

African seismic zone (AF). The northern areas of Africa (AF-a) are affected by crustal interplate earthquakes, due to the subduction of the African plate under the Mediterranean microplate. Divergent crustal earthquakes along the Red Sea and the African Rift (AF-b) dominate the Eastern African areas. The western African zones are almost immune from earthquakes.

European seismic zones (EU). Europe has one of the most complex seismic zones. The first zone (EU-a), the Mediterranean one, including the south of Spain and Italy, is dominated in the southern part by crustal subduction interplate earthquakes (where the African plate subducts the Mediterranean microplate), with the exception of the south of Sicily and Crete, where some intraslab earthquakes can occur. The northern zone is located along the Pyrenees and Alps, where the Central Mediterranean and Adria microplates subduct the Eurasian plate giving rise to crustal collision interplate earthquakes. Another very important zone is the Vrancea one, characterized by intraslab earthquakes, where the Black Sea microplate subducts the Pannonian microplate. The most active seismic zone in the Balkans is Greece, with many intraplate earthquakes to a very active network of faults. The Anatolia zone is characterized by the presence of strike-slip earthquakes along the North Anatolian fault and collision earthquakes in the south part of Anatolia. Central Europe (EU-b) presents many intraplate earthquakes especially along rifts in a very diffuse seismic zone. Finally, Northern Europe (EU-c) has the same intraplate earthquakes, due to the influence of the Middle Atlantic ridge, creating diffuse seismic areas. The eastern seismic part of Europe is dominated by the Ural zone (EU-d), where a collision between Europe and Siberia produced a chain of mountains.

Asian seismic zones (AS). The Arabian and Caucasian zones (AS-a) are dominated by collision earthquakes, produced by the contact between the Arabian and the Eurasian plates. The North of India and the Tibet zones (AS-b) are characterized by the presence of crustal collision interplate earthquakes, due to the contact between the Indian and the Asian plates along the Himalayan mountains. Peninsular India (AS-c) is affected by intraplate earthquakes. Another very important seismic zone is the Chinese plateau (AS-d), where intraplate earthquakes are the dominating events. This diffuse zone is the result of the collision of some ancient blocks. A very large seismic zone is situated in South Asia, where the majority of earthquakes are due to the subduction of the Indian plate under the Pacific plate (AS-e). The last zone, which is a very important and complex one, is the Japanese area (AS-f), where all types of earthquakes can occur: crustal interplate, intraplate, and intraslab earthquakes act in different zones of this area, being a function of the existing fault type, due to the simultaneous subduction of many tectonic plates under the Eurasian plate.

Australian seismic zone (AU) is characterized by intraplate earthquakes only, being situated far from the tectonic plate boundaries, inside of the Indo-Australian plate.

3.3 STRONG SEISMIC REGIONS

3.3.1 Earthquake types in strong seismic regions

Strong earthquakes are produced by

- *Crustal interplate sources*, the earthquake type being in function of the epicenter-to-site distance, giving rise to near-field or far-field earthquakes (Figure 3.10), with very different seismic characteristics.

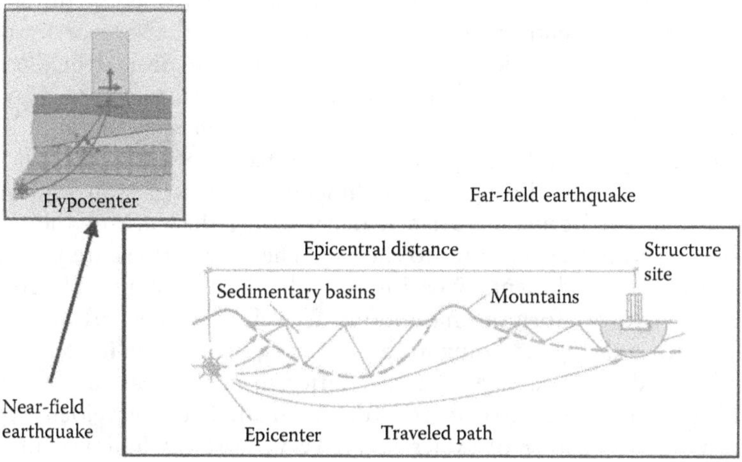

Figure 3.10 Crustal interplate earthquakes: near- and far-field earthquakes.

- *Intraslab sources* producing only one earthquake type (Figure 3.11), having only the characteristics of far-field ground motion, the effects of near-field disappearing due to the great distance to the surface.

Crustal interplate sources produce *crustal earthquakes* (subduction or strike-slip types). In the near-field area, the earthquake characteristics

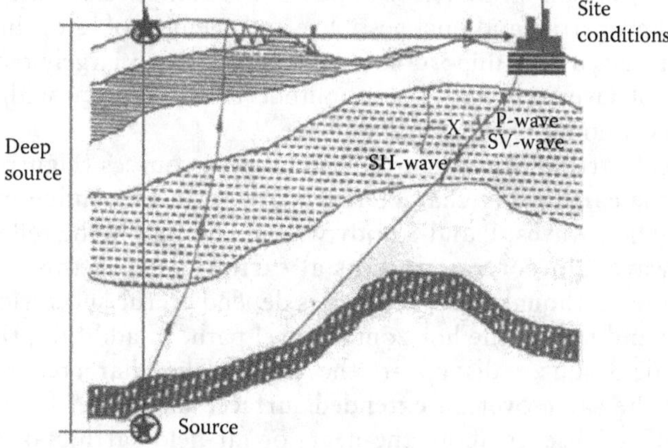

Figure 3.11 Intraslab earthquakes. (Anastasiadis, A., Mosoarca, M., Gioncu, V. 2012: *Journal of Constructional Steel Research*, 69(2012), 176–191; after Mazzolani, F.M., Mandara, A. 1992: Nuove strategie di protezione sismica per edifici monumentali. Editioni10/17, Salerno.)

rely on the propagation path of the body P and S waves, while in the far-field area, they depend on the propagation path of the surface L and R waves, as well as on the local soil conditions. In case of near-field earthquakes, the characteristics of ground motions are influenced only by the layers and soil stratification under the site. In contrast, in the case of far-field earthquakes, the propagation path effect depends on the percentage of the path travel through soft sediments. Deviations from a uniform horizontal layered crust model can occur along the path of the waves propagating from the epicenter to the site. These deviations are produced by a collection of sedimentary basins with alluviums, separated by irregular basement rock, forming mountains as well as geological and topographical irregularities. Owing to these effects, the recorded acceleration, velocities, and displacements are very different in the near- and far-field sites. Generally, the near-field records are related to the velocity pulse with a reduced number of cycles and a very short earthquake duration (produced by P and S waves), while the far-field records are characterized by many acceleration cycles and a long earthquake duration (produced by L and R waves).

Even though it is well known that the body waves, P–S, are characterized by high frequencies, and the surfaces waves, L–R, by low frequencies and, in addition, by the fact that the higher frequencies attenuate more rapidly than the lower ones as far as the distance increases, in the intermediate sites the earthquake characteristics are dominated by the far-field earthquake characteristics. It should be mentioned that for this earthquake type the site soil conditions play a leading role, in many cases being more important than the earthquake type or traveled path.

Secondly, the soil profile is composed by multilayers with different mechanical properties and thickness. The arrangement of layers in an alternative manner is a very important factor that has been largely recognized; the nature of layers changes the amplitude and frequency with a direct effect on the amount of structural damage.

Subcrustal earthquakes are produced by deep sources (Figure 3.11). In this case, the earthquake characteristics are given as a function of both L and R surface waves, P and S body waves, and finally the reflected and refracted waves. In contrast to crustal earthquakes, in this mentioned situation, the earthquake characteristics depend on the succession of vertical layers and not on the horizontal travel path. In addition, the effects of a near-field source disappear, the earthquake characteristics being practically the same over an extended surface. Therefore, for this type, it is difficult to discuss about the near- or far-field earthquakes. In any case, owing to the great depth of the source, the main characteristics correspond very well to the ones of far-field earthquakes, such as duration, number of cycles, influence of soil, site conditions, and so on.

3.3.2 Structural problems for near-field earthquakes

3.3.2.1 Main characteristics of near-field earthquakes

As these earthquakes are produced by faults situated in the earth's crust, the main effects are concentrated in the so-called *near-fault zones*, assumed to be within a distance of about 20–30 km from the fault rupture. Since it is very difficult to establish the exact distance between the site and the faults, all areas where interplate faults exist must be considered potentially situated in near-fault zones. Within this zone, the ground motions are significantly influenced by the rupture mechanism.

- *Subduction-thrust faults* (Figure 3.12a and b), which accommodate the differential motions between the heavier down-going oceanic plate and the lighter continental plate. So, these faults are the result of the contact between the top of the oceanic plate and the bottom of the newly formed continental block. Thrust-subductions occur when

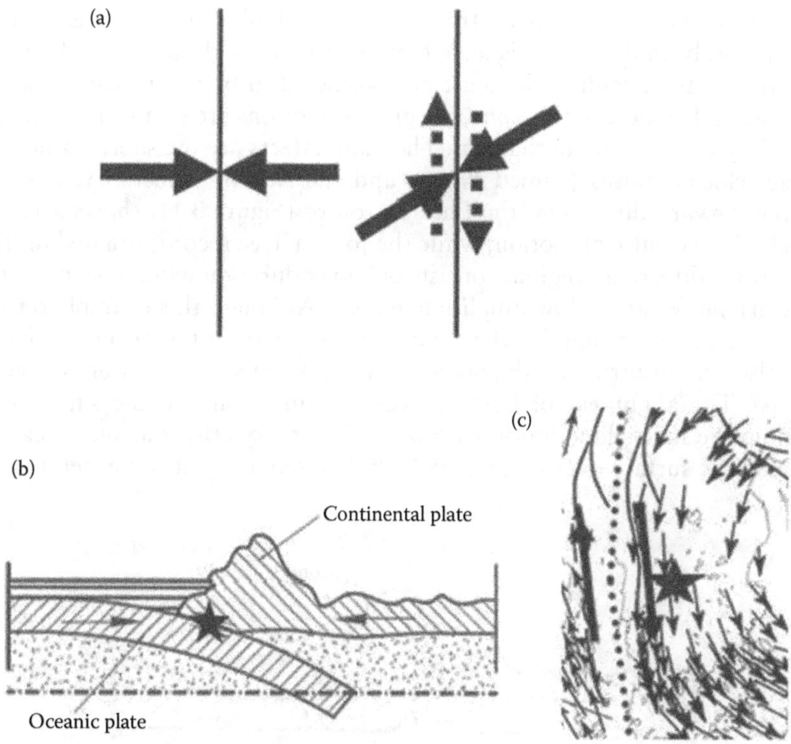

Figure 3.12 Subduction and strike-strip earthquakes.

the tectonic forces are normal to the fracture line. Some areas of the world are prone to such faults, including Alaska, the western coast of Canada, North Africa, Southern Europe, Iran, and Indonesia.

- *Strike-slip faults* (Figure 3.12a and c), which occur when the plate in contact moves parallel to the fault because the tectonic force is oblique. In this case, the two plates just slide against each other. The most famous faults are the Californian San Andreas fault, the North Anatolian fault, and the New Zealand Alpine fault.
- *Collision-thrust faults* (Figure 3.13), which occur in zones where two continental plates, with the same weight, are in contact. Along their boundaries, massive deformations of plate edges occur and the crusts buckle, wrinkle, and uplift into mountains. The most well-known faults are the ones of the collision between the Indian and the Asian plates, producing the Himalayan mountains, as well as the Italian and European plates that gave rise to the Alps and Apennines mountains.

As these earthquakes are produced by faults situated in the earth's crust, the main effects are concentrated in the near-field zones, assumed to be at a distance of about 20–40 km from the rupture faults. Since it is very difficult to establish the exact distance between the site and the faults, all areas where interplate faults exist must be considered to be potentially situated in near-fault zones. In this zone, the ground motions are significantly influenced by the rupture mechanism. The main effects are the short-duration, large-velocity pulses (named *flings*), and the *rupture directivity*. Owing to the forward directivity, the Lucerne record (Figure 3.14) shows a large brief velocity pulse of motion, while the Joshua Tree record, situated in the backward directivity region, consists of a long duration without significant velocity pulses and a low amplitude record. Although this example refers to a strike-slip earthquake, the conditions required for forward directivity are also met in thrust earthquakes (resulting from subduction or collision faults). The alignment of both the rupture direction and the slip direction up the fault plane produces rupture directivity effects at sites located around the surface exposure of the fault. Consequently, it is a general case

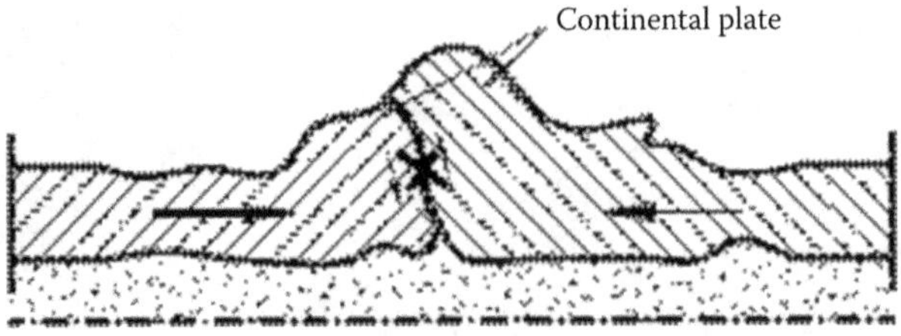

Figure 3.13 Crustal collision earthquake.

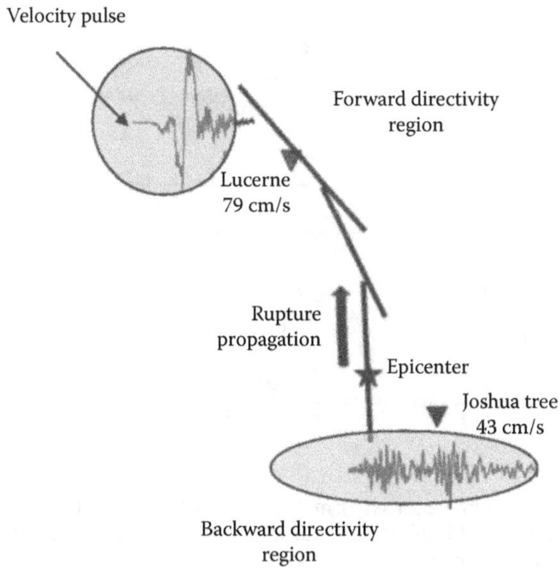

Figure 3.14 Directivity effect during the 1992 Landers (USA) earthquake. (Gioncu, V., Mazzolani, F.M. 2011: *Earthquake Engineering and Structural Design*. Taylor & Francis, London.)

when all sites located near the surface exposure of crustal faults experience forward rupture directivity for an earthquake occurring on these faults.

Regarding *large-velocity pulses*, this characteristic is very clear by examining the acceleration and velocity records from the TCU068 Station (Figure 3.15), recorded very near to the fault of the Chi-Chi (Taiwan) earthquake. While the acceleration record (with some exception around 35 s, where a small acceleration pulse can be observed) is very difficult to classify, the velocity record shows a very clear velocity pulse, with a long period (over 5 s). The same characteristics were observed during the Northridge and Kobe earthquakes. At the same time, the peak velocity during the Chi-Chi earthquake was very high, over 400 cm/s, which is the largest ever recorded in the world. During the Northridge earthquake, the maximum recorded velocity was about 200 cm/s (Station Rinaldi).

The main characteristics of *subduction-thrust faults* (including collision faults, as a peculiar case with vertical rupture surface) and *strike-slip faults* were studied by Aagaard (2000) (Gioncu and Mazzolani, 2011). An FEM model (using the potential of a parallel grid system with 256 computers) of the rupture process for the two fault types shows very important differences in ground motions, which are very difficult to be observed from the recorded values. The velocities that occurred were larger in the case of strike-slip faults in comparison with thrust faults. The thrust fault pulse

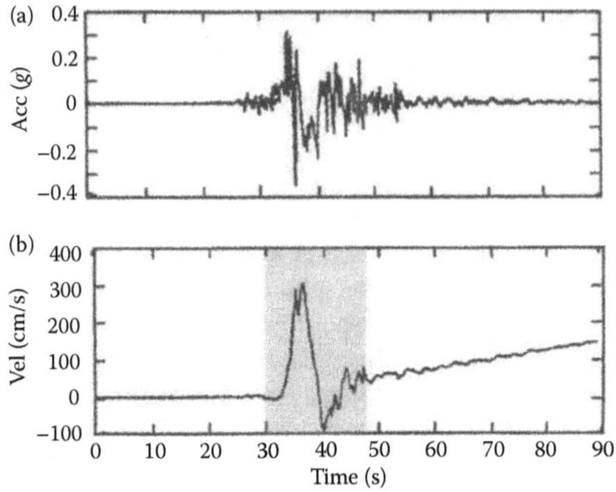

Figure 3.15 Time–history records from the 1999 Chi-Chi earthquake. (Gioncu, V., Mazzolani, F.M. 2011: *Earthquake Engineering and Structural Design*. Taylor & Francis, London.)

has a single important cycle, which occurred in the rupture direction, while the strike-slip fault pulse has two or three important cycles, produced normally in the direction of the rupture surface.

These aspects led to some very important differences in spectra and radiated energy. The acceleration amplifications are very high for strike-slip faults, also in the field of long structural periods, while being moderate for thrust faults. The radiated energies from strike-slip earthquakes are larger than those predicted by the Gutenberg–Richter relationship, while the ones from thrust faults are smaller than those predicted by this relationship. One can notice that, for the same magnitude, the radiated energies in the case of strike-slip faults are sometimes larger than the ones resulting in the case of thrust faults, showing that the last few are the most damaging among the interplate earthquakes (Figure 3.16). Fortunately, there are reduced numbers of strike-slip faults in the world. But, unfortunately, the earthquakes and the damage produced by the San Andreas fault are the most popular, giving the wrong impression that all faults, indifferently from the mechanism type, can produce the same severe damaging effects.

Another interesting phenomenon observed from the near-source earthquakes is the increase in vertical accelerations. It is generally acknowledged that vertical motions are smaller than horizontal ones. Recent observations during the last important earthquakes have demonstrated that the vertical accelerations in the near-source areas are sometimes greater than the horizontal ones, due to direct propagation of P-waves (Figure 3.17).

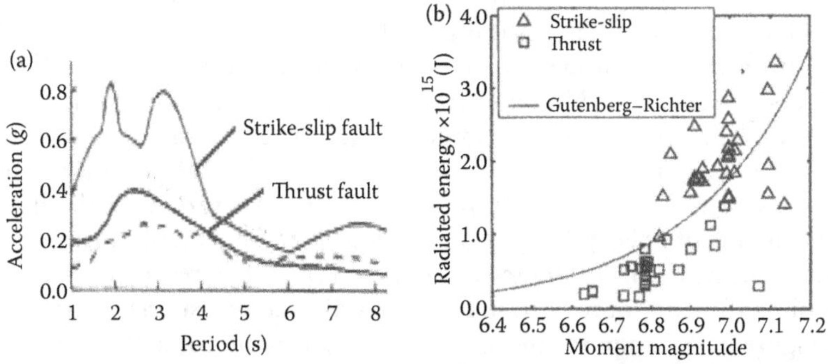

Figure 3.16 Differences between trust and strike-slip earthquakes: (a) acceleration spectra; (b) radiated energy. (Gioncu, V., Mazzolani, F.M. 2011: *Earthquake Engineering and Structural Design*. Taylor & Francis, London; after Aagaard, B.T. 2000: *Finite-Element Simulation of Earthquakes*. PhD Thesis, California Institute of Technology, Pasadena, California.)

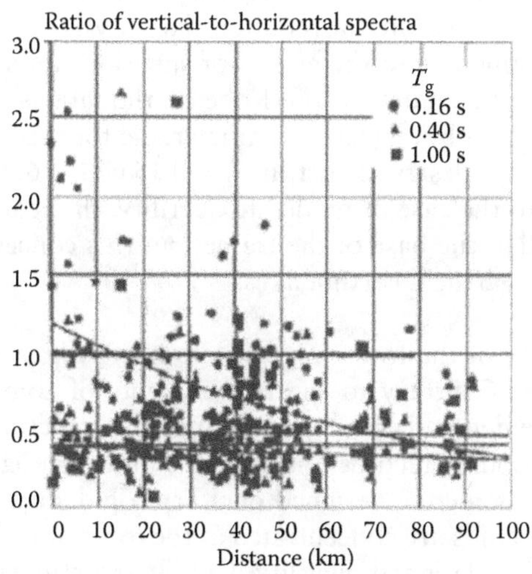

Figure 3.17 Vertical-to-horizontal component ratios during the Northridge earthquake. (Gioncu, V., Mazzolani, F.M. 2011: *Earthquake Engineering and Structural Design. Taylor & Francis, London.)*

3.3.2.2 Characteristics of structural responses

The structural design is influenced by the effects of very peculiar *characteristics of near-fault ground motions*. The first problem is due to a *reduced number of long velocity duration pulses*. The number of acceleration pulses is always greater than the number of velocity pulses, but the latter is dominant in respect of structural behavior. At the same time, the number of yield reversals of structures is higher than the action pulses. This number depends on the magnitude, which is always small (Mateescu and Gioncu, 2000). So, the near-field impulse-type earthquakes may force structures to absorb and dissipate a large amount of input energy with very few large yield cycles, determined as a consequence of high peaks of incremental velocity, which rise in the fault vicinity. As the structure must dissipate a great amount of energy (higher for strike-slip earthquakes) during a very short time, the *ductility demands are very large*. The response of framed structures to near-source ground motions also shows special characteristics. Figure 3.18 compares the story ductility demand for a 20-story structure subjected to near-source Californian and Japanese ground motions, in comparison with the ductility determined from the UBC 97 American code. Two cases are analyzed (also considering the earthquake type) (Alavi and Krawinkler, 2001):

- Frame with moderate designed ductility $V = 0.4W$ ($q = 2.5$) (Figure 3.18a) (V being the base shear force and W the structure weight). One can see that the maximum required ductility is concentrated at the top of the frame due to the effects of superior vibration modes. The larger ductility demand for the Kobe earthquake is the consequence of the multiple pulse, which is characteristic for this earthquake.
- Frame with high designed ductility $V = 0.15W$ ($q = 6.7$) (Figure 3.18b). In contrast to the case of moderate ductility, the required ductility is concentrated at the base of the frame, but this concept is not recommended for near-field earthquakes.

The particularity of the frame response to near-source ground motions is again prevalent. Contrary to the requirements of common codes, the distribution of the demands over the height of the structure is highly non-uniform for near-source records and depends on the designed ductility of the structure. For moderate designed ductility (rigid structure), the maximum ductility demands are concentrated at the top stories of the structure. In contrast, for high designed ductility (flexible structures), the maximum ductility demands are concentrated at the first few stories of the structure. Therefore, the severity of near-source ground motions leads to ductility demands, which are significantly larger and having less uniform distribution than those for the far-source records, which are represented in the common code provisions. At the same time, different ductility demands must be considered in structural design for thrust and strike-slip earthquakes.

Owing to the characteristics of the pulse seismic actions, developed with great velocity, and especially due to the lack of important restoring forces, the ductility demand could be very high, so the potential of the inelastic properties of structures for seismic energy dissipation has to be carefully

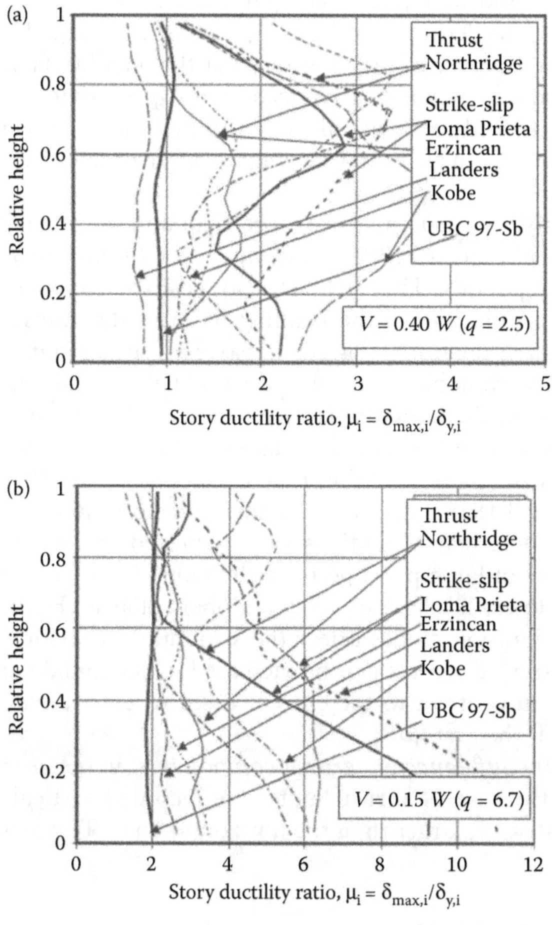

Figure 3.18 Ductility demands for near-source ground motions: (a) moderate ductility frame; (b) high ductility frame. (Gioncu, V., Mazzolani, F.M. 2011: *Earthquake Engineering and Structural Design*. Taylor & Francis, London; after Alavi, B., Krawinkler, H. 2001: Effects of near-fault ground motions on frame structures. Blume Earthquake Engineering Center, Department of Civil and Environmental Engineering, Stanford University, Report No. 138.)

examined. The *short duration* of the ground motions is a very important factor. A balance between the severity of the ductility demand, due to the pulse action, and the effects of the short duration must be seriously analyzed.

These aspects have an important consequence on the establishment of the reduction factor q, because, due to the short duration, the structure does not have enough time to dissipate a corresponding input seismic energy. Therefore, these factors must have different values in the case of near-source ground motions, in comparison with the most common far-source cases. Considering also that the near-field earthquakes produce in many cases the fracture of sections and that the ductility factor is limited, it is most rational to choose a reduced factor q for sizing the structures situated in near-field areas mainly with limited ductility demand and to concentrate the design effort in reducing the unfavorable effects of connection fractures.

Consequently, the structures are more rigid and the behavior of frames is dominated by a great *influence of superior vibration mode,* especially for short pulse periods. This fact can also have some unfavorable effects, introducing irregularities in the bending moment diagrams, especially at the middle high of the frame (Figure 3.19a). This situation induces a strong reduction of the available ductility in the middle or in the top parts of the frame, even when the structure is designed according to the normal code provisions. As the required ductility has a maximum just in these locations, the local collapse of the building may occur due to insufficient ductility (see also Figure 3.18a). This was a common phenomenon during the Kobe earthquake, where many buildings were damaged just in the middle stories. In the case of large pulse periods, the reduction of available ductility occurs at the lower levels, producing a global collapse (Figure 3.19b). So, the superior vibration modes affect the structural behavior and both the drift and ductility demands can be increased in the middle and top levels of structures, in contrast to the current cases where the main demand is concentrated at the first floor.

Regarding the *influence of vertical components*, in the near-fault zones, especially in the case of thrust faults, the recorded vertical components were, in many cases, larger than the horizontal ones. Two aspects must be

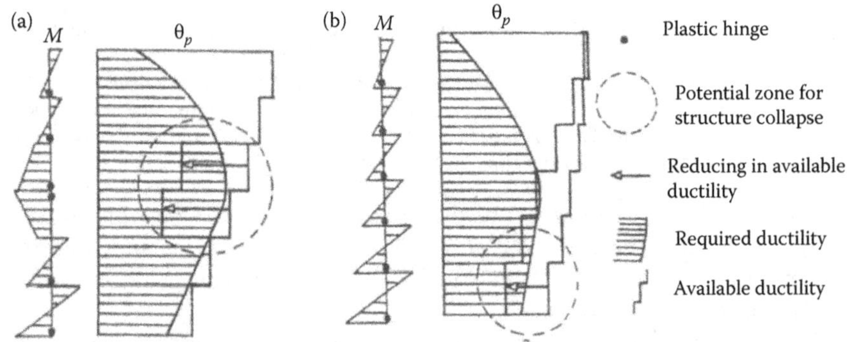

Figure 3.19 Required and available ductility: (a) reduced pulse periods; (b) large pulse periods. (Gioncu, V., Mazzolani, F.M. 2011: *Earthquake Engineering and Structural Design*. Taylor & Francis, London.)

considered. The first is related to the fact that the axial force in columns increases, especially at the top stories and interior columns, due to the vertical components. The second problems is related to the asynchronical vertical movements of frame foundations, due to the delay in seismic wave propagation, which can be important in the case of very short periods of vertical ground motions. At the same time, different ductility demands must be considered in the structural design for thrust and strike-slip earthquakes.

There is also one very disputable question: the *influence of strain rate* on structural response. In case of large velocities, as recorded during the Northridge and Chi-Chi earthquakes, the increasing of yield stress can reduce the difference between itself and the ultimate strength, with the contemporaneous decreasing of the ultimate elongation. So, steel structures behave as if made of brittle materials with a strongly reduced capacity of seismic energy dissipation. Therefore, the design methodology must verify the influence of strain rate (Gioncu and Mazzolani, 2002). The main problem refers to the question whether the velocities are sufficiently high to produce important effects. About this problem, see Chapter 5.

3.3.3 Structural problems for far-field earthquakes

3.3.3.1 Main characteristics of far-field earthquakes

Far-field earthquakes are the results of

- *Interplate faults* (Figure 3.10), the corresponding earthquakes occurring at a distance far from the source and the characteristics are influenced by the traveled path and site soil conditions.
- *Intraslab faults* (Figure 3.11), which are located in a subduction plate in a zone situated under the contact between two plates and under the earth's crust. The earthquake characteristics are influenced by the layers between the source and the surface, as well as by the site soil conditions.

The common main characteristics of these two far-field types are the large number of cycles, long duration of ground motions, and great sensibility to site soil conditions (Figure 3.20). In the last case, a pulse in acceleration occurs (Figure 3.20b) due to the resonance of natural and soil vibration modes.

There are two well-known cases, in which the influence of soil conditions at far distance has a great influence on the ground motion characteristics: the 1977 Vrancea and the 1985 Mexico City earthquakes. In the first case, owing to the deep fault position, the affected areas around the epicenter are very large and the effects of bad site soil conditions

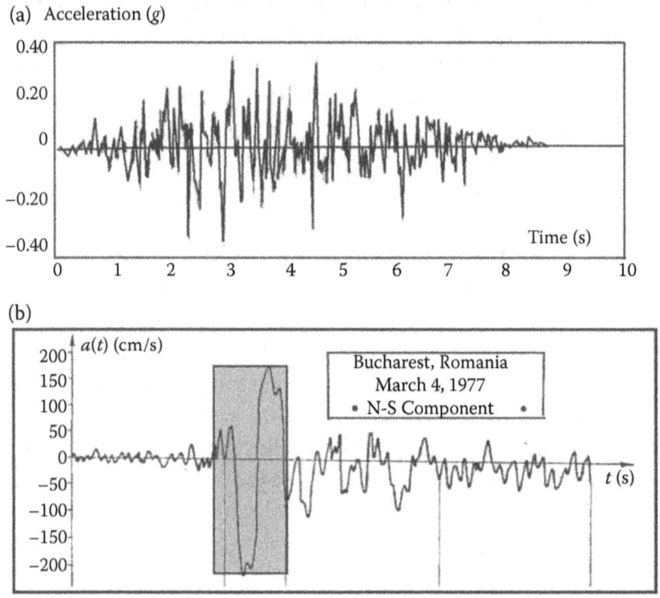

Figure 3.20 Recorded acceleration types: (a) good site soil conditions; (b) bad site soil conditions. (Gioncu, V., Mazzolani, F.M. 2002: *Ductility of Seismic Resistant Steel Structures*. Spon Press, London.)

have had a very catastrophic action on the RC buildings of Bucharest, situated at 160 km from the epicenter. In the case of the Mexico City earthquake, the most affected area is situated at 350 km from the fault and the damaging effects were produced because of the very bad soil conditions, Mexico City being erected on an ancient lake filled with alluvium material.

3.3.3.2 Characteristics of structural responses

The effects of reduction of ductility capacity due to the *accumulation of plastic deformations* (known also as *low cycle fatigue*) due large number of cycles (Figure 3.21) are the main aspects in case of far-field earthquakes and, consequently, they must be under design control. In many cases, the simple constructional requirements, similar to the ones provided by the present codes, do not assure sufficient safety against local failure and more detailed procedures for determining the available ductility, considering the actual number of important pulses and their type, should be required (Gioncu and Mazzolani, 2002). The number of pulses until fracture due to the accumulation of plastic deformation is presented in Figure 3.22, in

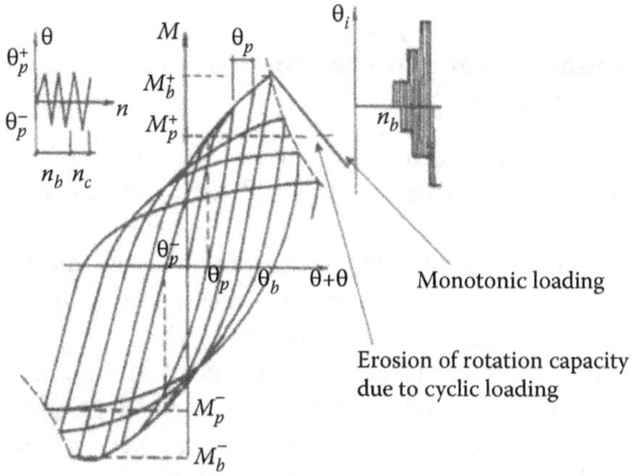

Figure 3.21 Ductility erosion due to cyclic loading. (Gioncu, V., Mazzolani, F.M. 2002: *Ductility of Seismic Resistant Steel Structures*. Spon Press, London.)

function of the yield ratio (yield stress over tensile strength). One can see that, for low yielding ratios, the number of fracture pulses is out of the range of practical values and, consequently, for far-field earthquakes, only the reduction of available ductility remains a problem to be verified. For details about this problem, see Chapter 5.

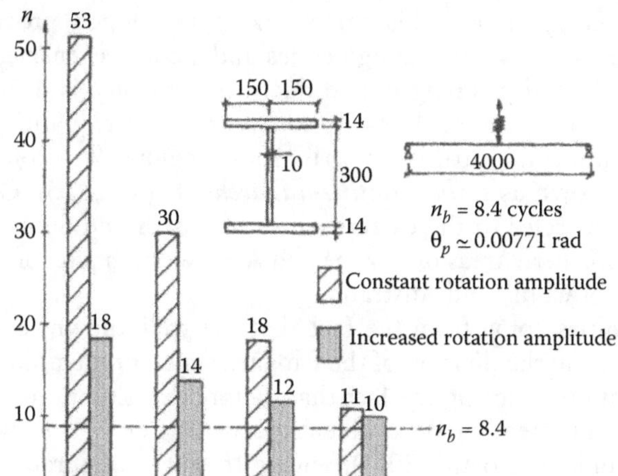

Figure 3.22 Number of fracture pulses in function of yield ratio. (Gioncu, V., Mazzolani, F.M. 2002: *Ductility of Seismic Resistant Steel Structures*. Spon Press, London.)

3.4 LOW-TO-MODERATE SEISMIC REGIONS

3.4.1 Earthquake types in low-to-moderate seismic regions

Earthquakes represent a risk in many parts of the world, particularly in the western North and South America, Japan, China, and the lands surrounding the Mediterranean basin, to mention just some of the higher seismic risk regions. But what does it mean regarding the seismic risk in regions with low-to-moderate seismicity, which represent a large area of the world? That is a question that has no clear answers.

The earthquake types are interplate (produced along the tectonic plate borders) and intraplate (occurring in the interior of tectonic plates) earthquakes. In spite of knowing that intraplate earthquakes are characterized by low-to-moderate magnitude, it can be observed that many very devastating earthquakes occurred in areas where no previous events have been recorded and where the current knowledge would suggest the existence of quiescent areas. There are few, if any, areas of the world that are immune to earthquake effects. Unfortunately, these zones can be very densely populated and the buildings are not prepared for the occurrence of earthquakes. Therefore, due to this situation, a great difference can exist between seismic hazard and seismic risk so that even a moderate earthquake can turn into a very devastating event (see the Tangshan—China and the Bhuj—India earthquakes).

The relative motions between the tectonic plates, occurring at their boundaries (producing interplate earthquakes), have been well understood and easily quantified. For this earthquake type, design procedures are included in the main seismic design codes and a comprehensive methodology, including rigidity, strength, and ductility demands, is developed. No comparably simple or agreed description yet exists for quantifying active intraplate continental earthquakes, in some regions far from the plate boundaries, known as *stable continental areas* (Figure 3.23). One can see that these areas refer to the eastern areas of North and South American Continents, northern areas of Europe and Asia, western part of Africa, and eastern areas of China and Australia.

The difficulties come from the fact that the well-known tectonic plate theory starts from the division of the Pangaea supercontinent into the present configuration, ignoring the fact that, before this situation, some cycles of gathering and spreading of continents occurred (the Wilson cycle theory, see Gioncu and Mazzolani, 2011), with different configurations that are known today. Therefore, many of the mountain belts and the diffuse seismic zones from the inland of the present continents are the result of this pre-Pangaea history. In addition, the tectonic plates are not perfectly rigid and some inland faults exist in the zones with crust weakness. Produced

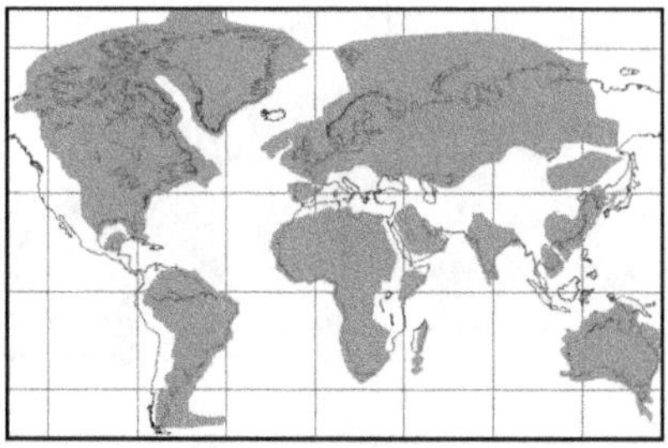

Figure 3.23 Stable continental areas. (Gioncu, V., Mazzolani, F.M. 2011: *Earthquake Engineering and Structural Design*. Taylor & Francis, London; after Mooney, W.D., Schulte, S., Detweiler, S.T. 2004: *26th Seismic Research Review, Trend in Nuclear Explosion Monitoring*, 439–448.)

exclusively in the earth's crust, the faults in diffuse zones due to the fracture of crust have different features in comparison with the tectonic plate boundaries: no divergence and no subduction. Fracture types can be classified as follows:

- Fracture in the compressed crust
- Fracture in the tension crust

The differences between these fracture types are presented as follows:

- *Fracture of weak compressed crust* (Figure 3.24). Active plates and microplates are stressed by both forces applied to the adjacent plates and forces applied to the plate boundaries due to the motions of

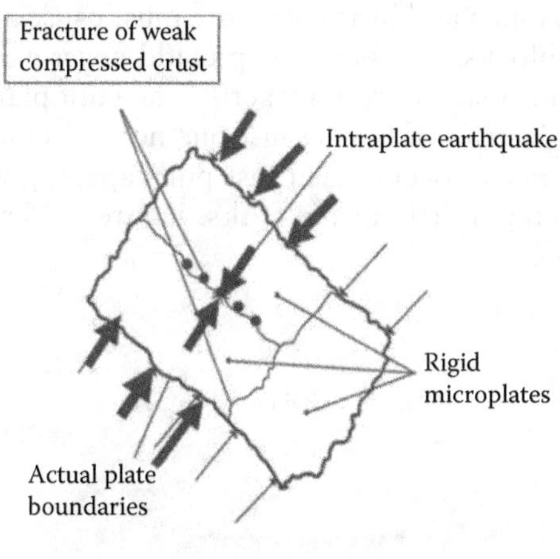

Figure 3.24 Fracture of weak compressed crust. (Gioncu, V., Mazzolani, F.M. 2011: *Earthquake Engineering and Structural Design*. Taylor & Francis, London.)

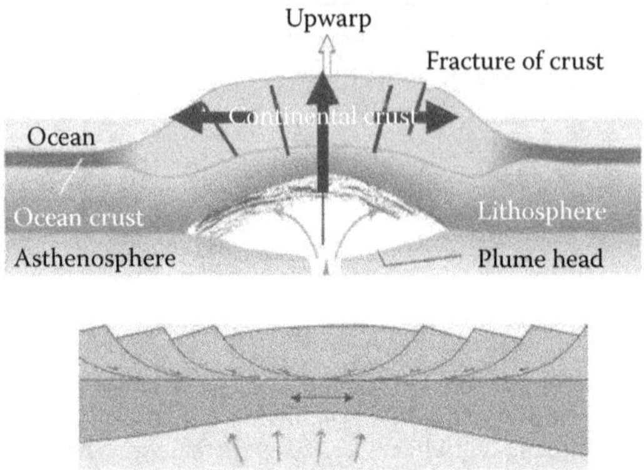

Figure 3.25 Tensioned crust caused by rising mantle plume. (Gioncu, V., Mazzolani, F.M. 2011: *Earthquake Engineering and Structural Design*. Taylor & Francis, London.)

plates. If the compressed crust presents a weakness, a fracture of the crust occurs, forming a reverse fault. In the case of low angles (average <45°), the fault is called a *thrust fault* and it often consists of fault planes parallel to sedimentary layers, forming ramps.

- *Fracture in the tension crust* (Figure 3.25). In the presence of crust weakness, the plume rising from the mantle produces tension stresses and fractures in the continental crust. So, a *normal fault* forms when the crust is stretched, tensioned, or pulled apart. After a certain amount of elastic deformation, the tension in the crust becomes too much and brittle fracture must cause a gap between the two crustal blocks. But as this gap could never exist in nature, the two blocks must stay in contact across the fault plane. To do this, in many cases, instead of having just one normal fault, a whole series of fault planes can occur: the crust pulls apart, giving rise to a *rift fault* with two parallel normal faults. Figure 3.26 shows the formation of *graben*.

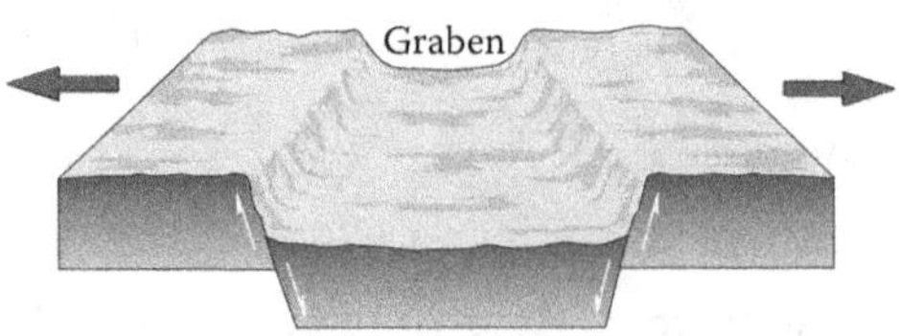

Figure 3.26 Formation of a graben.

3.4.2 Low-to-moderate earthquakes in European seismic areas

Earthquakes are widespread in Europe. The most destructive events occurred in the Mediterranean countries, particularly Greece, Italy, and Turkey, situated in the subduction zone between the African and Eurasian crustal plates. Other Balkan countries, such as Albania, Croatia, Macedonia, Serbia, and the western region of Romania, have experienced major crustal earthquakes due to the contact between the Adria and Dinarides crustal plates.

At the same time, there are low-to-moderate earthquakes, occurring in the so-considered *stable continental areas,* which generally do not produce casualities; however, owing to very high urbanization and the existence of nonengineered constructions, which are unprepared against earthquakes, they can lead to very important damage. The origin of these earthquakes is not easy to be explained, belonging to the geological history of these zones. This so-called *stable continental European area* is presented in Figure 3.27, being composed by a part of Spain, France, Belgium, the Netherlands, the United Kingdom, Germany and Scandinavia, western part of Romania, Russia, and so on.

In European countries, there are two different situations, in function of the number of earthquake types that occurred in their territories. The most significant areas influenced by only *one earthquake type*, the low-to-moderate one, are the borders among the Netherlands, Belgium, France, and Germany (Figure 3.28), where the Lower and Upper Rhine's grabens are capable of producing earthquakes of magnitude up to M 6.0 (BGHRC, 2003). These areas are typically classified as relatively stable continental regions, the seismic activity being low to moderate. The seismicity of Belgium is complicated

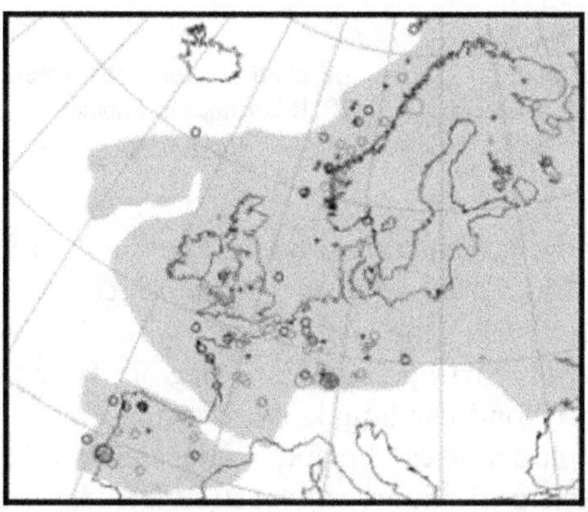

Figure 3.27 European continental stable areas. (From Gioncu, V. 2008: In *Seismic Risk 2008, Earthquakes in North-Western Europe* (ed. H. Degee), 11–12 September 2008, Liege, 249–260, 261–272; after Mooney, W.D., Schulte, S., Detweiler, S.T 2004: *26th Seismic Research Review, Trend in Nuclear Explosion Monitoring*, 439–448.)

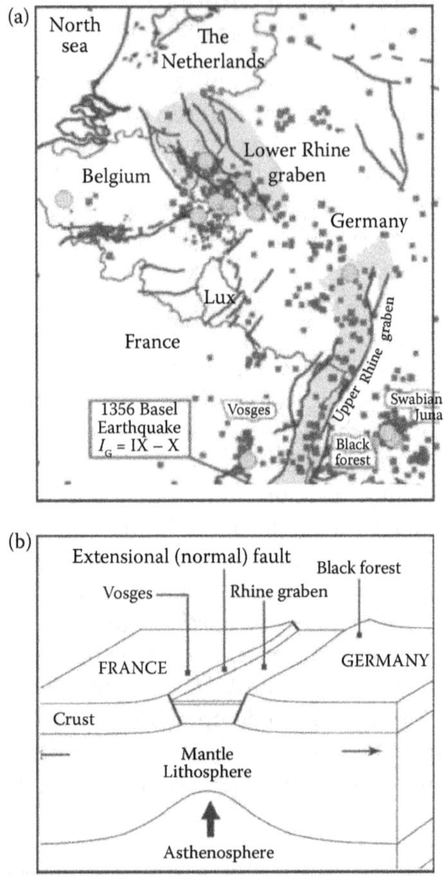

Figure 3.28 Rhine's graben: (a) graben and earthquakes; (b) formation of upper graben. (Gioncu, V., Mazzolani, F.M. 2011: *Earthquake Engineering and Structural Design.* Taylor & Francis, London.)

by the presence of the earthquakes produced by the Roer graben, the northwestern branch of the Rhine graben system, which seems to be the most active seismic area. During the last 300 years, the Roer graben has experienced at least seven earthquakes with magnitude ranging from 5.0 to 5.5. The 1992 Roermond earthquake with a magnitude M 5.3 struck the Netherlands. This moderate earthquake is considered to be one of the major historical seismic events in this area and it stimulates the activity of seismologists and engineers to estimate the true seismic hazard of this region. Are the normal faults along the Roer graben capable of producing an earthquake larger than the Roermond event? The answer is yes, the estimated maximum magnitude implying a magnitude M 6.4 (Camelbeek et al., 1996).

The UK territory is not generally associated with important earthquakes; however, about 20–30 earthquakes are felt by people each year and a few hundred smaller earthquakes are recorded by sensitive instruments. The largest known British earthquake occurred in 1931 in the English Channel, having a magnitude M 6.1, due to an existing sea graben. The studies show that a magnitude of M 5.0 occurs every 10–20 years and the largest possible earthquake is around M 6.5 (EB, 2008).

From the history of the Norwegian earthquake mentions, it is learnt that the maximum magnitude of M 5.8–6.0 occurred in northern Norway in 1819 and this event was considered to be the strongest one recorded in Northern Europe. It is anticipated that the maximum earthquake magnitude expected around Norway may reach M 7.0 (NORSTAR, 2004).

Another very important situation can be observed in countries that are influenced by *two earthquake types*. The southern area of Spain and Portugal are influenced by the subduction of the African plate beneath the European plate and by the collision between the Iberian plate and the European plate in the northern part. France is influenced by the eastern earthquakes, where the Italian plate collides with the European plate, along the Alps mountains, producing important earthquakes.

The Romanian earthquakes represent a clear example of this situation, where the eastern zone is influenced by the Vrancea intraslab earthquakes and the western zone by the intraplate earthquakes (Figure 3.29). The depth distribution outlines three enhancing zones of seismic activity: the crustal intraplate domain (0–40 km) in the western area of Romania and in the upper part (60–100 km); the Vrancea intraslab earthquakes in the lower part (100–180 km). While in the first case low-to-moderate earthquakes occur, whose magnitude does not exceed the value of M 5.6, in the second case, very strong earthquakes are present, with maximum expected magnitudes up to M 7.8.

In the case of countries with two earthquake types, the main problem is related to the necessity of having two different design methodologies: one for the very strong earthquakes and the other for low-to-moderate earthquakes. Unfortunately, the first case is the dominant one and code rules refer mainly to the most damaging earthquakes, but these rules are also extended to the second case, only with some very modest reduction in demand.

3.4.3 Main characteristics of low-to-moderate ground motions

The main characteristics of ground motions refer to the maximum foreseeable magnitude, duration, amplification (due to resonance effects), and number of characteristic pulses. Situated in crustal zones, the ground motions must have the features of near-source earthquakes, but with some modified

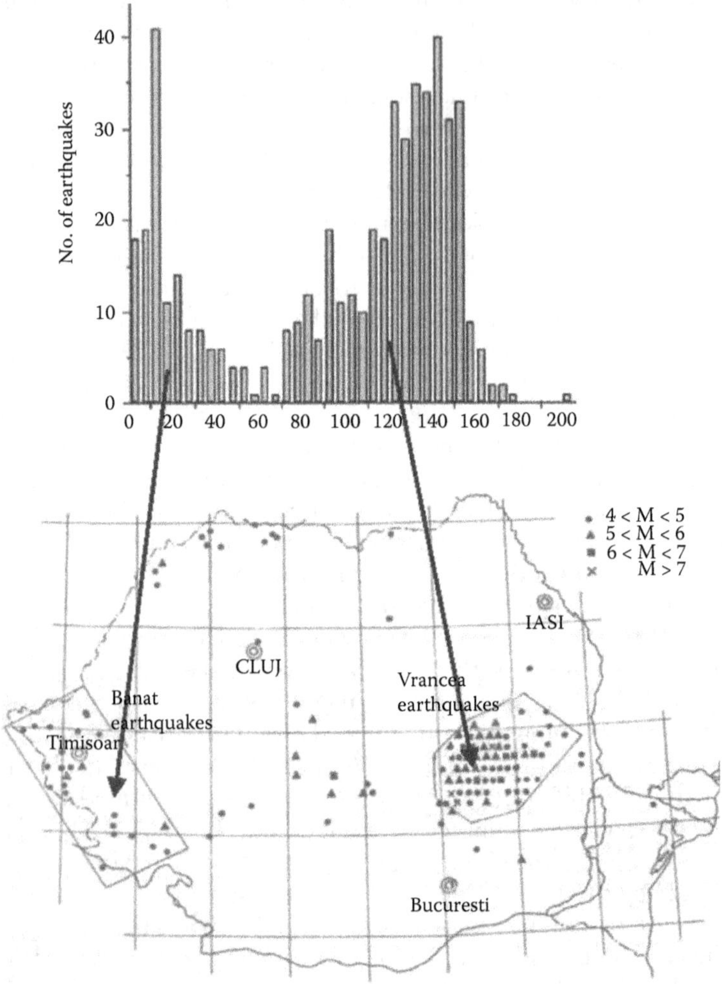

Figure 3.29 Romanian earthquakes: location of the two different seismic areas and earthquake number as function of depth. (Gioncu, V. 2004: Development and design of seismic-resistant steel structures in Romania. In *Behaviour of Steel Structures in Seismic Areas, STESSA 94* (eds. F.M. Mazzolani, V. Gioncu), Timisoara, 26 June–1 July 1994. E&FN Spon, London, 3–27.)

characteristics. The principal characteristics of low-to-moderate seismic regions, compared to high seismicity areas, can be summarized as follows:

- There are many uncertainties in delineating the seismic source zones, the geometry and location being not well known, due to the lack of fault traces on the ground surface. The historic seismicity record, at the same site, is too short compared to the recurrence periods of interplate earthquakes, where the frequency of occurrence is high. Therefore, the seismic sources and hazards are less readily defined.

- Earthquakes with large magnitudes are very rare events. The probability of large intraplate earthquakes is 2% in 50 years, whereas in comparison the probability of large interplate earthquakes is 10% in 50 years. This is mainly due to their reduced rupture areas in comparison with the interplate rupture areas. The territory of the United States is the best example to compare the occurrence of these two earthquake types. While the eastern part is dominated by intraplate earthquakes, the interplate earthquakes are most frequent in the western part. Figure 3.30 shows the return periods of peak accelerations in some cities of eastern and western United States. One can see very important differences between the hazard curves for the two selected cities, which are Memphis in eastern United States (intraplate earthquakes) and San Francisco in western United States (interplate earthquakes). The difference in accelerations is very high ($a_{mem} = 0.12g$, $a_{sf} = 0.51g$, MEM/SF = 24%), but it decreases for a return period of 2500 years (MEM/SF = 81%). One can notice that, if a return period

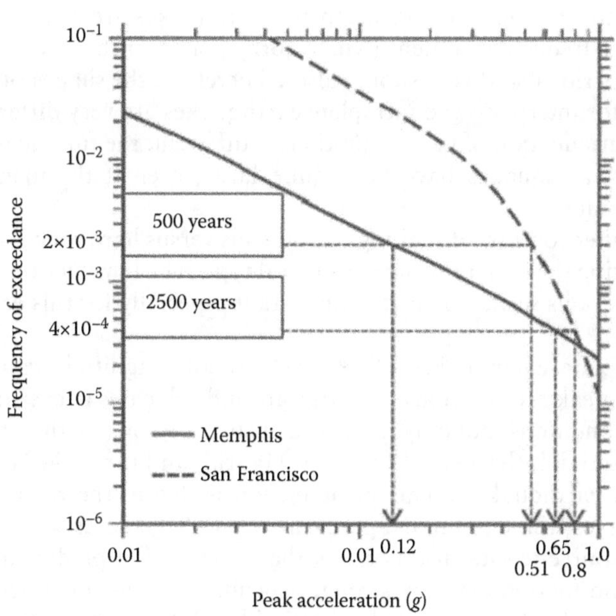

Figure 3.30 Peak accelerations for intraplate and interplate earthquakes. (From Gioncu, V., Mazzolani, F.M. 2011: *Earthquake Engineering and Structural Design*. Taylor & Francis, London.)

of 500 years gives rise to low or moderate earthquakes, the magnitude can be higher in some exceptional cases.

- The same observations result by examining the European earthquakes recorded in the so-called stable regions. As documented in the historical records, these regions usually experienced earthquakes with a magnitude below M 6 and, only in a relatively small number of cases, a magnitude M 6.5. The majority of earthquakes have a magnitude less than 4.5. But this observation does not exclude the possibility that some more important earthquakes could have occurred in the previous millennium. This remark is also applicable to the future.
- Buried thrust or rift fault types, caused by the compression or tension of the crust, produce the great majority of intraplate earthquakes. Only in very rare cases, these earthquakes result from a strike-slip fault. Local factors, such as strength of crustal rock, stress concentration, stress drop, and level of stress, play a dominant role in the earthquake process.
- The duration of intraplate earthquakes is shorter than that of interplate earthquakes. Intraplate records show that the strongest ground motion has a duration of just a few seconds. The motion starts with one or two relatively large-amplitude cycles, followed by five or six cycles of reduced amplitude, and then rapidly decays toward the end.
- The ground motions for intraplate earthquakes are characterized by higher-frequency content (which corresponds to short natural vibration periods) and very short pulses. Therefore, the shapes of acceleration for interplate and intraplate earthquakes are very different.
- The attenuation of ground motions is different: the full areas of intraplate earthquakes have been quite large, even if the magnitude is moderate.
- A higher contrast of shaking on soft soils versus hard rock is observed, showing the high influence of site soil type. The liquefaction of noncohesive soils seems to be very important especially for this earthquake type.
- Intraplate earthquakes of relatively small magnitude can produce large peak accelerations and large ground velocities in case of bad site soil conditions, but only in the near-source region (see the 1396 Basel, Switzerland, the 1811–1812 New Madrid, and the 1994 Northridge, USA, earthquakes). This phenomenon is due to the reduced depth, being less than 10 km in approximately 80% of cases.
- For source-to-site distance less than 10 km, the predicted levels of ground motion are very strong, depending on the focal depth of the earthquake. Considering the threshold of 0.2*g* for damaging motions, it can be observed that only those motions with focal depths less than 10 km far from the surface would be expected to generate potential damage on a limited surface around the epicenter.

- In the near-source zone, intraplate earthquakes also depend on the source type and the rupture propagation. From the observation, there is some evidence that the thrust faults produce higher ground motions than the normal faults for comparable hypocentral depths and equivalent magnitudes. In the same conditions, the magnitude of thrust faults can be more than three times the one corresponding to normal faults.
- The rare occurrence of severe seismic loading at a site situated in *stable continental regions* suggests that the philosophy for the design of buildings must be different from the one used in areas where interplate earthquakes occur.

Owing to these characteristics, only the near-source of the intraplate earthquakes can be very damaging. Unfortunately, the lack of intraplate records at a short distance from the epicenter encouraged the use of interplate earthquake records, even if it is known that there are important differences between these two near-source earthquake types.

Therefore, a high priority is to also create a network of seismic stations in a stable continental zone, where some earthquakes have been experienced, especially along the existing rifts that are already known. At the same time, it is necessary to elaborate special codes for the structural design of buildings located in areas with low-to-moderate earthquakes, because the use of the same codes elaborated for strong earthquake areas could be satisfactory from the safety side, but unsatisfactory from an engineering ethical point of view.

3.4.4 Structural design problems in the low-to-moderate seismic regions

The design philosophy in regions of low-to-moderate seismicity should be different from that in high seismicity regions. Differences in magnitude, duration, return periods, pulse type and number of important pulses, pulse periods, and so on justify the use of a different approach. Many buildings not designed for protection against earthquakes are present in these regions and the use of seismic rules, developed for high seismicity zones, for strengthening them is not economically convenient. However, completely changing the usual methodology is not a reasonable solution. In contrast, a reasonable way should be to adjust the existing rules to take into account the special characteristics of low-to-moderate earthquakes. In terms of design procedure, it means in practice to choose proper spectra and rational reduction factors.

The spectra for crustal fracture earthquakes are very different in comparison with the spectra of interplate earthquakes. A comparison among the corresponding spectra in the U.S. eastern territory (intraplate earthquakes)

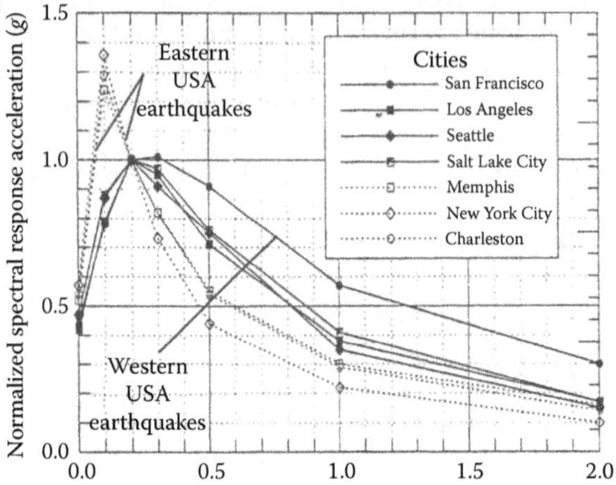

Figure 3.31 Comparison among spectra in the United States. (Gioncu, V., Mazzolani, F.M. 2006b: In *Behaviour of Steel Structures in Seismic Areas, STESSA 2006* (eds. F.M. Mazzolani, A. Waida), Taylor & Francis Group, London, 121–127; Gioncu, V., Mazzolani, F.M. 2011: *Earthquake Engineering and Structural Design.* Taylor & Francis, London.)

and western territory (interplate earthquakes) is presented in Figure 3.31. One can see that the difference consists of an important amplification of peak acceleration for very short periods (0.1–0.2 s).

It is very important to notice this great difference between intraplate and interplate earthquake characteristics, which must be reflected into the elastic spectra. In particular, this difference is referred to as the natural period of ground motions. For intraplate earthquakes, characterized by low-to-moderate earthquakes and short ground periods, an important increase in amplification is noticed for short structure periods. At the same time, the corner periods are much reduced for the intraplate earthquakes (0.3–0.5 s), in comparison with the interplate earthquakes, for which the natural periods are longer (0.5–1.5 s), especially in function of site soil conditions (Figure 3.32). Therefore, it is a crucial problem to use an adequate spectrum type, different from the one used in common codes (Figure 3.33). Only EUROCODE 8 introduces different spectra for earthquakes with magnitude $M < 5.5$ corresponding to low-to-moderate seismic area and earthquakes with magnitude $M > 5.5$ for strong seismic area. At the same time, it is very important to notice that in EUROCODE 8 the effects of soil conditions are given only for $M > 5.5$; for low-to-moderate earthquakes, weak soils produce an increase in acceleration amplification and some modification of corner periods (Figure 3.34).

Another very important problem is related to the *behavior factor q*, which reflects the capability of a structure to dissipate seismic energy

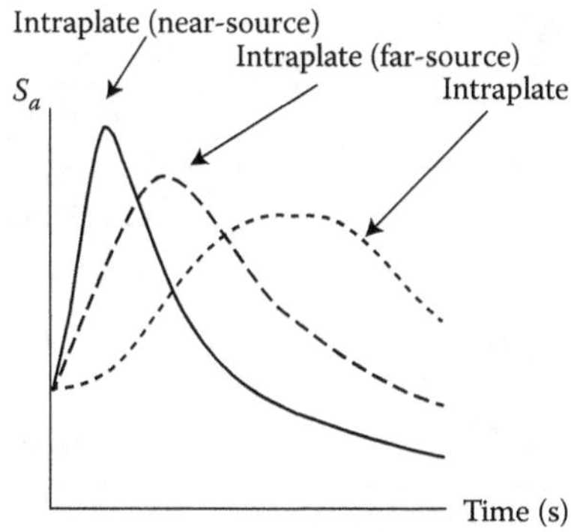

Figure 3.32 Influence of earthquake type on spectra. (Gioncu, V., Mazzolani, F.M. 2011: *Earthquake Engineering and Structural Design*. Taylor & Francis, London.)

through inelastic behavior, reducing the elastic spectra (Figure 3.35) and the seismic actions correspondingly. It characterizes the hysteretic behavior of the structural elements under severe events. It is often known as a general *ductility factor*, indicating the ability of the structure to absorb energy under reverse cyclic loading. The values of q range from 1.0 (for very brittle systems) to 6.0 (for very ductile systems), the maximum value being established for very ductile elements and for an important number of

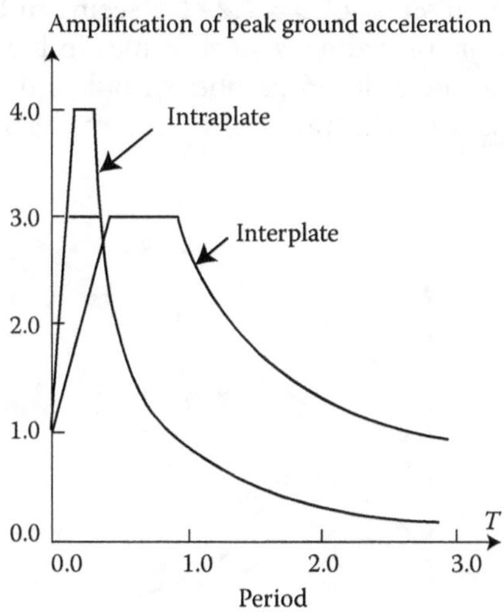

Figure 3.33 Normalized spectra for intraplate and interplate earthquakes. (Gioncu, V., Mazzolani, F.M. 2011: *Earthquake Engineering and Structural Design*. Taylor & Francis, London.)

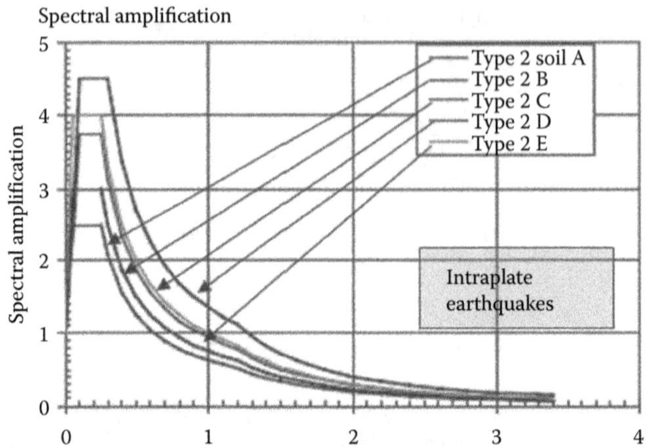

Figure 3.34 Elastic response spectra for soil classes. (Gioncu, V., Mazzolani, F.M. 2011: *Earthquake Engineering and Structural Design*. Taylor & Francis, London.)

cycles. As was shown in the above sections, the main characteristics of low-to-moderate earthquakes are short duration, short periods, and reduced number of important cycles. In this situation, the proposed behavior factor for strong earthquakes, characterized by long duration, medium and long periods, and large number of important cycles, must be revised. For this purpose, the maximum value of $q = 2.2–2.5$ seems to be reasonable as a *limited ductility factor* for ordinary steel frames in low-to-moderate seismicity regions, where the reduced number of pulses does not allow large seismic energy dissipation. This range of values was also proposed by Kelly and Zona (2006).

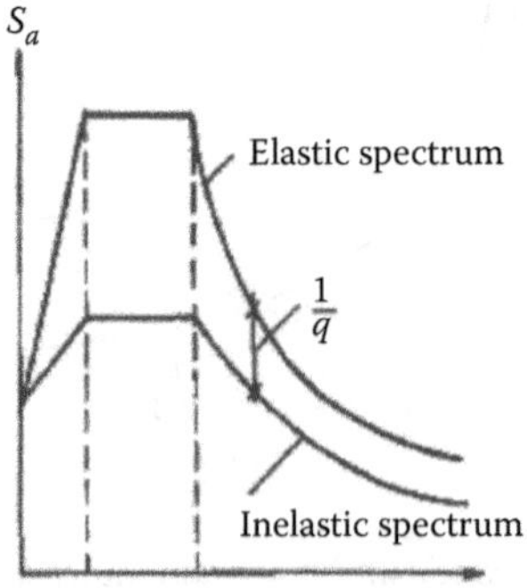

Figure 3.35 Inelastic response spectra.

3.5 PROPOSALS FOR IMPROVING THE NEW CODE PROVISIONS

3.5.1 Two topics for new codes

Earthquakes produced in urban areas have demonstrated that the economic impact of damage, due to the loss of function and business interruption, is huge and that, as a consequence, the damage control must become a more explicit design consideration. Recent earthquakes have demonstrated that the behavior of structures depends on the earthquake type (Gioncu and Mazzolani, 2006). Therefore, in order to improve the seismic design, the new codes must consider the following two topics:

- Performance-based design
- Specific design and performance evaluation approaches for structures in function of the earthquake type

As presented in Section 3.1.3, the implementation of these topics in the codification is constrained by the need to be *simple,* but *reliable*, to be assimilated, and not rejected, by the engineering community, which is generally conservative. For this reason, it must preserve the existing format of the present codes, the new aspects representing only some diversification of the principal design data.

3.5.2 Performance-based design

In recent years, a new design philosophy for building codes has been discussed within the engineering community. New building codes for earthquakes must adopt a performance-based design framework. The goal of a performance-based design procedure is to produce structures that have predictable seismic performance under multiple levels of earthquake intensity. Performance-based earthquake engineering implies design, evaluation, and construction of engineering facilities, whose performance under common and extreme loads responds to the various needs and objectives of owners, users, and the society.

In order to comply with these requirements, it is necessary for strong earthquake areas to pass from the two levels of the present codes to the three-level approach (Gioncu and Mazzolani, 2002). Two levels can be used only for low-to-moderate earthquake areas. The three-level approach is illustrated in Table 3.1.

- *Serviceability limit state,* under frequent low earthquakes, for which the structure must work in an elastic field and the nonstructural elements must remain undamaged. To guarantee their integrity, the interaction between structural and nonstructural elements must be evaluated by means of an elastic analysis, whose result is the reduction

Table 3.1 Three-Level Seismic Approach

Limit state	*Earthquake intensity*	*Avoided failure*	*Verification*	*Analysis method*	*Analysis object*
Serviceability	Low	Nonstructural elements	Rigidity	Elastic	Story drift
Damageability	Moderate	Local collapse	Strength	Elastoplastic	Section capacity
Ultimate[a]	Severe	Global collapse	Ductility	Kinematics	Rotation capacity

[a] Not necessary for low-to-moderate earthquake area.

of the interstory drifts. The basic verification refers to the structure rigidity by controlling the story drifts, which must be framed within the prescribed limits.

- *Damageability limit state,* under occasional moderate earthquakes, where the nonstructural elements are damaged (so the interaction of structural-to-nonstructural elements is neglected or just partially considered) and a very limited structural damage occurs. In this case, the analysis follows the elastoplastic methodology. The moderate damage to structural elements can be repaired without great technical difficulties. The basic verification refers to the structure members, by analyzing their section capacities.
- *Ultimate limit state (survivability),* under severe earthquakes, where all nonstructural elements are completely damaged and a global mechanism is formed. The analysis follows kinematics procedures. The basic verification considers the available ductility, which is expressed by the rotation capacity of plastic hinges. It must mention that this verification is not required for steel structures in low-to-moderate earthquake areas, because these structures can never arrive in the situation to form a collapse mechanism.

The application of performance-based design itself does not provide an improved or more effective structural performance. The checks are made according to the design methodology, which just guarantee the reliability given by a set of drawings and instructions for the builder. The improvement of the quality of the built product will depend on the development of advanced methodologies incorporating the new achievements in the demand and response capacity of structures. As has been shown in this chapter, the demand cannot ignore the earthquake type and the response capacity, which directly depends on this demand.

3.5.3 Influence of earthquake type

Despite using very advanced design methodologies, most of the seismic design work cannot describe the actual structure behavior, as very significant differences exist between the predicted and the actual response. The main reason for these differences is based on the neglect of the influence of ground motion types. The code provisions classify the earthquakes only in function of their magnitudes and specific spectra. Therefore, a structure, designed for a given magnitude and soil conditions, is supposed to behave in the same manner, regardless of whether it is situated in near-field or far-field sites or in zones influenced by interplate, intraplate, or intraslab sources. This is clearly not true. This is a great mistake on the part of the present codes, the structure behavior being very different for these sites, even if the magnitudes are the same. Therefore, a correct seismic design methodology must take into account these differences.

The first requirement for the next generation of codes is to complete the map of the seismic zonification in function of ground motion accelerations, by indicating on the map the *earthquake source type*:

- Interplate crustal source
  - Subduction (thrust)
  - Convergence
  - Strike-slip
- Intraplate crustal source
  - Crust fracture
- Intraslab subcrustal source

The second step for the next generation of codes is to consider in seismic design the main particularities of each earthquake type.

The main problem is to select a *set of spectra* corresponding to earthquake type, normalized to the acceleration, for the above three earthquake types and for the three limit states (Figure 3.36). The values in this figure, concerning the amplifications and corner periods, are given just as an example, the exact values being established for each case (Gioncu and Mazzolani, 2002). One can see that the spectra for the limit states are different, in contrast to the pattern of the present code spectra. For the serviceability limit states, the spectra are characterized by high amplification for short periods, with a steep reduction in the amplification of vibration periods exceeding the corner period. For the damageability limit states, the amplification is not so high and the reduction is more soft. For the ultimate limit states, no amplification occurs because the structure is transformed into a plastic mechanism, which works as the ground motions impose, without any interaction due to the lack of any vibration mode.

The differences depending on the earthquake types are very large. The interplate and intraslab earthquakes are in the first category (Figure 3.36a), being characterized by longer corner periods (especially for strike-slip and intraslab earthquakes). The amplification is higher for strike-slip and especially for intraslab earthquakes in comparison with the thrust earthquakes,

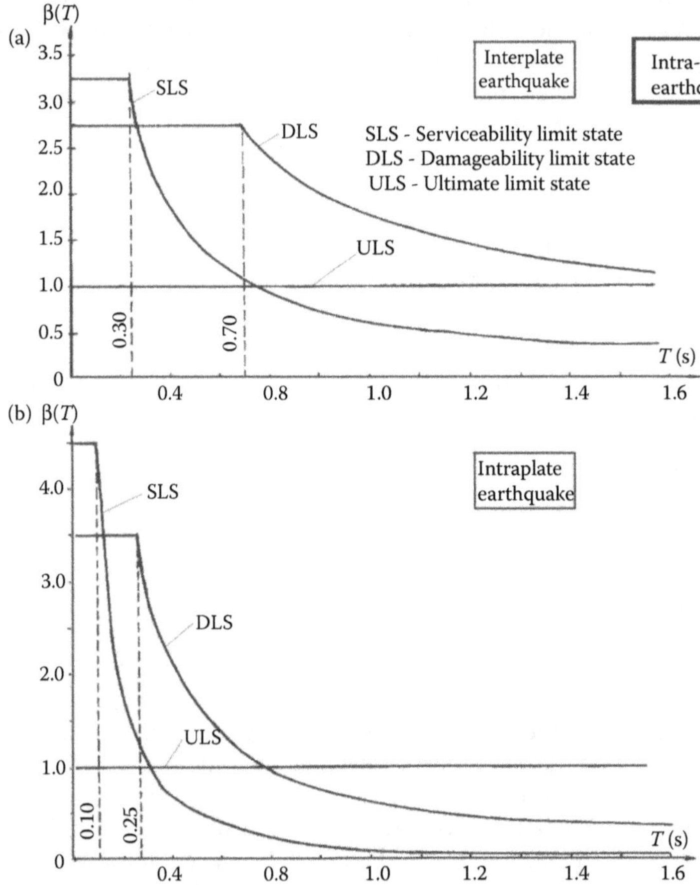

Figure 3.36 Design spectra: (a) interplate earthquake; (b) intraplate earthquake. (Gioncu, V., Mazzolani, F.M. 2002: *Ductility of Seismic Resistant Steel Structures*. Spon Press, London; Gioncu, V., Mazzolani, F.M. 2011: *Earthquake Engineering and Structural Design*. Taylor & Francis, London.)

due to the presence of a higher number of cycles. In the case of subduction, convergence, or intraplate, especially for crust fractures, the patterns of spectra are very different (Figure 3.36b). These spectra are also different for the different limit states.

A very important step in improving the next code generation is the topic dealing with the *different design philosophy* in function of the earthquake type. The seismic-resistant structures are usually designed relying on their ability to sustain high plastic deformations. The earthquake input energy is dissipated through the hysteretic behavior of plastic hinges during a high number of cycles of seismic loading. According to this design philosophy, the structure may be designed for lower forces than those that it has actually to resist. This seismic design philosophy has proved to be valuable during many great earthquakes and it corresponds to the methodology, which commonly considers the energy dissipated in the damaged members before the joint failure. From more recent circumstances, it has been questioned whether the current philosophy, based on the dissipation of seismic input energy by means of plastic hinges, should continue to be included in the modern codes. In fact, the recent earthquakes that occurred in near-source areas produced a widespread and unexpected brittle fracture of joints with very little or no evidence of plasticization of members, in contrast to the common assumptions in the code provisions. As a consequence, the critical question is whether the amount of seismic input rotations is also effective in the case of near-source earthquakes. Among the causes analyzed to explain these brittle fractures, the effect of high velocity of the pulse seismic loading, producing important strain-rate effects in the structural material, is considered by the specialists to be the main influencing factor (Gioncu and Mazzolani, 2002). Therefore, there are two different types of member behavior influenced by the earthquake type (see also Chapter 5):

- In the case of pulse seismic loading (Figure 3.37a), characteristic of near-source, thrust, and partial strike-slip earthquakes, the great velocity induces a very high strain rate with the consequence of increasing the yield stress, reducing the ductility and producing the brittle or ductile fracture of members or joints at the first or second cycle.
- In the case of cycle loadings (Figure 3.37b), characteristic of far-field and intraslab earthquakes on soft soils, an accumulation of plastic deformation occurs, producing a degradation of the structural behavior.

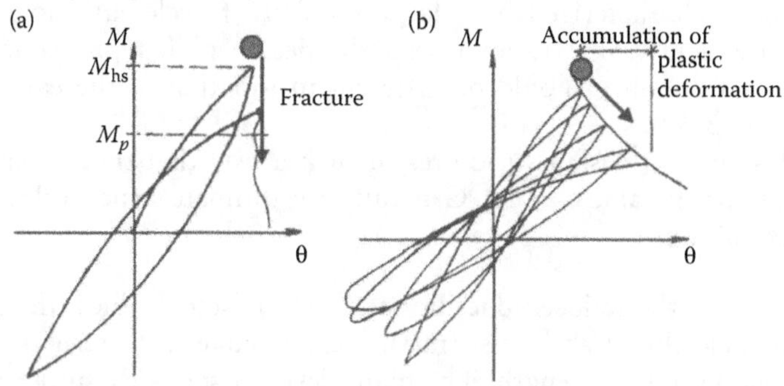

Figure 3.37 Member behavior under seismic actions: (a) pulse seismic loading; (b) cyclic seismic loading. (Gioncu, V., Mazzolani, F.M. 2002: *Ductility of Seismic Resistant Steel Structures*. Spon Press, London.)

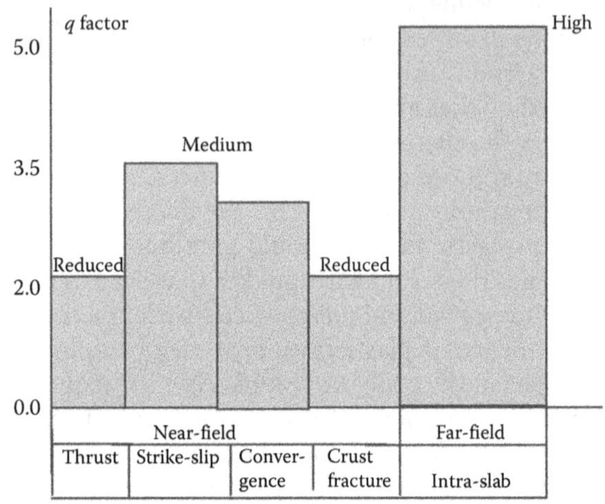

Figure 3.38 Proposed q factors in function of earthquake type. (Gioncu, V., Mazzolani, F.M. 2011: *Earthquake Engineering and Structural Design.* Taylor & Francis, London.)

Figure 3.38 presents a proposal for the values of the reduction factor q, in function of the earthquake type. This proposal refers to framed structures and normal soil conditions; for other structure types and bad soil conditions, the values can be modified accordingly. Only 1–2 pulses and short durations for thrust sources, 3–4 pulses, and moderate durations for strike-slip sources characterize the crustal near-source earthquakes. As a consequence, the seismic energy dissipation is reduced for thrust sources and it is the medium for strike-slip sources.

In the case of crustal convergence sources, where both thrust and strike-slip earthquakes can occur, a medium reduction factor is proposed. A higher value of the reduction factor is proposed for far-field and intraslab earthquakes, characterized by a large number of cycles and long duration. Consequently, it is rational that the design philosophy concerning strength and ductility should be different in function of the earthquake type (Figure 3.39).

It is desirable to design structures in such a way that they behave in a known and predictable manner. Generally, the ultimate structural response can be as follows:

- Owing to the reduced ductility and low q factor values (the case of near-field thrust and crust fracture earthquakes), the structure must develop a large strength. The main design issue is the protection of structural members against brittle fracture, which can be produced by the first or second cycle, due to the effect of the strain rate and a minimum ductile rotation.

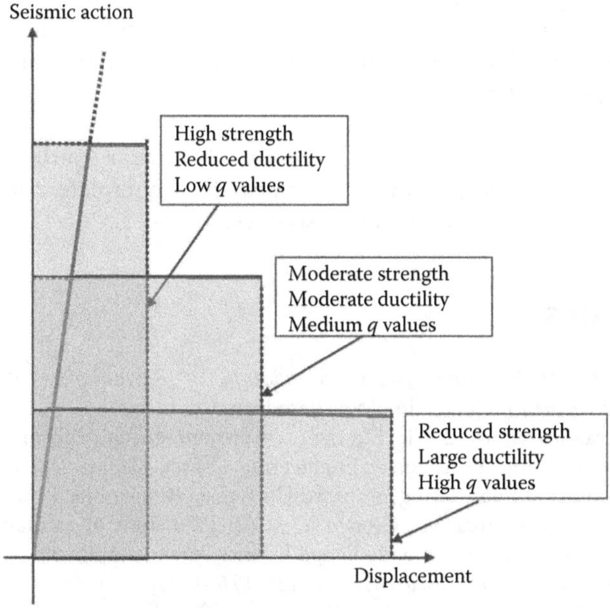

Figure 3.39 Different conceptions on strength, ductility, and q reduction factor values. (Gioncu, V., Mazzolani, F.M. 2011: *Earthquake Engineering and Structural Design.* Taylor & Francis, London.)

- For moderate ductility and medium q factor values (the case of near-field strike-slip or convergence earthquakes), the member fracture occurs after a reduced number of pulses. The design must protect the structural members against brittle fracture, by assuring a corresponding ductile rotation.
- For large ductility and high q factor values (the case of far-field and intraslab earthquakes), the main issue is the protection of structural members against large accumulation of plastic deformations, by assuring the structural members of large available ductile rotation.

Another very important aspect in seismic design is the *effect of superior vibration modes*, producing a change of the seismic forces distribution, in comparison with the distribution corresponding to the first vibration mode. There are three possible different situations, in function of the earthquake types:

- For *far-field and intraslab crustal earthquakes*, the effects of superior vibration modes are reduced, the first mode being dominant. These effects can produce some change in the seismic forces distribution only in case of very unusual flexible structures.

- For *near-field crustal earthquakes,* the effects of superior modes must be considered practically in all cases. The consideration of superior vibration modes reduces the plastic moment distribution in the frame, giving rise to favorable effects.

Regarding the *effect of vertical components* in near-source areas, they are very important only in the case of near-field interplate crustal earthquakes, especially for normal-thrust sources.

REFERENCES

Aagaard, B.T. 2000: *Finite-Element Simulation of Earthquakes.* PhD Thesis, California Institute of Technology, Pasadena, California.

Alavi, B., Krawinkler, H. 2001: Effects of near-fault ground motions on frame structures. Blume Earthquake Engineering Center, Department of Civil and Environmental Engineering, Stanford University, Report No. 138.

Anastasiadis A., Mosoarca M., Gioncu V. 2012: Prediction of available rotation capacity and ductility of wide-flange beams: Part 2: Applications. *Journal of Constructional Steel Research*, 69(2012), 176–191.

Bala, A., Radulian, M., Popescu, E. 2003: Earthquakes distribution and their focal mechanism in correlation with the active tectonic zones of Romania. *Journal of Geodynamics*, 36, 129–145.

BGHRC, Benfield Grieg Hazard Research Centre. 2003: A rift at the heart of Europe, http://www.benfieldhrc.org/activities/issues/issues1/pdf.

Camelbeeck, T., Meghraoui, M. 1996: Large earthquakes in Northern Europe more likely than once thought. *EOS*, 77(42), 405–409.

EB, Earthquakes Booklet. 2008: Earthquakes in UK, http://www.earthquakes.bgs.ac.uk/earthquakes/education/booklet.uk.htm.

EUROCODE 8 (EN 1988-1-1). 2004: Design of earthquake resistant structures. Part 1–1: General rules and rules for buildings.

FEMA 267. 1997: Interim guidelines advisory. Report No. SAC-96-03.

GB5011. 2001: *Code for Seismic Design of Buildings.* China Architecture & Building Press. China.

Gioncu, V. 2004: Development and design of seismic-resistant steel structures in Romania. In *Behaviour of Steel Structures in Seismic Areas, STESSA 94* (eds. F.M. Mazzolani, V. Gioncu), Timisoara, 26 June–1 July 1994, E&FN Spon, London, 3–27.

Gioncu V. 2008: Structural design problems in the low-to-moderate seismic regions. Part 1. Seismic risk and hazard; Part 2. Building's vulnerability. In *Seismic Risk 2008, Earthquakes in North-Western Europe* (ed. H. Degee), 11–12 September 2008, Liege, 249–260, 261–272.

Gioncu, V., Anastasiadis, A., Mosoarca, M.: Influence of earthquake type on the local ductility of steel beams. Part 2: Near field earthquakes. *Engineering Structures*, (submitted for publication).

Gioncu, V., Mazzolani, F.M. 2002: *Ductility of Seismic Resistant Steel Structures.* Spon Press, London.

Gioncu, V., Mazzolani, F.M. 2006a: Influence of earthquake types on the design seismic-resistant steel structures. Part 1. Challenges for new design approaches. In *Behaviour of Steel Structures in Seismic Areas, STESSA 2006* (eds. F.M. Mazzolani, A. Waida), Taylor & Francis Group, London, 113–120.

Gioncu, V., Mazzolani, F.M. 2006b: Influence of earthquake types on the design seismic-resistant steel structures. Part 2. Structural responses for different earthquake types. In *Behaviour of Steel Structures in Seismic Areas, STESSA 2006* (eds. F.M. Mazzolani, A. Waida), Taylor & Francis Group, London, 121–127.

Gioncu, V., Mazzolani F.M. 2011: *Earthquake Engineering and Structural Design.* Taylor & Francis, London.

IBC. 2006: *International Building Code,* Edition 2006. International Code Council, ICC.

Kelly, D.J., Zona, J.J. 2006: Design tips for steel in low or moderate seismicity regions. *Modern Steel Construction*, February, 20–25.

Mateescu, G., Gioncu, V. 2000: Member response to strong pulse seismic loading. In *Behavior of Steel Structures in Seismic Areas, STESSA 2000* (eds. F.M. Mazzolani, R. Tremblay), Montreal, 21–24 August 2000, Balkema, Rotterdam, 55–62.

Mazzolani, F.M., Mandara, A. 1992: Nuove strategie di protezione sismica per edifici monumentali. Editioni10/17, Salerno.

Mooney, W.D., Schulte, S., Detweiler, S.T 2004: Intra-plate seismicity and the determination of nuclear test events using an updated global earthquake catalogue. *26th Seismic Research Review, Trend in Nuclear Explosion Monitoring*, 439–448.

NEHRP. 2008: Strategic program for the national earthquake hazards reduction program (2009–2013). FEMA, NIST, USGS report 2008.

NORSTAR. 2004: Earthquakes in Norway, http://norstar.no/seismology/general/earthquakesinnorway.pdf.

SAC. 1996: Connection test summaries. Report No. SAC 96-02.

Sato, T., Murono, Y., Wang, H.B., Nishimura, A. 2001: Design spectra and phase spectrum modeling to stimulate design earthquake motions: a case study through design standards of railway facilities in Japan. *Journal of Natural Disaster Science*, 23(2), 89–100.

UBC. 1997: *Uniform Building Code*, Edition 1997. International Council of Building Officials, ICBO.

USGS (nd): Major tectonic plates of the world, http://geology.er.usgs.gov/eastern/plates.html.

Chapter 4

New generation of steel structures

4.1 INTRODUCTION

Steel structures for multistoried buildings can be classified according to the three well-known main classical categories, that is, moment-resisting frames (MRFs), concentric-braced frames (CBFs), and eccentric-braced frames (EBFs). In the last few decades, theory and practice have been devoted to improving the behavior of classical systems by means of more sophisticated details (Section 4.2) as well as to proposing more advanced systems for seismic-resistant schemes (Section 4.3) and beam-to-column connections (Section 4.4), for improving structural effectiveness and energy absorption capacity.

The damage produced in the beam-to-column connections of the perimetral MRFs during the 1994 Northridge earthquake recalled the attention on the advantage of the use of the so-called dog-bone system for protecting the connection. It is worth remembering that the dog-bone idea was proposed for the first time by André Plumier in 1990, within the activities of the Committee TC13 of the European Convention for Constructional Steelworks. It was necessary to analyze the effects of a strong earthquake to discover the beneficial effect of the application of this idea both for repairing damaged frames and for the new ones, in order to guarantee the formation of a global collapse mechanism, which is the most critical issue in the design of MRF. After these first experiences, the new generation of dog-bone, the so-called reduced beam section (RBS), has been fully investigated, but not yet completely codified (Section 4.2.2).

As is well known, the EBF system was proposed for the first time at the University of Berkeley, California, by Popov et al. in the 1970s. This system was internationally codified and many applications followed for new building structures. Today, the research attention is concentrated on the detail of the *link*, which is the key element of the whole system. Studies of interest are devoted to proposing removable links, which can be substituted after the damage occurs. A noncommon typology is the so-called inverted Y, which has been recently reevaluated as a very effective solution. Also, the use of EBF in the seismic upgrading of existing RC buildings has been successfully proposed (Section 4.2.1).

The CBF are the most common typology for seismic-resistant steel structures and their codified design rules are universally recognized, even if there are still some weak points to be improved. For instance, the common hypothesis to neglect the compression diagonal in X-bracings is full of uncertainties, which are not clarified.

These problems have been completely removed by the introduction of buckling restrained braces (BRB), which eliminate the difference between tension and compression members. In addition, a big advantage of BRB consists of the possibility of having only one diagonal for each structural mesh, which reduces the functional limitation of the system in the layout of the building. Starting from the BRB solution, which was patented for the first time in Japan, many other solutions have been proposed until now, basically following two configurations for the cross-section: mixed section (steel core in a steel tube filled with cast concrete) and all-steel section (steel core restrained by an external steel box). The details of the core and the restraining element are today the object of many research activities, faced with improving both the execution aspects and the energy absorption properties. Interesting applications of BRB are also performed for the seismic upgrading of the existing RC buildings (Section 4.2.3).

In the field of new solutions for bracings, particular attention is devoted to the use of special metallic panels—the so-called metal plate shear walls (MPSW). Their use was proposed for the first time in the United States for the seismic upgrading of the existing RC buildings after the Loma Prieta earthquake in 1989. For new steel structures, shear walls can be used as the only bracing system inside a pinned structure or they can be added to MRF structures, according to a dual configuration system. After the Kobe earthquake of 1995, all the main new buildings in Japan were equipped with special damping devices. In particular, the MPSW systems were frequently applied in multistoried steel buildings. The metal used was a special *low-yield steel* with a yield point of about 100 MPa, which guarantees the fully plastic behavior before the start of the early local buckling phenomena. This material is produced only in Japan, but not in Europe or the United States. The transfer of this technological system in other countries is based on the fundamental idea of substituting low-yield steel with aluminum, in particular, pure aluminum with a special heat treatment, which produces the reduction of the elastic limit to 40 MPa and increases the ultimate elongation to about 50–60%, that is, to obtain the ideal material for this kind of application. The research activity on MPSW developed in the last few years covers all types of metal panels to be used for bracing purposes: sandwich panels, unstiffened panels, and stiffened panels. With respect to the structural mesh, the configuration can be full bay, partial bay, or small panels in X-brace configuration. These systems have been studied both analytically and experimentally. Interesting applications have been performed for seismically retrofitting existing RC buildings. It is worth noticing that both the design rules and constructional details of these systems based on MPSW have not yet been internationally codified, despite the significant number of applications (Section 4.3).

A new solution for connections in MRF was proposed in the United States, based on the observation of damage after the 1994 Northridge earthquake. In fact, during this earthquake, steel MRFs, which are usually designed to dissipate energy through the formation of plastic hinges at the beam edges, did not perform in such a way, due to the unexpected brittle fracture in the nodes. As an effective alternative, the research team of the Leigh University in Bethlehem, USA, proposed the posttensioned energy-dissipating (PTED) connection system, which provides both self-centering and energy dissipation. Many theoretical and experimental activities have been developed both in the United States and Europe, but no codification has been set up as yet (Section 4.4).

4.2 IMPROVING EXISTING SOLUTIONS

4.2.1 Advanced eccentric-braced systems

The EBF is a hybrid lateral force-resisting system (Figure 4.1). In fact, it can be considered to be the superposition of two different framing systems: the MRF and the concentrically braced frame. EBFs can also combine the main advantages of these two conventional framing systems and minimize their respective disadvantages. In general, EBFs possess high elastic stiffness and a stable inelastic response under cyclic lateral loading, together with excellent ductility and energy dissipation capacity.

The key distinguishing feature of an EB system is that at least one end of each brace is connected to the beam in such a way as to isolate a segment of the beam called a "link." EBF arrangements are shown in Figure 4.1, where in each framing scheme the links are identified by a bold segment. The four EBF arrangements presented here are usually known as a split-K-braced frame, a D-braced frame, a V-braced, and an inverted-Y-braced frame.

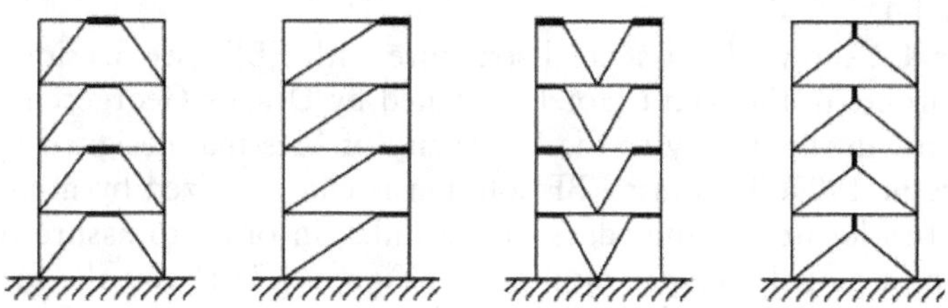

Figure 4.1 EBF configurations.

Research on the behavior of EBFs started in the second mid-1970s (Roeder and Popov, 1978) and continues to date. All these studies have confirmed the reliability of EBFs in resisting horizontal actions. Consequently, the number of civil applications is increasing day by day. Eccentrically braced frames in buildings, which provide a stiff and ductile lateral load-resisting system, typically include segments of beams, the so-called links, which are destined to yield and plastically deform in shear and bending.

The behavior of links in eccentrically braced frames for new buildings has been studied widely (Hjelmstad and Popov, 1983; Malley and Popov, 1983a,b; Manheim and Popov, 1983; Kasai and Popov, 1986a,b; Popov and Engelhardt, 1988; Foutch et al., 1989; Roeder et al., 1987; Foutch, 1989; Ricles and Popov, 1993), but their use is now also becoming a viable method to retrofit RC structures and protect bridges. Two examples of the use of shear links for bridge retrofitting are the Richmond–San Rafael Bridge (Itani et al., 1998) and the tower of the new San Francisco–Oakland Bay suspension cable bridge (McDaniel et al., 2003). Research on the application of shear links for seismic retrofitting of RC structures was recently carried out by Ghobarah and Abou Elfath (2001), Mazzolani (2006a,b). In particular, this system reveals that it has good reversibility because, in case of damage, it is easy to rehabilitate after an earthquake. The link is conceived to act as a seismic fuse, in such a way that the whole surrounding structure remains an elastic field. For this reason, a detachable link offers the actual possibility of easy removal and substitution.

Experimental investigations on removable bolted links have demonstrated the technological feasibility of this solution. The performance of short removable links and the possibility of easy replacement make them attractive for retrofitting RC structures. Very short links, which assure an elastic behavior of the connections, are preferred, due to the much easier replacement of damaged links. Concentration of damage in the removable link (performing like passive energy dissipation devices) may be accomplished by the capacity design principles, including fabrication of the link from steel with a lower yield strength than that of ordinary structural steel.

Some examples of structures with EBF systems are given in Figures 4.2 through 4.4.

Figure 4.2 shows the first steel structure with EBFs (see Section 2.2.4): (a) in Europe, in the Mara Hotel, designed by Dragos Georgescu (1996); (b) in Italy, in the library of the University of Messina, designed by Aldo D'Amore in 1997. This last EBF solution is characterized by a couple of vertical ties located at the edges of the links, in order to assure the formation of a global collapse mechanism. Figure 4.3a shows the facade of the Century Tower in Tokyo, designed by Norman Foster in 1991, which also demonstrates how mega-EBFs are used for characterizing the building

Figure 4.2 (a) Hotel Mara in Sinaia, Romania; (b) University of Messina, Italy.

Figure 4.3 (a) Century Tower, Tokyo (1991); (b) multistoried building in San Diego (California, USA).

(a)

(b)

Figure 4.4 (a) Multistoried building in Alikahya (Turkey); (b) Bilgi University in Istanbul (Turkey).

from an architectural point of view. A multistoried building with EBF—in San Diego, California—is mentioned in Figure 4.3b. Figure 4.4 illustrates two interesting applications of steel structures with EBFs in Turkey: one in Alikahaya (a), the second in Istanbul (b), being the new building of the Bilgi University, which received the ECCS award in 2005.

The main purpose of designing EBF is to restrict the inelastic action to the links and to design the framing around the links to sustain the maximum forces that the links can develop. Design using this strategy should ensure that the links act as ductile seismic fuses and preserve the integrity of the whole frame. For this reason, other components of the framing system (such as the diagonal braces, columns, and link connections) should be designed for the forces generated by the full yielding and strain hardening of the dissipative links. To this purpose, it is important to make explicit the distribution of internal actions in the EBF system and define a relationship between the frame shear force and link shear force, which depends only on the EBF configuration. The design actions in links can be calculated using equilibrium concepts.

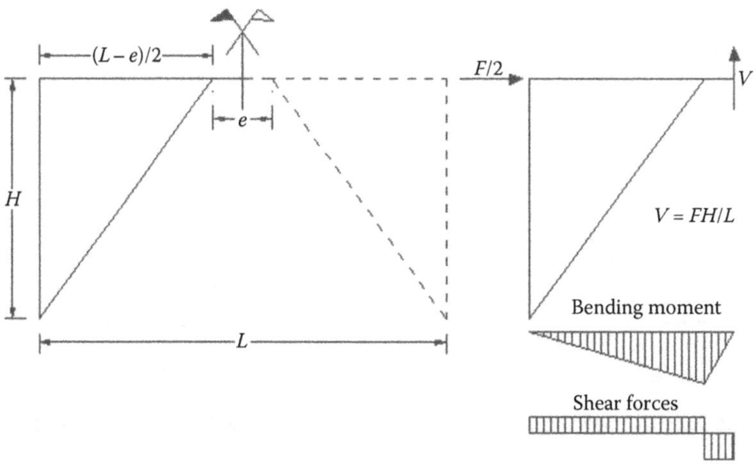

Figure 4.5 Design action in the "Link" for a split-K-braced EBF configuration.

For example, in a split-K-braced EBF (shown in Figure 4.5), assuming that the moment at the center of the link is equal to zero, the link shear force V can be expressed as

$$V = \frac{F \cdot H}{L} \tag{4.1}$$

where F is the lateral force, H is the interstory height, and L is the bay length.

In the case of an inverted-Y-braced EBF (Figure 4.6), assuming that the moment at the brace connections is equal to zero (i.e., in case of pinned braces), the link shear force V can be expressed as

$$F = V \tag{4.2}$$

where F is again the lateral force.

The link inelastic performance essentially depends on its length, cross-section properties, and web stiffener arrangement. For a given cross-section, the link length controls the yielding mechanism and the ultimate failure mode. Short links are mainly dominated by a shear mechanism; instead, a flexure mechanism controls the link response for long links. Intermediate links are characterized by combined shear and flexural yielding. Assuming perfect plasticity, no flexural–shear interaction, and equal link end moments, the theoretical dividing point between a short link (governed by shear) and a long link (governed by flexure) is a length of $e = 2M_p/V_p$, where $M_p = Zf_y$ is the plastic bending moment (Z being the plastic modulus and f_y the steel yielding stress) and $V_p = 0.55f_y \cdot (d - 2t_f) \cdot t_w$ (being d the depth of the cross-section, t_w the web thickness, and t_f the flange thickness).

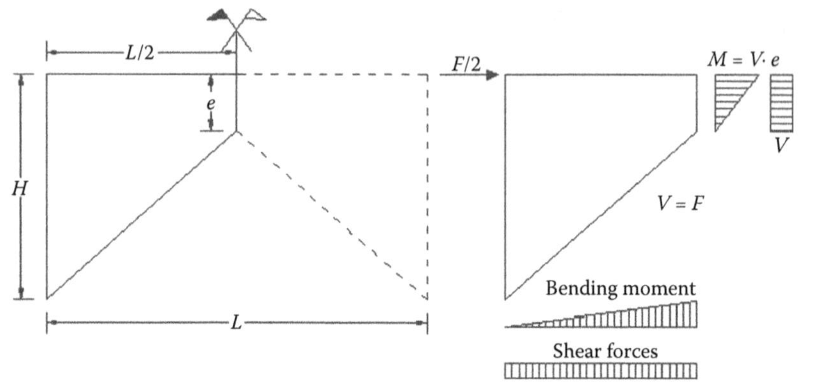

Figure 4.6 Design action in the "Link" for an inverted-Y-braced EBF configuration.

Experimental tests (Hjelmstad and Popov, 1983; Kasai and Popov, 1986a) indicate that the assumption of no $M–V$ interaction is reasonable, but the assumption of perfect plasticity is not correct. In fact, substantial strain hardening occurs in shear links. According to the first few tests performed in the 1980s, the average ultimate link shear force may reach the value of $1.5V_p$ (Kasai and Popov, 1986a; Popov and Engelhardt, 1988). Recent tests reported by Okazaki and Engelhardt (2007) seem to confirm the validity of such a value for the average overstrength exhibited by short links. However, other tests recently carried out on built-up shear links (Itani et al., 2003; McDaniel et al., 2003) showed larger overstrength factors, reaching values slightly larger than 2 in the case of normal carbon steel and up to 4 in the case of low-yield-point steel. In the tests reported by D'Aniello et al. (2005), overstrength factors larger than 3 have been exhibited by European-type wide-flange (HE) shapes. Anyway, the value currently accepted for the average overstrength of shear links is 1.5 and it forms the basis of modern steel seismic design codes, such as the AISC Seismic Provisions (2005). The effect of axial restraints on link performance was recently highlighted in Della Corte et al. (2007), showing that the tension axial force developing at large shear deformations, as a consequence of second-order geometric effects, may be nonnegligible and could represent one contributing cause for an overstrength factor larger than 1.5, which was measured in the tests carried out by McDaniel et al. (2003), Dusicka et al. (2004), and D'Aniello et al. (2005).

One implication of strain hardening is that both shear and moment yielding occur over a wide range of link lengths. The large end moments, combined with the significant strain gradient occurring in links, lead to very large flange strains. Kasai and Popov (1986a) estimated that the maximum link end moments can be assumed to be $1.2M_p$. Thus, from the link static, if the end moments are limited to $1.2M_p$ and the link shear is assumed to reach $1.5V_p$, the limiting link length is

$$e = \frac{2 \cdot (1.2M_p)}{1.5 \cdot V_p} = 1.6\frac{M_p}{V_p} \tag{4.3}$$

Then, the following equations can be used to classify the link mechanical response:

Shear (short) links:

$$e \leq 1.6\frac{M_p}{V_p} \tag{4.4}$$

Intermediate links:

$$1.6\frac{M_p}{V_p} < e < 2.5\frac{M_p}{V_p} \tag{4.5}$$

Flexure (long) links:

$$e \geq 2.5\frac{M_p}{V_p} \tag{4.6}$$

The ultimate failure modes of short and long links are quite different. In particular, inelastic web shear buckling is the ultimate failure mode of short links. This buckling mode can be delayed by adding web stiffeners (Figure 4.7a). Hjelmstad and Popov (1983) developed several cyclic tests to relate the web stiffeners spacing to the link energy dissipation; Kasai and Popov (1986) subsequently developed simple rules to relate stiffeners spacing and maximum link inelastic rotation up to the web buckling.

The work of Kasai and Popov (1986a,b) forms the basis of the current code provisions, such as the AISC Seismic Provisions (2005), leading to the following relationship between stiffener spacing (a), cross-section depth (d), and required inelastic shear deformation capacity (γ):

$$a = 30t_w - \frac{d}{5} \quad \text{for } \gamma = 0.08 \text{ rad} \tag{4.7}$$

$$a = 56t_w - \frac{d}{5} \quad \text{for } \gamma = 0.02 \text{ rad} \tag{4.8}$$

with a linear interpolation for intermediate values.

(a)

(b)

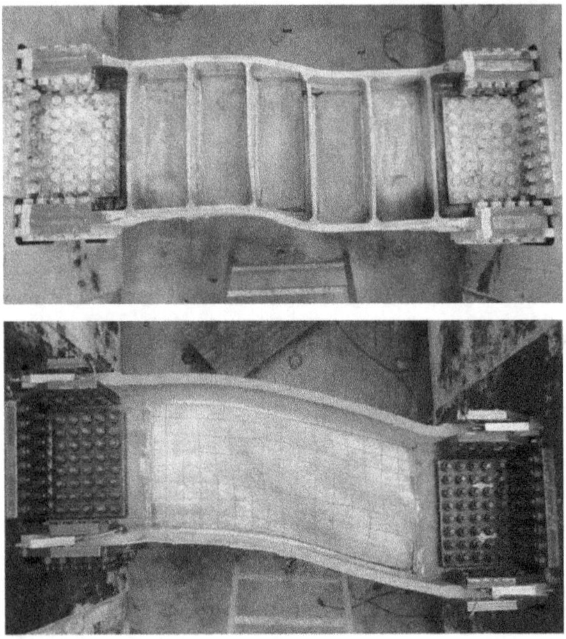

Figure 4.7 (a) Plastic deformation of stiffened short link. (b) Plastic deformation of unstiffened short links.

Test results carried out in the 1980s showed that the ultimate failure mode of short links is always the web fracture, which developed just after significant local web buckling took place, and it was activated in correspondence of those zones with very large strain concentration due to the buckled shape. In contrast, recent test results (McDaniel et al., 2003; Okazaki and Engelhardt, 2007) highlighted a failure mode not previously reported, consisting of a fracture developing prior to the buckling phenomena (Figure 4.8a). The fracture started consistently at the termination of a stiffener-to-web weld and then propagated either in a diagonal or in a horizontal pattern across the web. Shear links without stiffeners (Figure 4.8b) had lower plastic strain demands in the web as compared to those with stiffeners and consequently did not develop cracks until larger deformations were imposed. Localized plastic strains were also present in the stiffeners and the flanges of the links. The stiffeners developed localized plastic hinging at the connection to the flange, resulting in the cracks observed during the experiments (Figure 4.8c).

As reported by Dusicka et al. (2004), numerical models developed to investigate the plastic strain demands on different components of steel built-up links (Figure 4.9) show a consistent correlation between the location of the initial cracking during the experiments on shear links with

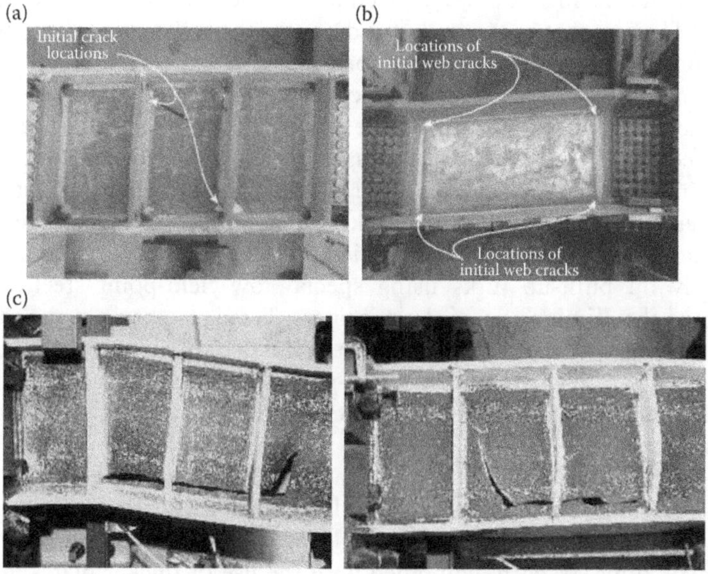

Figure 4.8 Short link web fracture. (a) Location of initial crack in stiffened link. (Dusicka, P., Itani, A.M., Buckle, I.G. 2004: In *Proceedings of the 13th World Conference on Earthquake Engineering*, Vancouver, B.C., Canada, 1–6 August, Paper No. 522.) (b) Location of initial crack in an unstiffened link. (Dusicka, P., Itani, A.M., Buckle, I.G. 2004: In *Proceedings of the 13th World Conference on Earthquake Engineering*, Vancouver, B.C., Canada, 1–6 August, Paper No. 522.) (c) Web fractures after testing for stiffened shear link. (Okazaki, T., Engelhardt, M.D. 2007: *Journal of Constructional Steel Research*, 63, 751–765.)

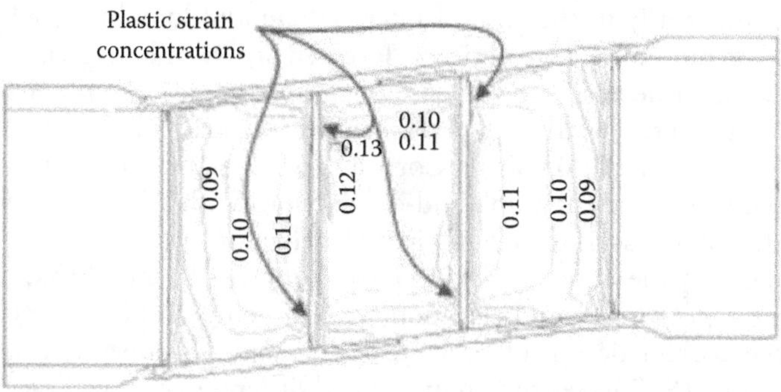

Figure 4.9 Plastic strain distribution in the web of built-up shear links. (Dusicka, P., Itani, A.M., Buckle, I.G. 2004: In *Proceedings of the 13th World Conference on Earthquake Engineering*, Vancouver, B.C., Canada, 1–6 August, Paper No. 522.)

stiffeners and the location of localized plastic shear strain in the numerical models. The increase in strain demand occurred consistently at the ends of the stiffener-to-web welded connection.

This indicates that the onset of cracking in the web observed during the experiments was likely caused by the combination of the influence of the heat-affected zone from welding and the plastic strain concentrations caused by the link deformations. The contours of the plastic shear strain showed a lower demand at the ends of the link length as compared to the midlength.

Besides, Dusicka et al. (2004) carried out an experimental and numerical study on built-up links using special low-yield-point steel, which permitted the elimination of the stiffeners. From the strain demand perspective, removing the stiffeners from the link length eliminates the localized plastic strains at the stiffener-to-web welds. As a result, the initial cracking and ultimate failure mode occurred at significantly higher link deformations.

The work presented by Rai and Wallace (2000) is based on the same concept of using a material with a low elastic limit; they proposed the use of aluminum for fabricating the shear links as energy dissipaters. Again, the low elastic limit strength, combined with the use of thicker plates, allows stiffeners to be avoided, because of the web local buckling being delayed at large plastic deformations.

Innovative solutions for eccentrically braced frames have also been studied by Ghobarah and Ramadan (1994) and by Stratan and Dubina (2002), who proposed the use of a bolted flush end-plated connection. The idea is that the bolted connection permits easy replacement of the shear links, after a damaging earthquake. Experimental results presented by Stratan and Dubina (2002) suggest that the end-plate connections should be designed in such a way as to avoid flexural plastic deformation of the end plates; otherwise, the removal of the link may be quite difficult. Furthermore, experimental results highlight the difficulty in keeping the bolts in an elastic range, especially in the case of intermediate and long links, where the flexural demand on the connection is larger. The use of bolted end connections becomes a necessity in the case of seismic upgrading of existing RC structures. In fact, several tests performed on RC structures equipped with EBs (D'Aniello et al., 2005; Della Corte and Mazzolani, 2006) showed that it is fundamental to design link end-connections with a large safety factor to provide a stable and reliable dissipative behavior.

Moreover, in the case of seismic retrofitting of RC buildings, another important aspect to adequately adopt steel bracing within the RC structure is the correct design of brace-to-RC structure connections. Maheri and Sahebi in 1997 have studied and tested different types of brace-to-RC structure connections. The best solution is to adopt gusset plates bolted to RC beam-to-column joints by means of bolts passing through drilled holes in the RC members or laterally, as shown in Figure 4.10. Analogously,

Figure 4.10 Diagonal brace-to-RC end connection. (Della Corte, G., Mazzolani, F.M. 2006: In *Proceedings of the 2nd FIB Congress*, Naples, Italy, 5–8 June.)

link-to-RC beam connections can be made by bolting link end plates to the RC structure by means of bolts passing through holes drilled in the RC beams, as shown in Figure 4.11.

An example of seismic retrofitting of RC structures by means of EBF systems has been developed within the ILVA-IDEM project (Mazzolani, 2006a), where EBFs have been applied for seismic upgrading of an existing two-storied RC structure, which has been subjected to the pushover cyclic test (D'Aniello et al., 2005; Della Corte and Mazzolani, 2006). In fact, these tests have been carried out within the context of a wider experimental research activity, all having the purpose of evaluating several innovative technologies for the seismic retrofitting/upgrading of an existing RC structure (Mazzolani, 2006b, 2008), which was subdivided into six modules. Each module has been retrofitted by using different techniques (see Figure 4.12): (1) base isolation with neoprene bearings, (2) steel BRBs, (3) composite material C-FRP, (4) eccentric steel braces, (5) shape memory alloys (SMA) braces, and (6) metallic shear panels (Figure 4.12).

The details of the RC structure, which has been equipped with the inverted Y EBF (n. 4 in Figure 4.12) and then tested, are shown in Figure 4.13. Three lateral loading tests have been performed and each test was characterized by shear links with bolted end-plate connections. Different design criteria have been applied in the design of the connections: (1) in the first test, no capacity design criteria have been considered; (2) in the

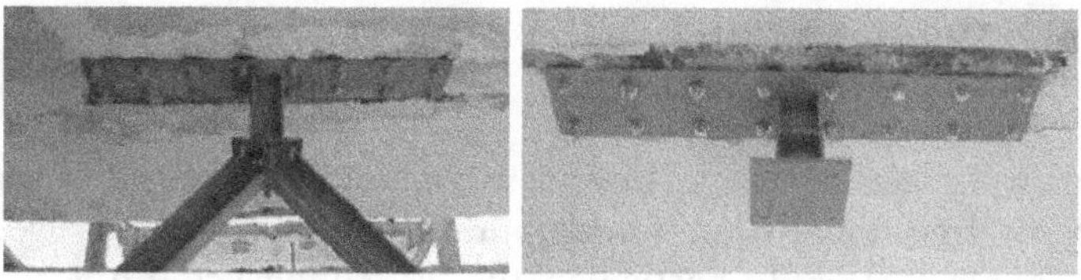

Figure 4.11 Link-to-RC beam connection. (Della Corte, G., Mazzolani, F.M. 2006: In *Proceedings of the 2nd FIB Congress*, Naples, Italy, 5–8 June.)

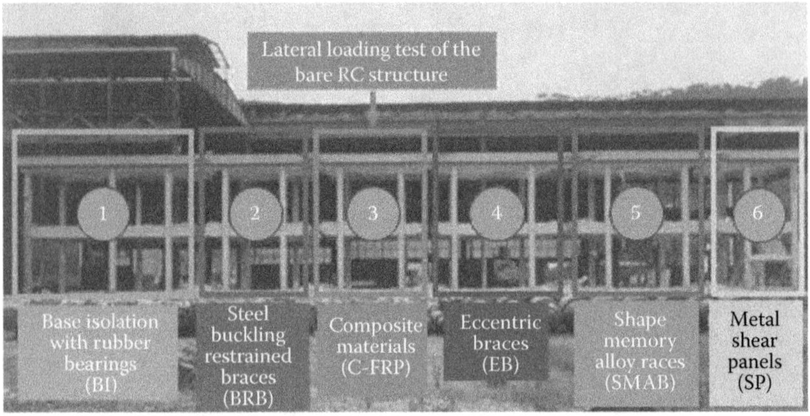

Figure 4.12 The systems experienced for retrofitting an existing RC structure, according to the ILVA-IDEM project. (Mazzolani, F.M. 2006a: *Seismic Upgrading of RC Buildings by Advanced Techniques: The ILVA-IDEM Research Project*. Polimetrica International Publisher, Monza, Italy.)

second test, a capacity design criterion has been applied, with a link shear overstrength factor equal to 1.89; and (3) in the third test, a design criterion similar to the one adopted for the second test has been implemented, but with a larger overstrength factor equal to 2.84. Test results suggest that shear links made with the European HE type of hot-rolled steel profiles are characterized by overstrength factors appreciably larger than the value (1.5) coming from tests on American wide-flange profiles.

As shown in Figure 4.14, the maximum measured increase of lateral strength, with respect to the original capacity of the bare structure, was 4.32 times in the first test, 6.37 times in the second test, and 6.21 times in the third test. This improvement of lateral capacity was essentially related to the link overstrength.

Figure 4.15 highlights the plastic deformation that occurred in the EB links during the test and testifies to the large local ductility that this system can guarantee.

4.2.2 Dog-bone systems

4.2.2.1 General concept

The Northridge seismic event (USA, 1994) caused unexpected brittle fractures in the traditional beam-to-column, welded flange-bolted web connections of steel MRF. In particular, the full confidence in the strength, stiffness, and ductility of such common moment connections was seriously undermined: most connections provided small inelastic deformations and

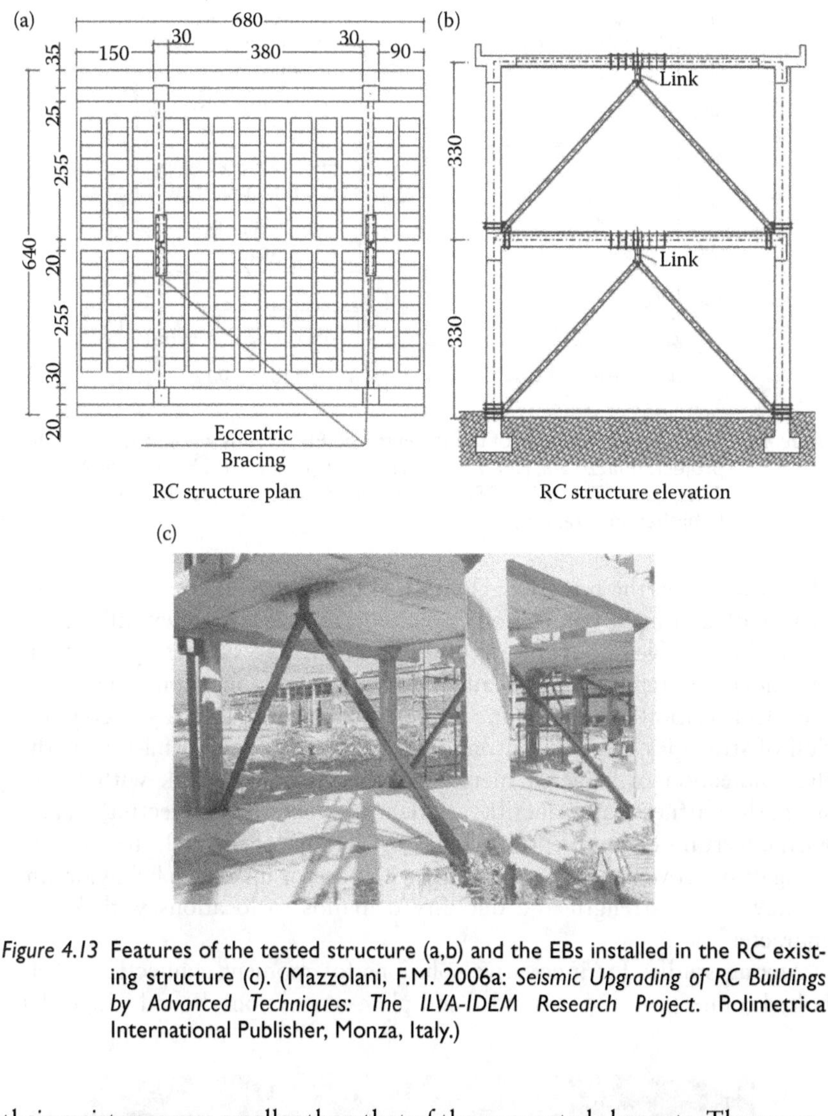

Figure 4.13 Features of the tested structure (a,b) and the EBs installed in the RC existing structure (c). (Mazzolani, F.M. 2006a: *Seismic Upgrading of RC Buildings by Advanced Techniques: The ILVA-IDEM Research Project.* Polimetrica International Publisher, Monza, Italy.)

their resistance was smaller than that of the connected elements. The experimental investigations that arose on typical American connections pointed out that the welds did not allow the development of large inelastic flexural strain into plastic hinges, limiting their local ductility in terms of rotational capacity (SAC, 1997; Roeder, 2002). Major causes were recognized, such as the bad quality of groove welds and the contribution of the concrete slab to the beam strength, which produces a large tensile strain in the bottom flange (Uang et al., 2000). Further to the evaluation of the pre-Northridge moment connections, aimed at going deeply into the reasons for such lack

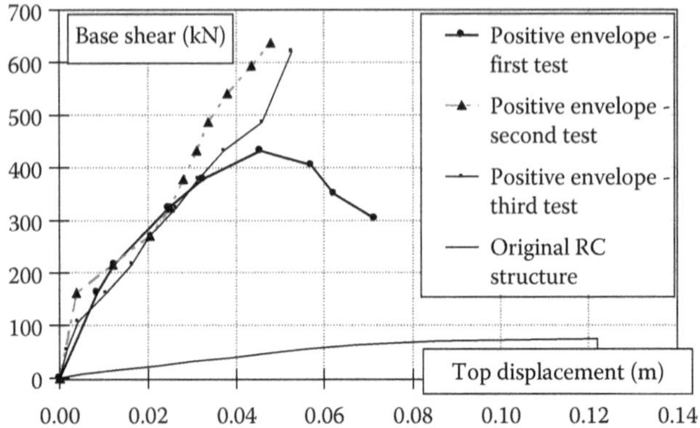

Figure 4.14 Lateral-load response of the three tested EB systems within the ILVA-IDEM project. (Mazzolani, F.M. 2006a: *Seismic Upgrading of RC Buildings by Advanced Techniques: The ILVA-IDEM Research Project*. Polimetrica International Publisher, Monza, Italy.)

of performance, the research activity had two aims: (1) the definition of a methodology for the connection evaluation, in order to achieve all the necessary knowledge about the strength, ductility, and stiffness peculiarities for understanding, predicting, and controlling the connection behavior; (2) the identification of new connection types, characterized by a combination of strategies, such as, on the one hand, the insertion of a fuse within the connection or adjacent members, it being a weak link with known strength, stiffness, and ductility, which protects any connecting element with uncertain or unacceptable behavior; on the other hand, the strengthening or improvement of connections with poor or unknown behavior aims at moving the strength and ductility demands to locations with known characteristics.

Among studies for innovative solutions for moment connections, field-bolted connections (i.e., extended end-plate connections, bolted flange plate

Figure 4.15 Link shear deformation during the test.

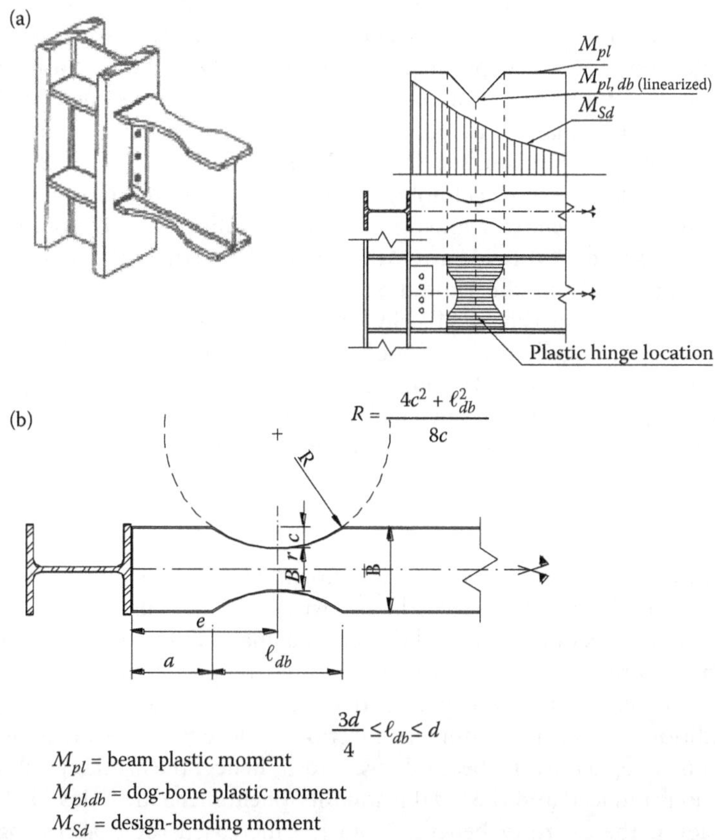

Figure 4.16 The dog-bone features: (a) beam-to-column connection; (b) shape of "dog-bone."

connections, bolted T-stub connections), and field-welded connections (i.e., welded flange-welded web connections, haunch connections, cover plate connections), the so-called RBS connections have been analyzed (Roeder, 2002).

The RBS connection, namely *dog-bone,* is a structural detail, which meets both the above-mentioned conception strategies; it consists of the introduction of a ductile fuse and the improvement of connections as well. Actually, RBS allows the plastic hinge from the beam ends to be shifted by a controlled weakening of the beam itself in a relevant position (Figure 4.16), hence ensuring a better structural performance under cyclic loads at both the local and global levels. In fact, the available ductility of the beam-to-column connections increases and, consequently, the seismic behavior of steel MRF enhances. Moreover, the flange force, which can be transmitted to the connection to the panel zone and to the column continuity plates, is reduced due to the small local reduction in the beam cross-section and in the overall lateral stiffness.

The dog-bone idea was proposed by Plumier (1990) and first applied by Georgescu (1996), at the base of the columns of the steel structure of the Mara Hotel in Sinaia, Romania, in order to guarantee the formation of a global collapse mechanism (see Section 2.2.4).

However, the type and technology of the RBS beam-to-column connection, as well as its behavior under cyclic load conditions, were comprehensively investigated worldwide after the Northridge earthquake (Chen et al., 1996), particularly in the United States (Uang et al., 2000; Engelhardt et al., 1996), within the FEMA/SAC project (Roeder, 2002; Carter and Iwankiw, 1998; Engelhardt and Sabol, 1997; Bruneau et al., 1998; Kunnath and Malley, 2002). The RBS connection was proposed either for optimizing the seismic behavior of new steel MRF or as an upgrading or retrofitting system for existing steel MRF in seismic areas.

4.2.2.2 Dog-bone design

RBS protects the node from plastic deformations, producing a ductility enhancement. At the local level, the inelastic deformations in the beam do not involve connection welds and the stress demand at the beam-to-column connection is reduced. As a consequence, the strength requirements for the column flange, continuity plates, and panel zone are less stringent than for traditional beam-to-column connections. Moreover, the cross-section reduction is applied to the beam flange width; hence, the *b/t* flange slenderness ratio is reduced and the local instability phenomena are delayed. At the global level, the ductility benefits from the increased rotational capacity; moreover, dog-bones encourage the attainment of more dissipative collapse mechanisms, making it easier for the capacity design for members to be achieved.

The RBS profile should be curve shaped (Figure 4.16). In fact, the tapered element shall not present any geometrical discontinuities, in order to avoid the trigger of cracks during inelastic excursions. Some advantages can then be achieved from both behavioral and economical points of view: first, the curved RBS behaves with the largest rotational capacity with respect to polyline-shaped solutions; second, curve tapering is the most economical technology. The flange area reduction by either constant-size or variable-size drilled holes (Zekioglu et al., 1997) induces a less performing strain concentration in beam flanges with respect to more extended and uniform plastic deformations, like a fuse in the whole tapered-cut RBS region. Recently, considering that a disadvantage of the RBS solution could be the difficulty in repairing beams damaged by an earthquake, Balut and Gioncu (2003) suggested realizing a "dismountable dog-bone," which is bolt connected to the current beam, as it is common for the dissipative links of the eccentric bracing-framed systems. In some cases, welded rib plates at the beam and column flanges are erected to favor the shifting of plastic deformation from the beam-to-column interface to the RBS zone and to reduce stresses at the beam-to-column welds (Zekioglu et al., 1997).

The main parameters affecting the design of RBS connections are (Figure 4.17) the amount of cross-section resistance reduction ($m_{db} = M_{pl,db}/M_{pl}$) and the beam flange reduced width (B_r); the RBS midsection location with respect to the beam-end (e); and the RBS segment length (l_{db}). The amount of the section plastic modulus reduction depends on both the location and the bending moment distribution due to the design loads. It has to be noted that a weakening of the beam is feasible for steel MRFs because of the member overstrength due to the drift limitations prescribed by the codes. The distance e is approximately equal to $(1/3 - 1/2) \times d$, where d is the beam height, because at that distance, according to the classical St. Venant principle, the effects of stress concentrations can be neglected. The l_{db} length should be determined for avoiding a local mechanism in the reduced portion and having a large dissipative zone (Anastasiadis and Gioncu, 1998).

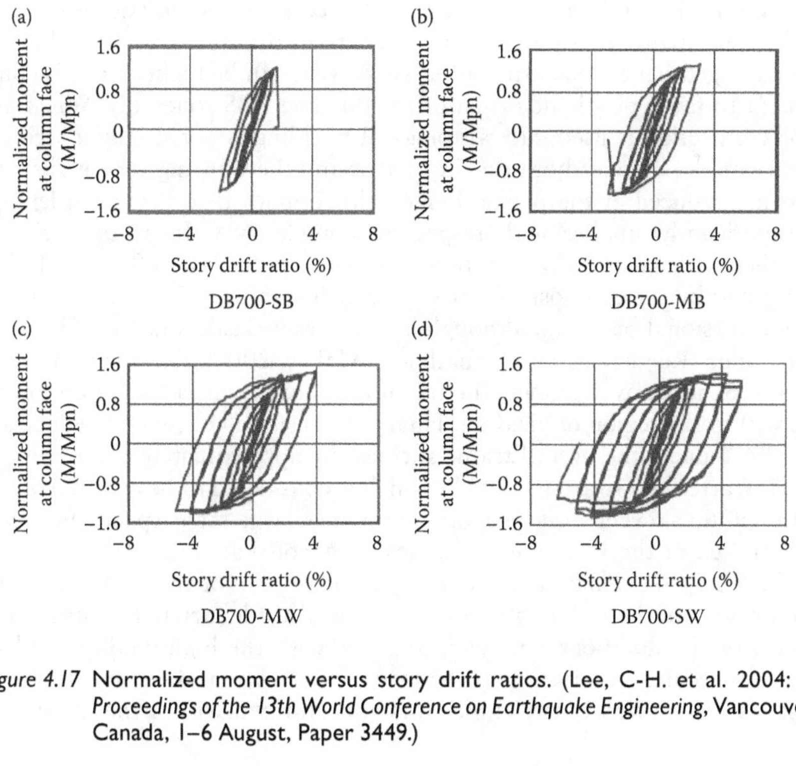

Figure 4.17 Normalized moment versus story drift ratios. (Lee, C-H. et al. 2004: In *Proceedings of the 13th World Conference on Earthquake Engineering*, Vancouver, Canada, 1–6 August, Paper 3449.)

4.2.2.3 Experimental activity

After the Northridge seismic event (1994), a wide experimental activity was developed in the United States, within the FEMA/SAC steel program (1995–2000), involving the scientific community and the industries. Such activity focused on the seismic design and evaluation of steel moment resisting frames (MRF) (Roeder, 2002; Kunnath and Malley, 2002). In particular, an extended part was dedicated to testing and assessing the pre- and post-Northridge connections. Regarding the RBS connections, a set of more than 10 full-scale specimens was tested under cyclic loads (Jones et al., 2002; Brandon and Uang, 2002; Gilton and Uang, 2002). The tests were supplemented by analytical studies aimed at parametrical analyses, in order to go more deeply into details of the RBS connection behavior. The main results of the investigations were evidenced afterward. RBS connections are able to achieve the SAC minimum plastic rotation of 3% (up to 0.05 rad), to prevent brittle weld fractures, and to reduce the strain concentration at the beam-to-column interface. Owing to flange reduction, local buckling of the RBS web could occur, so more stringent web slenderness limits at the RBS zone should be provided. The weak panel zone could give a significant contribution to the energy dissipation, but the extent of panel zone shear yield deformation should be controlled to prevent the excess of ductility demand on the weld and weld access region; in addition, even if the weak panel zones induce cost reduction, the cost of substitution of the damaged panel zone after the seismic event can be higher, so it is convenient to favor plastic deformations within the RBS zone only. When an RBS connection is used, lateral-torsional buckling is potentially a serious failure mode, because both the St. Venant and the warping torsional stiffness are reduced at the plastic hinge. This implies that strain hardening is significantly smaller with respect to completely welded connections, together with greater deterioration in resistance. Such an effect could be restrained by the composite floor slab, which stabilizes the beams against lateral-torsional buckling, although it increases inelastic strains in the bottom flange (Roeder, 2002; Kunnath and Malley, 2002). However, the slab does not cause any detrimental effect if the shear connectors are not present within the region of yielding. Finally, the usual flange width reduction of 40% reduces the overall frame stiffness by approximately 5% (Roeder, 2002; Carter and Iwankiw, 1998), which is considered to be acceptable.

In 1996, a cyclic test program with five large-scale specimens was undertaken at the University of Texas Ferguson Laboratory (Engelhardt et al., 1996) to evaluate the suitability of dog-bone for the 2000 moment connections of the American Stores Company headquarters building. The effects of the dog-bone cut type, together with the high-quality welded beam-to-column connection at the face of the column, were investigated. Test results showed that yielding was concentrated within the dog-bone region. At large plastic rotations, the specimens exhibited flange and web local buckling within the plastic hinge region; however, even at a plastic rotation equal to 0.02 rad, the degree of local buckling was still very slight and hardly noticeable. Considering that analytical studies of steel moment frames affected by the Northridge earthquake evidenced that most buildings experienced maximum plastic joint rotations of the order of 0.005–0.008 rad, it can be concluded that local buckling at the dogbone should not be a concern. A cost analysis was carried out, pointing out the economic advantage of RBS connections with respect to other post-Northridge connection solutions.

A shaking table test was performed by Chen et al. (1996, 1997) in Taiwan on two single-story steel frames, such as a traditional MRF and one equipped with RBS connections; the latter behaved better than the traditional one. In fact, plastic hinges formed at the RBS connection and dissipated more energy with respect to the traditional scheme, where stress was concentrated at the welds with consequent crack trigging. Practically the same elastic stiffness resulted for both frames.

Full-scale moment connection tests were carried out at the Charles Lee Powell Structural Laboratories at the University of California at San Diego (Zekioglu et al., 1997) on six specimens with tapered-cut flange RBS; in all the cases, welded vertical rib plates at the beam-to-column interface and welded web shear plates were used. Test results demonstrated a good performance of the RBS connections, confirming all the experimental evidence of other U.S. investigations.

In 1997, an experimental campaign was conducted at the University of California at San Diego (Uang et al., 2000) to investigate the effectiveness of RBS or welded haunch techniques for rehabilitation of the pre-Northridge steel moment connections. Six specimens were tested, three of them incorporating the floor slab. Test results showed that the presence of only RBS at the beam bottom flange, with the removal of steel backing and weld tabs, did not prevent brittle fractures at the top flanges, even after considering the floor slab contribution and removing the top flange steel backing. To have a ductile behavior, it was necessary to also replace the top flange weld. Lateral-torsional buckling under positive bending was prevented due to slab-bracing effects to the top flange, but flange local buckling still developed, whereas under negative bending, significant local buckling could interact with lateral-torsional buckling.

Recently, experimental tests on eight full-scale specimens of RBS steel moment connections were carried out in Korea (Lee et al., 2004), aiming at evaluating the effects of panel zone strength and the type of beam-to-column, both web bolted (*B*) and welded (*W*) connections, on the seismic performance of RBS connections. RBS was designed according to the FEMA 350 (2000) recommendations. These results show that welded web specimens develop satisfactory levels of ductility, as required for special moment frames, mostly in the cases of strong (S) panel zones with respect to medium (M) ones, whereas bolted web specimens perform poorly because of the premature brittle failure of the beam flange at the weld access hole (Figure 4.17).

4.2.2.4 Numerical activity

The influence of dog-bones on the seismic response of moment resisting frames can be evaluated by means of static inelastic pushover analyses, and it is expressed in terms of the main factors characterizing structural behavior, such as maximum bearing capacity, global ductility, and collapse mechanism. The main results and conclusions are depicted below. Numerical investigations focus on two main steps:

1. Definition of a design strategy of steel MRF modified by means of dog-bones
2. Assessment of the reliability of the RBS connection system for the seismic retrofitting or upgrading of steel moment resisting frames

The main results and conclusions depicted are summarized below.

4.2.2.4.1 Definition of a design strategy of steel MRF modified by means of dog-bone

This aspect was investigated by Faggiano and Landolfo (2002), Faggiano et al. (2003), and Montuori and Piluso (2000). The most simple design procedure for MRF with RBS, which seems to be more appropriate for seismic upgrading of existing structures, can be exemplified by means of the following steps (Faggiano and Landolfo, 2002; Faggiano et al., 2003):

1. Design of steel MRFs according to traditional methods
2. Insertion of appropriately designed dog-bones in the beams

The latter step consists of two more phases. The first one is the macrodesign of the dog-bone, such as the estimation of the plastic moment reduction ratio m_{db}, associated to the dog-bone position along the beam, e. Three different approaches were proposed:

- The *FEMA* relationship, which derives from experimental tests on RBS beam–column assemblages
- The Montuori–Piluso (*MP*) relationship, which derives from a simplified analytical formulation
- The Faggiano et al. relationship, which simply synthesized the two above-mentioned formulations

Such relationships for the dog-bone macrodesign are

$$e_{\text{FEMA}} = a + (l_{db}/2) \quad (FEMA)$$

$$e_{MP} > \sqrt{2 \cdot (1 + m_{db}) \cdot \frac{M_{pl}}{q}} + 2\sqrt{\frac{M_{pl}}{q}} \quad (MP) \tag{4.9}$$

$$e_{MP} = e_{FEMA} \cdot \zeta = 1.5 \times 0.75 \cdot d \cdot \zeta \quad \text{(Faggiano et al.)}$$

where q is the vertical load distributed uniformly along the beam; the coefficient ζ can be easily deduced from the abaci presented in Figure 4.18a, as a function of IPE profiles (W_x), vertical loads Q, and plastic moment reduction ratio m_{db}.

Montuori and Piluso proposed a design abaqus for dog-bone as well (Figure 4.18b). It appears that, for a given value of M_{pl} or W_x and q, an increase in m_{db} causes a decrease in e; similarly, for a given value of m_{db}, an increase in q causes a decrease in e.

The second step is the microdesign of the dog-bone, such as the determination of the dog-bone midsection flange reduced width, B_r, and design of the constructional detail (Figure 4.19).

A more refined design methodology, which seems more appropriate for seismic design of new structures, was set up by Mazzolani and Piluso (1996). They extended to MRF modified by RBS, the design procedure based on the kinematic theorem of limit design, and on the concept of the mechanism equilibrium curve, thus being able to assure a collapse mechanism of global type (Mazzolani and Piluso, 1996; Faella et al., 1998; Montuori and Piluso, 2000). The method leads to both the failure mode control and the local ductility control. It has been noted that the use of dog-bone connections can induce a saving in the columns' total weight up to 30%, corresponding to a saving in the total structural weight up to 20%. Moreover, as far as e/L increases, there is an increase in the columns' plastic moments and the structural weight.

4.2.2.4.2 Assessment of the reliability of the RBS connection system for the seismic retrofitting or upgrading of steel MR frames

The main remarks and conclusions of some recent numerical investigations aimed at the evaluation of the influence of dog-bone connections on the structural behavior of MRF are synthesized hereafter (Faggiano and Landolfo, 2002; Faggiano et al., 2003).

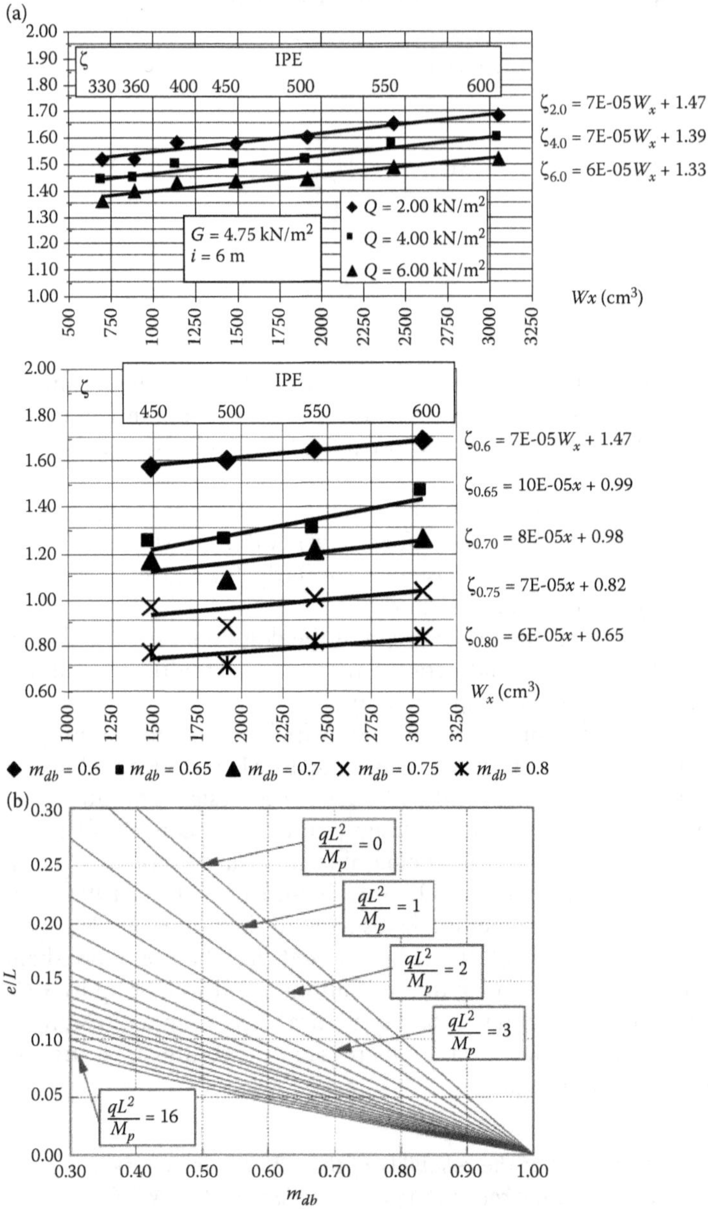

Figure 4.18 Dog-bone macrodesign abaci: (a) Faggiano et al. (2002, 2003); (b) Montuori and Piluso (2000).

$$B_r = \frac{[k_1 + 2\cdot B\cdot k_2]\cdot m_{db} - k_1}{2\cdot k_2}\ ;\ k_1 = t_w\,(h - 2t_f)^3$$

$$k_2 = 3t_f\cdot h^2 - 6t_f^2\cdot h + 4t_f^3$$

$$B_r^* = m_{db}\times B$$

No web contribution to bending
(a scatter of about 15–20% in excess)

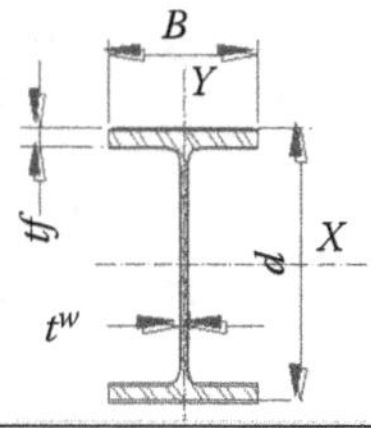

Figure 4.19 The flange reduction for the IPE profiles.

Multistoried steel MRF configurations have been analyzed in two successive numerical investigation campaigns:

- The first one (Faggiano and Landolfo, 2002): different bay spans ($L = 6.0$, 9.0, 12.0 m) and a fix amount of the RBS plastic moment reduction ratio $m_{db} = 0.6$, it appearing the most significant and commonly used value, were assumed.
- The second one (Faggiano et al., 2003): the impact of both the number of stories and m_{db} ratios on the global behavior of MR frames is examined; 2 bays (spanning $L = 9.0$ m), 5 and 8 stories steel MRF are studied, considering different amounts of plastic moment reduction ratios (m_{db} equal to 0.6, 0.65, 0.7, 0.75, 0.8).

Structures are designed by means of traditional methods (*BASE*), modified with both the *FEMA* dog-bones and the *MP* dog-bones, according to Equation 4.9. The design data and the geometrical configurations are shown in Figure 4.20.

Static inelastic pushover analyses are performed by using the computer program SAP2000 Non-linear 7.12. The P–Δ effects are considered and the inelastic behaviors of dissipative zones are defined by concentrated rigid-plastic hinges at the structural members' end. Dog-bones are modeled as "nonprismatic section frame" elements; for the sake of simplicity, the curve profile is linearized.

A performance-based comparison between the different RBS design criteria is made on the basis of the pushover behavioral curves (i.e., top sway displacements Δ versus base shear V, see Figure 4.21). The main results are synthesized afterward:

- The RBS elements are able to protect the nodes from plastic deformations; in fact, the inelastic deformations in the beam do not involve the connection welds, the RBS undergoing inelastic deformations instead of the beam ends.

(a)

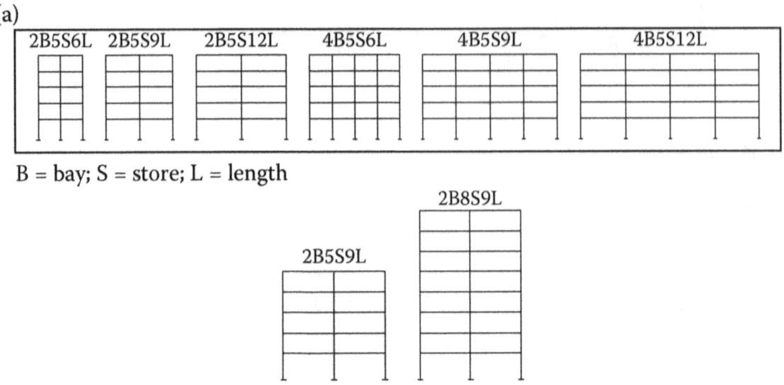

(b)

The design data	(Eurocodes 3 and 8); Steel grade S275	
Vertical loads	ULS design	SLS design
DL 4.75 (kN/m^2)	$q_d = 4$	$\nu = 2$
LL 2.00 (kN/m^2)	Soil A	$(d/h)_{\text{lim}} = 0.006$
	ξ (%) = 5	
	PGA ($\times g$) = 0.35	

Figure 4.20 Case studies: (a) first numerical investigation; (b) second numerical investigation.

- The number and the length of bays do not affect the RBS effectiveness.
- The presence of dog-bones has a negligible influence on both the structural weight and the overall lateral flexibility, whereas the overall bearing capacity of the modified frames is reduced.
- The ductility coefficient of frames modified by RBS, $\mu = \Delta_{0.03}/\Delta_y$ (where Δ_y is the top sway displacement at the activation of the first plastic hinge, this situation being assumed as the first yielding condition; $\Delta_{0.03}$ is the top sway displacement at the attainment of the inelastic rotation equal to 0.03 rad in the most engaged plastic hinge, this situation being assumed as a conventional collapse condition), matches the design requirements, it being always larger than the design q-factor ($q_d = 4$).
- Small amounts of cross-sectional resistance reduction, such as $m_{db} = 0.75 - 0.8$, are unsuitable, because RBS should be located very close to the beam ends, thus involving the beam-to-column connection.
- The global ductility always increases.
- As far as the collapse mechanism is concerned, it can be observed that the beam plastic hinges only form in the dog-bone elements.

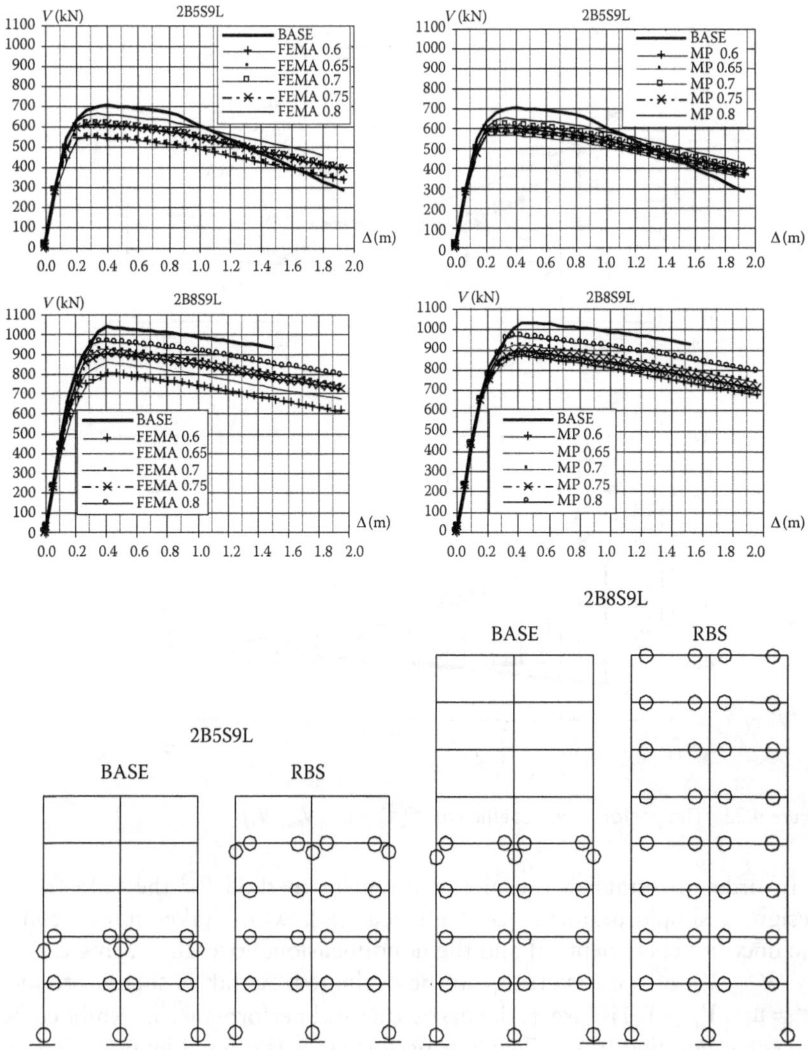

Figure 4.21 The pushover behavioral curves V(kN) versus Δ(m) and typical collapse mechanisms with and without RBS.

- RBS induces a decrease in both the plastic hinge number in the column and the extent of plastic rotations.
- RBS allows the attainment of failure modes, which involve a higher number of stories, resulting in a higher dissipation capacity of structures.
- The RBS position along the beams has no influence on the failure modes.

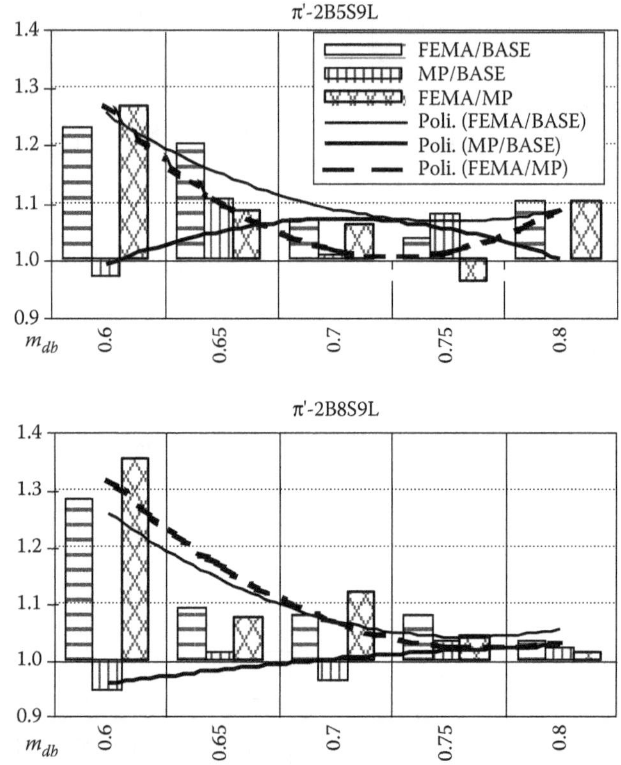

Figure 4.22 The performance coefficient $\pi'(\pi' = \mu \times V_{max}/V_d)$.

In order to establish the most convenient method for the RBS frame design, a simple performance coefficient (π'), which takes into account the ductility coefficient (μ) and the nondimensional overall bearing capacity (V_{max}/V_d) of the structure, can be evaluated. As far as such a product ($\pi' = \mu \times V_{max}/V_d$) is larger, the best structural performance is exhibited by the corresponding frame. The best performance is offered by $m_{db} = 0.6$, as suggested by *FEMA* case (Figure 4.22).

4.2.2.5 Further developments

The improvement of the MR-frame performance induced by RBS connections has been appreciated by means of an extensive program of both experimental and numerical investigations. For new structures, the cross-section reduction is applied at the beam flanges, both superior and inferior, obtaining in any case-symmetric double-T-shaped cross-sections. Nevertheless, the traditional RBS design evidently applies with some difficulties to enhance the ductility properties of existing structures, because the primary structure is partially or fully inaccessible to further interventions. Actually, in case of retrofitting, the current section reduction should be obtained just at the lower beam flange, obtaining a nonsymmetric constructional detail, whose behavior is not yet fully investigated. Basic aspects concerning the use of nonsymmetric RBS connections as moment-resisting retrofitting systems should be examined from the point of view of the design methodology and the geometric characteristics of the weakening detail. Another item to be elucidated is that the actual ductility category of the reduced beam cross-section should be considered, in order to calibrate the conventional collapse situation to the actual enhanced rotational capacity. Eventually, the RBSs connection effectiveness as a seismic upgrading or retrofitting system should be validated by means of extensive dynamic time–history seismic analyses.

4.2.3 Buckling-restrained-braced systems

4.2.3.1 Criticism to classical concentric braces (CBs)

The use of steel CBs for the seismic protection of structures is a well-established technique. However, ordinary steel CBs are characterized by a limited ductility capacity under cyclic loading. In fact, their response is not symmetric and exhibits substantial strength degradation, owing to buckling of the braces under strong compression (see Figure 4.23a and c).

(a) (b) (c) (d)

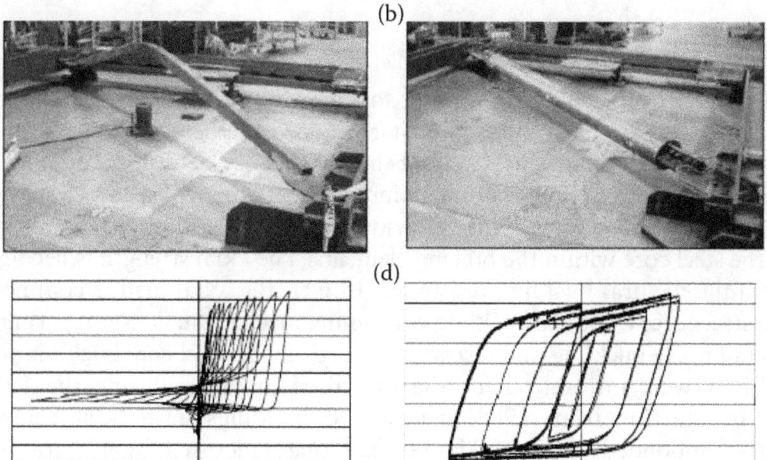

Figure 4.23 Comparison between classic and buckling-restrained brace: (a) classic brace (CB) buckled out in compression; (b) buckling-restrained brace (BRB) without buckling; (c) hysteresis behavior of classic brace (CB); (d) hysteresis behavior of buckling-restrained brace (BRB).

In addition, this complex response originates actual distributions of internal forces and deformations in the overall structure, which substantially deviate from the prediction based on conventional models (Jain and Goel, 1979; Khatib and Mahin, 1987). Besides, ordinary CBs, made, for example, in common double T sections, are difficult to be tailored in terms of required strength and stiffness, thus giving rise very often to some "weak" or "soft" stories. Finally, out-of-plane buckling of braces could cause severe damage to nonstructural components.

These problems in the seismic design of CBs have long been investigated. For example, research has been carried out in the United States, Canada, and Japan, in order to limit the flexural slenderness of the brace to a small value, so as to be compatible with the ductility requirements associated with the adopted strength reduction factor. Research has also been addressed to solve the problem of damage concentration in a few weak stories, by means of alternative structural schemes, such as the "Zip" frame configuration proposed by Khatib and Mahin (1987). The substitution of bare steel braces with composite ones has also been suggested for improving the available ductility (Liu and Goel, 1987).

When ordinary braces are used for strengthening existing constructions, such as when they are added to existing RC frames, the problem of the introduction of additional axial forces in the columns of the existing frames also needs to be considered. This problem is obviously exasperated when the brace slenderness is kept low in order to limit the strength degradation of the brace under compression.

4.2.3.2 BRB concept and details

BRBs have been conceived in order to solve the above problems. As shown in Figure 4.23b and d, BRBs are characterized by the absence of buckling, which allows a stable hysteretic behavior, different from traditional CB braces. In addition, BRBs permit an independent design of stiffness, strength, and ductility properties. This behavior is achieved by limiting the buckling of the steel core within the bracing elements. The axial strength is decoupled from the flexural buckling resistance. In fact, the axial load is confined to the steel core, while the buckling restraining mechanism does not permit the overall brace buckling and restrains the high-mode steel core buckling.

BRBs were introduced for the first time in Japan during the 1980s, where, starting from 1988, nearly 300 buildings have been equipped with "unbonded" braces (Figure 4.24) manufactured by Nippon Steel Corporation. After several tests carried out in 1999 at the University of California, Berkeley, the technology has also been implemented in the United States by utilizing BRBs for the seismic retrofitting of the UC Davis Plant and Environmental Sciences. The BRB technology is currently undergoing strong development, with a growing number of buildings using

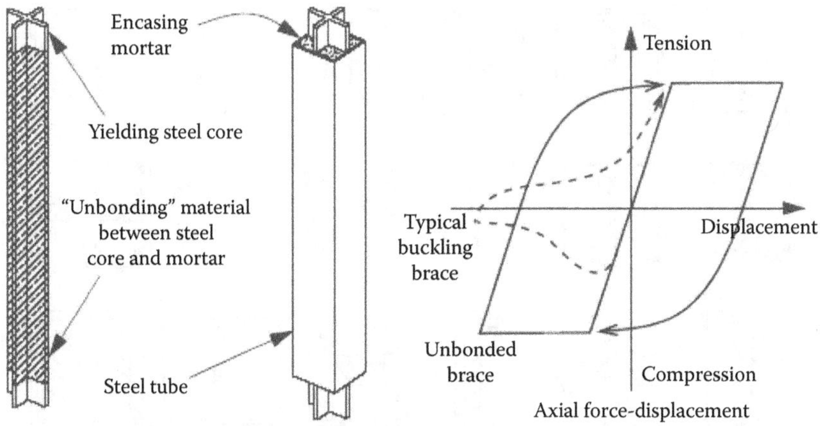

Figure 4.24 Features of typical "unbonded" brace.

buckling-restrained braces as a primary lateral force-resisting system. This strong development is also testified by several research studies that are ongoing in the United States, Taiwan, and obviously Japan (Tsai et al., 2004a; Sabelli and Aiken, 2004; Wada and Nakashima, 2004).

Different types of BRBs (Figure 4.25) have been studied, all based on the same concept of using tubes or other elements for restraining lateral displacements while allowing axial deformations of the core. In the most classical form, the restraining element is a steel tube filled with concrete. A layer of unbonding material is placed at the contact surface between the core plates and the filling concrete, this version therefore being called the "unbonded brace."

The unbonding material allows the brace to slide freely inside the buckling restraining unit and lets the transverse expansion of the brace to take place when the core yields in compression.

"Only-steel" solutions have also been proposed, with two or more steel tubes in direct contact with the yielding steel plates. In the latter case, the restraining tubes can also be connected by bolted steel connections, thus allowing an easy inspection and maintenance during the lifetime or after a damaging earthquake (Tsai et al., 2004a). An adequate gap size between the brace and the restraining tubes is also required in case of "only-steel" BRBs, in order to provide the necessary space for relative deformation between both member components.

Yielding of this special type of bracing occurs when the plastic strength of the core steel plates is achieved. The axial stiffness is determined by the combination of two springs in series having the axial stiffness of the internal core and terminal tapered plates. The length and size of the latter can be independently fixed to some extent. In any case, the possibility of avoiding

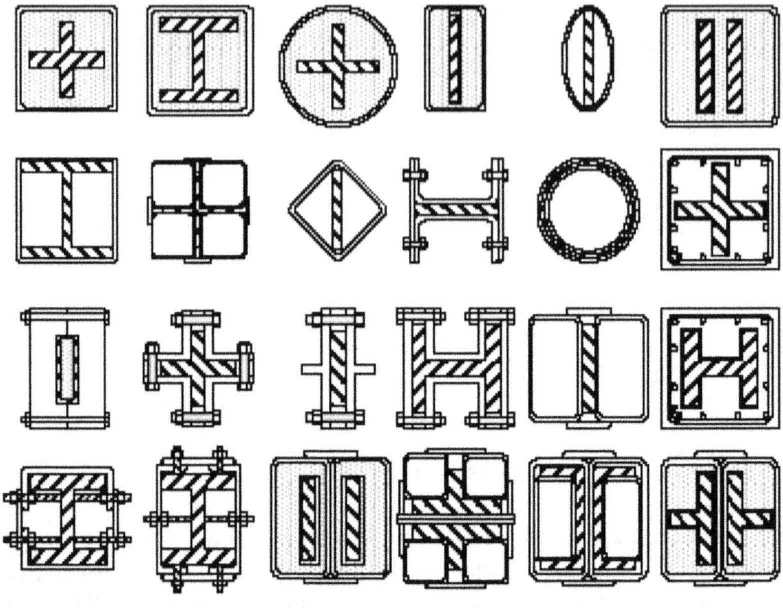

Figure 4.25 Different types of BRBs. (Tsai, K.C. et al. 2004a: In *Proceedings of the 13th World Conference on Earthquake Engineering*, Vancouver, Canada.)

compression buckling allows very slender steel plates to be used as the core of the BRB, with a relatively low plastic strength and without impairing the system ductility. In this way, the yielding of the BRB can be regulated to very low interstory drifts, thus permitting the dissipative action to be activated soon.

Low cycle fatigue (failure) characteristics have been shown to depend on a variety of factors, including the restraining mechanism used, material properties, local detailing, workmanship, loading conditions and history, and so on. Inelastic deformation (ductility) capacities are generally quite large, with cumulative cyclic inelastic deformations often exceeding 300 times the initial yield deformation of the brace before failure (Nakamura et al., 2000).

Anyway, as mentioned above, the basic principle, which characterizes the BRB response, is based on the possibility of decoupling of axial-resisting and flexural-resisting aspects in the compression field. In fact, a steel core plate has to resist axial stresses, while buckling resistance is provided by an external sleeve, which may be made of steel, concrete, or composite. Figure 4.26 shows the parts that form a common BRB-bracing member.

It is possible to divide the core into three zones:

- The yielding zone, with a reduced cross-section area, inside of the zone of lateral restraining provided by the sleeve (zone C)

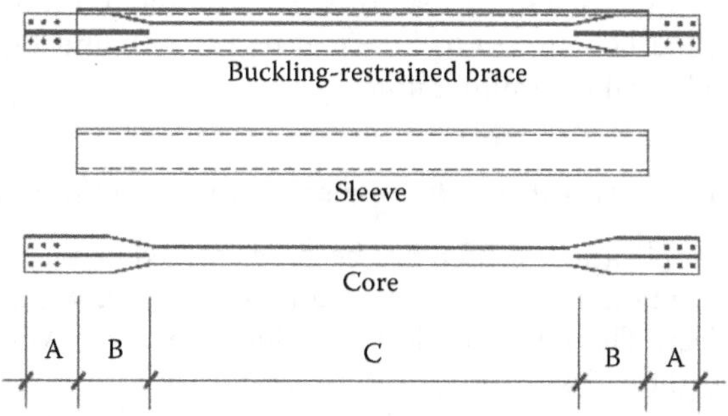

Figure 4.26 Schematic view of the BRB component elements.

- The transition zones, with a larger area than the yielding zone, but similarly restrained (zone B)
- The connection zones, which extend outside the sleeve and connect to the frame by means of gusset plates (zone A)

In order to prevent the instability of the lateral restraint (sleeve) and to permit the full axial yielding of the steel core, the end-connections of these devices have to be able to transfer forces to the core without the development of a significant stress state in the sleeve. Furthermore, the end-connections have to be designed to avoid modes of overall instability of the bracing member, as shown in Figure 4.27.

In addition, special end details must be designed in order to permit inelastic deformations of the steel core.

The lateral strength of the BRB device is closely related to the lateral stiffness of the support element. Chen (2002) suggested that the nominal limit strength in compression (N_{BRB}) can be calculated according to the following relationship:

$$N_{BRB} = \frac{N_E}{1 + N_E\delta_0/M_y} \tag{4.10}$$

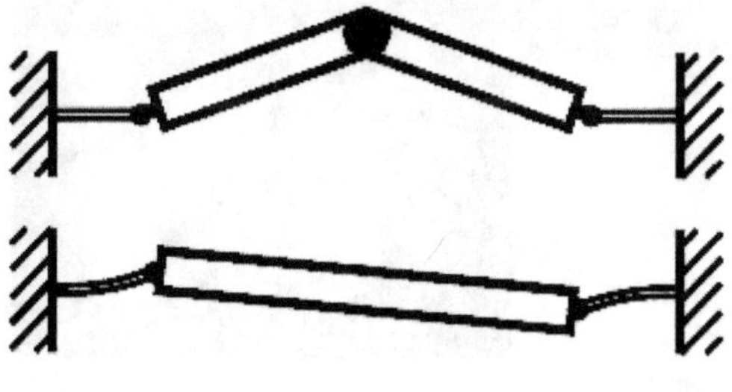

Figure 4.27 Typical modes of overall instability for bracing members.

where N_E is the Euler buckling load of the restraining unit, δ_0 is an initial crookedness, usually assumed to be equal to $L/1000$, and M_y is the yield moment of the lateral restraining unit.

4.2.3.3 Applications of BRBs in new and existing buildings

An important application of BRB has been performed in a newly built area located in the Ropongi quarter of Tokyo (Figure 4.28), within the Tokyo Midtown Project. The tallest building in this area is the Midtown tower with a total height of 248 m.

The seismic-resistant structure of this building is based on the use of BRB in a dual configuration of MRF and CBF (Figure 4.29); the braces are based on the Japanese patented system, like an "unbonded" type inside of a tubular steel sleeve, which looks very stocky.

In the United States, an example is that of the Wallace F. Bennett Federal Building (Salt Lake City, Utah, USA), shown in Figure 4.30. This eight-storied RC building is an important office building built in the early 1960s and then seismically retrofitted by BRBs, which are placed externally, in such a way as to characterize the architectural appearance of the renovated building (Figure 4.30).

In Italy, BRBs have been adopted for seismic protection of one building of the University of Ancona (Figure 4.31); the BRB is of an "unbounded" type, patented by the Italian FIP company.

Recently, the BRB braces have been used in the seismic retrofitting of the Deutsche Bank building in Naples (Figure 4.32). The original structure of this building was a steel skeleton, whose internal part was suspended from the roof for allowing a free space at the ground floor. During the 1990s, its

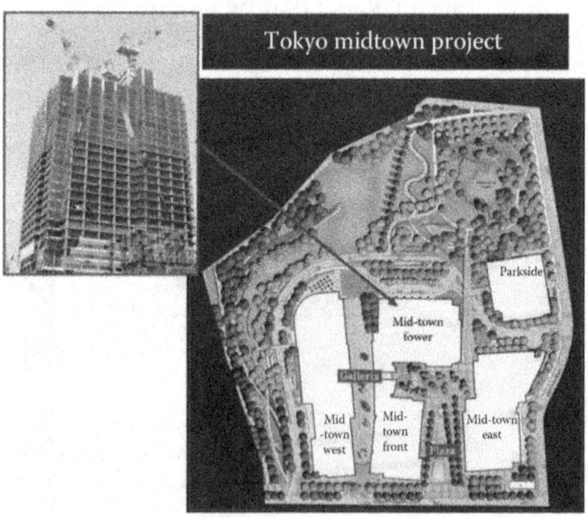

Figure 4.28 The Midtown Tower in Tokyo.

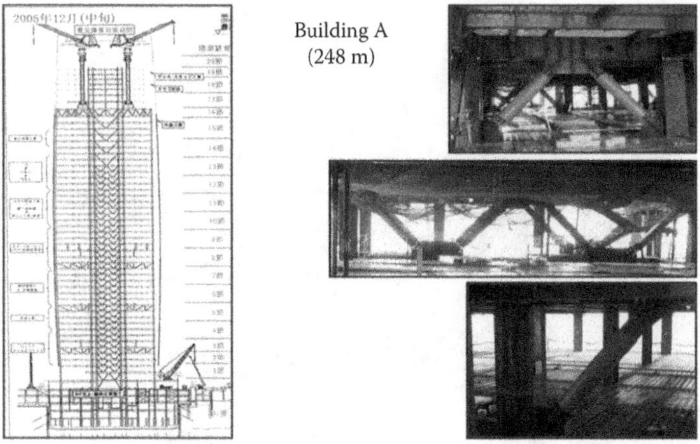

Figure 4.29 The BRB structural system in the Midtown building in Tokyo.

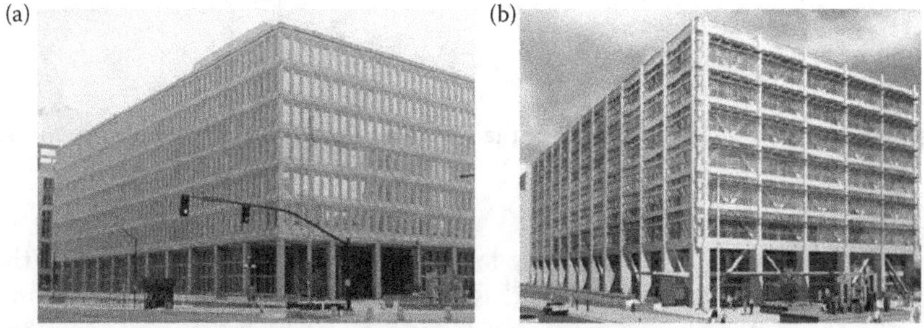

Figure 4.30 Wallace F. Bennett Federal Building (USA): before (a) and after (b) seismic retrofitting.

Figure 4.31 University of Ancona (Italy).

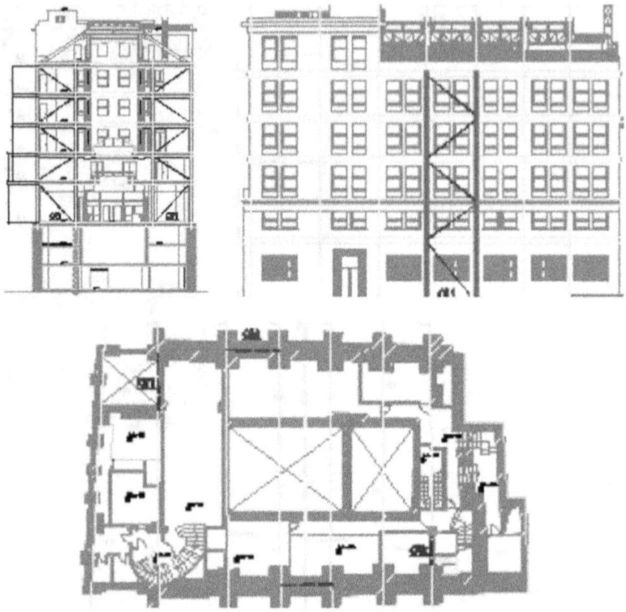

Figure 4.32 The seismic retrofitting of the Deutsche Bank building in Naples by means of the BRB-bracing system.

load-carrying capacity for gravity loads was increased by substituting the original ties with prestressed high-strength steel cables. Seismic retrofitting has been done by introducing two couples of BRB bracings along the facades (Figure 4.32). The BRB used type is "all-steel"; one diagonal per mash is connected to the steel columns by bolted connections (Figure 4.33).

Figure 4.33 The BRB brace in the Deutsche Bank building in Naples and the connection details.

4.2.3.4 Seismic upgrading of existing RC buildings

The simplicity of this system permits the adoption of BRBs for the seismic protection of existing structures. In particular, BRB systems are also a very suitable technique for retrofitting RC structures, because they can get a judicious modification of the structural properties of an existing building, such as lateral strength, stiffness, and ductility, improving its performance in case of future earthquakes. Moreover, these systems reveal good reversibility, because they allow easy substitution after the earthquake if they are damaged, since these devices are reported to be replaceable. In particular, in case of detachable BRBs, according to the "all-steel" typology, it is possible to design them to be opened, in order to control their inside condition after each seismic event.

Recent full-scale tests of "all-steel" BRBs attached to an existing two-storied one-bay RC structure (Della Corte et al., 2005; Mazzolani, 2006a,b, 2007) have shown the excellent performance of this type of device, but they also emphasize the need to carefully design end-brace connections against local buckling failure modes. In these experimental tests, BRBs have been applied for seismic upgrading of an existing two-storied RC structure, which has been subjected to a cyclic pushover test within the ILVA-IDEM project (Mazzolani, 2006a).

The features of the RC structure, which was equipped with BRBs and then tested, are shown in Figures 4.34 and 4.35. Two types of BRBs have been tested (Figure 4.36): in the first type, the restraining sleeve was made by using two external rectangular tubes, connected together by welded steel batten plates (Figure 4.36a); the second is of a detachable type, again

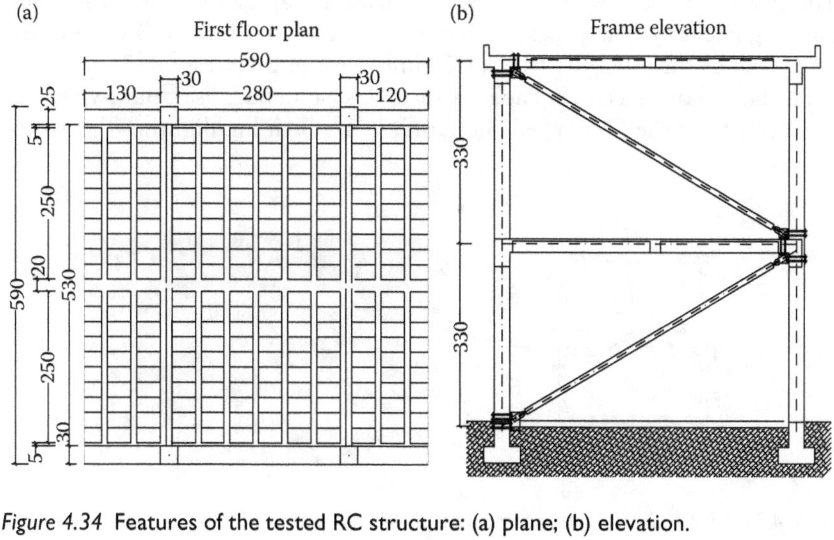

Figure 4.34 Features of the tested RC structure: (a) plane; (b) elevation.

Figure 4.35 The RC structure equipped with BRBs.

being made of two restraining rectangular tubes, but joined together by means of bolted steel connections (Figure 4.36b). In both cases, the internal yielding core has a rectangular steel plate.

The test results are encouraging about the possibility of using this special type of braces for improving the seismic resistance of existing RC structures. In Figure 4.37, the lateral-load response of the two tested BRB systems is compared in terms of the envelope curve corresponding to the positive loading direction. Besides, the behavior is also compared with the results of a previous pushover test, which can be considered to be representative of the bare RC structure response. As can be observed, the load-carrying capacity is appreciably increased (five times for the BRB test no. 2), as well as the lateral stiffness (12 times for the BRB test no. 2).

The lateral displacement capacity of BRB system no. 2 is about two times the capacity of the first type. Therefore, it is worth highlighting the excel-

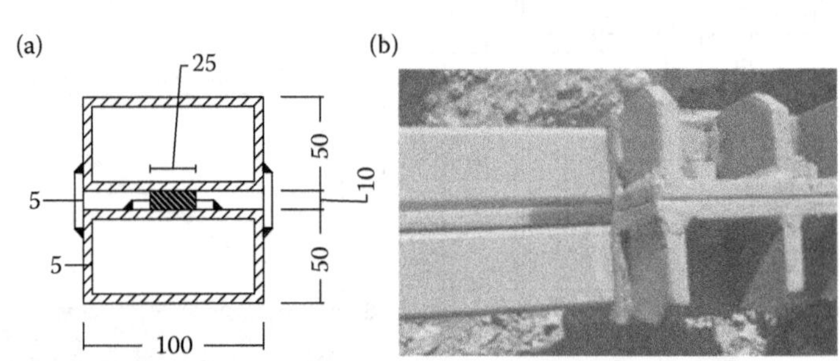

Figure 4.36 BRB sleeves: (a) nondetachable sleeve; (b) detachable sleeve.

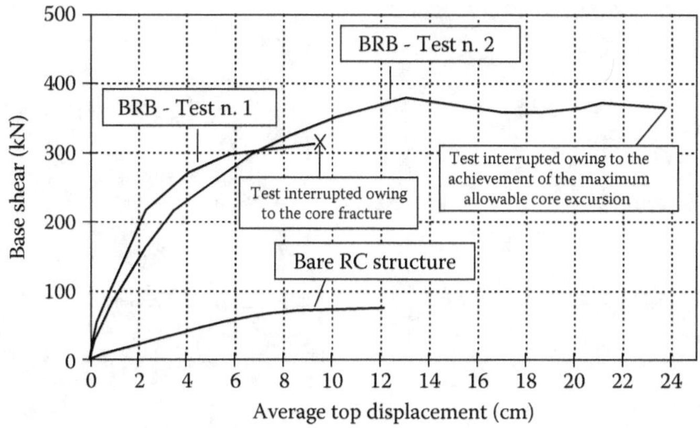

Figure 4.37 Lateral-load response of the two BRBs tested within the ILVA-IDEM project.

lent experimental response of BRB type 2, both in terms of load-bearing capacity and ductility.

The pictures of Figure 4.38 summarize the damage pattern evidenced during the tests. Figure 4.38a and b shows the dark visible parts of the BRB core of the second and the first types, respectively; they highlight the relative displacement between the internal core and the restraining tubes. The relative displacements are developed when the BRB works either in tension or in compression.

Figure 4.38c shows the initial inelastic higher buckling mode of the inner core, which is expected as a normal response of this system, because of the presence of an inner clearance of 1 mm for each side of the yielding core. In particular, the local buckling in the weak axis direction occurs at the end of the core plastic region. This phenomenon becomes very apparent at the maximum first-story drift reached during the test, corresponding to the end of the core free length working stroke.

In Figure 4.38d–f, the local buckling of the internal core of one BRB placed at the first story at the maximum interstory drift reached during the test (about 5.6% of the story height) is shown. It is worth noting that the beneficial energy dissipation effect due to this phenomenon is not present in the "unbounded" BRB types, where it is impossible to know what happens inside of the external sleeve.

Figure 4.38g and h illustrates the local buckling failure of the end-plate connection during the compression of one BRB at the first story. This unexpected phenomenon occurred at just one location and it may be attributed to some damage produced in the gusset plates during the erection of the BRBs. In fact, on that occasion, the gusset plates were forced and deformed, thus introducing geometric imperfections and losing some parts of the restraining effect against out-of-plane rotation of the BRB. This implied that the

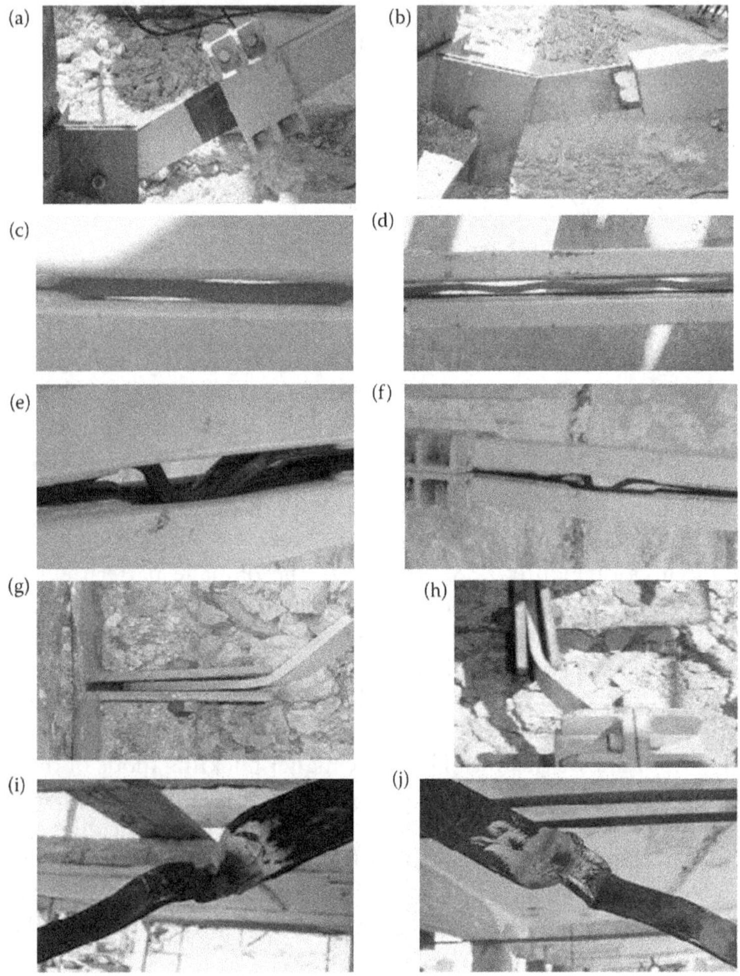

Figure 4.38 Damage patterns of BRB tested within the ILVA-IDEM project.

buckling length of the BRB end portion increased significantly and, consequently, its local buckling capacity decreased rapidly. Another confirmation of this consideration is the fact that the local buckling phenomenon occurred during the third cycle in compression. This again underlines that the undesired end-connection failure mechanism was induced by the local imperfections produced by the damage during the erecting phase of the braces. However, the maximum lateral deflection of the end portion was measured during the test and it was about 85 mm. It is important to highlight that the other braces behaved stably in compression up to the maximum interstory drift of 5.6% at the first floor.

It can be remarked that the problem of local buckling of the BRB end-portions has also been emphasized by previous experimental research carried out by Tsai et al. (2004b) and Chen et al. (2004). Furthermore, it has again been confirmed during the tests given by D'Aniello et al. (2007).

Figure 4.38i and l shows the localization of plastic flexural strain at the transition section between the reduced core and the end tapering. This phenomenon was essentially due to the combined effect of sliding friction between the core and the restraining tubes and the small thickness (hence, the low flexural stiffness) of the tube walls.

A new series of full-scale inelastic cyclic static tests of novel "all-steel" BRBs applied to an existing RC building has been recently developed within the international PROHITECH project on "Eathquake Protection of Historical Buildings by Reversible Mixed Technologies" (Mazzolani, 2009). In order to demonstrate the reversibility of the intervention, the examined BRBs "all-steel" type were designed to be demountable, thus allowing one to control the condition of the devices after seismic events and, if necessary, to easily replace the yielded steel core. In addition, the second scope of the research was to demonstrate the possibility of hiding the brace inside of the gap between the two masonry panels of the external claddings.

The tested building has a rectangular plan (18.50 × 12.00 m) and two levels, with the first and second floor heights equal to 4.60 and 8.95 m, respectively (Figure 4.39).

Before testing the building with BRBs, it was initially tested twice under horizontal cyclic loads in transversal direction, the first time in its original condition, in order to evaluate the contribution of nonstructural components (external cladding, internal partition walls) on the overall carrying capacity of the RC skeleton. All damage was concentrated at the ground floor and practically nothing happened at the upper level.

After repairing both the external claddings at the ground floor by means of C-FRP in the form of near surface mounted bars (Della Corte et al., 2005;

Figure 4.39 The RC building tested within the PROHITECH project.

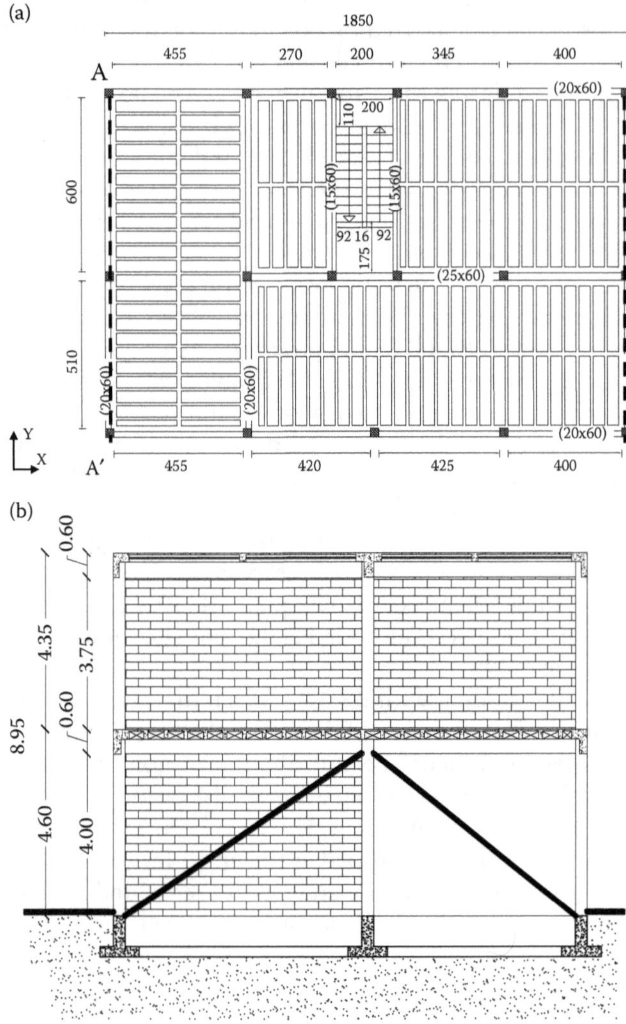

Figure 4.40 The building equipped with BRBs: (a) structural plan; (b) brace configuration.

D'Aniello et al., 2009) and the RC columns, the BRBs were mounted in the two couple of mashes at the ground floor in the loading direction. After these two initial tests, the RC structure was partially repaired and recentered.

Figure 4.40a shows the plan location of the braces at the ground floor (highlighted with the dashed lines, axis A–A′), while Figure 4.40b illustrates the vertical arrangement of the BRBs. Three different BRB systems have been designed and tested separately and subsequently. Particular attention in designing the novel BRBs was paid to details for locating them inside the free space usually in between the facing and the backing of masonry infill walls commonly used for cladding of RC buildings. In order to demonstrate the feasibility of hiding the device inside the claddings, for the first test of this experimental series, the masonry cladding was reconstructed in one bay. Figure 4.41 shows some intermediate erection phases (a, b, and c) and the final configuration before testing with the BRB located inside of the masonry cladding (d).

The building equipped with BRBs was tested by applying a horizontal cyclic inverted triangular force distribution up to the development of a clear collapse mechanism. The test setup consisted of a reacting steel structure with a push–pull system made of six hydraulic jacks used for applying the load in two opposite directions. Figure 4.42 illustrates a view of both the reacting frame and the loading jacks used for applying the lateral force.

The jacks have a maximum stroke of ±30 cm and a maximum load equal to 496 kN in compression and 264 kN in tension (corresponding to a total force of 2976 kN and 1584 kN, respectively, in push and in pull action). They have been connected to a hydraulic pump by means of a circuit in order to always guarantee the same pressure in all the jacks. The latter have been located at a height of 7.31 m and spaced horizontally 3.64 m on centers. The lateral load has been transferred to the two slabs of the building through a steel-trussed structure (Figure 4.42b).

(a) (b) (c) (d)

Figure 4.41 Intermediate erection phases (a,b,c) and the final configuration (d).

Figure 4.42 Reacting structure (a) and hydraulic jacks (b).

Story displacements have been monitored by using a Zeiss-Trimble S10 station (Figure 4.43a) and reflecting prisms (Figure 4.43b). Six measuring points (i.e., six reflecting prisms) have been fixed, corresponding to the two building levels, three at the first floor and three at the second floor (Figure 4.43c). These points allow the average story translation and rotation about the vertical axis to be measured.

The concept of the novel device descends from the previous experience matured within the ILVA-IDEM project (Mazzolani, 2006a,b), where two types of all-steel BRBs have been studied and tested (see Figures 4.34 through 4.37).

Figure 4.44 shows the geometry of the new tested BRB prototypes (henceforth called types A, B, and C). The restraining sleeve was designed

Figure 4.43 Location of the measuring station (a) and reflecting targets (b), which are located in six points, three per floor (c), allowing the measurement of the average story translation and rotation about the vertical axis.

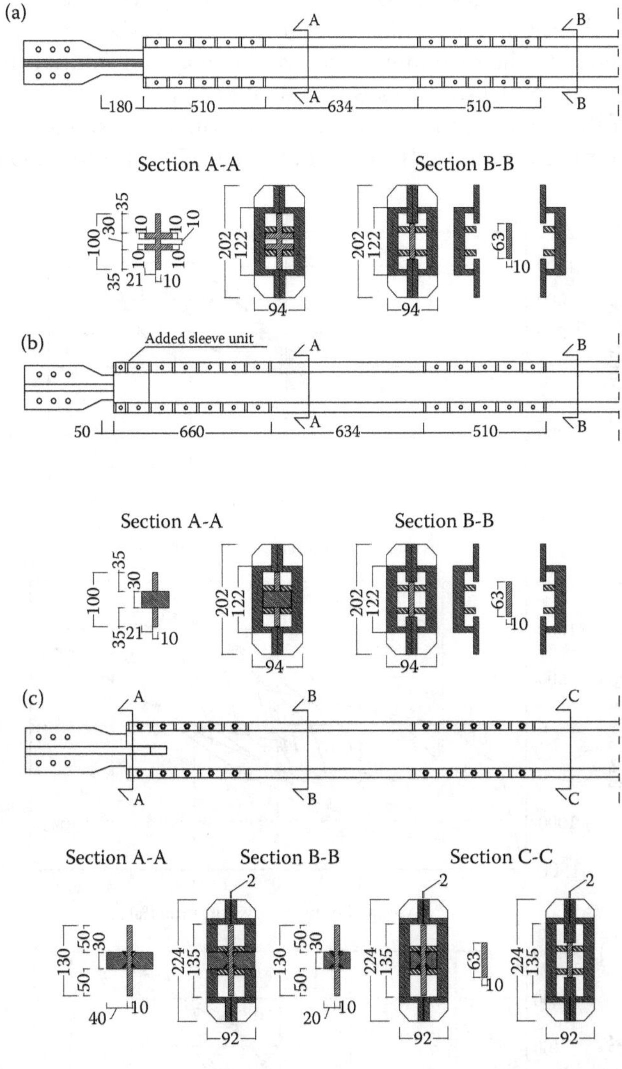

Figure 4.44 **Geometry and details of BRB type A (a), type B (b), and type C (c).**

with a minimum ratio N_E/N_y equal to 2.1 (Mazzolani, 2009), N_E being the Eulerian critical load of the sleeve and N_y the core plastic strength. It is worth noting that the main differences among the tested prototypes are

1. The inner clearance between the yielding core and the restraining sleeve (1 mm per core side in type A and 2 mm in types B and C)
2. The different arrangement of the unrestrained end portions of the core

More details on the design assumptions of these devices can be found in D'Aniello et al. (2008, 2009). The experimental results showed a satisfactory global response of the examined systems. Anyway, substantial differences were recognized among the investigated BRB prototypes.

In case of BRB type A, the tested structure exhibited a stable response up to a maximum interstory drift ratio of about 1.25% (Figure 4.45a), which

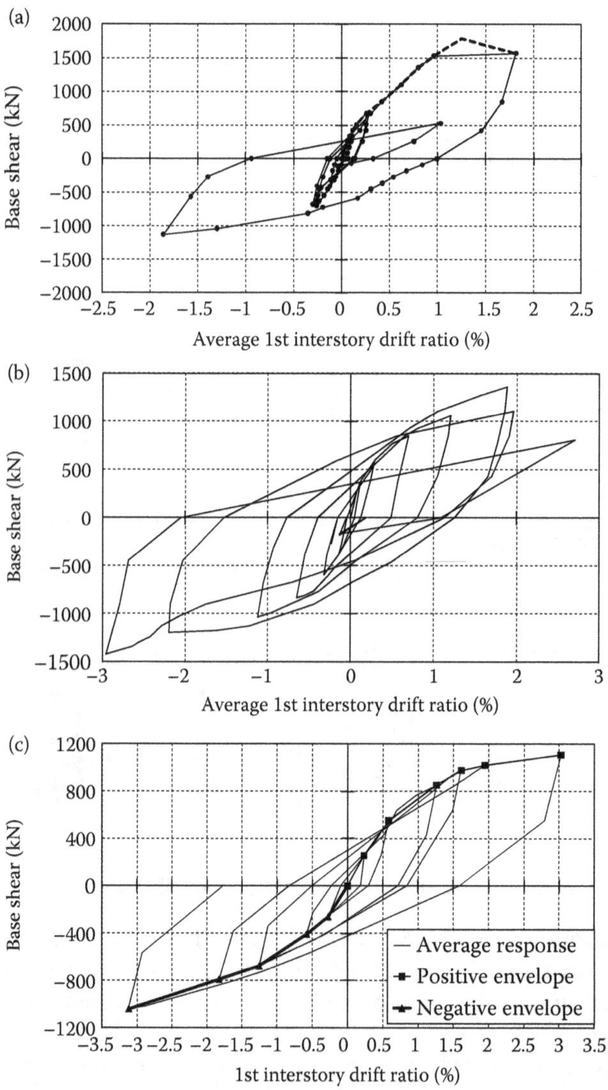

Figure 4.45 Experimental cyclic response of BRBs: (a) prototype A; (b) prototype B; (c) prototype C.

corresponded to a local-distortional buckling failure of the unrestrained end portion of the steel core. The corresponding ductility was $\mu = \theta_{max}/\theta_y = 6.94$. For a larger deformation demand, the local-distortional buckling failure of the unrestrained nonyielding end-plate produced strength degradation.

BRB type B showed a satisfactory overall performance, characterized by stable and symmetric hysteresis loops. As can be noted in Figure 4.45b, the tested device showed a symmetric response in the interstory drift range of ±1.5%, with a minimum ductility capacity of about $\mu = 10.50$. In particular, the experimental test was completed in conjunction with the end of the core free stroke.

BRB type C showed a better overall response, characterized by the complete efficiency as a ductile fuse up to the design value of the maximum interstory drift ratio. Indeed, as highlighted in Figure 4.45c, the tested prototype showed a fully stable and symmetric response in the design interstory drift range (±2%) and allowed the maximum allowable interstory drift of ±3% to be reached in a stable manner. The tested structure reached a maximum interstory drift of about 3.12% and a maximum ductility $\mu = 20.8$. Moreover, it is worth noting that the loading program applied to this prototype was more severe than the one used for the other symmetric response in the design interstory devices. Indeed, fully reversed interstory drift ratios of 1.0%, 1.5%, and 3.0% and continuing with cycles at 1.5% up to the core fracture were imposed.

The possibility of hiding the devices within the space between the two facades of masonry infill walls and the excellent performance experienced by BRB type C confirmed the effectiveness of these devices, which can be used as a viable system for the seismic retrofitting of existing RC structures.

4.3 NEW SOLUTIONS OF BRACING SYSTEMS

4.3.1 Shear wall systems

MPSWs represent a very appealing system to improve the seismic performance of new and existing steel and RC structures. Generally speaking, an MPSW is a lateral load-resisting system composed by metal plates connected to columns and girders of a primary surrounding structure.

As stated by Astaneh-Asl (2001), the connection between the steel plates and the two lateral boundary columns behaves like a vertical cantilever plate girder that is able to resist horizontal story shear and overturning moment due to lateral actions. The plates act as the web of the resulting girder, the columns form its flanges, and the horizontal floor beams act as transverse stiffeners.

The plates can be arranged in one or more bays along the height of the primary structure, which in case of steel structures can be simply pin-jointed

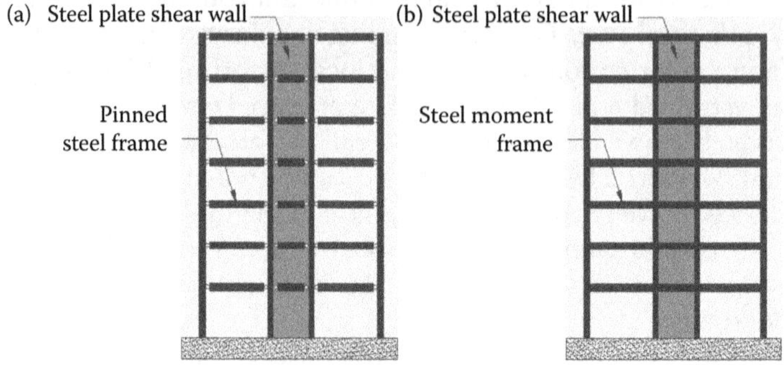

Figure 4.46 Typical steel structural systems with shear walls: (a) standard structural system (SSS); (b) dual structural system (DSS).

or moment resisting (Figure 4.46). In the first case, the system is defined as a "standard structural system" (SSS), and in the latter case, as a "dual structural system" (DSS).

In the last 40 years, many studies on the performance of MPSWs have been undertaken and new design criteria (CSA, 2001), based on the assumption that shear plates can offer a suitable postbuckling strength, have been introduced. Such studies highlight the advantages that such a type of lateral load-resisting system can provide with respect to the traditional systems (see Table 4.1).

In particular, it has been clearly pointed out that the assessment of safety of buildings equipped with shear plates must be carried out considering a design ultimate limit state not corresponding to the out-of-plane buckling of the infill panels; otherwise, an adequate performance of a structure with MPSWs would consist in applying heavily stiffened thick plates, but considering their postbuckling resources given by tension field-resisting mechanisms. In this case, the use of MPSWs can be considered economically competitive with respect to other seismic strategies, as well as to the most employed RC shear walls (Bruneau et al., 2005).

On the other hand, the need of a suitable theory for interpreting the postbuckling behavior of steel plates was felt even previously. On this purpose, one of the most significant conclusions of the VIII Congress AIPC in

Table 4.1 Performances of the Most Popular-Braced Systems Compared to MPSWs

	Stiffness	*Ductility*	*Strength*
MRF	Poor	Excellent	Good
CBF	Excellent	Poor	Good
EBF	Good	Excellent	Good
MPSWs	Good	Excellent	Excellent

New York during the 1960s stated: "The linear theory of buckling is not an adequate base for designing structures with thin plates and it has to be surrogated with a non linear theory taking in account the post-buckling resources of the same plates. It is necessary to consider the maximum in plane tensions and deformations bearable by the plates after that buckling phenomenon occurred .... There is a lack of mathematical approaches which allow to establish a calculation method which keep into accounts all the influencing parameters.... A new series of experimental and numerical analyses which include the post-buckling behaviour have to be carried out...."

It is only since the 1980s that many experimental and numerical studies on the postbuckling behavior of MPSWs have been carried out in order to provide useful rules and design criteria to be adopted in national and international standards. They were based on the assumption that a shear plate can provide an effective initial stiffness, thus limiting interstory drifts value under low seismic actions, it being also able to develop a tension field-resisting mechanism. Once the buckling phenomena are triggered off, the shear plate behaves in a very ductile way (Figure 4.47), with a significant amount of energy dissipated under cyclic loads.

In particular, the stiffening and the dissipative function of structures equipped with MPSWs, under both static and dynamic actions, have been deeply investigated, besides comparing their performances with the ones offered by different seismic devices. Moreover, accurate finite element models have been implemented in order to identify the main behavioral phenomena acting in a shear panel; various metal materials and geometrical configuration, that is, characterized by different location of stiffeners or by holes wisely made on the base plate, have been considered as well.

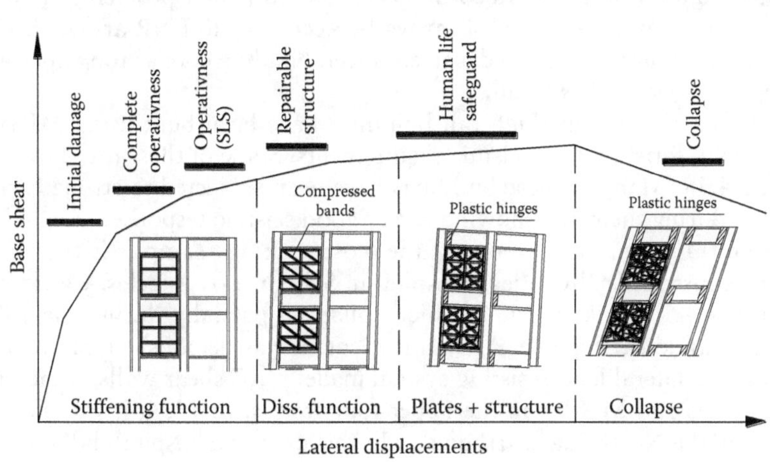

Figure 4.47 Nonlinear behavior of a DSS equipped with MPSS and its performance levels.

Beyond the above structural aspects, MPSWs began to be considered convenient even with respect to the most common RC shear walls, due to the fact that

1. No concrete casts are necessary, with both a significant speeding up of the erection process and a reduction of the construction costs; moreover, the use of controlled metals derived from industrial processes allows a major quality control and, thus, inferior safety levels to be adopted in the design phase.
2. The structural lightness produces minor stresses on both columns and foundations, a reduction in the input seismic load, as well as less occupancy of space, with evident advantages from the architectural point of view.
3. Inspection of the resisting structural elements, which can be necessary after a significant seismic event, is easier.
4. MPSWs can be applied even in hard environmental conditions, in which RC shear walls may not be economical, that is, in very cold regions.
5. Compared to RC shear walls, MPSWs can be much easier and faster to be erected, when they are used in the seismic retrofitting of an existing building.

MPSWs also began to be considered economically convenient in the beliefs of both designers and contractors. From this point of view, it is very worth noting what happened in 1978 to a 53-storied high-rise building erected in Tokyo. As referred by Astaneh-Asl (2001), "...the structure was initially designed using reinforced concrete shear walls. However, according to Engineering News Record (1978), due to patent problem, the R/C walls were converted to steel shear walls. According to ENR article (ENR, 1978), the contractor rejected a steel-braced building core as too expensive compared to steel shear wall."

A lot of steel middle-high tall buildings have been built using MPSWs in order to satisfy severe seismic requirements. Some of these are shown in Figure 4.48. Many of these buildings have been subjected to strong earthquakes during their life, showing a very good seismic response.

From this point of view, one of the most striking examples is presented by the six-storied Olive View Hospital in Sylmar, Los Angeles, California (Figure 4.48a), built in 1964, which collapsed partially during the 1971 San Fernando earthquake. After this event, it was demolished and rebuilt adopting a lateral load-resisting system made of RC shear walls in the first two stories and MPSWs in the upper four stories. In 1994, it was shaken again by the Northridge earthquake. In this case, the hospital did not sustain any serious damage.

Figure 4.48 Steel buildings with MPSWs: (a) Sylmar Hospital, Los Angeles. (www.air-worldwide.com, source USGS.) (b) The Nikken Sekkey building in Tokyo. (Photo by F.M. Mazzolani.) (c) Hyatt Regency Hotel, Dallas. (http://dallasregency.hyatt.com.) (d) Kobe office building, Kobe. (Photo by Shindo, Hiroyuki.) (e) U.S. Federal Courthouse, Seattle. (Art Anderson Associates.) (f) The Century, San Francisco. (Seilie, I.F., Hooper, J.D. 2005: *Modern Steel Construction, AISC*, 45(4), 37–43.)

As explained by Astaneh-Asl (2001), the California Strong Motion Instrumentation Program (CSMIP) had instrumented the Sylmar hospital before the Northridge earthquake. The acceleration at roof level exceeded 2.3*g* while the ground acceleration was recorded at about 0.66*g*. This amplification was due to the high stiffness of the structures, but, nevertheless, only slight damages to the nonstructural parts have been registered.

Another significant example, often cited in the current literature, is the one related to a 35-storied office building in Kobe, Japan (shown in Figure 4.48d), built in 1988 and subjected to a strong excitation during the 1995 Kobe earthquake. The structural system was given by the three basement levels made of RC shear walls, composite walls at the first and second floors, and steel shear walls from the second floor to the top. As discussed in Astaneh-Asl (2001), after the earthquake, the structure presented no significant damages. In fact, only local buckling of stiffened plate shear walls at the 26th story has been detected, with a permanent roof drift of 225 mm in the northerly direction and 35 mm in the westerly direction.

Finally, the first applications in Japan of DSSs equipped with dissipative shear panels have to be mentioned (Figure 4.48b). New steel buildings were erected in the harbor area of Tokyo, after the 1995 earthquake, by using stiffened shear panels located around the stair-case/elevator area. These panels are made of a special low-yield steel with a yield point of 100 MPa. In this case, the base shear panels have been conceived as dampers, acting as a fuse and providing a passive protection to the whole structure. This function allows the steel elements of the primary structure to remain in the elastic field, even under a strong earthquake.

4.3.2 Behavior of metal shear panels

4.3.2.1 General concept

Shear walls made of metal plates can be subdivided into two main typologies:

1. Walls able to improve the strength and the stiffness of the primary structure
2. Walls able to dissipate the input seismic energy

Both these typologies of MPSW systems can also be characterized in terms of geographical areas where they have been mainly studied and developed. Thus, thin unstiffened MPSWs mainly having a stiffening and strengthening function have been principally promoted in the United States and Canada, whereas stiffened dissipative shear panels have been mainly proposed in Japan and Italy. Also, the possibility of employing lightweight cladding panels for seismic purposes has been largely investigated in Europe, as a development of the well-known stressed skin design approach, which was mainly addressed to corrugated sheeting used as roofing of industrial buildings (Davies and Bryan, 1982).

In the following paragraph, an overview on the aforementioned typologies is given, in order to pick out the most important past and present developments in the state-of-the-art MPSWs. Prior to this, as a first step, it is necessary to give a synthetic description of the most important theoretical concepts related to the behavior of both stiffened and unstiffened thin shear plates, as well as the main concepts regarding the modeling of shear plates belonging to the MPSWs system.

4.3.2.2 Theoretical issues

When thin metal plates are subjected to a growing shear in-plane force, buckling phenomena occurring before the metal yielding usually have an important influence on the system response. This trivial consideration is a simple extension in the two-dimensional field of the well-known concept of buckling of a one-dimensional slender beam. However, it must be pointed out that shear plates and beams are very different as far as the postbuckling behavior is concerned. In fact, while for a beam element an equilibrium bifurcation is detectable, in the case of a shear plate the stable postcritical behavior allows further strength resources up to the plate failure (Figure 4.49a, where the three curves a, b, and c, respectively, refer to growing imperfection levels).

As shown in Figure 4.49b, a pure shear-resisting mechanism acts in the prebuckling phase, with equal compressive and tensile principal stresses, developing large hysteretic loops in case of both the cyclic loading history and yielding achievement. Then, if instability is triggered along the principal direction in compression, a postbuckling behavior is developed. It is characterized, under growing shear, by increasing tensile stress, whereas the compression stress remains substantially constant (see Figure 4.49c), thus entailing a rotation of the stress state (Höglund, 1997).

The critical tangential stress for rectangular plates subjected to pure shear is given in Equation 4.11:

$$\tau_{cr} = k \cdot \frac{\pi^2 \cdot E}{12 \cdot (1 - \nu^2)} \cdot \left(\frac{t}{d}\right)^2 \tag{4.11}$$

where E is the elasticity modulus, ν the Poisson factor, t the thickness of the plate, and d the characteristic transversal dimension, which usually depends on the maximum free deformable length detectable on the plate. Lastly, k, which is defined as the plate factor, depends on both the aspect ratio b/d and the boundary conditions. Many authors have provided useful

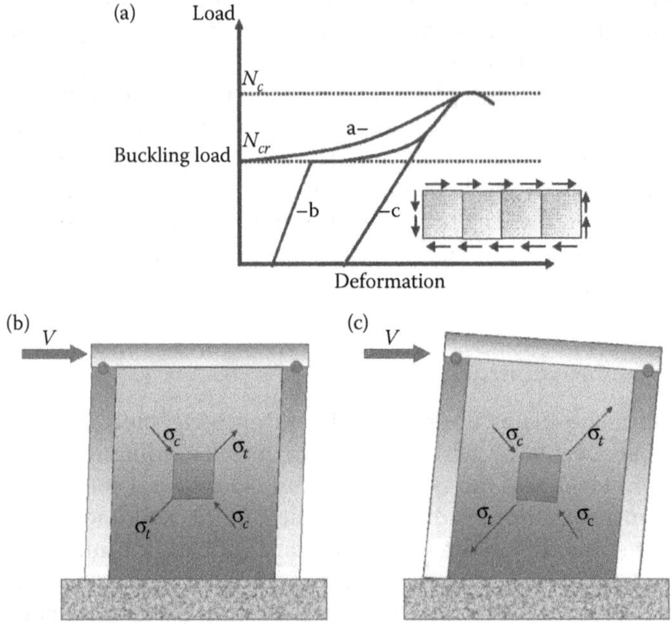

Figure 4.49 (a) Postbuckling behavior of shear plate; stress state on the web of a not buckled (b) and of a buckled (c) beam web subjected to shear.

indications about the exact value to be adopted for this parameter, considering several constraint conditions along the edges and different values of the aspect ratio. Among these, Timoshenko (1915, 1936) dealt with plates with several geometrical shapes, while Skan and Southwell (1924), Seydel (1933), and Stein and Neff (1947) studied cases with different in-plane geometries and simply supported boundary condition at their edges. The results obtained by these authors are summarized in Figure 4.50a and interpretation by the following equation was found (Stein and Neff, 1947):

$$k = 5.34 + \frac{4.00}{\alpha^2} \tag{4.12}$$

Then, Budiansky and Connor (1948) provided the same type of curve (Figure 4.50b) with reference to clamped shear plates, also on the basis of the results obtained by Skann and Southwell (1924), who for infinite clamped plates found a k value equal to 8.98 as well as the one obtained by Cox (1933) and Iguchi (1938). In such a case, the analytical curve representing the obtained results is

$$k = 8.98 + \frac{5.60}{\alpha^2} \tag{4.13}$$

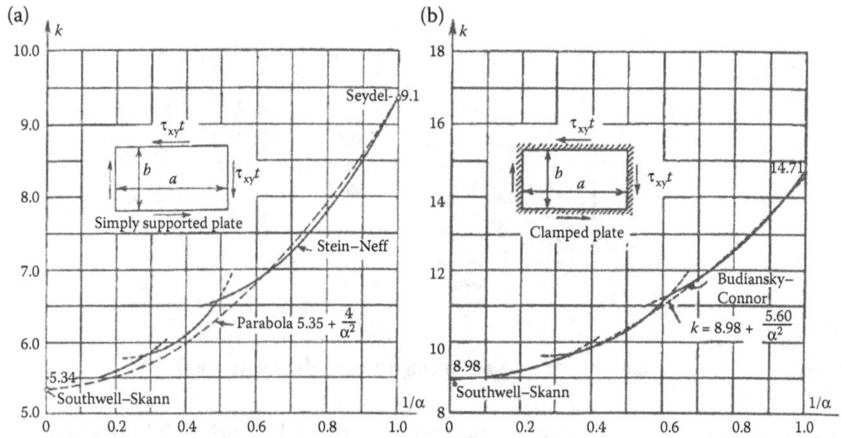

Figure 4.50 Values of the plate factor k for (a) simply supported and (b) clamped plates.

The above formulations have been generalized by Stein and Fralich (1949) considering both the case of aspect ratios equal to 1 and 2 for simply supported multistiffened plates. In these cases, the *k* factor has been expressed as a function of the flexural stiffness of transversal ribs.

However, the above results are to be considered acceptable only if buckling phenomena occur in the elastic field; otherwise, the expression (4.11) must be corrected by means of a coefficient η as shown in the Equation 4.14:

$$\tau_c = k\eta \frac{\pi^2 E}{12 * (1 - \upsilon^2) * \lambda^2} = \eta \tau_{cel} \tag{4.14}$$

where the coefficient η, on which many authors have been focused their attention, can be generally expressed as in Equation 4.15:

$$\eta = \frac{E_{red}}{E} \tag{4.15}$$

In both cases (buckling occurring either in the elastic or plastic field), after τ_c is achieved, a tension field mechanism develops. This aspect was first clarified by Wagner (1931) for the web of girders. He supposed that, after the plate buckling, a tension band with a slope angle arises. He determined the value of this angle (Figure 4.51), in case of a very thin web, when the girder flanges had enough stiffness to not interfere with the tension field mechanism.

Also, Basler and Thurlimann (1959) provided some interesting remarks, solving practical problems regarding the rules for the design of shear

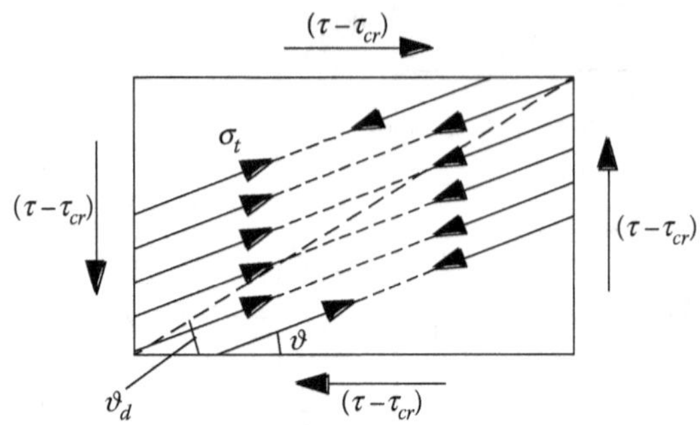

Figure 4.51 Tension field mechanism according to the Wagner theory.

plates, accounting for their stable postbuckling behavior. Their method was adopted by the AISC provisions and took into consideration only the flexural stiffness of the transversal web, neglecting the flange stiffness, thus presuming that buckling phenomena remain concentrated in each panel portion and that the rotated tension field anchors itself only on the transversal ribs, allowing the whole girder working as a truss. Successive authors developed more refined theories based on different hypotheses on the anchorage of the tension field, as well as on the formation of plastic hinges on the flanges. The methods of AARAU (Herzog), CARDIFF (Rochey et al.), KARLSRUHE (Steinhardt et al.), LEIGH (Chern and Ostapenko), OSAKA (Komatsu), PRAHA-CARDIFF (Skaloud and Rockey), STOCCOLMA (Höglund, base method of EC9 Provision), TOKIO (Fujii et al.), GÖTEBORG (Bergfelt), ZURIGO (Dubas), and TRIESTE (Mele), synthesized in Figure 4.52, have been therefore developed (Ballio and Mazzolani, 1983).

These methods are valid only under the hypothesis of a tension field developed between two successive transversal ribs. This happens only if the transversal stiffeners are characterized by an adequate flexural stiffness (Figure 4.53a); otherwise, a global buckling could develop and the yield bands could involve two or more panel portions (Figure 4.53b).

4.3.2.3 Shear panels modeling

On the basis of experimental tests, Thorburn et al. (1983) introduced two analytical models in order to determine the stiffness offered by steel plates under increasing shear loads, when the compression forces are negligible and the plates are arranged in a frame with hinged beam-to-column joints.

The first modeling approach is known as the *strip model* and it is shown in Figure 4.54a. In this method, the panel is represented by a series of

Method	Resisting mechanism	Boundary conditions
AARAU (HEZOG)	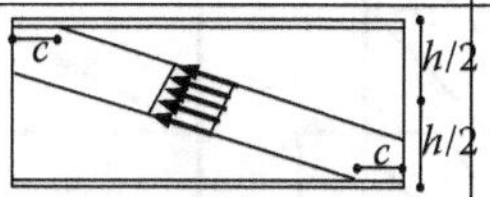	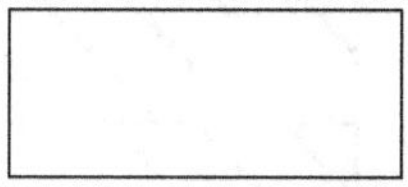
CARDIFF (Rockey et al.)	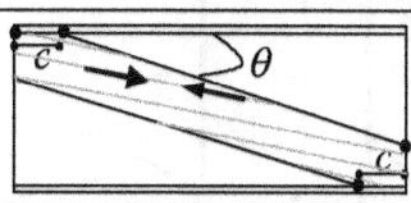	
KARLSRUHE (Steinhardt et al.)	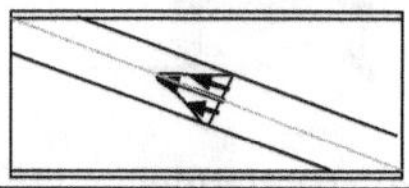	
LEHIGH (Chern and Ostapenko)	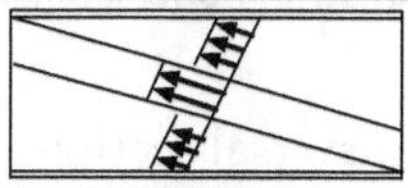	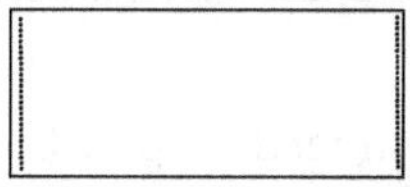
OSAKA (Komatsu)	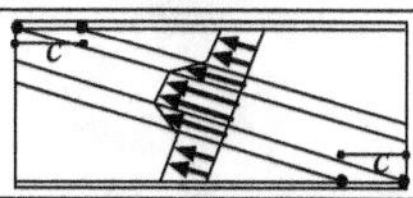	
PRAHA-CARDIFF (Skaloud and Rockey)	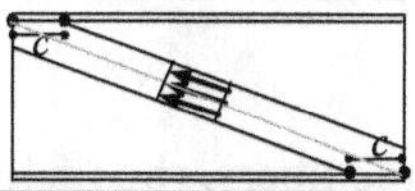	
STOCKOLM (Höglund)	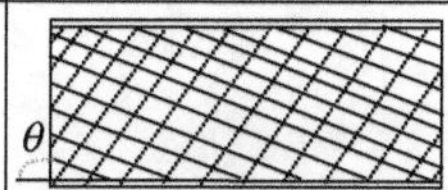	
TOKYO (Fujii et al.)	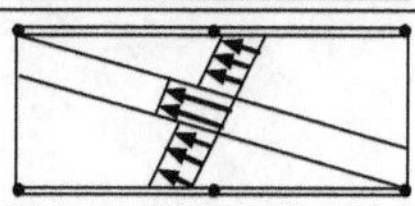	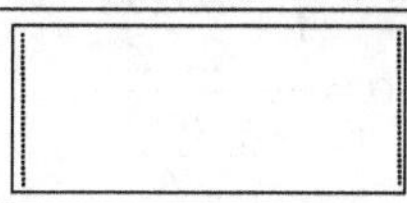

Figure 4.52 Tension field methods based on different modalities of anchorage of the yield band. (Ballio, G., Mazzolani, F.M. 1983: *Theory and Design of Steel Structures*. Chapman & Hall, London, New York.)

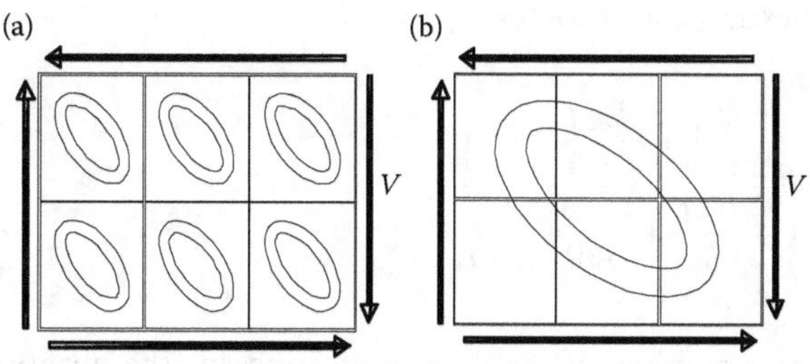

Figure 4.53 Local (a) and global (b) buckling phenomena due to inadequate second moment of area of transversal and longitudinal intermediate stiffeners.

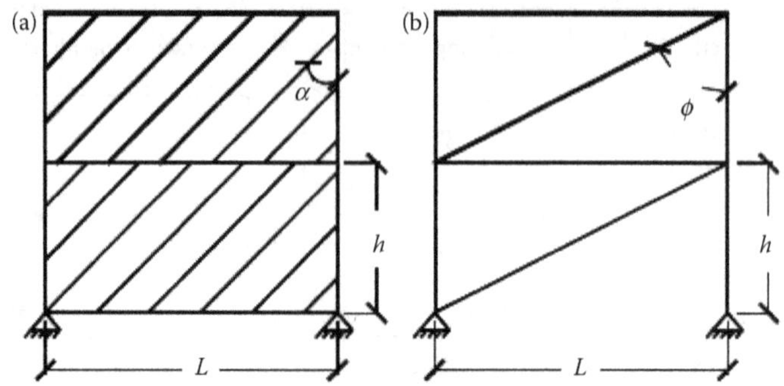

Figure 4.54 Shear panels modeling: (a) strip model and (b) equivalent brace model.

inclined hinged strips whose transversal section A_s is determined as a function of the plate thickness t:

$$A_s = \frac{(L\cos\alpha + h \cdot \sin\alpha)}{n} \cdot t \tag{4.16}$$

and, in case of rigid boundary members, the inclination angle α of the strip is given by

$$\tan^4\alpha = \frac{1 + \dfrac{L \cdot t}{2 \cdot A_c}}{1 + \dfrac{h \cdot t}{A_b}} \tag{4.17}$$

in which A_b and A_c are the beam and column cross-section areas, respectively; L the length of the beam, and h the height of the frame interstory. This expression was subsequently modified by Timler and Kulak (1983) in order to take into account the flexibility of the column, by introducing its moment of inertia I_c; it becomes

$$\tan^4\alpha = \frac{1 + \dfrac{L \cdot t}{2 \cdot A_c}}{1 + \dfrac{h \cdot t}{A_b} + \dfrac{h^4 \cdot t}{360 \cdot I_c \cdot L}} \tag{4.18}$$

As far as the number of strips n is of concern, the authors suggested assuming a minimum number of 10 for obtaining refined results. Starting from the strip model theory and applying a plastic analysis on the panel, it is possible to develop equations that allow the ultimate shear strength V to be obtained for assigning to the plate of the MPSWs system (Equation 4.19).

$$V = \frac{1}{2} \cdot f_y \cdot t \cdot L \cdot \sin 2\alpha \tag{4.19}$$

The second and more simplified model, known as the *equivalent brace* (Figure 4.54b), schematizes thin panels by means of a single diagonal hinged brace in each frame field.

The authors showed that, for infinitely rigid columns, the panel thickness (t) and the area A of the single equivalent diagonal are related by the following relationship:

$$A = \frac{t \cdot L \cdot \sin(2 \cdot \alpha)}{2 \cdot \sin\phi \cdot \sin 2\phi} \tag{4.20}$$

where L is the beam length, ϕ the acute angle of the brace with respect to the column, and α the inclination angle expressed in Equation 4.17.

Many authors also focused their attention on the models that must be assumed for reproducing the shear panel's hysteretic behavior. In particular, with reference to the geometrical models based on the use of an equivalent diagonal, analytical relationships have been suitably proposed by Nakashima (1995), for compact dissipative low-yield-strength shear panels, and by De Matteis et al. (2011), for pure aluminum shear panels. The last authors proposed a Ramberg–Osgood relationship to be calibrated on the basis of experimental analyses.

Finally, it is worth mentioning the work of Chen and Jhang (2006), who proposed a two-force strip model based on different strength-deformation behaviors in tension or compression, to be established on the basis of the slenderness of the panels to be modeled (Figure 4.55).

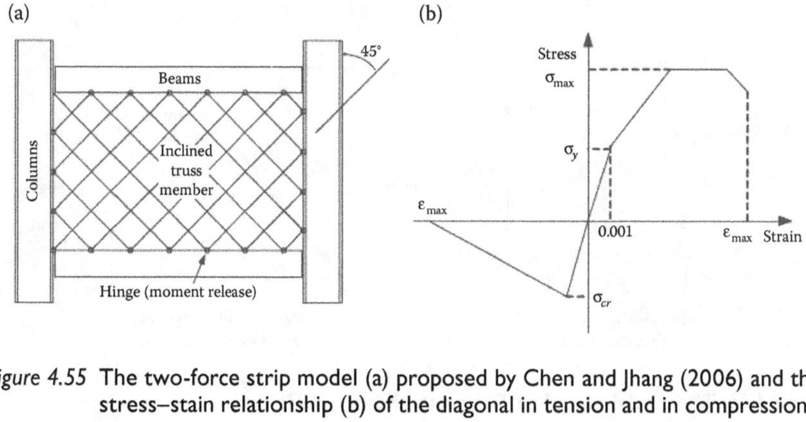

Figure 4.55 The two-force strip model (a) proposed by Chen and Jhang (2006) and the stress–stain relationship (b) of the diagonal in tension and in compression.

4.3.3 Type of shear panels

4.3.3.1 Thin plates

Unstiffened thin plates in shear are usually employed in MR structures, as part of an MPSWs system, in order to satisfy the serviceability limit state requirements by means of a significant increasing of lateral stiffness.

In addition, a fair hysteretic function is usually provided by these systems, as their dissipative response is generally characterized by stable hysteretic cycles, which are however degraded by pinching effects due to buckling phenomena.

Many studies on this type of devices have been performed in order to detect the MPSWs contribution to the improvement of the whole structure. The first research program was carried out by Mimura and Akiyama (1977), who developed a method to predict the behavior under monotonic and cyclic loadings of slender panels. In particular, the behavior of the global system (frame + panel) in terms of force–displacement response was assumed to be the sum of the single contributions offered by the panel and the external frame.

However, the most significant scientific outcomes on the behavior of unstiffened shear plates date back to 1983 (Thorburn et al., 1983; Timler and Kulak, 1983), where two research reports of the University of Alberta in Canada exposed analytical studies and experimental full-scale tests (Figure 4.56), clarifying that shear walls can also work adequately after the occurrence of buckling phenomena, establishing that the behavior of MPSWs could be adequately modeled by means of strip models, as described in Section 4.3.2.

Tromposch and Kulak (1987) conducted an experimental activity successively on large-scale steel shear panels, which was very similar to the ones tested by Timler and Kulak (1983). The main differences were due to the different dimensions of the single panel, in the use of bolted, instead of welded, beam-to-column connections, in the presence of larger beam

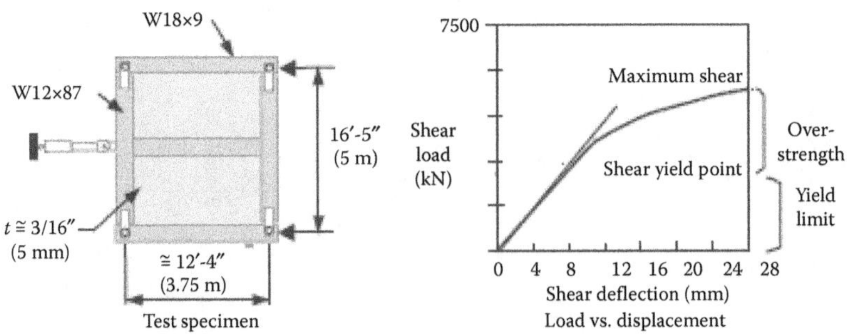

Figure 4.56 The model tested by Timler and Kulak (1983).

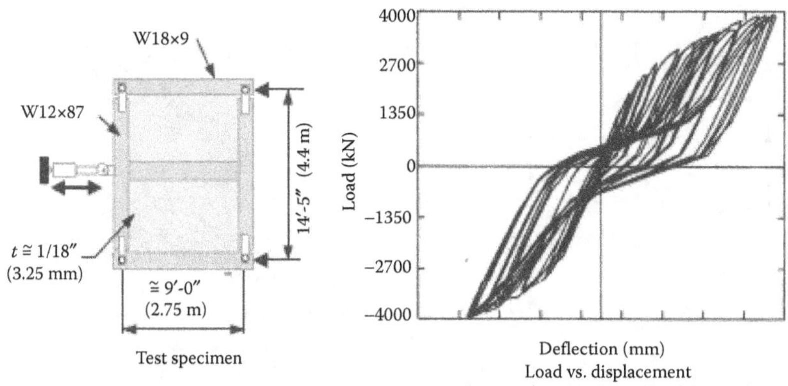

Figure 4.57 The model tested by Tromposch and Kulak (1987).

profiles, in the column preloading, in order to simulate the effects of gravitational loads, and in the adopted loading history, considering both the cyclic and monotonic loading tests (Figure 4.57).

Elgaaly et al. (1993) performed an experimental investigation carried out on five single-bay three-storied steel frames, modeled in a length scale 1:4, in order to detect the effects that the slenderness ratio of the shear plates and the possible types of beam-to-column connections (hinged or moment resisting) played on the shear wall behavior. On the basis of these results, they applied both a finite element model and a model based on a revision of the equivalent strips method to simulate the experimental tests.

Xue and Lu (1994) carried out an analytical study on three-bays–12-stories MRFs, having the intermediate bay stiffened by shear walls. The object of this research study was represented by the evaluation of the results produced by the use of different beam-column and plate–frame connections on the performances of the global system.

Driver et al. (1997) carried out a large-scale test of multilevel shear walls, having moment-resisting beam-to-column connections, in order to better identify the elastic stiffness, the first yielding, the ductility and the energy absorption capacity of the system, investigating into the stability of the hysteresis cycles and into the factors that caused the failure mechanism of the wall (Figure 4.58).

Rezai (1999) performed the first experimental dynamic test with a shaking table on shear walls realized with steel panels (Figure 4.59) with the main purpose of appraising the reliability of the prescriptions provided by the Canadian code for this structural typology. A four-level and one span steel frame, having a width of 918 mm and an interstory height of 900 mm, was considered, while the shear panels were characterized by a

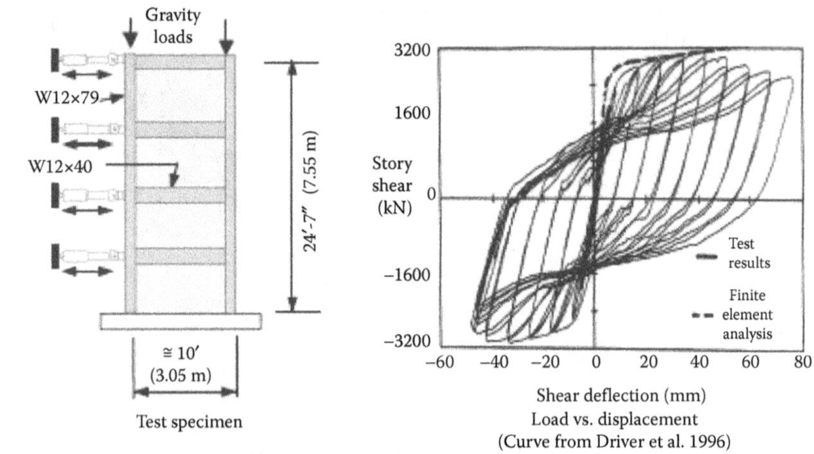

Figure 4.58 The model tested by Driver et al. (1997).

thickness of 1.5 mm, being connected to 2.5 mm steel plates welded to the frame members.

Lubell et al. (2000) tested two shear wall systems, applying them to a quasi-static cyclic according to the provisions specified by the ATC 24 protocol: the first system having four-level panels and the second one composed by a single shear plate (Figure 4.60). In all the examined systems, the panels had a thickness of 1.5 mm, a base of 900 mm and a *b*/*h* ratio equal

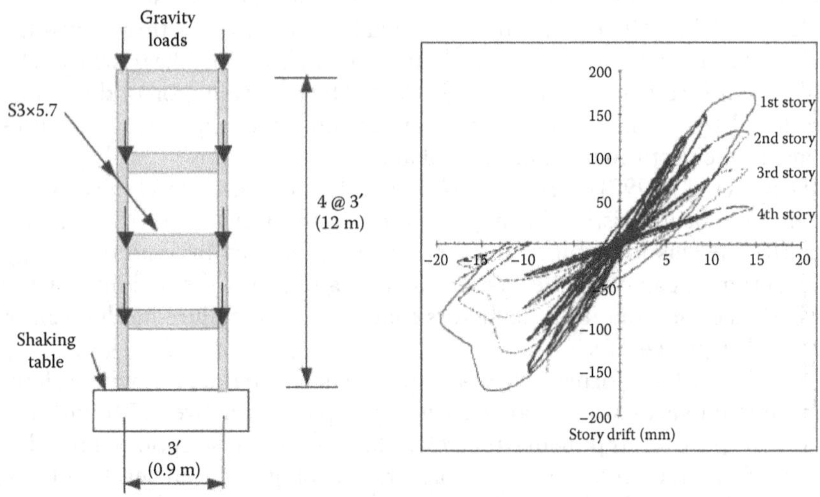

Figure 4.59 The model tested by Rezai (1999).

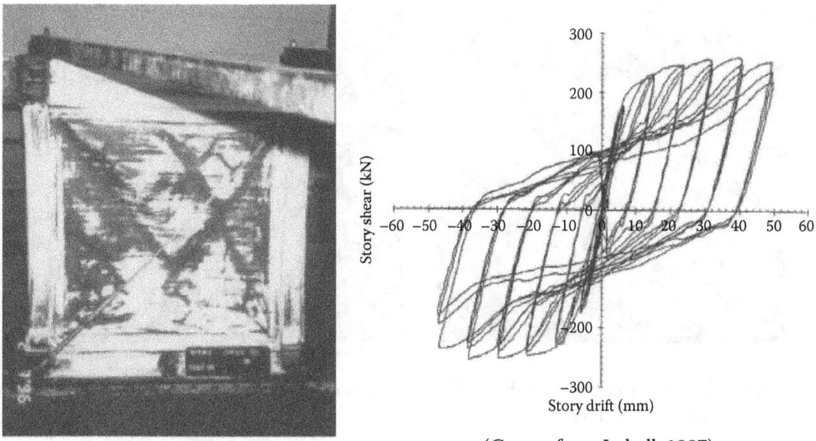

Test specimen

(Curves from Lubell, 1997)
Load vs. displacement

Figure 4.60 The model tested by Lubell et al. (2000). (Photo: Courtesy of C. Ventura.)

to 1, while the beam-to-column connections were of rigid type. The one-level system was loaded up to seven times the yielding structural displacement and exhibited a collapse of the contrast frame due to the excessive out-of-plane displacement of the upper part of the system.

Sabouri-Ghomi and Roberts (1991) and Sabouri-Ghomi et al. (2003) developed a general method for the analysis of shear walls characterized by different configurations of panels, with and without openings, with and without flexural stiffeners. The proposed analytical model, known as the plate–frame interaction (PFI), could easily be applied in the engineering practice, because it allows the design of both the panel and the frame members, without considering the interaction existing between the two structural components.

Astaneh-Asl and Zhao (2002) carried out research finalized to the study of the shear panels behavior under cyclical loads. The experimental test was performed on a system composed by a panel connected both to a composed steel–concrete column (filled type) and to steel beams and columns. The shear panel was endowed with a stiffening plate, located at midheight, and connected by means of bolts to the steel column. The panel was 9.5 mm thick, while the thickness of the stiffening plate was 6.4 mm.

Alinia and Dastfan (2006) investigated into the influence of the surrounding member rigidities on the elastic shear buckling and postbuckling behavior of infill panels. It has been shown that shear walls, surrounded by beams and columns, should not be considered as simply supported. It has also been shown that the torsional stiffness of supporting members is highly effective in increasing the elastic buckling load, but it does not

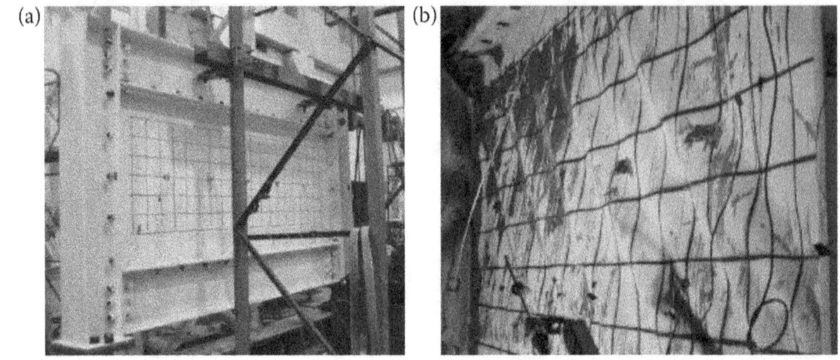

Figure 4.61 Test on a shear unstiffened steel plate by Bruneau et al. (2005): (a) before test; (b) after test.

affect the postbuckling strength. In addition, it has been established that the beam-to-column connection type has no significant effect on the panel's behavior.

Finally, a significant example of experimental tests on the stiffening light-gauge steel shear panel provided in Bruneau et al. (2005) is presented. An SPSW test specimen utilizing a light-gauge infill (thickness of 1.0 mm, 0.0396 in) panel has been statically tested (Figure 4.61). The infill was found to provide approximately 90% of the initial stiffness of the system, while a better dissipative performance has been shown by other systems, that is, perforated steel shear panels able to realize stable hysteretic cycles without significant pinching effects.

4.3.3.2 Dissipative shear panels

In the last few decades, new panel typologies with a suitable dissipative capability have been set up in order to improve the seismic performance of steel and RC buildings, using them as adequate sacrificial devices (dampers). Observing Equation 4.21 of energy balance, where E_i is the total input energy due to an earthquake, E_e is the elastic energy, E_k is the kinetic aliquot, E_{hs} and E_{hd} are the dissipated energy for hysteresis by the bare structure and the additional dampers, respectively, while E_v is due to viscous phenomena.

$$E_i = E_e + E_k + E_{hs} + E_{hd} + E_v \tag{4.21}$$

It is obvious that the rational increase of the E_{hd} term can entail a suitable reduction of the sum of the other aliquots and therefore of the damageability of the primary structure, which can therefore be considered to be passively controlled. As a consequence, the application of energy-absorbing shear panels leads the whole structure to a favorable situation, where the main structural members remain in the elastic field, while the plasticity is conveniently concentrated in a few parts. Obviously, the aforementioned energy-absorbing panels are really effective only if they are characterized by adequate rigidity and dissipative capability.

The above considerations broadly represent the basic concepts of the flexible–stiff mixed structural form application (Akiyama, 2006), which can be considered to be one of the most important seismic design strategies, explicitly taking into account the inelastic deformation of the structure.

The selected shear panel typologies, which are chosen to play a role of dissipative system for the structure where they are installed, must substantially satisfy the well-known basic concepts and requirements usually employed for the capacity design application. For this reason, the use of low-yield stress point material (in order to have the so-called LYSW) or the resort to a weakening technique of the system, for instance, by means of holes applied in the base plate, has been introduced.

It is clearly desirable that the dissipative capability offered by these kinds of systems does not also degrade for high shear strain values. Therefore, in order to avoid the pinching effects due to buckling phenomena, the application of ribs is necessary to transform a slender configuration into a compact one (Figure 4.62).

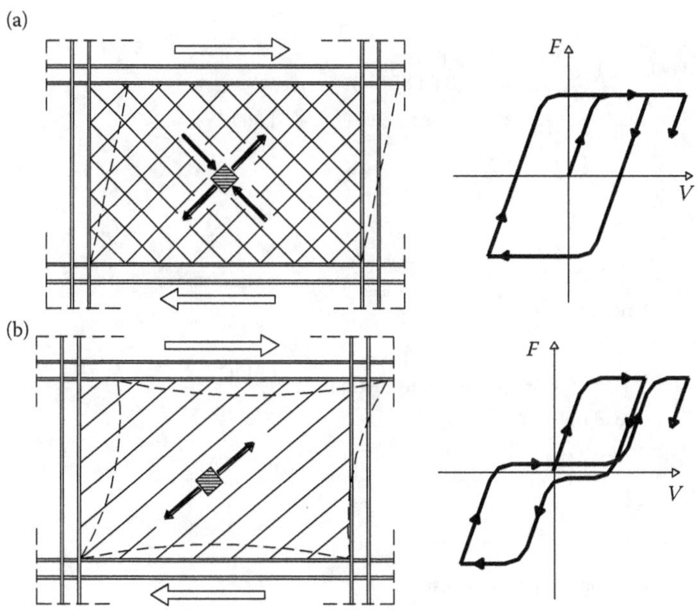

Figure 4.62 Shear energy dissipation mechanisms: (a) pure shear (compact shear panels); (b) tension field (slender or semicompact shear panels).

As far as the use of low-yield base material is concerned, many studies, especially in the Japanese area, have been carried out. In particular, the first applications in using low-yield steel date back to 1991–1992 (Tamai et al., 1991; Izumi et al., 1992).

However, the first and most significant work on strain-hardening behavior of shear panels made of low-yield steel has been carried out by Nakashima (1995) and Nakashima et al. (1994, 1995). Experimental tests (Figure 4.63) have been performed in order to detect the cyclic behavior of six compact shear panels, characterized by a different width-to-thickness ratio as the test variable and involving significant strain hardening. They were made of low-yield steel characterized by a nominal yield stress of 120 MPa and a large ductility with an ultimate elongation of almost 50%.

It was found that strain hardening during cycles with the same deformation amplitude, as well as under increasing deformation, was significant.

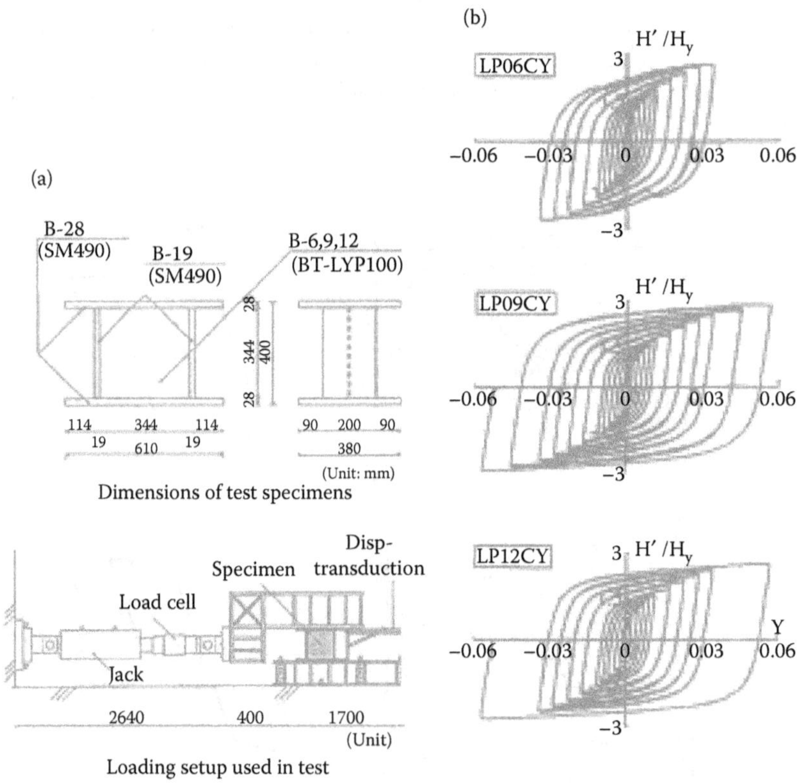

Figure 4.63 Experimental tests carried out by Nakashima (1995): (a) test apparatus; (b) shear force versus drift relationships.

Moreover, the behavior of tested shear panels has been accurately simulated by a multisurface model by means of appropriate choice parameters. Also, some numerical models have been based on either the bilinear or the Ramberg–Osgood model, combining adequately both the isotropic and the kinematic hardening components.

Nakagawa et al. (1996) carried out a very interesting experimental campaign in order to investigate both the hysteretic behavior of some LYSW configurations, characterized by different rib arrangements (Figure 4.64a), and the performance of a full-scale three-storied steel frame equipped with LYSWs (Figure 4.64b).

Tanaka et al. (1998) described the application of some LYSWs to about 10 buildings for the purpose of detecting both response and damage parameters of the main frames during large earthquakes, while in Tanaka and Sasaki (2000), it has been proved that LYSWs with a width-to-thickness ratio less than 40 are able to exhibit a very excellent hysteretic performance (Figure 4.65). Moreover, empirical equations to estimate maximum strength and allowable deformation of the panel have been proposed for the seismic response design purpose. Finally, the tested panel's behavior have been simulated using a skeleton-shift model.

Katayama et al. (2000) performed examinations of shear panels, with width-to-thickness ratio equal to 42, under static and dynamic loading, in order to investigate their energy dissipation capacity also for different strain-rate values.

In particular, it has been established that the shear stress increase in LYSW is mainly caused by strain hardening, but it is also to relate to the strain rate, whose maximum contribution is 20% and no more than 6% for maximum drift angles, when the effect of the strain rate becomes smaller as strain hardening increases. As a consequence, in any case, the energy dissipation capacity under dynamic loading is larger than that for static loading.

Another interesting technique to have dissipative shear walls, especially considering that low-yield materials are not always available in the whole world market, is that of weakening the base plate by means of application of holes. In this manner, the capacity design requirement can be satisfied by means of geometrical intervention.

One of the first applications was proposed by Roberts and Sabouri-Ghomi in 1992. They carried out a series of quasi-static cyclic loading tests on shear plates with centrally placed circular openings. All the panels exhibited stable S-shaped hysteretic loops, with strength and stiffness decreasing linearly with an increase in the size of openings.

Another interesting study was proposed by Bruneau et al. (2005), in which a shear panel with circular 200-mm-diameter openings has been statically tested (Figure 4.66).

At the end of this synthetic roundup on dissipative systems, another appealing solution was proposed by Hitaka and Matsui (2003), in which

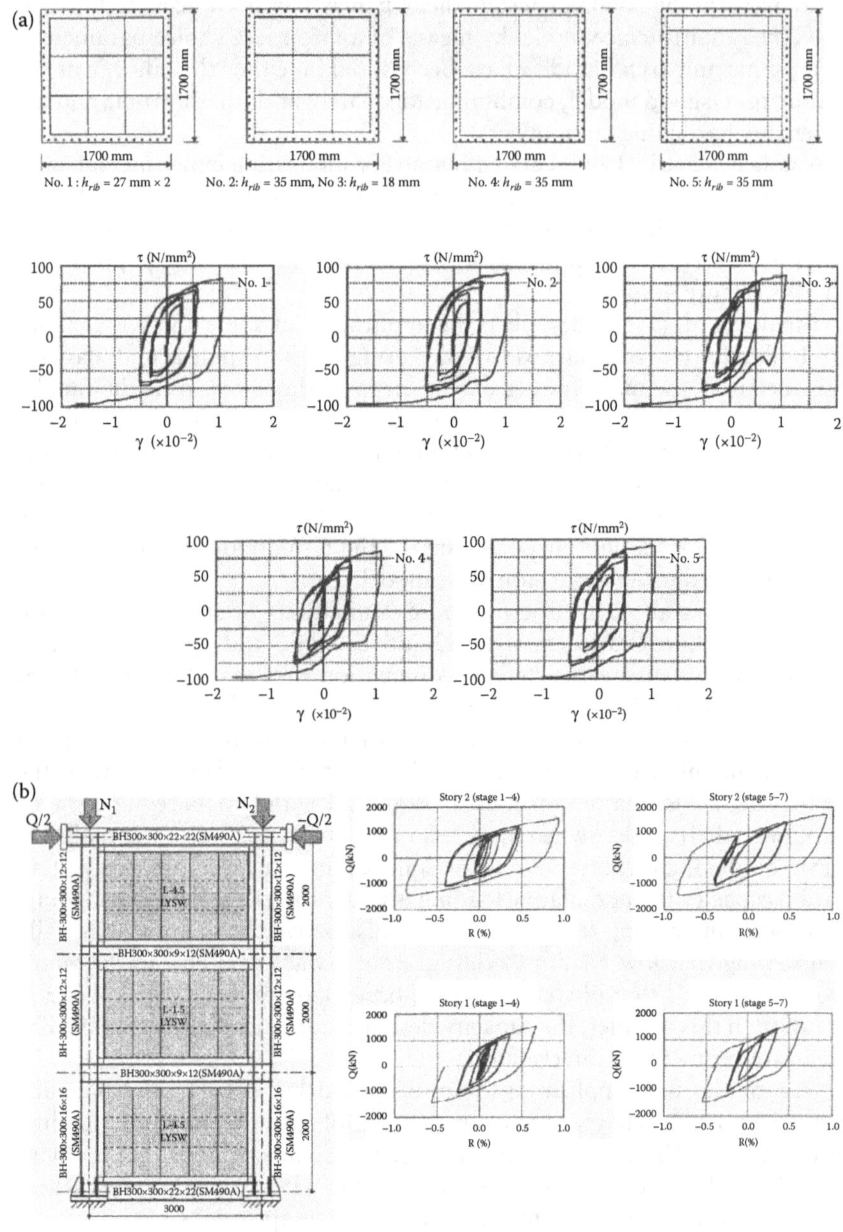

Figure 4.64 Experimental tests carried out by Nakagawa et al. (1996): (a) shear–load test on stiffened panels with different stiffener arrangements; (b) tests on a full-scale three-story steel frame equipped with LYSWs.

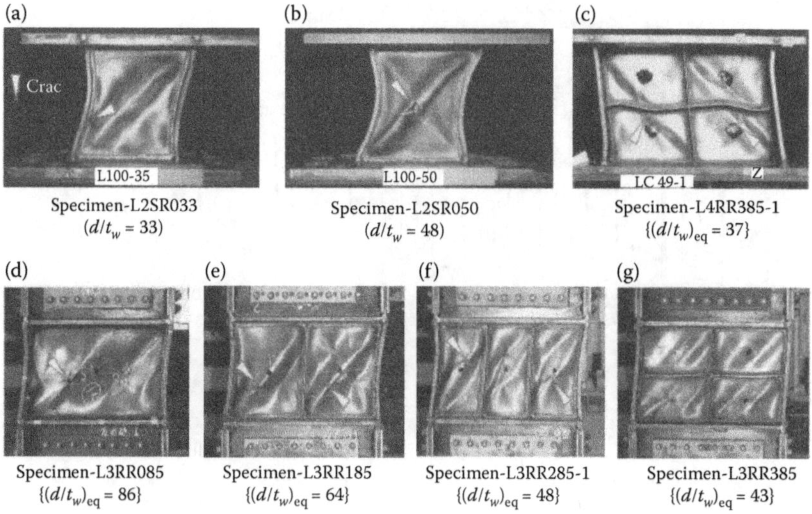

Figure 4.65 Experimental tests carried out by Tanaka et al. (2000).

MPSWs are equipped with shear panels with slits, without the need for heavy transverse stiffeners (see Figure 4.67). In this case, the shear panel behavior is determined by the portions of the plate included between the slits. These behave as a series of vertical links, which undergo large flexural deformations relative to their shear deformation. Forty-two silted plates of one-third of the full scale have been subjected to both monotonic and cyclic static lateral loads.

The obtained results have shown that shear plates with slits are generally characterized by large ductility and a fair dissipative capability, having

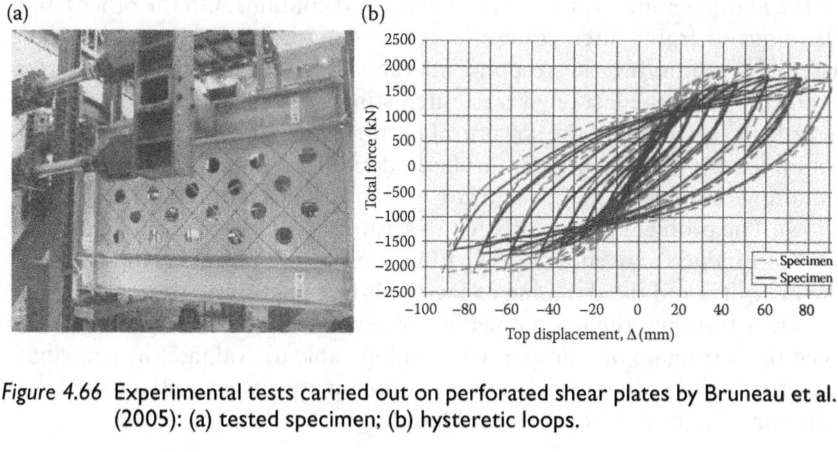

Figure 4.66 Experimental tests carried out on perforated shear plates by Bruneau et al. (2005): (a) tested specimen; (b) hysteretic loops.

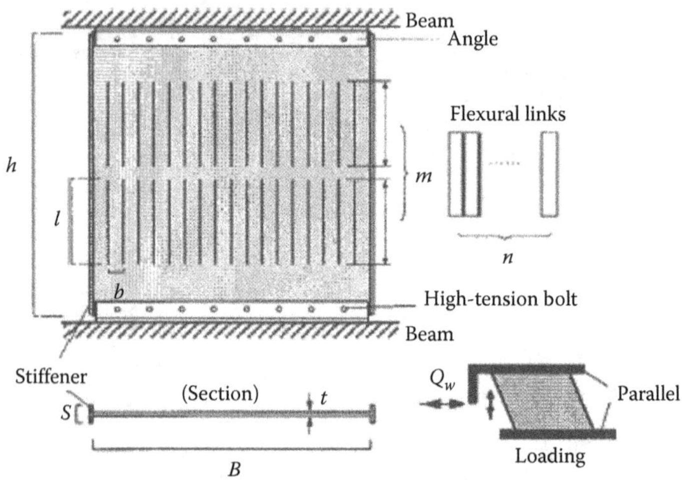

Figure 4.67 Shear panels with slits. (Adapted from Hitaka, T., Matsui, C. 2006: *Proceedings of the 5th International Conference on Behaviour of Steel Structures in Seismic Areas–STESSA 2006*, 241–246.)

performance levels that can be adjusted by changing the slits configuration (interval, length, and numbers of layers of slits). Moreover, it is worth noting that prior to the onset of out-of-plane deformations, the hysteresis response of the shear wall could be simply predicted by models based on the mechanisms of link bending and shear plate deformation.

4.3.3.3 Lightweight sandwich shear panels

For many years, the design of moment-resisting steel frames has been carried out by considering the resisting action of the whole structure provided by the main members only, that is, beams and columns. On the other hand, claddings-to-frame interaction affects the structural behavior of the whole assembly, modifying the response of the single components (De Matteis et al., 1996). In the last few years, interesting surveys have been performed in order to clarify how the structural performance of buildings under horizontal loads may benefit from properly designed external cladding panels, simultaneously developing enclosure and insulating purposes (De Matteis, 1998). The problem is currently felt in seismic areas, where such an interaction can induce new energy-dissipating sources, by changing the stiffness, the ductility, and the damping capacity of the structure.

In this framing, the main objective of several studies in recent years has been the setting up of a proper methodology able to evaluate the contributing effect of cladding panels and to contemplate such an effect in global structural analyses of steel-framed buildings.

An example is provided by the studies performed by Mazzolani et al. (1997) in the framework of the CREA research project, in which experimental and numerical analyses have been carried out (De Matteis and Landolfo, 1996, 1999a,b).

The performed experimental analysis dealt with the response of cladding sandwich panels inserted into pin-jointed frames submitted to monotonic and cyclic shear loading. Testing specimens are made of three subpanels, each one of dimension 100 × 250 cm. The whole panel assembly has an overall dimension of 300 × 250 cm. The columns of the frame are made of steel box sections, composed by welding two double channel profiles, and perfectly pin-joined to the ground. Horizontal members are made of two built-up steel channel sections. Beam-to-column connections are perfectly hinged. Three panel typologies are considered (Figure 4.68).

In all cases, sandwich panels are made of external steel sheets, with a thickness of 0.6 mm, and insulating polyurethane core having a total thickness of 40 mm.

In case of typology A, embossed sheets with slight stiffening ribs are used, while typologies B and C are composed by completely flat sheets. For types B and C, in order to improve the behavior of connections, panel edges are reinforced by means of internal cold-formed channel profiles, which are connected to the sheets through either a special 25 mm width biadhesive band all around the panel (typology B) or aluminum rivets of 6.3 mm diameter (typology C). The connections between the subpanels are made of steel self-drilling screws of 6.3 mm diameter and 200 mm spacing. As far as the panel-to-external frame connections are concerned, the shear diaphragm is fixed between the webs of the two channel sections composing the horizontal beam by means of steel bolts (Figure 4.69). Two rows of class 8.8 bolts, 8.0 mm diameter and 110 mm spacing, are used.

Both the monotonic and the cyclic loads have been transmitted to the top beam by a hydraulic jack (Figure 4.70), which is able to impose a maximum lateral displacement of 200 mm in monotonic tests and 100 mm

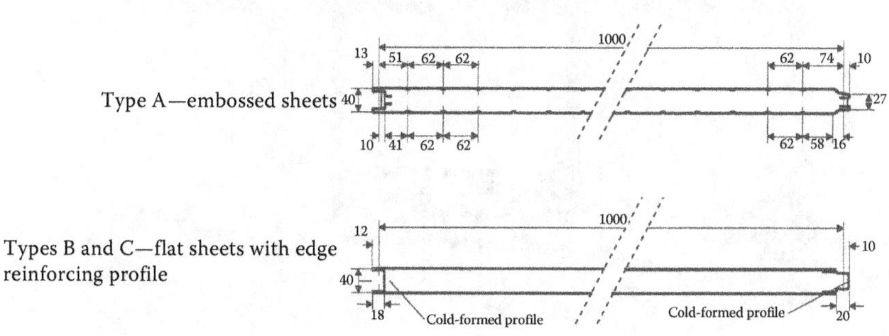

Figure 4.68 The sandwich panel typologies studied by Mazzolani et al. (1997).

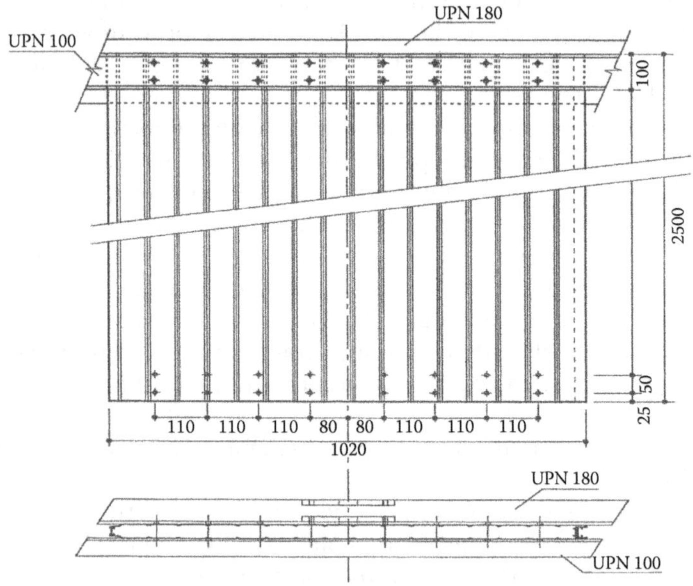

Figure 4.69 Panel-to-external frame connection. (Mazzolani, F.M., Landolfo, R., De Matteis, G. 1997: Analysis of the contributing effect of building panels on steel structure resistance to seismic and aeolian phenomena, Consortium CREA Technical final report. European Community Commission, Executive Committee F6 "Steel Structures," Agreement no. 7210–SA/421.)

Figure 4.70 The testing apparatus.

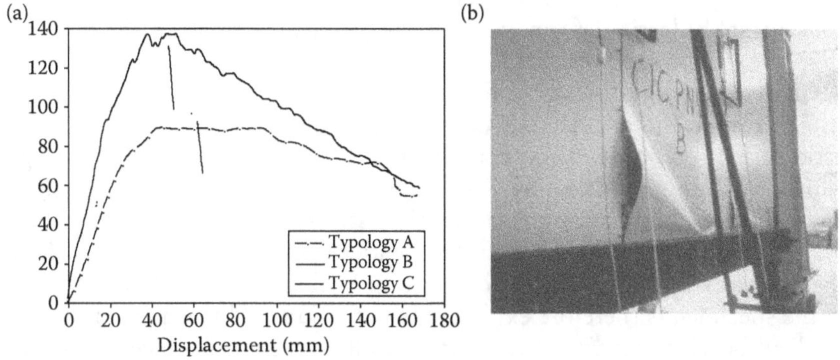

Figure 4.71 Experimental monotonic curves (a); collapse mode of the panel typology B (b).

in cyclic tests. Resistive transducers have been used for registering the displacements.

For each panel typology, two tests have been carried out under both monotonic and cyclic loading. The results obtained in case of monotonic loading, in terms of shear force versus shear panel deformation, are summarized in Figure 4.71a, where the average curves obtained for the three analyzed panel typologies are depicted.

The observed failure mechanisms are strongly related to the typology of the panel, being particularly influenced by the presence of the internal cold-formed profile (specimen typologies B and C).

For specimen typology A, the collapse has been caused by the failure of connectors. In particular, tearing of sheeting corresponding to both panel-to-panel screws and panel-to-external frame bolts is noticed. Bolted connections collapsed due to the stripping of the panel sheeting, starting from the ones located at the external corners of the assembly. As far as the screw connections are concerned, besides a remarkable deformation of the sheeting holes causing the slipping between adjacent subpanels, the shear breaking of some connectors in the central part of the connecting lines has been noted.

In contrast, in case of specimen typology B, the collapse was essentially characterized by the buckling of the panel facing, which exhibits out-of-plane deformations involving large portions of the external sheeting. This phenomenon involves the compressed diagonal of every subpanel composing the assembly, initiating from a slight wrinkling phenomenon at panel corners. Such behavior is mainly due to the insufficient resistance of the biadhesive tape (Figure 4.71b).

The observed failure mechanism is also confirmed by the corresponding load–displacement curve, which shows a sudden decrease of the load once that peak value is attained (see Figure 4.71a). After the buckling of the sheeting, the internal panel reinforcing profiles represent the sole load-bearing system, which suddenly collapsed due to the development of concentrated plastic deformations in the elements located in the vertical edge of external subpanels.

Finally, typology C exhibited the best behavior, providing the highest strength value and an adequate deformation capacity. The high restraining action exercised by the rivets partially prevents the buckling of the sheeting, which involves only little portions located at the corners of each subpanel, due to the wrinkling phenomena. In this case, the collapse is essentially due to the failure of the connector. In fact, with respect to panel typology B, a slighter stress redistribution after sheeting buckling occurred. The failure of the specimen is therefore exploited by a tension field mechanism, where the load-bearing capacity is concentrated along the tensile diagonal of the panel assembly and caused by the collapse of panel-to-external frame bolt connections located at the corner of the tension-diagonal field. Also in this case, the occurrence of plastic deformations involving the internal panel cold-formed profiles is detected.

As far as the cyclic behavior is concerned, test results essentially provide the same failure modes evidenced under monotonic tests. The conventional limit of the elastic range (F_{el}) and the corresponding displacement (Δ_{el}) are deduced on the basis of monotonic load–displacement curves. Cyclic responses obtained for the tested panel typologies are represented in Figure 4.72. The obtained hysteretic cycles are very similar to every typology, evidencing the reduced energy adsorption capacity of the panel assembly. The main reason for such behavior is the scarce performance of the connecting system, which exhibits large ovalizations at the sheeting holes. The slip-type behavior of the system is evident in the shape of reloading and the unloading branches, which therefore cause a remarkable "pinching" of the hysteretic loops.

As experimentally stated, shear diaphragms exhibit a loss of stiffness before the achievement of the maximum load. In fact, experimental results give evidence of the occurrence of significant nonlinear behavior beyond

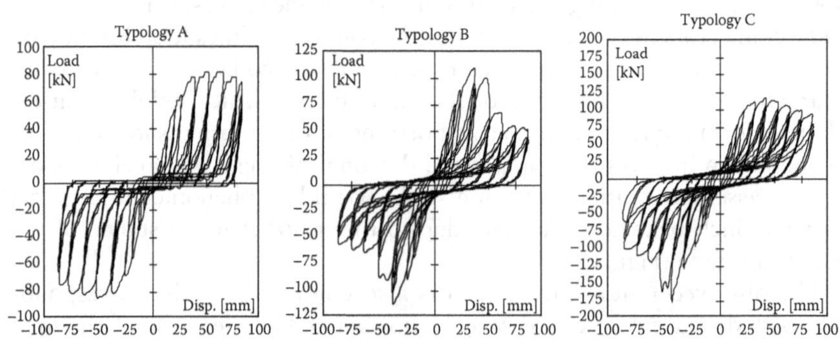

Figure 4.72 Cyclic test results.

a load level approximately equal to two-thirds of the maximum bearing capacity of the system. Besides, in order to exactly compute the energy adsorption capacity of the system, the behavior beyond the peak load has to be the object of investigation as well. In fact, in case of panels that do not exhibit a very brittle collapse mechanism, the plastic range could be considerable and the softening branch could account for the ductility of the system.

The examination of the obtained results points out that the global shear monotonic response of the panel is mainly characterized by three different paths: linear in the first stage of the loading process, nonlinear up to the maximum shear load F_{max} is reached, decreasing with a constant slope up to the collapse. By assuming the linear behavior limit at 2/3 F_{max} and by considering a parabolic trend in the range between such a limit and the maximum load, the mathematical equations of the multisection curve can be easily determined as follows (De Matteis and Landolfo, 2000):

$$0 \leq \Delta \leq \Delta_{el} \qquad F = k_0 \Delta \tag{4.22}$$

$$\Delta_{el} < \Delta \leq \Delta_{max} \qquad F = F_{max}\left[1 - \frac{1}{3}\left(\frac{\Delta_{max} - \Delta}{\Delta_{max} - \Delta_{el}}\right)^{n_0}\right] \tag{4.23}$$

$$\Delta_{max} < \Delta \leq \Delta_{ult} \qquad F = F_{max} - k_h(\Delta - \Delta_{max}) \tag{4.24}$$

where

$$k_0 = \frac{2}{3}\frac{F_{max}}{\Delta_{el}} \qquad k_h = \frac{F_{max} - F_{ult}}{\Delta_{ult} - \Delta_{max}}$$

$$\Delta_{rif} = \frac{3}{2}\Delta_{el} \qquad n_0 = 3\frac{\Delta_{max} - \Delta_{el}}{\Delta_{rif}}$$

The values of the fundamental parameters Δ_{el}, Δ_{max}, F_{ult}, F_{max}, and F_{ult} should be determined according to the results provided by the tests or reliable numerical simulations.

The application of the above mathematical model has been referred to tested sandwich panel typologies. In Figure 4.73, the comparison with the experimental results (in dashed lines) is shown. In the same figures, values for relevant parameters are also reported. The comparison shows a good capability of the proposed model to interpret the global behavior of the tested sandwich typologies.

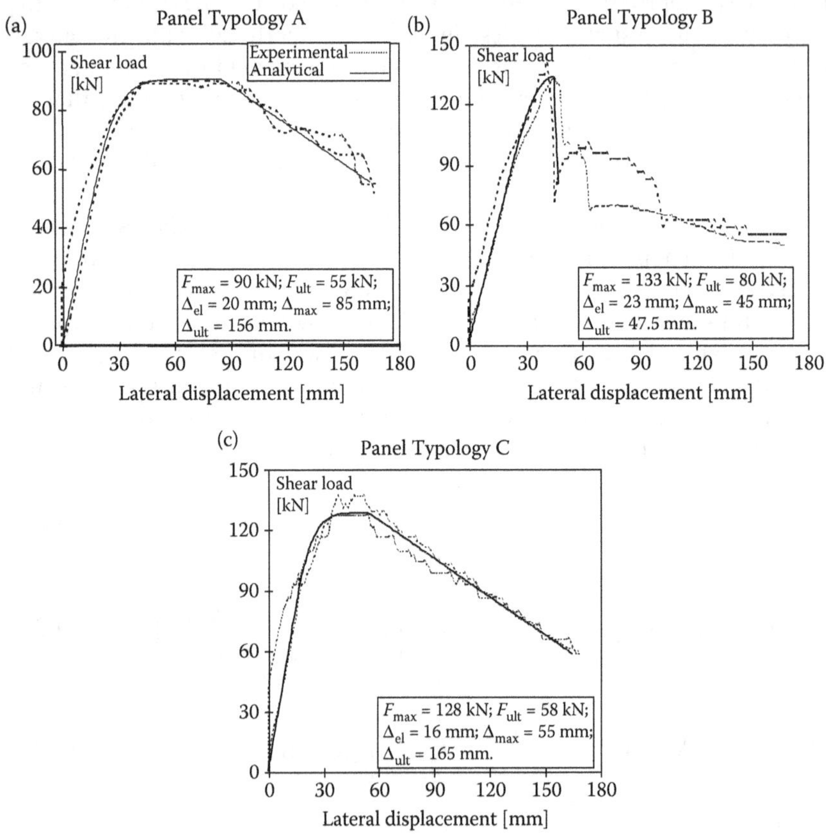

Figure 4.73 The application of the analytical model.

In particular, a proper selection of the fundamental parameters allows all the observed collapse mechanisms to be well reproduced:

1. Connector failure due to tearing of the thin sheets (panel Typology A)
2. Local buckling of panel sheeting with a sudden failure mechanism (panel Typology B)
3. Intermediate behavior with connector failure after local buckling of panel sheeting

In the first case, the model is able to account for the great displacements occurring at the maximum shear load, while in other cases it is able to reproduce the very brittle behavior of the system (2) and the one characterized by limited plastic excursions (3). Therefore, once the monotonic response of the panel has been established by either an experimental test or adequate numerical simulation, the above analytical model may be profitably used for a closed-form representation of the shear panel behavior.

In order to predict the monotonic response of the sandwich panel in shear, as an alternative to the experimental tests, a numerical FEM model, which is able to simulate the entire force–displacement relationship of a one-story shear wall system, has been set up (De Matteis, 2005). The model has been developed by means of the nonlinear code ABAQUS. It has been calibrated on the basis of experimental results concerning panel typologies A and C, which exhibited different collapse mechanisms. The panel typology B has been neglected due to its brittle failure mode, which makes such typology not advisable for structural purposes. In both the analyzed cases, numerical simulations have provided good results, emphasizing the model capability to properly interpret the actual behavior of the system with or without the panel sheeting instability phenomena.

The connection behavior has been represented through a suitable and properly developed analytical model based on a Ramberg–Osgood relationship. In particular, in order to describe the shear force versus shear displacement relationship of the connection, the following formula has been assumed:

$$\Delta = \frac{F}{k} + p_0\left(\frac{F}{F_0}\right)^n \tag{4.25}$$

where k is the initial stiffness, F_0 the reference load (usually F_0 is assumed as the conventional elastic limit), p_0 the residual displacement corresponding to F_0, while the exponent n interprets the shape of the curve and characterizes the postelastic behavior of the connection. Coefficient n has been determined by assuming as reference points the conventional elastic limit (F_0–Δ_0) and the point corresponding to the attainment of maximum connection-bearing capacity (F_c–Δ_c). The model has been calibrated with respect to some experimental results obtained by a wide experimental campaign (De Matteis and Landolfo, 1999a,b).

In order to apply the calculation procedure, the critical shear load of the system has been determined by assuming for sandwich core a finite space model with thickness equal to 40 mm, Young's modulus $E_c = 0.4$ N/mm^2, and Poisson's coefficient $\nu_c = 0.25$. The buckling analysis has provided critical load values equal to 70 and 80 kN for panel typologies A and C, respectively. The higher value for typology C is due to constraints exercised by internal reinforcing profiles on the unloaded panel edges.

In Figure 4.74, for both panel typologies A and C, the simulated F–Δ curves are provided. The comparison with experimental results emphasizes

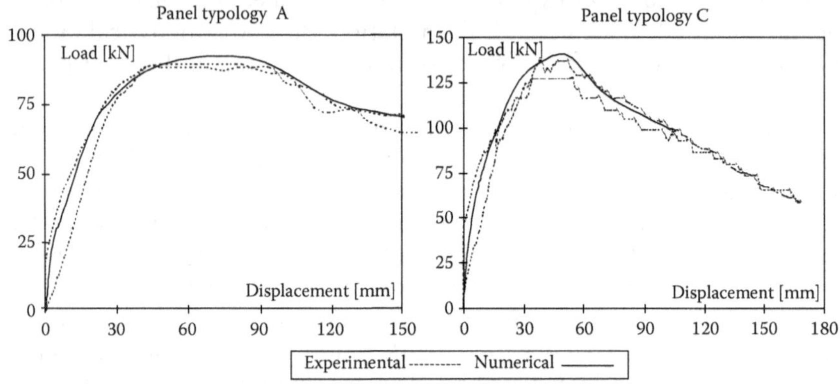

Figure 4.74 The application of the FE model.

the capability of the proposed finite element approach to interpret well the whole response of a sandwich shear wall system, also in case of the relevant buckling phenomena. In fact, all significant mechanical characteristics, namely, initial stiffness, ultimate strength, deformation capacity, as well as softening behavior, are correctly represented. Therefore, the proposed numerical approach seems to be a reasonable alternative to monotonic full-scale shear tests.

As far as the cyclic behavior is concerned, in the proposed model, a skeleton curve has been assumed as a reference curve, which is running from the moving point each time the displacement is greater than the maximum one ever reached during the previous loading history. The skeleton curve of the cyclic behavior rules the evolution of hysteretic loops by fixing the basic points of loading and reloading branches. In fact, all cycles are characterized by a strong "pinching" behavior. Basically, this is due to the connector slips, the most important damaging component being the panel sheeting in bearing. Beyond the elastic limit, the higher the shear load applied, the bigger the ovalization of sheeting holes produced. In addition to the elastic components, the subsequent unloading and reloading branches are mainly governed by the friction forces, which restrain free movements of mechanical fasteners inside plastic deformed holes. When contact between the connectors and sheeting is recovered, the main equilibrium path (skeleton curve) is reestablished.

In order to represent the hysteretic behavior, the unloading and reloading branches have been defined both for the positive and negative displacement ranges (Figure 4.75). The branches of the cyclic loops are described through the combination of three basic functions, constraining the curve to pass through a given point of the envelope P_i (F_i, Δ_i) as well as through the point

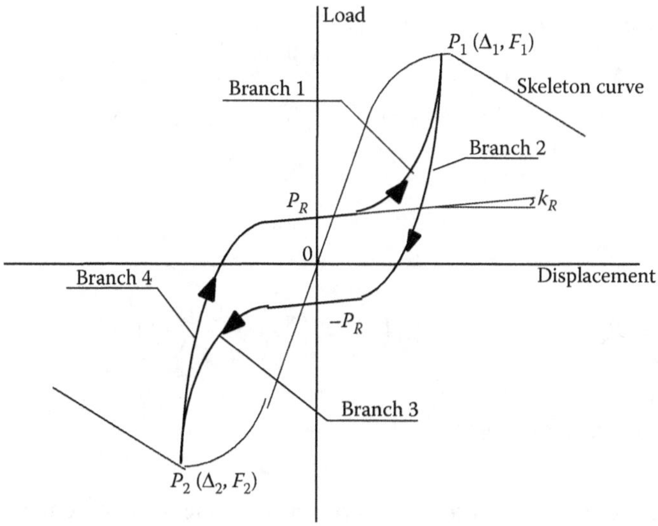

Figure 4.75 The proposed hysteretic model.

P_R at zero deflection. With reference to Figure 4.75, for each branch of the hysteretic loop, the following equations have been assumed:

$$\text{Branch 1:} \quad F = L(\Delta) + N_1(\Delta) \cdot \frac{\exp(a_1\Delta)}{(\Delta+1)^{SR}};$$

$$\text{with} \quad a_1 = \frac{\ln\left\{(\Delta+1)^{SR}\left[F_1 - L(\Delta_1)\right]/N_1(\Delta_1)\right\}}{\Delta_1} \tag{4.26}$$

$$\text{Branch 2:} \quad F = L(\Delta) + N_1(\Delta) \cdot \frac{\exp(a_2\Delta)}{(\Delta+1)^{(SR+SU)}} - 2P_R;$$

$$\text{with} \quad a_2 = \frac{\ln\left\{(\Delta+1)^{(SR+SU)}\left[F_1 + 2P_R - L(\Delta_1)\right]/N_1(\Delta_1)\right\}}{\Delta_1} \tag{4.27}$$

$$\text{Branch 3:} \quad F = L(\Delta) + N_2(\Delta) \cdot \frac{\exp(a_3|\Delta|)}{\left(|\Delta|+1\right)^{SR}} - 2P_R;$$

$$\text{with} \quad a_3 = \frac{\ln\left\{\left(|\Delta|+1\right)^{SR}\left[F_2 + 2P_R - L(\Delta_2)\right]/N_2(\Delta_2)\right\}}{|\Delta_2|} \tag{4.28}$$

$$\text{Branch 4:} \qquad F = L(\Delta) + N_2(\Delta) \cdot \frac{\exp(a_4 |\Delta|)}{\left(|\Delta| + 1\right)^{(SR+SU)}};$$

$$\text{with} \qquad a_4 = \frac{\ln\left\{\left(|\Delta| + 1\right)^{(SR+SU)} \left[F_2 - L(\Delta_2)\right] / N_2(\Delta_2)\right\}}{|\Delta_2|} \tag{4.29}$$

where

$$L(\Delta) = P_R + k_R \Delta; \qquad N_1(\Delta) = \frac{P_R}{2\Delta_1}\left(\frac{\Delta}{\Delta_1 - \Delta/2}\right)^2;$$

$$N_2(\Delta) = \frac{P_R}{2\Delta_2}\left(\frac{\Delta}{\Delta_2 - \Delta/2}\right)^2 \tag{4.30}$$

and F is the current shear load, Δ is the current shear displacement, P_1 and P_2 are the generic points of the skeleton curve in the positive and negative ranges, respectively, P_R is the load at zero deflection, and k_R is the slope of the slippage part in the unloading and reloading branches. Finally, SR and SU are two shape factors, which control the steepness of reloading and unloading branches, respectively. They may assume every positive value, while in case of a negative value, the compatibility with other parameters should be assessed.

The value of the parameters P_R, SR, and SU should be set on the basis of experimental results. In particular, P_R and k_R are related to the frictional resistance of the connecting system, while the coefficients SR and SU allow hysteretic cycles with different amplitude and shape to be considered. In particular, the steepness of the reloading branches increases as far as the SR value increases. On the other side, SU rules the difference in steepness between the unloading and reloading branches, and therefore it is strictly related to the amplitude of hysteretic loops.

In order to check the reliability of the proposed cyclic model, it has been applied to simulate some tests concerning panel typologies A and C. As a skeleton curve, the one corresponding to the envelope of the experimental cyclic curve, described by means of the analytical model presented above, has been considered. In Figure 4.76, the comparison of results is reported, being the experimental curves in dashed line.

The obtained F–Δ simulation curves are quite similar to the experimental ones, both above and below the peak resistance and for all ductility levels. Finally, it is important to evidence as the introduction of the shape factors SR and SU allow the panel behavior to be interpreted in different circumstances. In fact, both comparisons show the same degree of approximation, even if the response of the A and C panel typology is different in

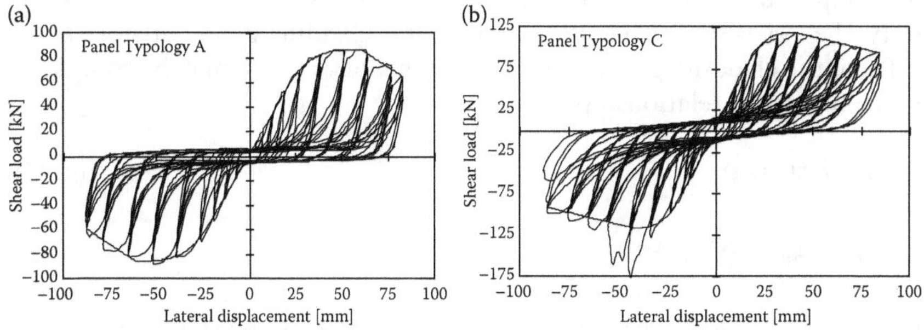

Figure 4.76 The application of the cyclic model: (a) panel typology A; (b) panel typology C.

the shape of the hysteretic curves. The better performance of panel typologies C, due to the collaboration of the internal reinforcing profile, is well simulated by using a coefficient SR = –1, contrary to the value SR = 1 used for typology A.

With reference to the whole structure, the attention is focused on the seismic design of moment-resisting steel frames endowed with structural lightweight cladding panels. The aim is to investigate the possibility to profit from their contributing effects on both the serviceability and the ultimate limit states of the structure. An original design procedure is based on the assumption that the external lateral force is resisted by a composite action. In fact, besides the intrinsic stiffness provided by the bare frame, the structural system benefits from the complementary stiffness provided by the cladding panels acting as vertical shear diaphragms. According to this assumption, in the first phase of the design procedure, in order to minimize the structural weight, the bearing frames are designed according to strength requirements only, that is, according to the conventional ultimate limit state provided by the seismic codes. Then, in the second phase, cladding panels are properly designed in order to provide the necessary complementary stiffness, allowing the covering of the discrepancy with the applied design limits for the interstory drift at the serviceability limit state.

The elastic stiffness to be supplied by the cladding panels has to be preliminarily evaluated at each story, assuming that the global external horizontal force is shared between the frame and the shear panels according to their relative stiffness. With this aim, the following interstory drift index can be defined:

$$\phi_i = \Delta_i / \Delta_{i,\mathrm{lim}} \tag{4.31}$$

where Δ_i is the actual interstory drift value at story i evaluated on the bare frame (neglecting the cladding action) and $\Delta_{i,\mathrm{lim}}$ is the design interstory drift limit imposed by the code at that story. The more the index F_i is close to unity, the less is the stiffness requirement for cladding panels at story i.

Then, the cladding panel stiffness at each story i ($k_{0,p,i}$) may be computed by the following relationships:

$$1 + c_i \psi_i = \phi_i \quad (4.32)$$

$$\psi_i = k_{0,p,i} / (V_i / \Delta_i) \quad (4.33)$$

where V_i is the design story shear load and c_i is the number of bays at the i story, where the cladding panels are inserted.

Once the required shear stiffness of the cladding panels is estimated at every story, the internal force distribution among the structural members and cladding panels can be determined by means of an elastic analysis under the predefined external action prescribed by the code. Then, the minimum elastic strength of the shear panels may be easily evaluated. In the applied procedure, it must be stated that the cladding panels should remain elastic under frequent earthquakes, that is, under lateral forces corresponding to the application of a serviceability limit state.

4.3.4 Pure aluminum shear panels

4.3.4.1 General concept

In the last decade, a significant research activity on dissipative pure aluminum devices has been carried out in Italy (De Matteis et al., 2007, 2008, 2009; Brando, 2007). In particular shear panels, made of pure aluminum and suitably reinforced by ribs in order to deliver shear buckling in the plastic deformation field, have been tested and numerically analyzed.

The implemented research has been carried out on three different levels. The first level is focused on the material behavior. The used alloy presents many innovative aspects (De Matteis et al., 2012), which need to be analyzed in order to correctly interpret the mechanical behavior under both monotonic and cyclic loading. An almost pure aluminum, namely, the alloy EN AW 1050A, has been selected. Moreover, in order to improve the mechanical features of the material, a special health treatment, consisting in an annealing, has been carried out. The choice of pure aluminum as the base material is justified by both the low yield strength and high ductility, which it is able to offer.

The second and third levels of research concern the cyclic behavior of shear panels. Two different panel typologies are taken into account by developing both experimental and numerical studies. The first panel typology, conceived for application in a framed structure with a full bay configuration, had in-plane dimensions 1500×1000 mm^2 and a thickness of

5 mm; the second panel typology, conceived to work in a bracing type configuration, had a square shape with side lengths of 500 mm and a thickness of 5 mm. In both cases, several types of plate slenderness ratios have been tested in order to evaluate their influence on the cyclic performance of the system. The obtained results have been used as the basis for the calibration of numerical models as well as to interpret the influence of the applied ribs on the structural performance of the tested systems.

In the following paragraph, an overview of the most significant results of this research is provided and discussed.

4.3.4.2 Innovation by pure aluminum

The proposed shear panels are fabricated with plates and stiffeners made of aluminum alloy AW1050 A H24, having a thickness of 5 mm. The chemical composition and nominal mechanical properties of the adopted material are listed in Table 4.2.

The preliminary uniaxial tensile tests on a wide quantity of specimens were not satisfactory, as they showed a higher conventional yielding strength (f_{02}) and a lower ductility than the nominal ones. To enhance the desirable ductile property and reduce the yield stress, a heat-treatment process has therefore been carried out. The specimens have been submitted to cycles of heat-treatment characterized by different phases with constant temperature, each one having a duration of 4 h. Successive tensile tests have shown a considerable improvement of ductile properties with a significant reduction in yield strength and an increase in ultimate elongation. In particular, a conventional yield stress f_{02} of 18 MPa and a ductility of around 50% have been registered.

For the sake of clarity, the comparison, in terms of strain–stress relationship, between the adopted pure aluminum alloy after the applied heat treatment, a common aluminum alloy (AW5154 A), and a typical low-strength steel is shown in Figure 4.77.

In addition, as already emphasized in previous studies (Nakashima, 1995, Katayama et al., 2000), the characterization of the cyclic behavior of a low-yield-strength material is very important, as the isotropic hardening

Table 4.2 Chemical Composition and Mechanical Properties of the Adopted Aluminum Alloy 1050A H24

Commercial denomination		*Impurities*
Aluminum AW 1050A		0.02% Cu, 0.40% Fe, 0.31% Si, 0.07% Zn, 0.02% TI, 0.02% others
	Mechanical Properties	
Tensile strength (MPa)	Yield strength (0.2% offset, MPa)	Elongation on 5 cm (%)
70–100	30–70	20–40

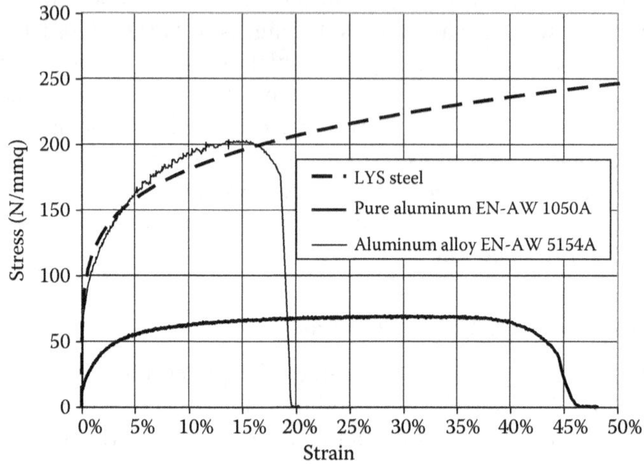

Figure 4.77 Stress–strain relationship between low-strength materials.

component could favorably influence the dissipative effect of the devices made of this type of material, since a reinflation of the hysteretic cycles is usually detectable. For this reason, specific tension-compression cyclic tests have been carried out on pure aluminum specimens equipped with a steel "jacket" able to inhibit out-of-plane deformations due to buckling phenomena in compression (see Figure 4.78).

The obtained results have shown an important dissipative behavior (see Figure 4.79a) characterized by full hysteretic cycles, a substantial isoresistance for each displacement level and the existence of an isotropic hardening component (see Figure 4.79b). However, as far as the cyclic degradation is concerned, the tests do not allow to yield final conclusions, since testing

Figure 4.78 Buckling-inhibited pure aluminum specimen.

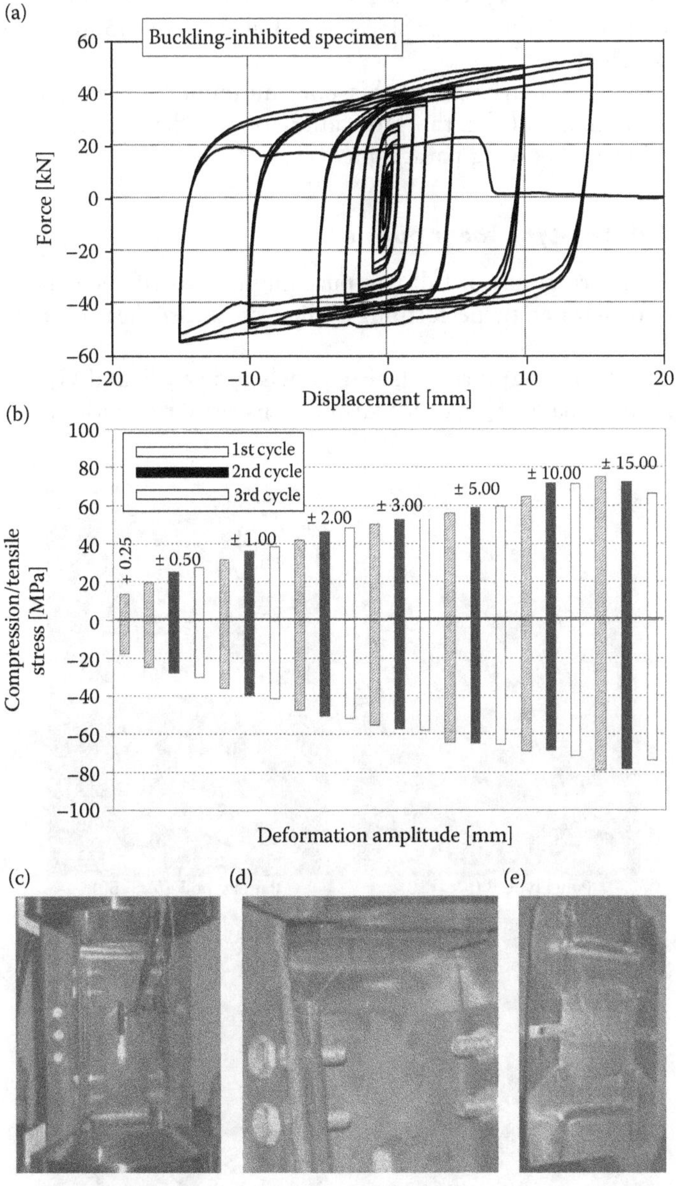

Figure 4.79 Buckling-inhibited pure aluminum specimen: (a) hysteretic behavior; (b) strength for each cycle; (c, d, e) collapse mode.

specimens showed unexpected local buckling phenomena, which invalidate the experimental results for displacements higher than +15 mm (see Figure 4.79c through e).

All the above remarks allow this pure aluminum to be considered as a convenient material for the fabrication of the devices, which will be described in the following paragraphs.

4.3.4.3 Full bay-type shear panels

The structural response of full bay pure aluminum-stiffened shear panels has been investigated by performing cyclic tests under shear on four different panels.

For the basic configurations (shear panels type B, F, and G; see Figure 4.80), stiffeners made of longitudinal and transversal ribs with a rectangular

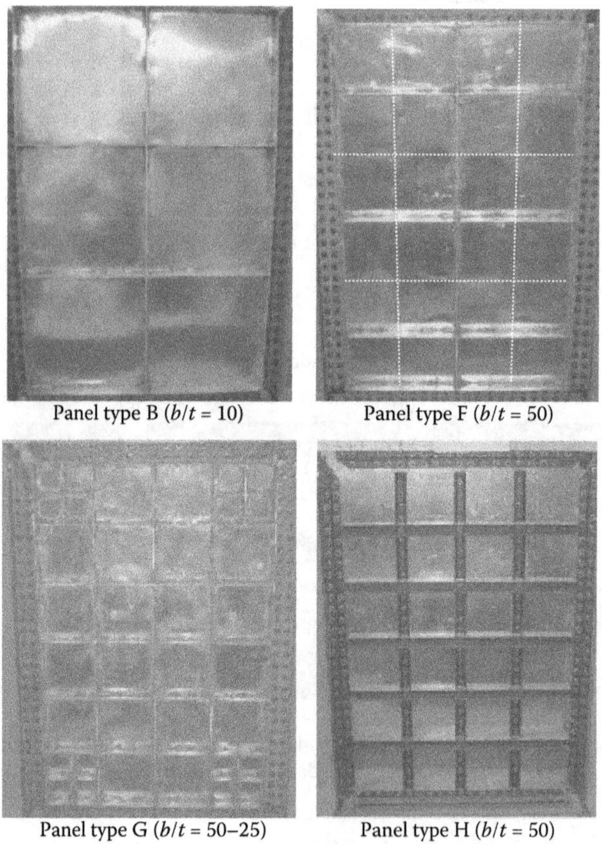

Figure 4.80 Geometrical configuration of tested full bay shear panels.

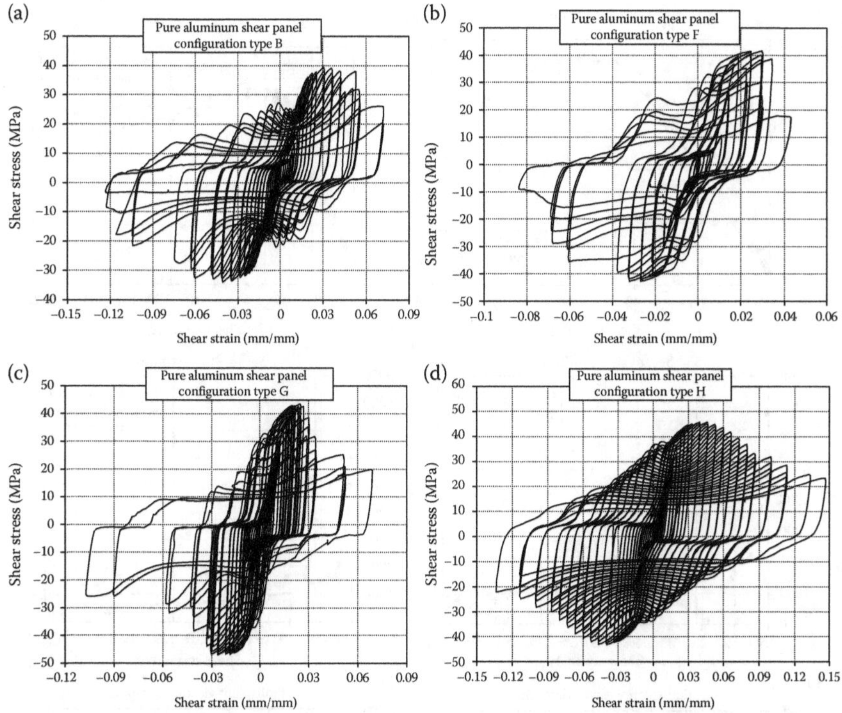

Figure 4.81 Hysteretic cycles provided by the tested full bay configurations.

cross-section (depth of 60 mm and thickness of 5 mm) have been used. Such stiffeners are connected to the base shear plate by means of welding. In addition, steel channel-shaped stiffeners, connected to the basic aluminum plate by means of bolted joints, have been considered (shear panel H).

The results of the experimental tests have been provided in terms of cyclic response (see Figure 4.81), determined as the relationship between the applied force (F) and the corresponding shear deformation (γ), putting in evidence that, as expected, the H typology, characterized by a lower slenderness, as well as by bolted stiffeners, performs in a better way. The response of tested specimens has also been characterized (see Figures 4.82 and 4.83) and compared in terms of a maximum hardening ratio (τ_{max}/τ_{02}, with $\tau_{02} = f_{02}/\sqrt{3}$), a secant shear stiffness (G_{sec}), and an equivalent viscous damping factor (ζeq) (De Matteis et al., 2008).

The experimental evidence highlights four main behavioral phases, whose corresponding shear strain range is also depicted in Figures 4.82 and 4.83:

- Phase 1: negligible buckling phenomena
- Phase 2: local buckling occurrence

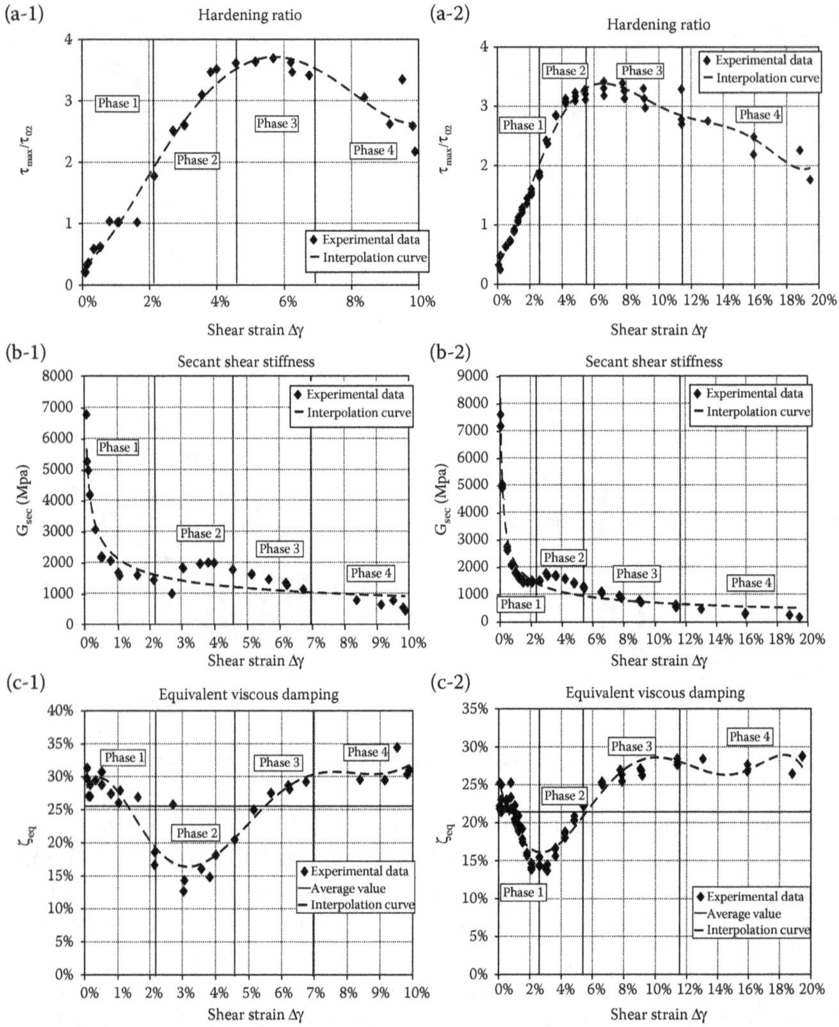

Figure 4.82 Cyclic performance of full bay shear panel type F (a-1, b-1, c-1) and type B (a-2, b-2, c-2).

- Phase 3: global buckling occurrence
- Phase 4: collapse phase

Based on the obtained results, the tested specimens have shown a very high ductility and a good structural performance in terms of strength, stiffness, and dissipative capacity, proving that the considered systems can be usefully adopted as passive seismic protection devices.

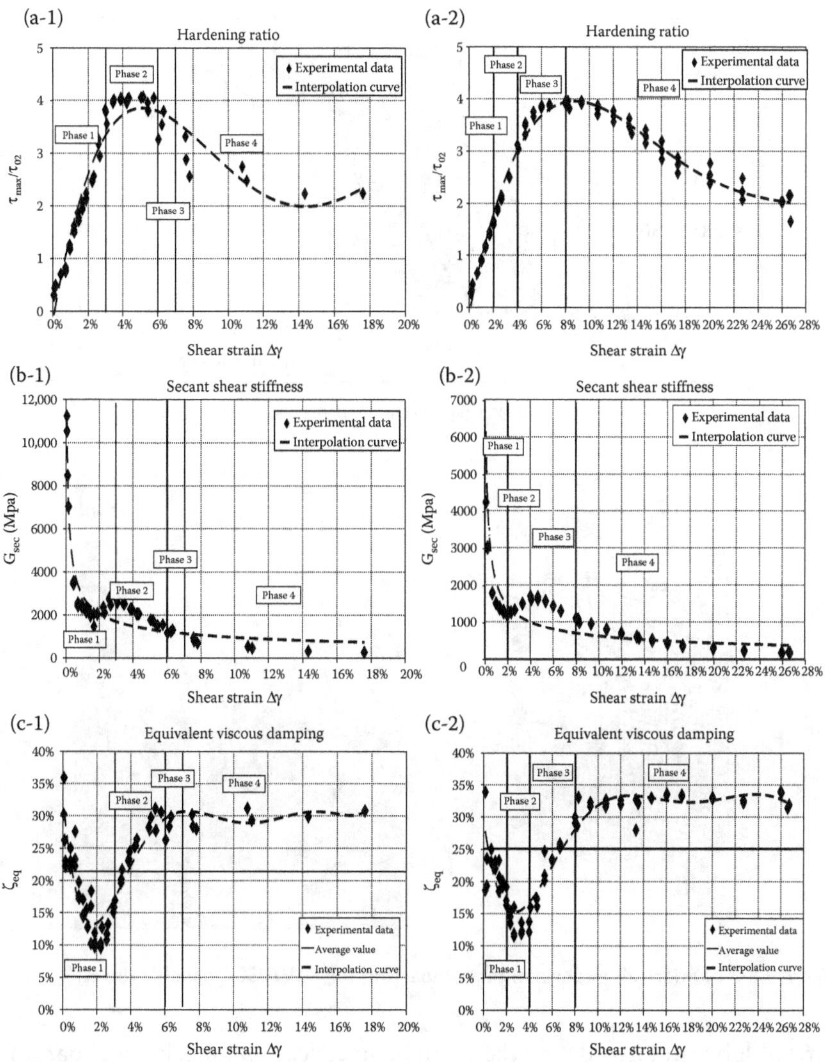

Figure 4.83 Cyclic performance of full bay shear panel type G (a-1, b-1, c-1) and type H (a-2, b-2, c-2).

In order to interpret the monotonic and hysteretic behavior of the tested systems, sophisticated FEM models have been set up. For example, the comparison between the numerical simulation and the experimental results is provided in Figure 4.84 for panel type B. It is clearly evident the proposed model is able to correctly interpret all the main behavioral phenomena of tested specimens. Therefore, the FEM model has been applied as a sort of

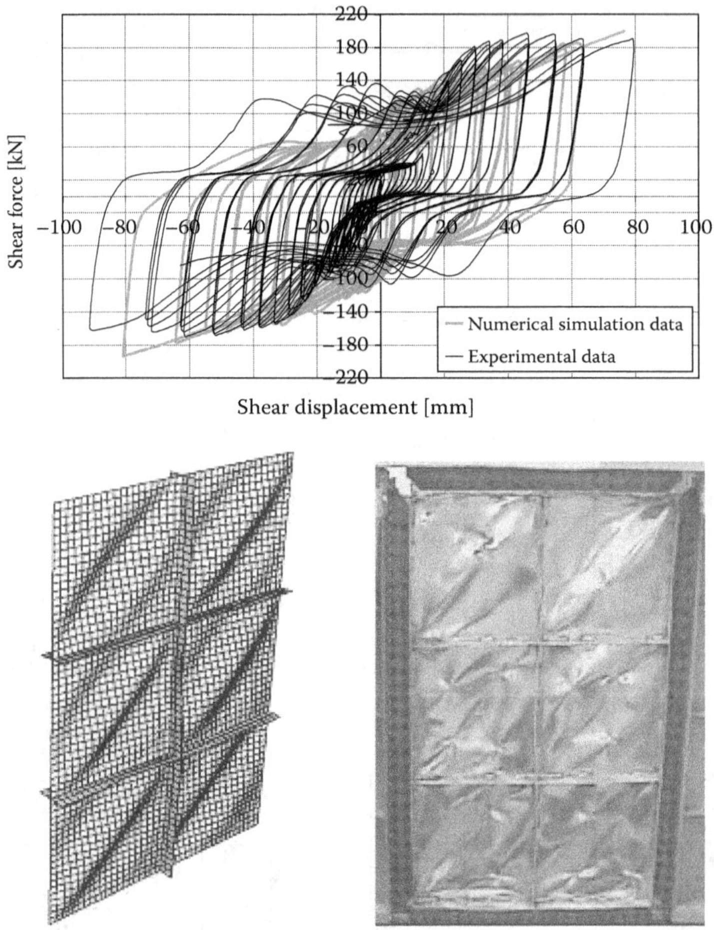

Figure 4.84 Numerical–experimental comparison for AW1050A panels type B.

virtual laboratory to define the main geometrical and mechanical parameters influencing the global response of the simulated systems.

In particular, a parametric study has been performed by varying the rib depth, determining the optimal geometrical configurations in terms of mechanical performance and fabrication costs. Major details on these aspects are given in Formisano et al. (2006).

4.3.4.4 Bracing-type pure aluminum shear panels

Bracing-type pure aluminum shear panels (BTPASPs) have also been proposed as an alternative and attractive solution for the passive protection

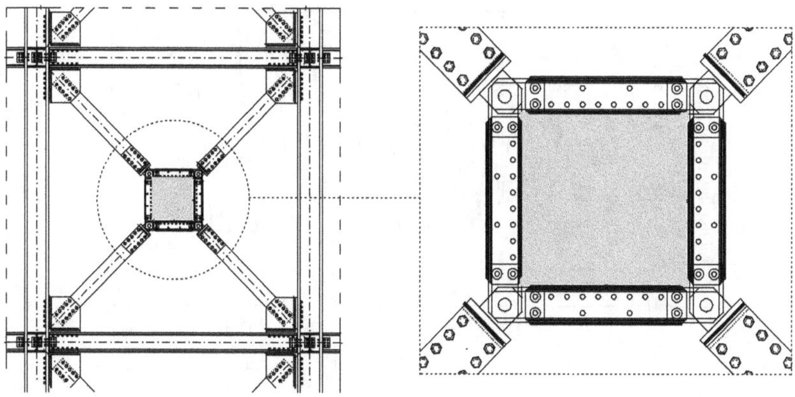

Figure 4.85 Steel frames equipped with BTPASP.

of new and existing structures (Figure 4.85) under low-intensity seismic events. In fact, on the basis of the aforementioned obtained results, full bay aluminum shear panels showed a good performance, in terms of both energy dissipation and damping capability, for medium-large lateral displacements, while some slipping phenomena have been observed for small inter-story drift levels. Thus, a rational reduction of the side length, so as to establish a favorable ratio between the interstory drift displacement of the relevant primary structure and the shear deformations of panels, has been implemented.

A wide experimental campaign on diagonally loaded BTPASPs, involving prototypes with different aspect ratios, is currently in progress, aiming at assessing the influence of the main behavioral parameters on the response of the system (De Matteis et al., 2011). The tested shear panels are characterized by global dimensions of 500 × 500 mm and a thickness of 5 mm. They are equipped with welded rectangular-shaped ribs, equally placed on the two faces of the panels, whose depth is 60 mm and which are made of the same material and having the thickness of the basic plates.

The testing apparatus is composed by a pin-jointed steel framework and linked to the panel edges by means of tightened steel bolts. Tested specimens are subjected to diagonal cyclic forces, according to the displacement history shown in Figure 4.86, where the ordinate indicates the applied diagonal displacement.

The cyclic response of two configurations, characterized by the *b*/*t* value equal to 100 ("type 1"), 50 ("type 2"), 33.3 ("type 3"), and 25 ("type 4"), respectively (see Figure 4.87), has already been critically studied.

The obtained results showed a good hysteretic performance, which was influenced by different collapse modes depending on the applied stiffener

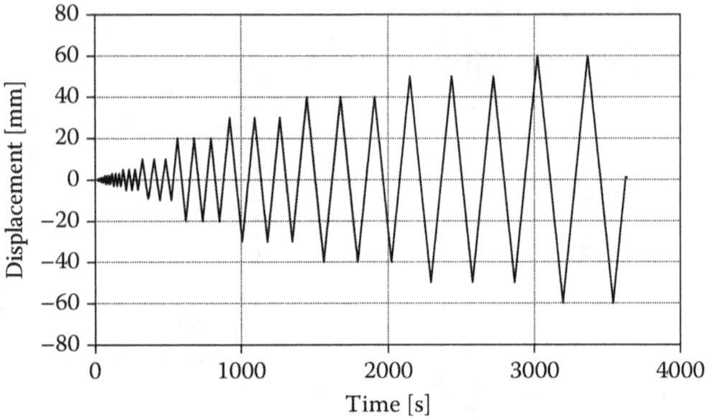

Figure 4.86 The displacement history on the tested specimens.

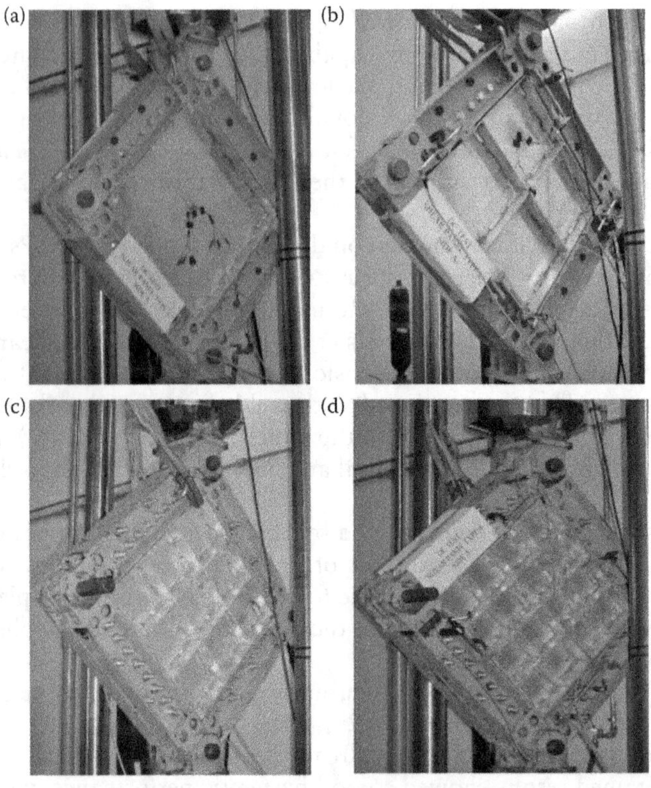

Figure 4.87 BTPASPs configurations type 1 (a), type 2 (b), type 3 (c), and type 4 (d).

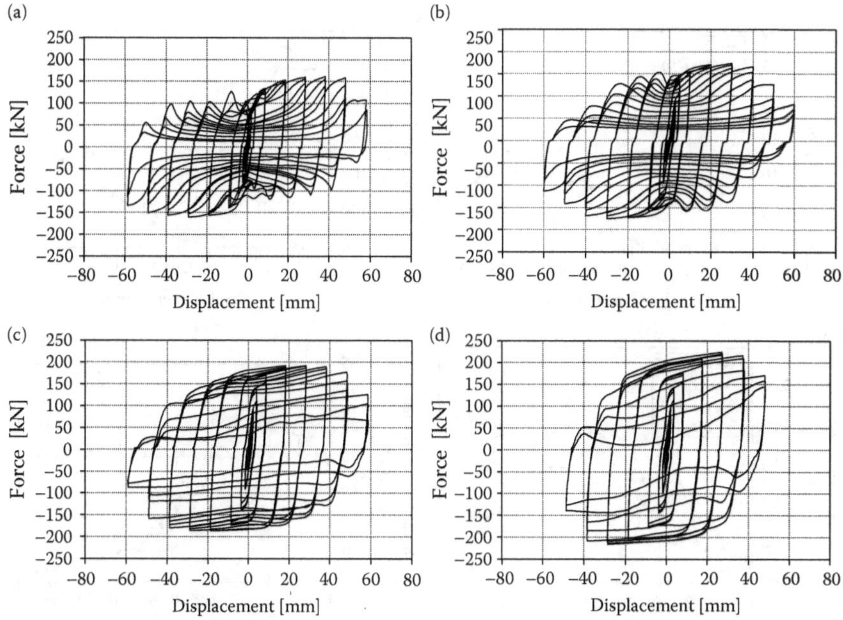

Figure 4.88 **BTPASPs hysteretic cycles: type 1 (a), type 2 (b), type 3 (c), and type 4 (d).**

configuration. In fact, the internal rib system acts as a sort of framework axially stressed, providing a resistant contribution to the panel and, therefore, a larger stress in the connecting system, which usually represents the weakest component of these devices.

In Figure 4.88, the obtained hysteretic cycles for the tested specimens, together with the relevant collapse modes, are provided, while in Figure 4.89 the dissipated energy per cycle, the cumulated dissipated energy, the equivalent viscous damping ratio, the hardening ratio, and the secant stiffness are illustrated.

The obtained results clearly emphasize that both panel configurations provide a good hysteretic performance, with fat hysteretic cycles also for high deformation levels. It is worth noting that the higher values of the equivalent viscous damping factor (about 50%) have been achieved for large shear strains.

It has also to be observed that the two tested BTPASPs "type 3" and "type 4" provided substantially the same dissipative behavior, even though the configuration "type 4" exhibited a larger peak strength. In contrast, it presents a lower ductility due to a premature collapse of the connection system, which is also highlighted by a quick reduction of the equivalent viscous damping factor starting from a diagonal displacement value of 50 mm. Hence, shear panel "type 3" seems to represent the optimum

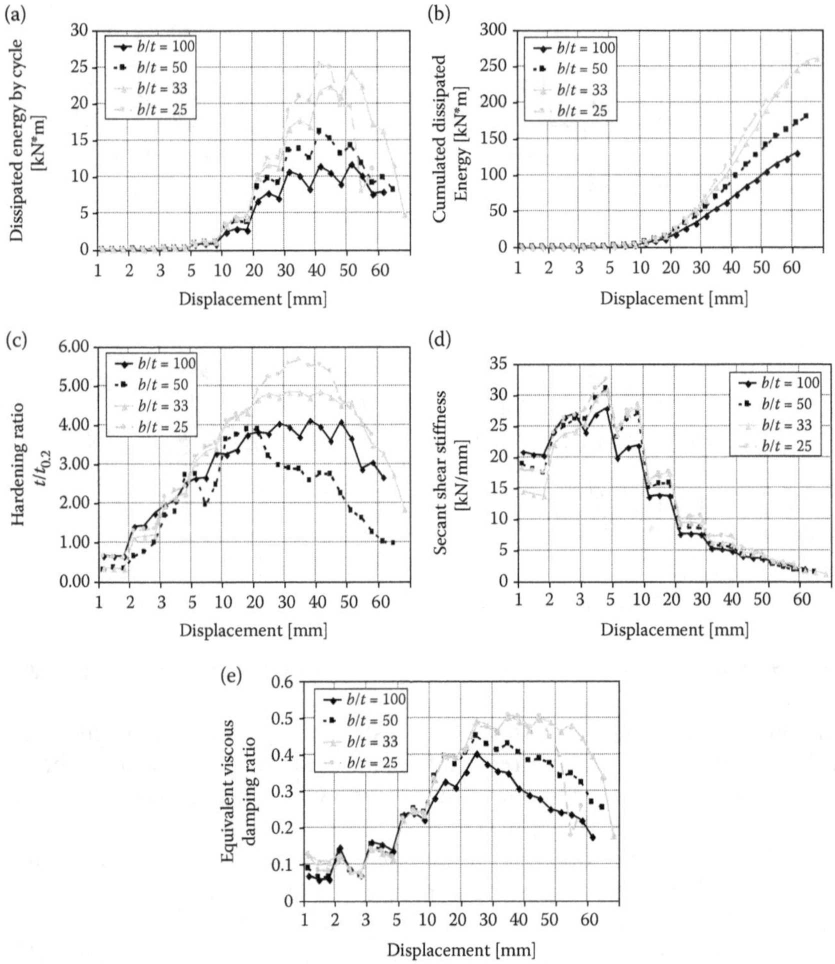

Figure 4.89 BTPASPs experimental results in terms of (a) dissipated energy per cycle; (b) cumulated energy; (c) hardening ratio; (d) secant shear stiffness; and (e) equivalent viscous damping factor.

configuration as a good compromise between reduced fabrication costs and good hysteretic behavior.

As for full bay-type shear panels, also in this case, FEM numerical models have been set up on the basis of experimental analyses (Brando and De Matteis, 2012). The four numerical models used for the bracing-type shear panels are shown in Figure 4.90.

According to the experimental layout, an external diagonal displacement has been statically applied to the top beam of the external pin-jointed frame

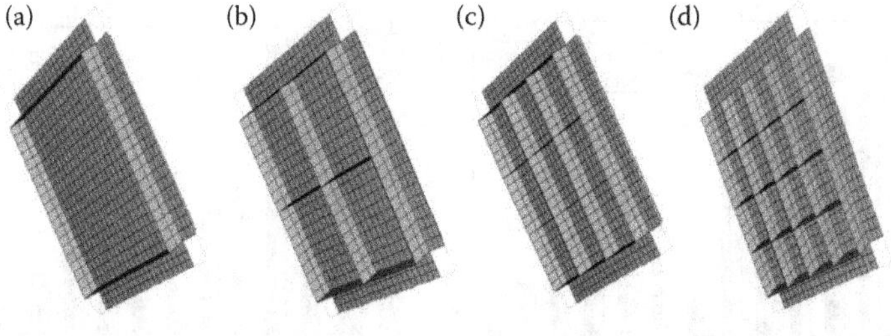

Figure 4.90 The BTPASPs FEM numerical models: type 1 (a), type 2 (b), type 3 (c), and type 4 (d).

of the FEM models. In Figure 4.91, the comparison is provided in terms of hysteretic cycles. It is to be noticed that only diagonal displacement demands ranging from −40 mm to +40 mm have been taken into account. In fact, when larger displacements are attained, the system response is influenced by both the failure of the perimeter connecting system and the fracture of the base plate, which are not contemplated in the numerical model. These aspects are evident from a careful inspection of the experimental hysteretic loops, from which it is possible to observe a contraction of the ±40.0 mm second and third cycles.

Figure 4.92 shows the comparison in terms of secant stiffness, while in Figure 4.93 the equivalent viscous damping measured at each shear strain demand is considered. The obtained results prove the reliability of the numerical model, which is able to capture all the main behavioral aspects of the system: namely, the strength, the stiffness, and dissipative features, including the pinching effects due to the buckling phenomena.

It is also evident that the initial slipping phenomena, which are unavoidable for the practical tolerance in every steel structure and whose entity has

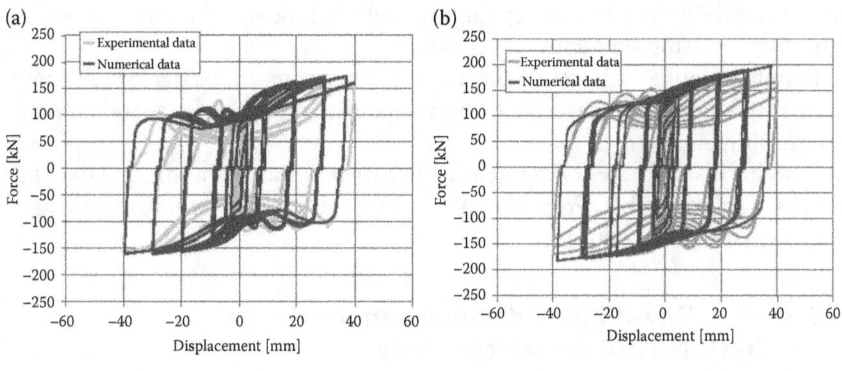

Figure 4.91 Comparison between numerical and experimental results in terms of BTPASPs hysteretic cycles: shear panel type 1 (a) and type 2 (b).

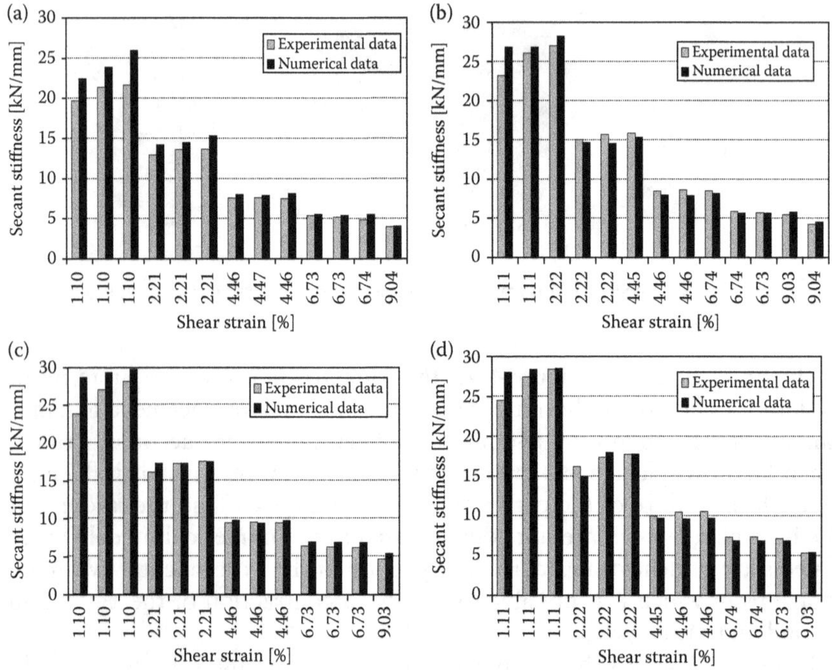

Figure 4.92 Comparison between numerical and experimental results in terms of BTPASPs secant stiffness: type 1 (a), type 2 (b), type 3 (c), and type 4 (d).

been interpreted during the calibration procedure of the numerical model, lead to a degradation of the dissipative response of the system for small displacement values.

On the other hand, for medium–high shear deformation levels, the dissipative capability reaches its maximum level, with peak values of the equivalent viscous damping factor of 40–45%.

Finally, Figure 4.94 also shows a comparison in terms of ultimate deformed shapes. For the sake of brevity, only two of the four studied panels are shown.

Also in this case, a good agreement between numerical and experimental results is recognizable. From the same picture, it is also possible to evidence the ultimate stress values.

4.3.5 Buckling-inhibited shear panels: A new hysteretic damper typology

An innovative configuration of the dissipative shear panel has been recently introduced (Brando et al., 2013). In this proposed new system, the main

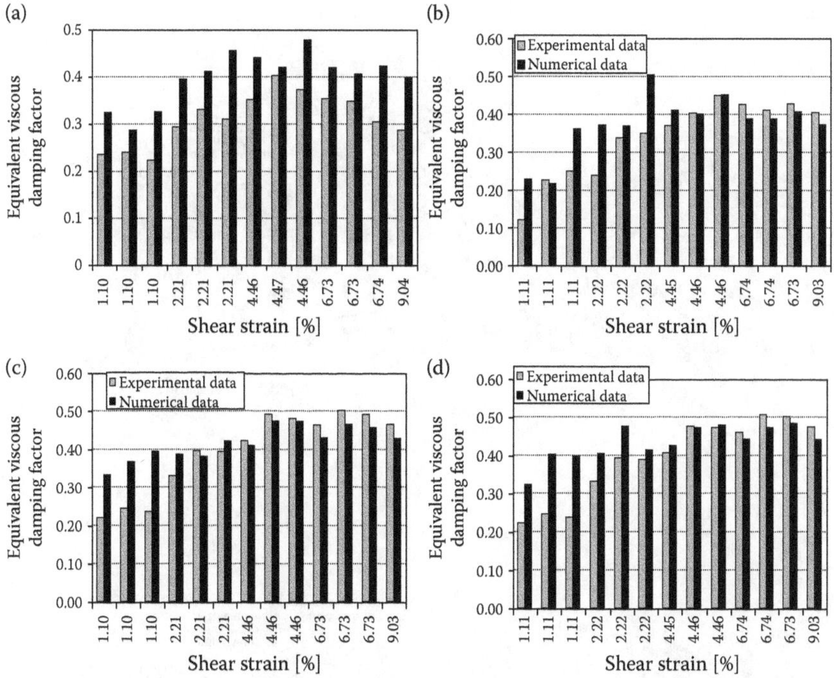

Figure 4.93 Comparison between numerical and experimental results in terms of secant stiffness: type 1 (a), type 2 (b), type 3 (c), and type 4 (d).

buckling modes of the base shear plate are inhibited by the application of external devices placed in parallel to the base plate itself and not directly loaded by the external forces.

In detail, in order to prevent shear buckling of the basic plate by restraining the out-of-plane displacements, two different solutions have been proposed. The first solution represents a "partially" buckling-inhibited

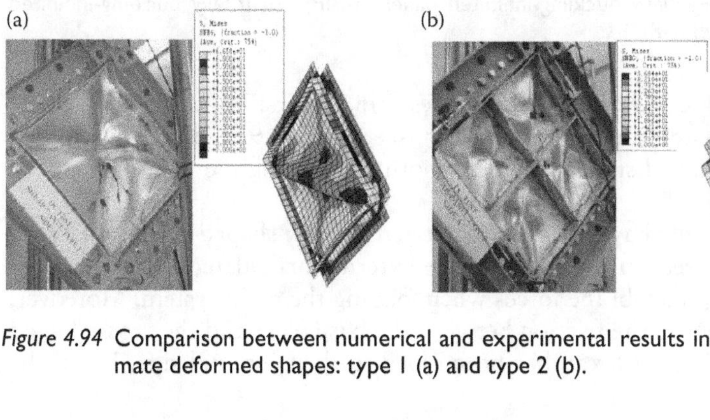

Figure 4.94 Comparison between numerical and experimental results in terms of ultimate deformed shapes: type 1 (a) and type 2 (b).

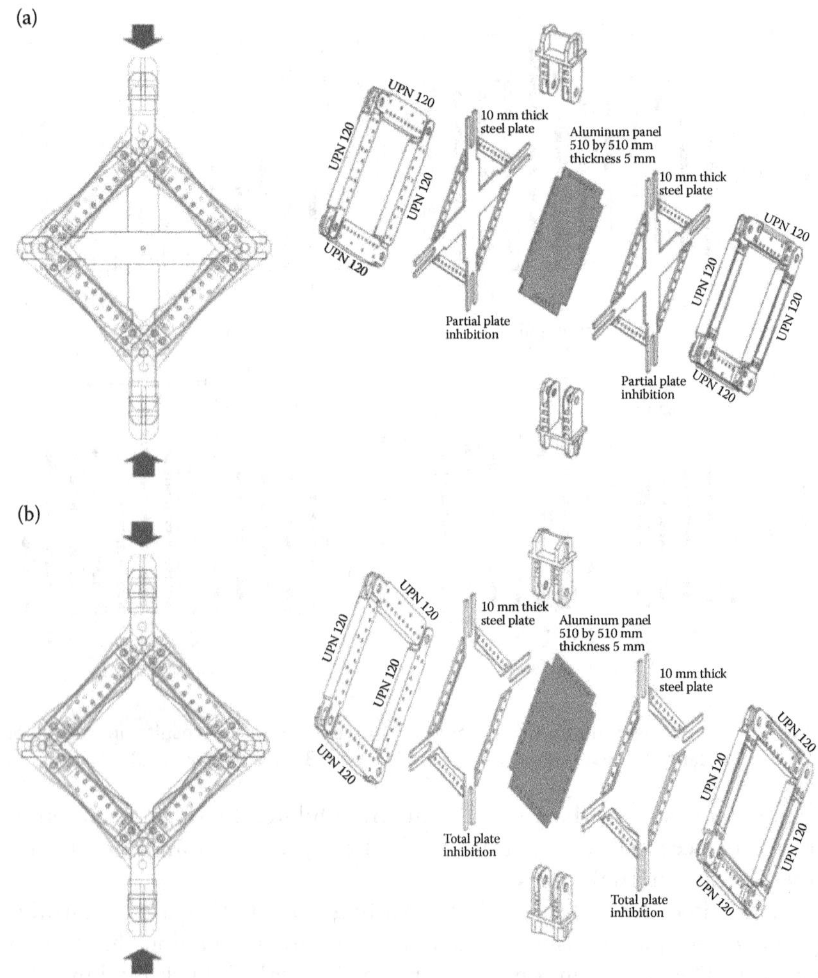

Figure 4.95 (a) Partially buckling-inhibited panel (p-BIP); (b) totally buckling-inhibited panel (t-BIP).

panel (p-BIP), conceived in order to restrain the first fourth critical modes of the base plate. It has been obtained (see Figure 4.95a) by arranging two thick cross-shaped steel elements at both sides along the diagonals of the plate.

These elements have been characterized by fork-shaped slotted end connections centered on the hinge of the external articulated frame in order to not develop membrane forces when loading the main system. Moreover, in order to reduce the friction between the base plate and the cross-shaped elements, a sheet of lexan has been glued to their internal side. It is to be pointed out that the partial buckling inhibition devices allow some secondary buckling phenomena developing along the medians of the triangular not restrained portions of the base plate.

The second solution represents a "totally" buckling-inhibited panel (t-BIP). It is conceived in order to restrain possible out-of-plane displacements of the entire base plate. The external devices, constituting the restraining system, are two octagonal-shaped steel plates, which are characterized by a thickness of 10 mm and are able to cover almost the entire aluminum plate (Figure 4.95b). Also in this case, a lexan sheeting has been employed in order to reduce the friction between the parts; in addition, slotted end connections have been used to accommodate in-plane movement of the buckling-inhibiting plate.

In order to prove the effectiveness of the above systems, experimental tests have been carried out on the two prototypes shown in Figure 4.96. These have been cyclically loaded by a diagonal force, according to the displacement history given in the ECCS provisions.

The obtained hysteretic responses are shown in Figure 4.97 in terms of shear stress–shear strain relationship. The large loops testify to the high dissipative capacity of the tested panels, as also for the high shear demands. Significant strength degrading effects can be registered only beyond a diagonal displacement of ±40.00 mm, corresponding to a shear strain demand of about ±9.00%.

The comparison between the obtained outcomes evidences that the specimen "t-BIP" performs better than the "p-BIP" one, the latter being affected by some secondary buckling phenomena developed during the

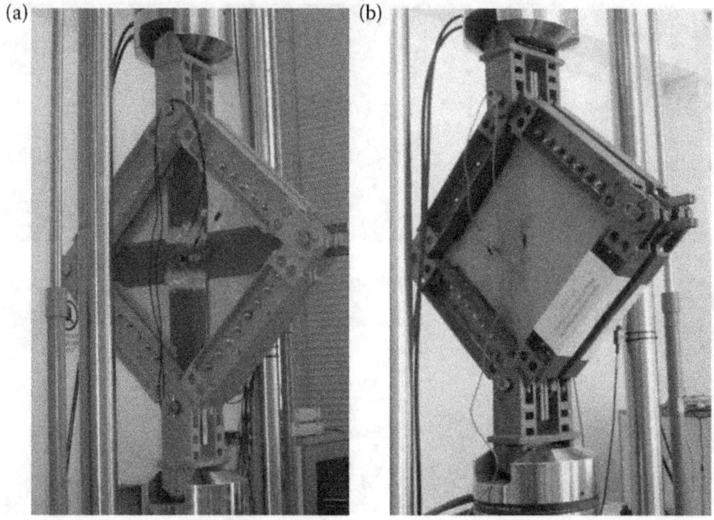

Figure 4.96 The tested partially (a) and totally (b) buckling-inhibited shear panels.

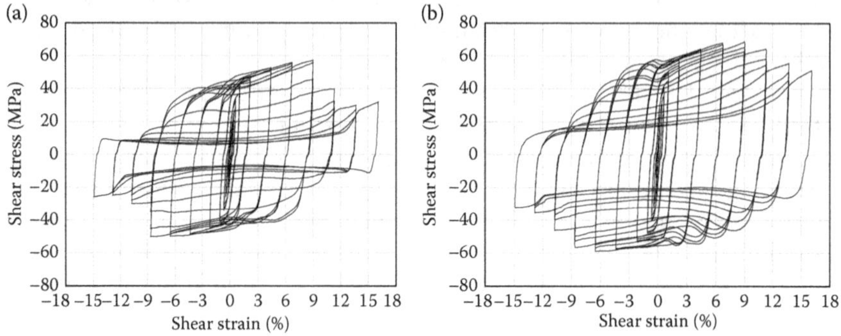

Figure 4.97 The obtained hysteretic cycle: (a) p-BIP and (b) t-BIP.

test. In addition, a confinement adjustment effect can be observed for the "t-BIP" specimen. This is similar to the one related to the "compressive adjustment factor" already evidenced for BRBs (AISC/SEAOC, 2001), leading to a general increase in strength for large deformation.

The collapse of the system arises when a diagonal displacement of ±60 mm (shear strain of 11.37%) is reached (see Figure 4.98).

At this stage, a pure shear failure is evident, with fractures completely developed along the edges of the plate, without plate tearing in the center of the panel and without any damage of the perimeter connecting system. The last remark allows us to state that the connections designed to not slip are properly dimensioned.

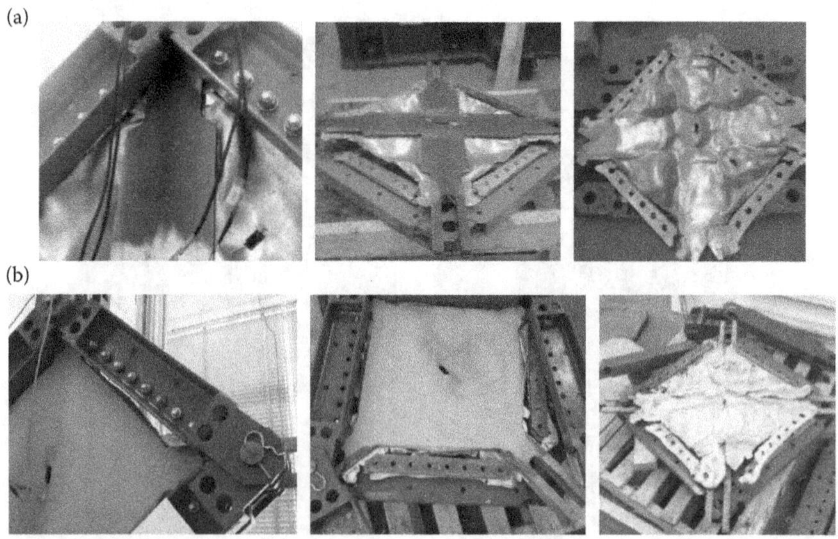

Figure 4.98 Collapse modes of tested specimen: (a) p-BIP and (b) t-BIP.

On the basis of the above tests, numerical models have been developed in order to carry out parametrical analyses. The outcomes of these analyses are going to be finalized. These will serve to provide useful design rules to be adopted in modern codes and guidelines.

4.3.6 Retrofitting of existing RC structures

The possibility of using metal shear panels for seismic retrofitting of existing buildings has been evaluated within the already-mentioned ILVA-IDEM research project, which developed experimental tests carried out on a full-scale gravity-load designed RC building, in order to evaluate different innovative seismic retrofitting techniques (Mazzolani, 2006). The original RC building is located in the Bagnoli district of Naples, where the ancient industrial site of Italsider was settled. It is a two-story building erected during the 1970s, serving as an office building. To perform the testing activity, it was preliminarily deprived of internal and external walls (Figure 4.99a) and then cut at both floors in order to obtain six separate structural modules (Figure 4.99b). In this section, the attention is paid to the use of both aluminum alloy and steel shear panels used as a retrofitting system of the substructure highlighted in Figure 4.99a.

It is worth noting that the considered RC substructure had been previously tested in transversal direction for evaluating the effectiveness of SMA brace. Therefore, it was affected by serious damages at the ends of columns due to the presence of cracks and the instability of the steel bars. After repairing the existing cracks by means of epoxy-resin injections, a couple of transversal X-braces were introduced in order to avoid the transversal sway of the first floor, when the structure is loaded in the longitudinal direction (Figure 4.100a). The contiguous RC substructure, placed beside the one under examination, was used as the retaining structure by inserting adequate steel V-shaped bracings (Figure 4.100b).

(a) (b)

Figure 4.99 The original RC building after the partition removal (a) and subdivision in separate modules (b).

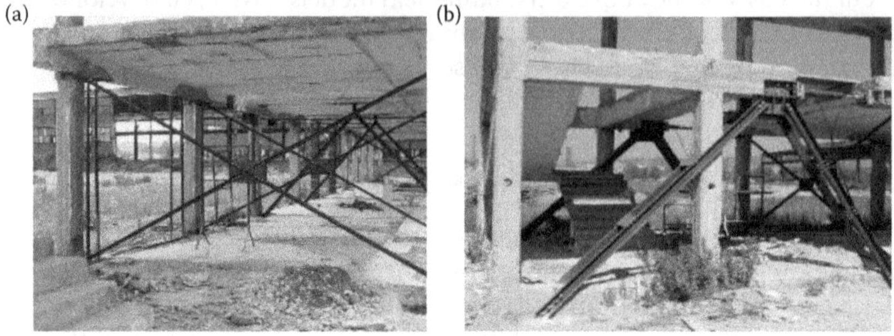

Figure 4.100 (a) X and (b) V bracings, as stiffening systems of the bare RC modules.

Hence, the bare RC structure has been preliminarily tested in a longitudinal direction, without reaching the collapse of the system, by means of a preliminary cyclic test, by using two hydraulic jacks able to apply a total force of 300 kN at the first floor of the structure.

The obtained test results provided useful information on the ultimate strength and the exact lateral stiffness levels offered by the bare RC structure, as it is evident from Figure 4.101, where the envelope curve of the cyclic test is depicted.

Such findings have been used for finalizing the retrofitting design of the substructure, according to the prescribed prerequisites, which should be guaranteed by the combined system (RC frame and metal shear panels).

The retrofitting design has been developed in the framework of the performance-based design methodology (Mistakidis et al., 2006), according to the procedures of the ATC-40 guidelines (1996). Therefore, the required strength and stiffness contribution provided by the shear panels have been defined,

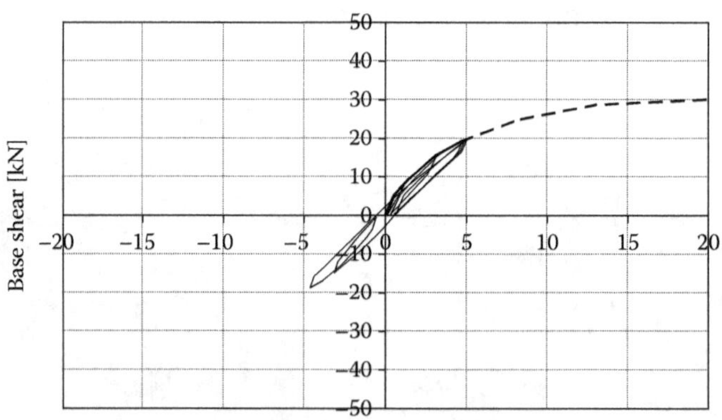

Figure 4.101 Results of the cyclic test performed on the bare RC substructure.

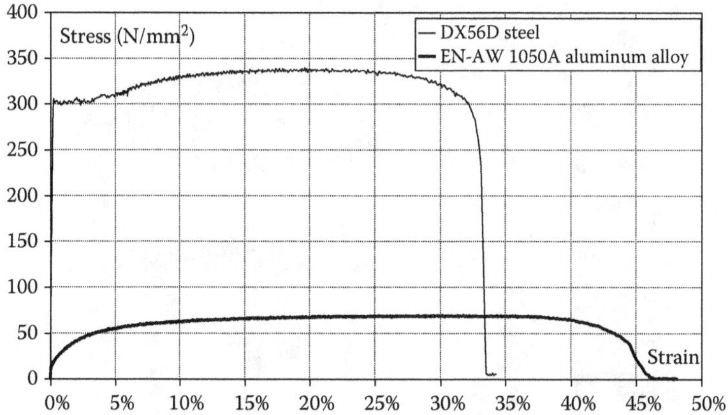

Figure 4.102 Mechanical characteristics of materials used for shear panels.

determining values of 270 kN and 15 kN/mm, respectively (Formisano et al., 2006). Subsequently, on the basis of the existing analytic simplified formulations (Sabouri-Ghomi et al., 2003), the shear panel configuration has been defined, obtaining the dimensions of b = 600 mm and d = 2400 mm, with a thickness t of 1.15 mm and 5 mm, in case of base material DX56D steel and EN-AW 1050 aluminum alloy, respectively (Figure 4.102).

In order to guarantee a *b/d* ratio lower than 0.8, which is indicated as the lower bound of the aspect ratio below which the development of a correct plastic mechanism is not achieved, the panel has been subdivided into six parts by means of the insertion of appropriate transversal rectangular fishplates with a thickness of 4 mm. Therefore, the choice of the better connection system between sheeting and fishplates has been carried out on the basis of ad hoc experimental tests (Formisano et al., 2006).

Also, a refined finite element model of the selected panels has been implemented by the ABAQUS nonlinear software, in order to correctly predict their performance under lateral loads (Figure 4.103a).

Furthermore, in order to confirm the validity of the proposed design solution and for evaluating the possible relative interaction problems between the RC structure and the added devices, a global analysis of the retrofitted structure, in which the shear plates have been modeled according to the "strip model" theory, has been performed by means of the SAP 2000 calculation program (Figure 4.103b) (De Matteis et al., 2007). Then, on the basis of the above numerical simulation, the design of the reinforcing intervention on the bare RC structure has been carried out, it being based on the strengthening of both the first level beams and the foundation beams by means of the coupled UPN 220 profiles, reinforcing plates, and threaded M16 bars (Figure 4.104). Finally, experimental cyclic tests on the

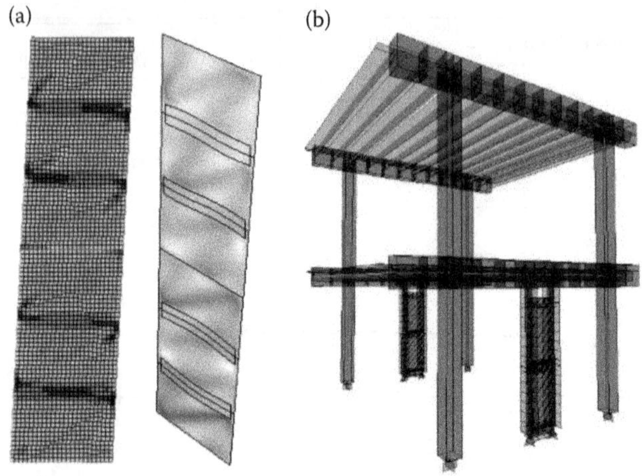

Figure 4.103 Final stress and deformation state of aluminum shear panels (a) and numerical model of the whole reinforced structure (b).

bare RC structure, upgraded with steel and aluminum shear panels, have been performed.

In both cases, a fully plastic behavior of single panel portions occurred at the end of the test, as shown in Figure 4.105, where the comparison with the initial plate configuration is also illustrated.

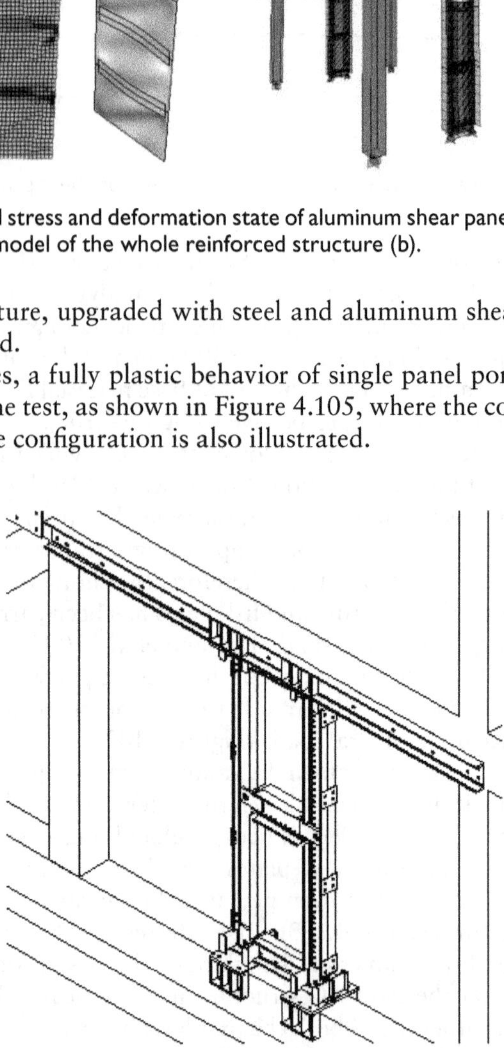

Figure 4.104 Reinforcing interventions on the original RC structure.

Figure 4.105 Initial and final deformed shape of steel (a) and aluminum (b) shear panels.

The comparison between the responses of the retrofitted structures with steel and aluminum shear panels, respectively, and the bare structure is given in Figure 4.106, in terms of envelope curves of the experimental tests. A significant improvement in strength (10 and 11.5 times with steel and aluminum panels, respectively), initial stiffness (2.5 and 2 times with steel and aluminum panels, respectively), and ductile capacity (interstory drift greater than 3.5% and 6.5% when steel and aluminum panels, respectively) can be observed.

Based on the above highlights, the performed experimental results showed that steel shear panels can be considered as a profitable system for improving the lateral load resistance of existing RC structures, whereas the

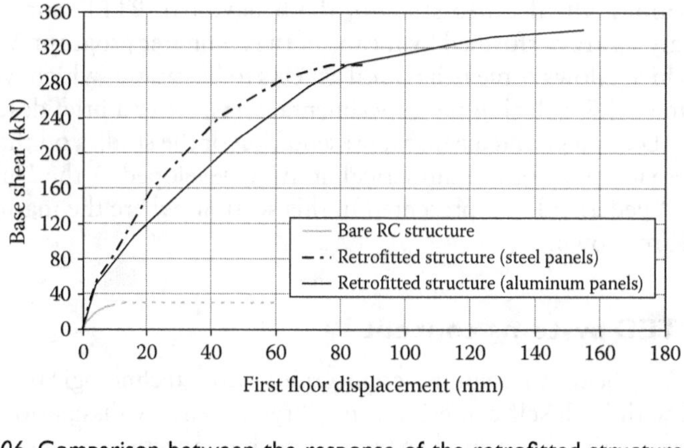

Figure 4.106 Comparison between the response of the retrofitted structures and the bare RC structure.

pure aluminum shear panels are also able to improve the ductility features and the dissipative capacity of the primary structure.

4.4 NEW SOLUTIONS FOR CONNECTIONS

4.4.1 Introduction

Steel MRFs are usually designed so that, in case of strong seismic events, a series of plastic hinges forms at the beam ends, which should provide the dissipation of the input seismic energy. Unfortunately, during the 1994 Northridge (USA) earthquake, unexpected brittle fractures occurred in the nodal areas of a lot of steel MRFs, thus preventing the formation of plastic hinges at the beam ends, with consequent loss of ductility. Consequently, many efforts have been made in order to improve the behavior of beam-to-column connections and a series of innovative systems has been proposed and tested (FEMA, 2000).

PTED connections represent an innovative solution, in which self-centering capability and energy dissipation capacity are associated (Ricles et al., 2001; Christopoulos et al., 2002a; Rojas et al., 2005; Wolski et al., 2006). A series of experimental studies has been carried out, allowing the validation of the PTED concept (Christopoulos et al., 2002a,b; Wolski et al., 2006; Ricles et al., 2002; Garlock et al., 2005). Moreover, some simplified numerical models have been set up, in order to analytically study the behavior of steel MRFs equipped with PTED beam-to-column connections (Ricles et al., 2001; Rojas et al., 2005; Christopoulos and Filiatrault 2003).

A campaign of numerical studies has been developed at the University of Naples "Federico II" (Esposto et al., 2005, 2006a,b; Esposto, 2007), whose principal aim is the detailed study of the behavior of PTED connections. These analyses are performed by means of the computer program ABAQUS (2004), which allows some behavioral aspects to be caught, which would be difficult to evaluate during the experimental tests. After a brief description of the PTED concept and a general presentation of the studies on the PTED systems, a summary of the numerical activity developed at the University of Naples "Federico II" is presented in this section, where the main results obtained are shown.

4.4.2 PTED systems concept

The PTED beam-to-column connections are technological systems endowed with both self-centering capability and energy dissipation capacity. Consequently, the cyclic behavior of such connections is flag shaped, as shown in Figure 4.107: when the seismic event occurs, dissipation of the

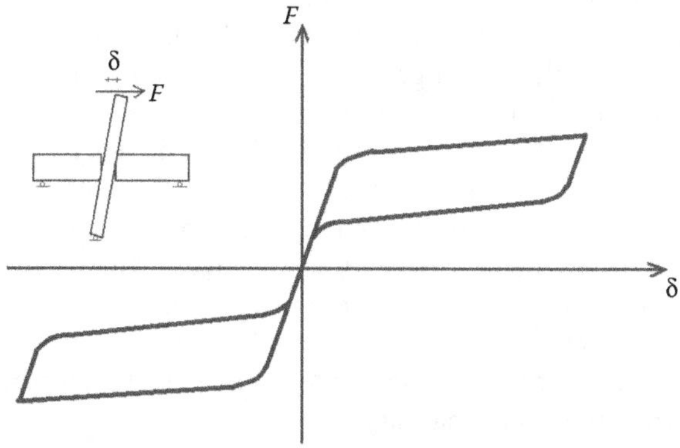

Figure 4.107 Flag-shaped cyclic behavior of PTED connections.

input seismic energy takes place, while at the end of the earthquake, the structure returns to the initial configuration.

The behavioral peculiarities of PTED systems are obtained thanks to the ad hoc subsystems. In particular, the self-centering capability is due to the high-strength steel bars or strands (PT-system), which are posttensioned in elastic range, whereas the energy dissipation capacity is due to the yielding or friction devices (ED-system). A schematic representation of a PTED beam-to-column connection is shown in Figure 4.108.

In steel frames equipped with PTED connections, the transmission of the internal forces between beams and columns is essentially due to the PT-action, which generates a uniform distribution of contact stresses in compression at the beam-to-column interface. Shear force is transmitted by the friction between beams and columns. For small values of the bending moment, beams and columns are in full contact and the connection behaves like a traditionally rigid one. When the bending moment increases, the contact stress in compression at the corner of the beam ends in tension becomes zero and a gap opens at the interface with the column. This

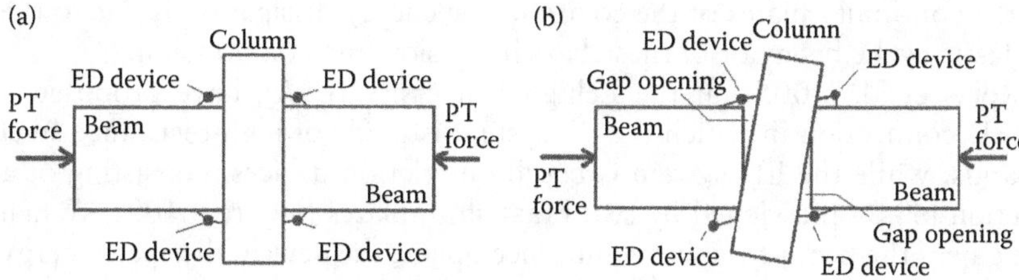

Figure 4.108 Schematic representation of the PTED connection behavior: (a) initial undeformed configuration; (b) deformed configuration.

causes the elastic elongation of the PT-elements and, at the same time, the ED-elements, after an initial elastic behavior, begin to dissipate energy, by deforming up to the plastic range or by friction. When the seismic action stops, the elastic returning action of the PT-system tends to bring the structure back to the initial configuration. Since the energy dissipation is guaranteed by the ED-system, the main structural elements are preserved from damage. Obviously, the PT-system must be designed to guarantee the full contact at the beam-to-column interface in the design conditions and the adequate friction for the shear resistance. Furthermore, it must behave in elastic range in case of severe earthquakes, because the breaking of PT-elements would lead to the collapse of the whole structure, due to the lack of shear transmission capability.

4.4.3 Studies on PTED systems: General framework

4.4.3.1 Technological solutions

To date, several types of PTED beam-to-column connections have been proposed, the main differences being the practical realization of the PT- and ED-systems, as shown in Figure 4.109. Ricles et al. (2001, 2002) and Garlock et al. (2005), at the Lehigh University (USA), have studied a PTED connection in which the PT-system is based on a series of high-resistance steel strands, while the ED-system is composed by bolted steel top-and-seat angles (Figure 4.109a). When the gap at the beam-to-column interface opens, the steel strands elongate in elastic range and the angles deform, undergoing plastic deformations in the stress concentration zones. In particular, three cylindrical plastic hinges form at each angle, one next to the column bolt washers and two in the angle fillet.

Christopoulos et al. (2002a,b), at the University of California in San Diego (USA), have studied a PTED connection in which the PT-system is based on a couple of high-resistant steel bars, while the ED-system is made of steel bars threaded in steel couplers and placed into steel confining cylinders, which prevent their buckling in compression (Figure 4.109b). Both the couplers and confining cylinders are welded to the beam flanges and/or to the continuity plates of the column. The energy dissipation is due to the inelastic cyclic behavior of these bars in tension and compression.

Rojas et al. (2005), at the Lehigh University (USA), have proposed a PTED connection in which the PT-system is made of high-resistance steel strands, while the ED-system is based on friction devices, consisting of a friction plate sandwiched by two brass shim plates (Figure 4.109c). When the gap at the beam-to-column interface opens, the devices dissipate energy by friction.

More recently, Wolski et al. (2006), at the Lehigh University (USA), have introduced a PTED connection type still endowed with high-resistant

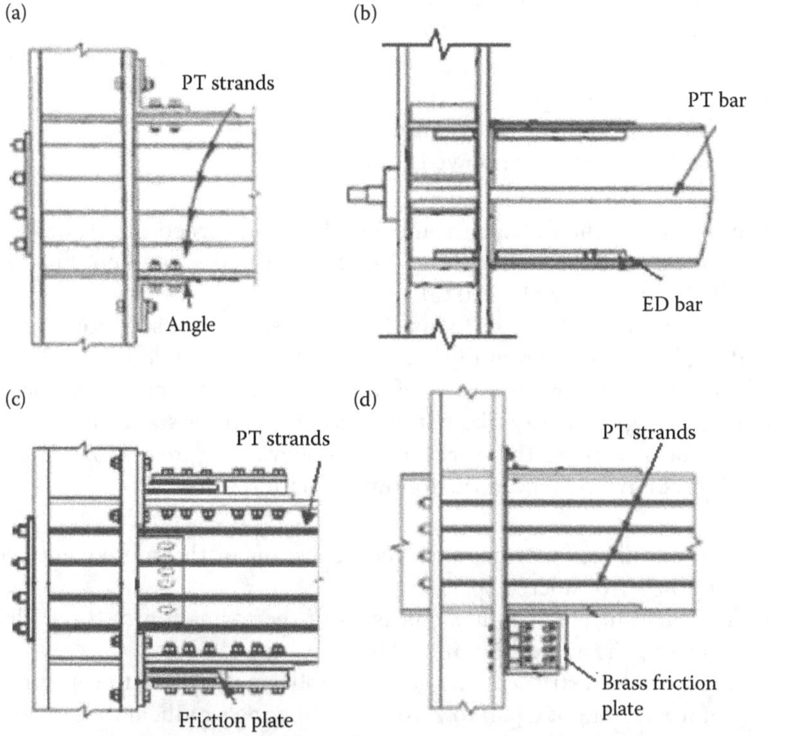

Figure 4.109 Technological solutions for PTED beam-to-column connections, characterized by different arrangements for PT- and ED-systems: (a) PT-steel strands and ED-bolted angles. (Ricles, J.M. et al. 2001: *ASCE Journal of Structural Engineering*, 127(2), 113–121.) (b) PT- and ED-steel bars. (Christopoulos et al. 2002a: *ASCE Journal of Structural Engineering*, 128(9), 1111–1120.) (c) PT-steel strands and ED-friction devices. (Rojas, P., Ricles, J.M., Sause, R. 2005: *ASCE Journal of Structural Engineering*, 131(4), 529–540.) (d) PT-steel strands and ED-top angle and bottom friction device. (Wolski, M., Ricles, J.M., Sause, R. 2006: In *Proceedings of 5th International Conference on the Behaviour of Steel Structures in Seismic Areas (STESSA 2006)*, Yokohama, Japan, 14–17 August, 481–487.)

steel strands, as PT-arrangement, but characterized by an ED-system composed of friction devices placed only below the beam bottom flange (Figure 4.109d). Such a technological solution has been proposed in order to avoid interferences between the friction devices and the composite floor slab.

4.4.3.2 Experimental studies

The behavior of some of the proposed PTED beam-to-column connections has been investigated by means of a series of experimental studies.

The characterization of the connection behavior has been made at different levels:

- ED-subassemblages
- Whole PTED beam-to-column connections
- PTED connections endowed frame

The studies on the ED-subassemblages have concerned the cyclic behavior of top-and-seat bolted angles (Garlock et al., 2003) and confined steel bars (Christopoulos et al., 2002a).

Garlock et al. (2003) have focused on the behavior of steel top-and-seat bolted angles, with the aim of determining how the angle size and geometry, together with the presence of washer plates, affect the connection stiffness, strength, energy dissipation capacity, and resistance to low cycle fatigue. The scheme of the test setup is shown in Figure 4.110a. The performed tests have evidenced the following main results:

- The initial stiffness has remained constant in the subsequent cycles for all the considered specimens.
- The yield mechanism has consisted in the formation of three plastic hinges in the angles (Figure 4.110b).
- The postyield stiffness has been sensibly smaller than the elastic one, but it has not got equal to zero, thus showing significant material and geometrical hardening.

The force–deformation cycles for some of the tested specimens are shown in Figure 4.110c.

Christopoulos et al. (2002a) have performed an experimental analysis on the confined steel ED-bars with the aim of assessing their tension–compression cyclic behavior, of verifying the capability of the coupler welds to

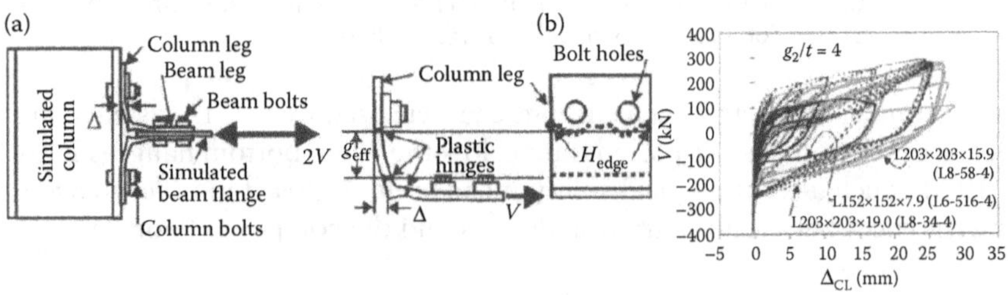

Figure 4.110 Tests on top-and-seat bolted angles: (a) specimen scheme; (b) yield mechanism; (c) cyclic behavior of some specimens. (Garlock, M.M., Ricles, J.M., Sause, R. 2003: *ASCE Journal of Structural Engineering*, 129(12), 1615–1625.)

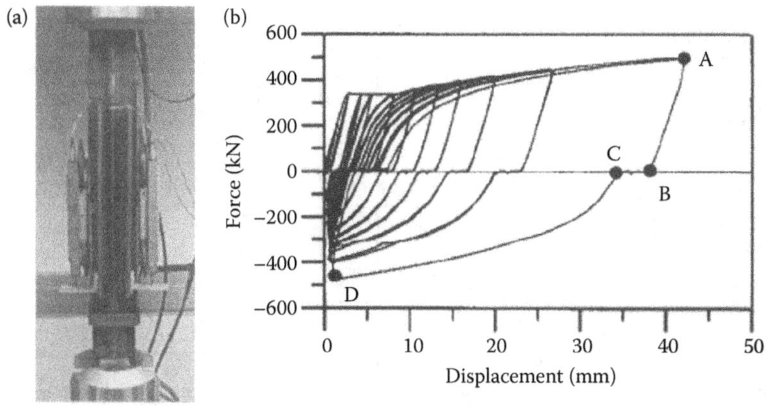

Figure 4.111 Test on confined ED-bars: (a) test specimen; (b) cyclic response. (Christopoulos, C. et al. 2002a: *ASCE Journal of Structural Engineering*, 128(9), 1111–1120.)

withstand the axial capacity of the ED-bars, and of evaluating the effectiveness of the confining cylinders in limiting the buckling of the ED-bars. The test specimen is shown in Figure 4.111a, whereas the force–displacement output of the test is shown in Figure 4.111b. The test results have shown the adequate stability and the good energy dissipation characteristics of the hysteretic behavior of the system.

Experimental analyses on the proposed whole PTED beam-to-column connections have been carried out by Ricles et al. (2002), Garlock et al. (2005), Christopoulos et al. (2002a), and Wolski et al. (2006). The connection effectiveness has been validated and the expected flag-shaped cyclic behavior has been evidenced. Moreover, the influence of some geometrical parameters has been also investigated.

Ricles et al. (2002) have carried out experimental tests on beam-to-column connections with PT-strands and ED-angles, each of them made of one column and two beams. In all the tests, the same beam size has been considered, whereas the steel wide flange or composite concrete-filled steel tubes have been used for the columns (Figure 4.112a and b). The experimental campaign has focused on the influence and on the PTED connections behavior, due to a series of parameters, namely, the presence of shim plates, of beam flange reinforcing plates, of features of the top-and-seat ED-angles, and of the posttensioning force. As an example, the influence on the connection behavior of the angle thickness is shown in Figure 4.112c, whereas the effect of the presence of the posttensioning force is evident in Figure 4.112d. These tests have evidenced that the shim and reinforcing plates are necessary to control the inelastic deformation of the beams, that the size and geometry of the angles influence the connection moment

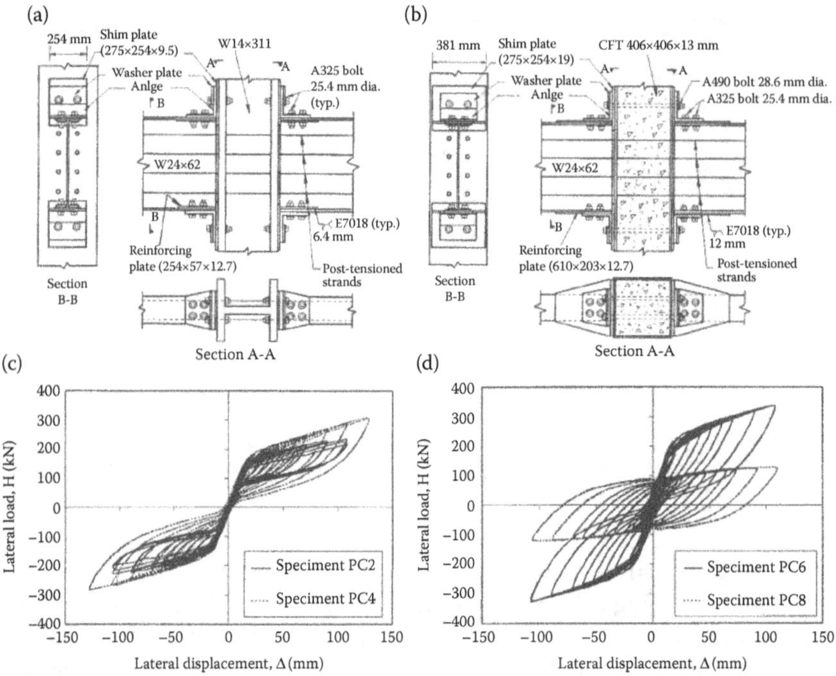

Figure 4.112 Tests on PTED connections with PT-strands and ED-angles: typical specimen with steel wide-flange column (a) and composite concrete-filled steel tube column (b); force–displacement curves, in function of the angle thickness (c); and presence of posttensioning force (d). (Ricles, J.M. et al. 2002: *ASCE Journal of Structural Engineering*, 128(7), 850–859.)

capacity and the energy dissipation capacity, and that the strands must be designed to remain elastic, in order to assure the self-centering and load-carrying capability of the system.

The studies by Garlock et al. (2005) are an expansion of the above-described work, since they have been related to similar PTED connections, focusing on different parameters. In particular, such studies have investigated the effects, on the connection behavior, of the value of initial post-tensioning force, of the number of strands used in the connection, and of the length of the beam flange reinforcing plates. The main results of this experimental campaign can be summarized as follows:

- Larger values of the initial posttensioning force increase the capability of the connection in achieving larger bending moments, but, at the same time, they can lead to undesirable beam flange buckling phenomena, with consequent reduction of the PT-force itself.

- The buckling phenomena can be avoided by using longer reinforcing plates and/or by using a smaller initial posttensioning force.
- The yielding of strands can be avoided by using a larger number of strands, which also allow a larger connection moment and greater ductility to be achieved.

Christopoulos et al. (2002a) have experimentally validated the connection with PT-bars and ED-bars by performing a test on an external beam-to-column connection. The detail of the test specimen node is shown in Figure 4.113a, whereas the force–displacement experimental cyclic curve is shown in Figure 4.113b. The main result of the experimental test carried out is that the proposed connection is able to undergo large deformations with energy dissipation characteristics, while keeping both the beam and column undamaged and without residual drift.

Wolski et al. (2006) have developed and experimentally studied a connection based on the use of steel strands, as a PT-system, and bottom flange friction devices, as an ED-system. In Figure 4.114a, the test setup is shown, whereas the moment–rotation response is plotted in Figure 4.114b. The test has shown the good energy dissipation provided by the connection, together with the full self-centering properties.

Finally, an experimental study has been carried out by Christopoulos et al. (2002b) on a half-scale three-column–two-beam simple steel frame equipped with PTED internal and external connections. The test specimen is shown in Figure 4.115a and the cyclic response, in terms of lateral force–lateral displacement, is plotted in Figure 4.115b. The test on the frame has confirmed the results obtained on the external beam-to-column connection, say the effectiveness of such system of providing both self-centering and energy dissipation capabilities.

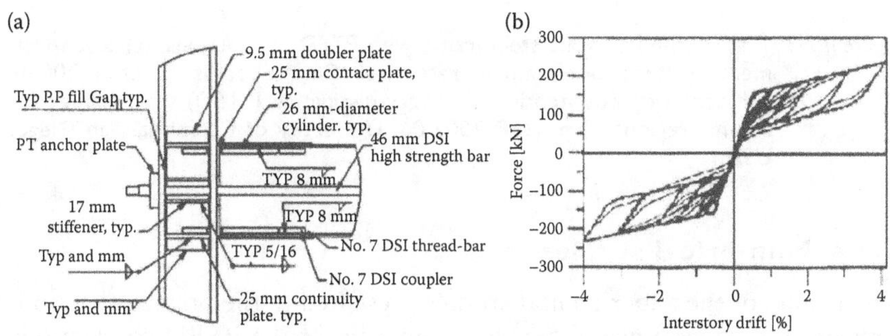

Figure 4.113 Experimental test on the connection with PT- and ED-bars: (a) specimen details; (b) force–drift relationship. (Christopoulos, C. et al. 2002a: *ASCE Journal of Structural Engineering*, 128(9), 1111–1120.)

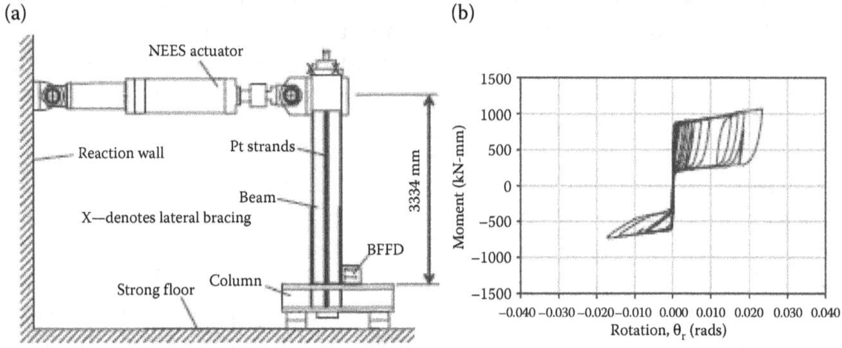

Figure 4.114 Experimental test on the connection with PT-strands and ED-bottom flange friction devices: (a) test setup; (b) moment–rotation cyclic response. (Wolski, M., Ricles, J.M., Sause, R. 2006: In *Proceedings of 5th International Conference on the Behaviour of Steel Structures in Seismic Areas (STESSA 2006)*, Yokohama, Japan, 14–17 August, 481–487.)

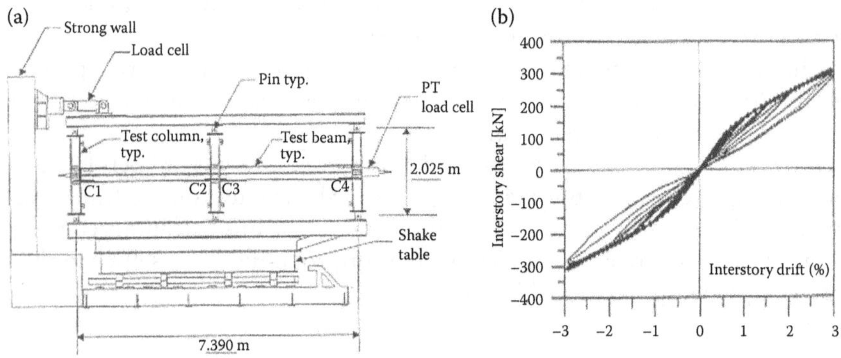

Figure 4.115 Test on the half-scale steel frame with PTED connections: (a) test specimen; (b) force–displacement response. (Christopoulos, C. et al. 2002b: Self-centering post-tensioned energy-dissipating (PTED) steel frames for seismic regions. Rep SSRP 2002/06, University of California, San Diego, USA.)

4.4.4 Numerical studies

The results of the experimental studies presented in the previous section, first of all, are very useful for the evaluation of the PTED connections behavior. Furthermore, they can be exploited for the setup and validation of a series of numerical models, allowing the investigation of very delicate aspects, which are difficult to measure by means of the experimental tests.

It is possible to subdivide the numerical studies on PTED connections into two main groups:

- Studies aimed at evaluating the seismic behavior of steel MRFs equipped with PTED beam-to-column connections
- Studies aimed at investigating in detail the local behavior of the connections

The numerical model setup by Ricles et al. (2001), based on fiber elements, has been developed and implemented with the aim of performing both the parametric inelastic static analyses of interior beam-to-column connections (Figure 4.116a) and the dynamic time history analyses of a whole steel MRF (Figure 4.116b). The parametric inelastic static analyses have focused on the effects, on the system behavior, of connection details such as the level of posttensioning force, the presence of shim and beam flange reinforcing plates, and the geometrical features of the angles. The analytical model has also been used for dynamic time history analyses of a six-story steel MRF equipped with PTED connections, which have proved to guarantee self-centering capability and adequate stiffness, strength, and ductility. Moreover, the analyzed frames with PTED connections have exceeded the performance of a frame with typical welded connections subject to the same earthquake records.

Christopoulos and Filiatrault (2003) have numerically modeled their connection by defining adequate equivalent springs and studied the seismic response of MRFs equipped with them. The analytical model of both internal and external PTED connections used for the numerical analyses are shown in Figure 4.117. Three buildings with 3, 6, and 10 stories,

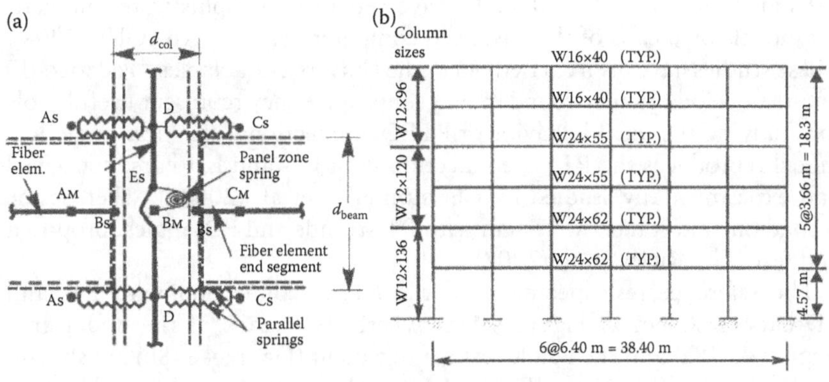

Figure 4.116 (a) Analytical model for internal PTED connection; (b) six-bay six-story frame subjected to the dynamic time history analyses. (Ricles, J.M. 2001: *ASCE Journal of Structural Engineering*, 127(2), 113–121.)

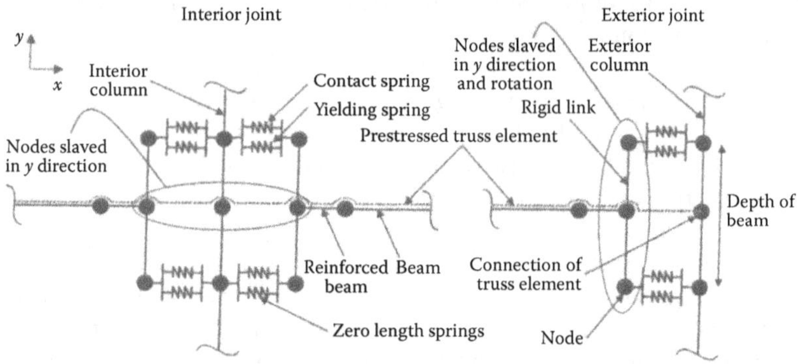

Figure 4.117 Analytical models of internal and external PTED connections with PT- and ED-bars. (Christopoulos, C., Filiatrault, A. 2003: In *Proceedings of 4th International Conference on the Behaviour of Steel Structures in Seismic Areas (STESSA 2003)*, Naples, Italy, 9–12 June, 511–517.)

respectively, have been considered in the analyses. First, they are designed as welded MRFs and, then, as equipped with the corresponding PTED connections. Also in this case, the results have confirmed the self-centering and energy dissipation capabilities of MRFs endowed with PTED connections.

The PTED system proposed by Rojas et al. (2005) has been studied by means of an analytical model based on fiber elements, somehow similar to that developed by Ricles et al. (2001), which has been used for performing inelastic static pushover and dynamic time history analyses. The seismic performance of the frames has proved to be satisfactory in terms of strength, story drift, local deformation, and self-centering capability.

Esposto et al. (2005, 2006a,b) have set up two sophisticated numerical models by means of the advanced computer program ABAQUS (2004). These studies have been carried out at the University of Naples "Federico II" and they belong to the second main group, since they represent useful tools for studying the local behavior of PTED connections. The first numerical model reproduces the PTED connection with PT- and ED-bars, conceived and experimentally studied by Christopoulos et al. (2002a), whereas the second one interprets the system with PT-strands and ED-angles, proposed and tested by Ricles et al. (2002).

The reference test specimen for the PTED connection with PT- and ED-bars is shown in Figure 4.118, where the details of the nodal area (Figure 4.118a) and the whole test arrangement (Figure 4.118b) are shown.

The details of the node (Figure 4.118a) show the presence of additional steel elements, with respect to the basic component parts of the connection: the continuity plates, which contribute to the transmission of the

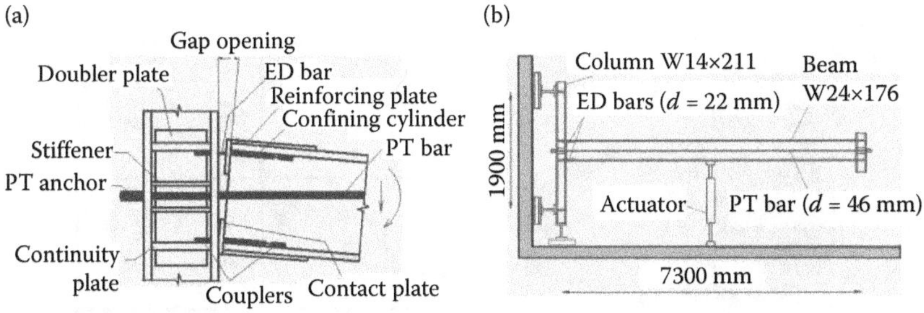

Figure 4.118 **Reference test specimen: (a) detail of the nodal area; (b) test arrangement. (Christopoulos, C. et al. 2002a: *ASCE Journal of Structural Engineering*, 128(9), 1111–1120.)**

beam flanges' forces to the column; the stiffeners, which support the column flanges next to the anchor of the PT-bars; the doubler plates, which increase the shear resistance of the column web; the contact plates, welded to the beams, which guarantee a smooth interface between beams and columns; the reinforcing plates, which are useful to prevent the beam flanges from yielding; the couplers, which provide the constraint for the ED-bars threaded into them; the confining cylinders, which prevent the ED-bars from buckling in compression, after their yielding in tension.

The PT-bars were made of high-strength steel (ultimate stress equal to 1030 MPa), and the ED-bars are made of elastic–plastic hardening steel (yield stress equal to 400 MPa), whereas all the other component parts of the tested PTED connection are made of a 345 MPa yield stress steel. Details on the size of the component parts and on the design of the test specimen can be found in Christopoulos et al. (2002a).

The experimental test consisted in a series of increasing vertical displacements imposed by the actuator at the midspan of the beam (Figure 4.118b).

The numerical model reproduces in detail the symmetrical half part of the specimen with respect to the middle plane of the beam and column's webs, in order to reduce the computational costs (Figure 4.119a). The finite element mesh is obtained by using tridimensional continuum elements (i.e., solid hexahedral elements) for all the components of the model. Such elements are useful for complex nonlinear analyses involving the contact, plasticity, and large deformations. Since contact interactions are present in the model, first-order finite elements are considered. In fact, in contact simulations, they allow correct calculation of the nodal loads, corresponding to a constant contact pressure, in contrast to second-order elements, which do not allow the univocal association of the nodal loads to the load case of a contact pressure at the element surface. Moreover, reduced integration has been used, in order to contain the computational efforts.

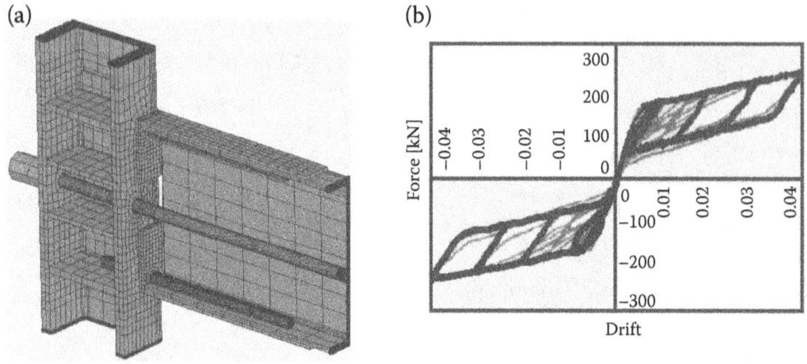

Figure 4.119 Numerical model of the PTED connection with PT- and ED-bars: (a) finite element mesh; (b) numerical versus experimental results. (Esposto, M., Faggiano, B., Mazzolani, F.M. 2006b: In *Proceedings of 8th International Conference on Computational Structures Technology*, Las Palmas de Gran Canaria, Spain, 12–15 September, Paper No. 117.)

Three different materials have been modeled: elastic–plastic structural steel, hardening steel, and high-resistance steel. The material properties, in terms of yield and ultimate true stress, and the connection component parts associated with each material are indicated in Table 4.3.

Special attention has been paid to the modeling of the interactions between the component parts of the connection. In particular, the interaction between the welded surfaces in the test specimen has been modeled by a tie constraint, so that no relative motion between the surfaces in contact is possible. As far as the contact interactions are concerned, two different formulations have been considered:

- A "penalty" tangential friction contact, for modeling the interaction between the column flange and contact plates
- A frictionless tangential contact, for modeling the interactions between the ED-bars and confining cylinders as well as between the PT-bars and the holes in the column flange they are located in

The numerical results faithfully reproduce the experimental ones, demonstrating that the numerical model is well calibrated, as shown in Figure

Table 4.3 Material Properties

Component parts	*Yield stress (MPa)*	*Ultimate stress (MPa)*
Column, anchors, continuity plates, stiffeners, doubler plates, contact plates, reinforcing plates, beam	345	345
ED-bars	400	700
PT-bars, couplers, confining cylinders	1030	1030

4.119b, where the comparison between the numerical and experimental results is presented.

The stress and strain state of the connection after the application of the PT-action and at a drift equal to 4% is shown in Figure 4.120. The main results obtained from the numerical analysis can be summarized as follows:

- At the PT-force application, stress concentrations are apparent in the column flange at the PT-anchor, in the column web, and in the beam next to the contact plates, as was expected (Figure 4.120a).
- At a drift equal to 4% (Figure 4.120b), the stress concentrations extend in the superior part of the model, at the beam-to-column contact plate, in the reinforcing plate, in the column web, and in the beam flange (where the maximum stresses are located at the end of the reinforcing plates).
- Contact plates and reinforcing plates undergo plasticization since the drift is equal to 0.5% and 3%, respectively.
- ED-bars show a stable cyclic behavior, able to adequately dissipate energy.
- Concerning the PT-bars, some local rotations at the PT-hole in the column flange occur at a drift equal to 4% (the PT-bar comes into contact with the PT-hole, causing a dangerous stress concentration in the PT-bar).

As previously mentioned, the second setup finite element model reproduces the PTED connection conceived and tested by Ricles et al. (2002). Preliminarily, a model of the ED-subassemblage, consisting in a couple of steel bolted angles, has been set up (Figure 4.121a). The model of the ED-subassemblage, set up by the software ABAQUS, has reproduced the L8-58-4 specimen described in Garlock et al. (2003), with the only difference that a half system has been considered, exploiting the symmetry with respect to the column web midplane. Accordingly, appropriate boundary

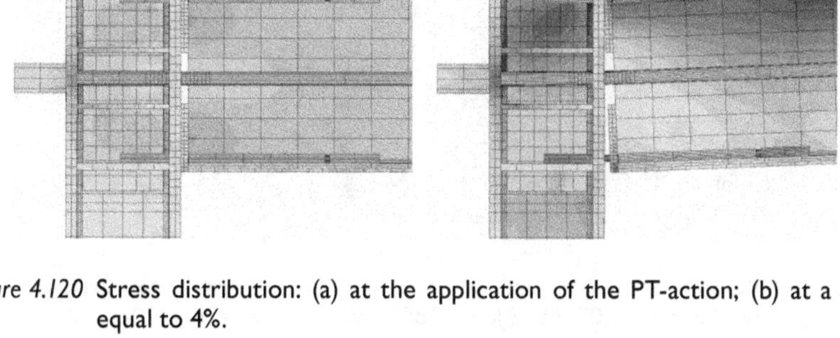

Figure 4.120 Stress distribution: (a) at the application of the PT-action; (b) at a drift equal to 4%.

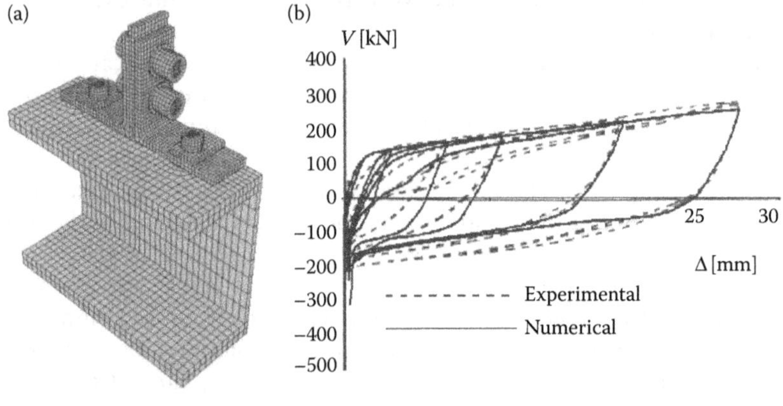

Figure 4.121 Numerical model of the ED-subassemblage: (a) finite element mesh; (b) numerical versus experimental results. (Esposto, M., Faggiano, B., Mazzolani, F.M. 2006a: In *Proceedings of 5th International Conference on the Behaviour of Steel Structures in Seismic Areas (STESSA 2006)*, Yokohama, Japan, 14–17 August, 299–304.)

conditions have been imposed, in order to prevent the out-of-plane displacements and the rotations around in-plane axes. The geometrical and mechanical characteristics of the specimen, together with the cyclic loading history, can be found in the reference paper. The test results are presented in terms of the V–Δ curve, where Δ is the applied top displacement and V is the corresponding measured vertical force. The comparison between the experimental and numerical results (Figure 4.121b) shows the good calibration of the model, which has been implemented in the whole PTED internal connection model. Details on this model can be found in Esposto et al. (2006a).

The PC4 specimen tested by Ricles et al. (2002) (Figure 4.122) has been modeled in detail, with the only difference being that only a half system has

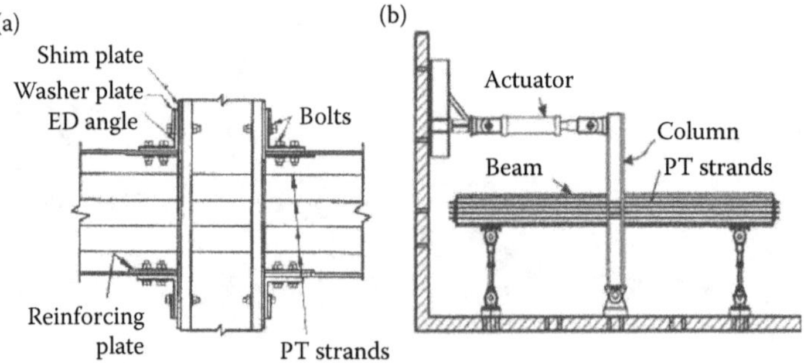

Figure 4.122 Reference test specimen: (a) detail of the nodal area; (b) test arrangement. (Ricles, J.M. 2002: *ASCE Journal of Structural Engineering*, 128(7), 850–859.)

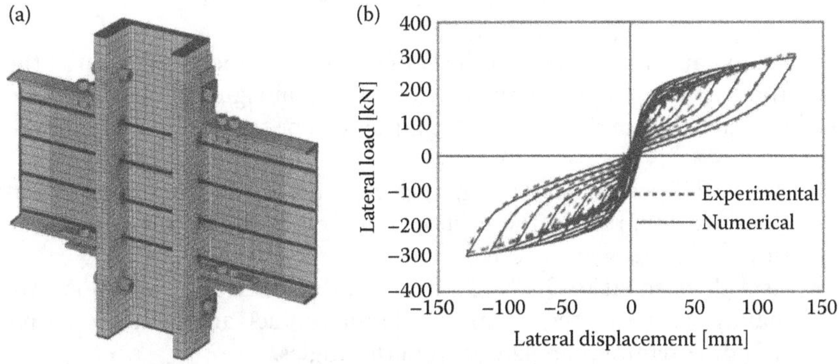

Figure 4.123 Numerical model of the PTED connection with PT-strands and ED-angles: (a) finite element mesh; (b) numerical versus experimental results. (Esposto, M., Faggiano, B., Mazzolani, F.M. 2006a: In *Proceedings of 5th International Conference on the Behaviour of Steel Structures in Seismic Areas (STESSA 2006)*, Yokohama, Japan, 14–17 August, 299–304.)

been considered, exploiting the symmetry (Figure 4.123a). As in the case of the ED-angle model, boundary conditions have been imposed accordingly. The geometrical and mechanical characteristics of the specimen, together with the cyclic loading history, can be found in the reference paper.

The criteria used for the selection and calibration of the finite element mesh of the model are the same used for the model shown in the previous paragraph.

The material properties for each component part of the model are given in Table 4.4. As for the previously described model, true values of stress and strain have been assumed. The interactions between the different component parts (weldings, contacts, etc.) have been modeled by using the same criteria indicated previously.

The numerical results reproduce the experimental ones with pretty good approximation. Some differences, in terms of initial stiffness and strength,

Table 4.4 Material Properties

Component parts	*Yield stress (MPa)*	*Ultimate stress (MPa)*
Column, washer plates and anchor plates	345	550
Beams	248	530
Shim plates and reinforcing plates	845	1030
ED-angles	263	567
PT-strands	1305	1865
Bolts	635	810

are apparent for the first cycles, whereas the last cycles have been successfully reproduced (Figure 4.123b).

The stress and strain states of the connection after the application of the PT-action and at a drift equal to 3.5% are shown in Figure 4.124. The main results obtained from the numerical analysis can be summarized as follows:

- At the PT-action application (Figure 4.124a), the stress concentrations are apparent in the column web and in the beams next to the shim plates, as was expected.
- At a drift equal to 3.5% (Figure 4.124b), stress concentrations are mainly visible at the beam-to-column contact areas, in the corresponding reinforcing plates and in the angles.
- The ED-angles have shown the expected yielding mechanism, corresponding to the formation of three cylindrical plastic hinges.
- The shim and reinforcing plates have always behaved in an elastic range.
- For large values of drift (3.5%), the strands have been in contact with the column flange holes.
- Beams have undergone plastic deformations for large values of lateral displacements (to the authors' knowledge, this condition has not been reached during the tests, but anyway it represents a possible scenario to be avoided, since it would lead to the reduction of the PT-force).

The presented numerical models appear to be useful tools for the detailed investigation of the connection behavior, allowing the analysis of several still unexplored aspects.

The first necessary step, still lacking at this time, is the comparison, in terms of cyclic behavior, between the PTED beam-to-column connections and the traditional welded ones. In this aim, the simplified model

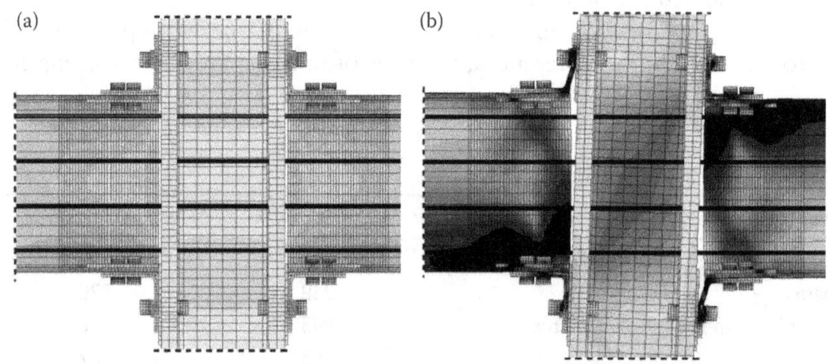

Figure 4.124 Stress distribution: (a) at the application of the PT-action; (b) at a drift equal to 3.5%.

assumption of considering only the half symmetrical part of the tested specimens must be removed, since the dissipative mechanism of the traditional beam-to-column nodes (formation of plastic hinges at the beam extremities) is not symmetric. A model of the PTED connection with PT- and ED-bars without the symmetry imposed condition has been set up and reproduced the experimental results very well.

Another important aspect to be investigated is represented by the evaluation of the capability of the PTED connections to resist vertical loads, also in combination with the seismic loads. In particular, the numerical analyses must be focused mainly on the following two problems:

- The connection capability of transmitting the vertical shear and the local stability of the beams. In fact, in the case of PTED connections, the transmission of the vertical shear from the beams to the columns is assigned to the friction at the beam-to-column interface.
- For this kind of connection, the local stability of the beams seems to be a problem of greater concern than in the case of traditional connections: the compression in the beams induced by the bending moment, due to the presence of vertical and/or seismic loads, is coupled with the initial compression state, which the beams are subjected to, due to the PT-action, with a consequently increasing risk of local instability occurrence.

In both cases, the simplified model assumption, consisting in the exploitation of the system symmetry, is not applicable, since the local instability phenomena are not symmetric. Some investigations should also concern the behavior of the columns, in the contemporary presence of vertical and seismic loads, since, when the gap opens at the beam-to-column interface, strong stress concentrations are present in the column web (Figures 4.112b and 4.116b).

In order to achieve an adequate optimization of the constructive detail for PTED beam-to-column connections, the influence of a series of geometrical parameters on the system behavior must be appropriately evaluated. This aim could be pursued by means of both experimental and numerical studies. However, considering that an experimental campaign can be much more expensive, in terms of economic costs, than a series of numerical analyses, the use of well-calibrated numerical models represents a powerful and relatively cheap tool of investigation.

The peculiarities of the PTED beam-to-column connections require specific design rules. Keeping this purpose in mind, the results of the numerical analyses on the PTED connections behavior, together with the already-performed experimental tests, will represent the basis of a proposal for codification of the design of steel MRFs equipped with such types of innovative nodes.

REFERENCES

ABAQUS, Inc. 2004: ABAQUS Standard User's Manual, version 6.5.

AISC 2004: 2005 *Seismic Provisions for Structural Steel Buildings*. American Institute of Steel Construction, Chicago, IL.

AISC/SEAOC 2001: Recommended Provisions for BRB, Structural Engineers Association of California: Seismology and Structural Standards Committee and American Institute of Steel Construction, Inc. 2001.

Akiyama, H. 2006: Prediction of seismic responses of buildings based on the energy concept. In *Proceedings of the 5th International Conference on the Behaviour of Steel Structures in Seismic Areas (STESSA 2006)*, Yokohama, Japan.

Alinia, M.M., Dastfan, M. 2006: Behaviour of thin plate shear walls regarding frame members. *Journal of Constructional Steel Research*, 63(7), 730–738.

American Institute of Steel Construction, Inc. (AISC) 2005: *Seismic Provisions for Structural Steel Buildings*. Standard ANSI/AISC 341/05, Chicago, IL, USA.

Anastasiadis, A., Gioncu, V. 1998: Influence of joint details on the local ductility of steel moment resisting frames. In *Proceedings of the Greek National Conference on Steel Structures*, Thessaloniki, Greece.

Astaneh-Asl, A. 2001: *Seismic Behavior and Design of Steel Shear Walls*. Steel TIPS Report, Structural Steel Educational Council, Moraga, CA, January.

Astaneh-Asl, A., Zhao, Q. 2002: Cyclic tests of steel shear walls, volume II—Appendices to final report. Final Report to the Sponsors, Report Number UCB/CEE-STEEL-01/01a, Dept. of Civil and Env. Engineering, Univ. of California, Berkeley, July.

Ballio, G., Mazzolani, F.M. 1983: *Theory and Design of Steel Structures*. Chapman & Hall, London-New York.

Balut, N., Gioncu, V. 2003: Suggestion for an improved 'dog-bone' solution. In *Proceedings of the 4th International Conference on the Behaviour of Steel Structures in Seismic Areas (STESSA 2003)*, Naples, Italy, 9–12 June, 129–134.

Basler, K., Thurlimann, B. 1959: Plate girder research. In *Proceedings AISC National Engineering Conference*, New York.

Brando, G. 2007: Experimental tests on bracing type pure aluminum shear panels. *Pollack Periodica*, 2(3), 73–84.

Brando, G., D'Agostino, F., De Matteis, G. 2013: Experimental tests of a new hysteretic damper made of buckling inhibited shear panels. *Materials and Structures/Materiaux et Constructions*, pp. 1–13. (Article in Press).

Brando, G., De Matteis, G. 2012: Experimental and numerical analysis of a multi-stiffened pure aluminium shear panel. In *Thin-Walled Structures*, ISSN: 0263-8231, October 2011, Elsevier, Oxford (UK), 49(10), 1277–1287, doi: 10.1016/j.tws.2011.05.007.

Brandon, C., Uang, C.-M. 2002: Cyclic response and design recommendations of weak-axis reduced beam section moment connections with deep columns. *ASCE Journal of Structural Engineering*, 128(4), 464–473.

Bruneau, M., Berman, J., Lopez Garcia, D., Vian, D. 2005: Steel plate shear wall buildings: Design requirements and research. Published on-line on http://www.eng.buffalo.edu/~bruneau/.

Bruneau, M., Uang, C.M., Whittaker A. 1998: *Ductile Design of Steel Structures*. McGraw-Hill, USA.

Budiansky, B., Connor, R.W. 1948: Buckling stresses of clamped rectangular flat plates in shear. NACA Tech. Note 1559.

Caccese, V., Elgaaly, M., Chen, R. 1993: Experimental study of thin steel-plate shear walls under cyclic load. *ASCE Journal of Structural Engineering*, 119(2), 573–587.

Canadian Standards Association (CSA). 2001: *Limit States Design of Steel Structures, CAN/CSA S16-01*. Willowdale, Ontario, Canada.

Carter, C.J., Iwankiw, N.R. 1998: Improved ductility in seismic steel moment frames with dog-bone connections. *Journal of Constructional Steel Research*, 46(1–3): Paper n. 253.

Chen, C.C. 2002: Recent Advances of Seismic Design of Steel Buildings, in Taiwan. *Proceedings of the International Training Programs for Seismic Design of Buildings Structures*. (http://www.ncree.org.tw/itp2002/07_RecentAdvancesOfSeismicDesignOfSteelBuildingsInTaiwan.pdf)

Chen, C.-H., Hsaio, P.-C., Lai, J-W., Lin, M.-L., Weng, Y.-T., Tsai, K.-C. 2004: Pseudo-dynamic test of a full-scale CFT/BRB frame: Part 2—Construction and testing. In *Proceedings of the 13th World Conference on Earthquake Engineering*, Vancouver, Canada, CD-ROM, Paper no. 2175.

Chen, S.J., Chu, J.M., Chou, Z.L. 1997: Dynamic behaviour of steel frames with beam flanges shaved around connection. *Journal of Constructional Steel Research*, 42(1), 49–70.

Chen, S.J., Jhang, C. 2006: Cyclic behaviour of low yield point steel shear walls. *Thin Walled Structures*, 44, 730–738.

Chen, S-J., Yeh, C.H., Chu, J.M. 1996: Ductile steel beam-to-column connections for seismic resistance. *ASCE Journal of Structural Engineering*, 122(11), 1292–1299.

Christopoulos, C., Filiatrault, A. 2003: Seismic demands on post-tensioned energy dissipating moment-resisting steel frames. In *Proceedings of 4th International Conference on the Behaviour of Steel Structures in Seismic Areas (STESSA 2003)*, Naples, Italy, 9–12 June, 511–517.

Christopoulos, C., Filiatrault, A., Uang, C.M. 2002b: Self-centering post-tensioned energy-dissipating (PTED) steel frames for seismic regions. Rep SSRP 2002/06, University of California, San Diego, USA.

Christopoulos, C., Filiatrault, A., Uang, C.M., Folz, B. 2002a: Posttensioned energy dissipating connections for moment-resisting steel frames. *ASCE Journal of Structural Engineering*, 128(9), 1111–1120.

Cox, H.L. 1933: Summary of the present state of knowledge regarding sheet metal construction. A.R.C.R & M. No. 1953.

Davies, J.M., Bryan, E.R. 1982: *Manual on Stressed Skin Diaphragm Design*, Granada Publishing Ltd., London.

D'Aniello, M., Della Corte, G., Mazzolani, F.M. 2005: Seismic Upgrading of RC buildings using Eccentric Braces: Full-scale experimental tests. In *20th CTA Conference—First International Workshop on Advances in Steel Constructions*, Ischia (Italy), 26–28 September.

D'Aniello, M., Della Corte, G., Mazzolani, F.M. 2007: A special type of buckling-restrained brace for seismic retrofitting of RC buildings: Design and testing. In *Proceedings of the 21st CTA Conference*, Catania, Italy, 3–5 October.

D'Aniello, M., Della Corte, G., Mazzolani, F.M. 2008: Experimental tests of a real building seismically retrofitted by special Buckling-Restrained Braces. *AIP Conference Proceedings*, 1020 (PART 1), 1513–1520.

D'Aniello, M., Della Corte, G., Mazzolani, F.M. 2009: Seismic protection of reinforced concrete buildings by means of "all-steel" buckling-restrained bracings. *Protection of Historical Buildings* (ed. F.M. Mazzolani), CRC Press, Taylor & Francis Group, London.

Della Corte, G., D'Aniello, M., Mazzolani, F.M. 2005: Seismic upgrading of RC buildings using Buckling Restrained Braces: Full-scale experimental tests. In *20th CTA Conference—First International Workshop on Advances in Steel Constructions*, Ischia (Italy), 26–28 September.

Della Corte, G., D'Aniello, M., Mazzolani, F.M. 2007: Inelastic response of shear links with axial restraints: Numerical vs. analytical results. *5th International Conference on Advances in Steel Structures, ICASS 2007*, 3, 651–656.

Della Corte, G., Fiorino, L., Mazzolani, F.M. 2008: Lateral-loading tests on a real RC building including masonry infill panels with and without FRP strengthening. *Journal of Materials in Civil Engineering*, 20(6), 419–431.

Della Corte, G., Mazzolani, F.M. 2006: Full-scale tests of advanced seismic upgrading techniques for RC buildings. In *Proceedings of the 2nd FIB Congress*, Naples, Italy, 5–8 June.

De Matteis, G. 1998: The effect of cladding panels in steel building under seismic actions. PhD Thesis, University of Naples, Italy.

De Matteis, G. 2005: Effect of lightweight cladding panels on the seismic performance of moment resisting steel frames. *Engineering Structures*, 27(11), 1662–1676.

De Matteis, G., Brando, G., Mazzolani, F.M. 2011: Hysteretic behaviour of bracing-type pure aluminium shear panels by experimental tests. *Earthquake Engineering and Structural Dynamics*, John Wiley & Sons, Chichester, England.

De Matteis, G., Brando, G., Mazzolani, F.M. 2012: Pure aluminium: An innovative material for structural applications in seismic engineering. *Journal of Construction & Building Materials*, 26(1), 677–686, doi: 10.1016/j.conbuildmat.2011.06.071.

De Matteis, G., Brando, G., Panico, S., Mazzolani, F.M. 2009: Bracing type pure aluminium stiffened shear panels: An experimental study. *International Journal of Advanced Steel Construction (IJASC)*, S.L. Chan, W.F. Chen and R. Zandonini (editors-in-chief), The Hong Kong Institute of Steel Construction, ISSN 1816–112X, doi 10.1556/Pollack.2.2007.3.7, 5(2), 106–119.

De Matteis, G., Formisano, A., Panico, S., Mazzolani, F.M. 2008: Numerical and experimental analysis of pure aluminium shear panels with welded stiffeners, In *Computer and Structures*, ISSN: 0045–7949, Pergamon-Elsevier Science Ltd, Oxford, England, doi: 10.1016/j.compstruc.2007.05.027, 86/6, 545–555.

De Matteis, G., Landolfo, R. 1996: Structural behaviour of sandwich panel shear walls: An experimental analysis. *Materials and Structures*, 32, 331–41.

De Matteis, G., Landolfo, R. 1999a: Mechanical fasteners for cladding sandwich panels: Interpretative models for shear behaviour. *Thin Walled Structures*, 35, 61–79.

De Matteis, G., Landolfo, R. 1999b: Modelling of lightweight sandwich shear diaphragms for dynamic analyses. *Journal of Constructional Steel Research*, 53(1), 33–61.

De Matteis, G, Landolfo, R. 2000: Diaphragm action of sandwich panels in pin-jointed steel structures: A seismic study. *Journal of Earthquake Engineering*, 4(3), 251–275.

De Matteis, G., Landolfo, R., Mazzolani, F.M. 1996: On the shear flexibility of corrugated shear panels. Steel structures. *Journal of Singapore Structural Steel Society*, 6(1), 103–112.

De Matteis, G., Mazzolani, F.M., Panico, S. 2007: Pure aluminium shear panels as dissipative devices in moment-resisting steel frames. *Earthquake Engineering and Structural Dynamics*, Wiley InterScience, 36, 841–859, doi: 10.21002/eqe.

Driver, R.G., Kulak, G.L., Kennedy, D.J.L. 1997: Seismic behaviour of steel plate shear walls. Struct. Engrg. Rep. No. 215, Dept of Civ. Engrg, University of Alberta, Edmonton, Alta., Canada.

Dusicka, P., Itani, A.M., Buckle, I.G. 2004: Evaluation of conventional and specialty steels in shear link hysteretic energy dissipators. In *Proceedings of the 13th World Conference on Earthquake Engineering*, Vancouver, B.C., Canada, 1–6 August, Paper No. 522.

Elgaaly, M., Caccese, V., Du, C. 1993: Postbuckling behaviour of steel-plate shear walls under cyclic loads. *ASCE Journal of Structural Engineering*, 119(2), 588–605.

Engelhardt, M.D., Sabol, T.A. 1997: Seismic-resistant steel moment connection: Developments since the 1994 Northridge Earthquake. Zandonini R. Elnashai A., Dexter R. (eds). *Progress in Structural Engineering and Materials*, 1(1): 68–77.

Engelhardt, M.D., Winneberger, T., Zekany, A.J., Potyraj, T.J. 1996: *The Dogbone Connection: Part II. Modern Steel Construction*, August: 46–55.

Esposto, M. 2007: FEM modeling of PTED beam-to-column connections for earthquake resisting steel frames. *Pollack Periodica, Akadémiai Kiadó*, 2(1), 101–112.

Esposto, M., Faggiano, B., Mazzolani, F.M. 2005: Numerical analysis of PTED beam-to-column connection systems for steel MRFs. In *Proceedings of 20th CTA Conference*, Ischia, Italy, 26–28 September, 427–434.

Esposto, M., Faggiano, B., Mazzolani, F.M. 2006a: Numerical vs. experimental results on a PTED beam-to-column connection for seismic resistant steel frames. In *Proceedings of 5th International Conference on the Behaviour of Steel Structures in Seismic Areas (STESSA 2006)*, Yokohama, Japan, 14–17 August, 299–304.

Esposto, M., Faggiano, B., Mazzolani, F.M. 2006b: Numerical modelling of PTED connections for steel moment resisting frames. In *Proceedings of 8th International Conference on Computational Structures Technology*, Las Palmas de Gran Canaria, Spain, 12–15 September, Paper No. 117.

Faella, C., Montuori, R., Piluso, V., Rizzano, G. 1998: Failure mode control: Economy of semi-rigid frames. In *Proceedings of the 11th European Conference on Earthquake Engineering*, Paris, France, 6–13 September.

Faggiano, B., de Cesbron de la Grennelais, E., Landolfo, R. 2003: Design criteria for RBS in MR frame retrofitting. In *Proceedings of the 4th International Conference on the Behaviour of Steel Structures in Seismic Areas (STESSA 2003)*, Naples, Italy, 9–12 June, 683–690.

Faggiano, B., Landolfo, R. 2002: Seismic Analysis and Design of Steel MR frames with "dog-bone" connections. In *Proceedings of the 12th European Conference on Earthquake Engineering*, London, U.K., Elsevier Ltd, Paper 309.

Federal Emergency Management Agency (FEMA) 2000: Recommended seismic design criteria for new steel moment frame buildings. Rep. No. FEMA 350, Washington, D.C., USA.

Formisano, A., Mazzolani, F. M., Brando, G., De Matteis, G. 2006: Numerical evaluation of the hysteretic performance of pure aluminium shear panels, In *Behaviour of Steel Structures in Seismic Areas (STESSA 2006)*, (eds. F.M. Mazzolani, A. Wada), Taylor & Francis/Balkema Publisher, London, ISBN 0-415-40824-5, 211–217.

Foutch, D.A. 1989: Seismic behavior of eccentrically braced steel building. *ASCE Journal of Structural Engineering*, 115(8), 1857–1876.

Foutch, D.A., Goel, S.C., Roeder, C.W. 1989: Seismic testing of full-scale steel building—Part I. *ASCE Journal of Structural Engineering*, 115(11), 2111–2129.

Garlock, M.M., Ricles, J.M., Sause, R. 2003: Cyclic load tests and analysis of bolted top-and-seat angle connections. *ASCE Journal of Structural Engineering*, 129(12), 1615–1625.

Garlock, M.M., Ricles, J.M., Sause, R. 2005: Experimental studies of full-scale post-tensioned steel connections. *ASCE Journal of Structural Engineering*, 131(3), 438–448.

Georgescu, D. 1996: Recent developments in theoretical and experimental results on steel structures. Seismic resistant braced frames. *Costruzioni Metalliche*, 1, 39–52.

Ghobarah, A., Abou Elfath, H. 2001: Rehabilitation of a reinforced concrete frame using eccentric steel bracing. *Engineering Structures*, 23, 745–755.

Ghobarah, A., Ramadan, T. 1994: Bolted link-column joints in eccentrically braced frames. *Engineering Structures*, 16(1), 33–41.

Gilton, C.S., Uang, C-M. 2002: Cyclic response and design recommendations of weak-axis reduced beam section moment connections. *ASCE Journal of Structural Engineering*, 128(4), 452–463.

Hitaka, T., Matsui, C. 2003: Experimental study on steel shear wall with slits. *ASCE Journal of Structural Engineering*, May, 586–594.

Hitaka, T., Matsui, C. 2006: Seismic performance of Steel Shear Wall with Slits integrated with multi-story composite moment frame. *Proceedings of the 5th International Conference on Behaviour of Steel Structures in Seismic Areas–STESSA 2006*, 241–246.

Hjelmstad, K.D., Popov, E.P. 1983: Cyclic behavior and design of link beams. *ASCE Journal of Structural Engineering*, 109(10), 2387–2403.

Höglund, T. 1997, Shear buckling resistance of steel and aluminium plate girders. *Thin-Walled Structures*, 29(1–4), 13–30.

Iguchi, S. 1938: Buckling of rectangular plates clamped on all four sides by shear stresses. *Proceedings of the Physico-Mathematical Society of Japan*, 20, 814.

Itani, A., Douglas, B.M., ElFass, S. 1998: Cyclic behavior of shear links in retrofitted Richmond-San Rafael Bridge towers. In *Proceedings of the 1st World Congress on Structural Engineering*, San Francisco, Paper No. T155–3.

Itani, A.M., Elfass, S., Douglas, B.M. 2003: Behaviour of built-up shear links under large cyclic displacement. *Engineering Journal, American Institute of Steel Construction*, 40(4), 221–234.

Izumi, M. et al. 1992: Low cycle fatigue tests on shear yielding type low yield stress steel hysteretic damper for response control, part 1 and 2. In *Proceedings of the Annual Meeting of the Arch Inst*, Tokyo, Japan, 1447–1450 (in Japanese).

Jain, A., Goel, S. 1979: Seismic response of eccentric and concentric braced steel frames with different proportions. UMEE 79R1, Department of Civil Engineering, University of Michigan, Ann Arbor, July.

Jones, S-L., Fry, G.T., Engelhardt, M.D. 2002: Experimental evaluation of cyclically loaded reduced beam section moment connections. *ASCE Journal of Structural Engineering*, 128(4), 441–451.

Kasai, K., Popov, E.P. 1986a: General behavior of WF steel shear link beams. *ASCE Journal of Structural Engineering*, 112(2), 362–382.

Kasai, K., Popov, E.P. 1986b: Cyclic web buckling control for shear link beams. *ASCE Journal of Structural Engineering*, 112(3), 505–523.

Katayama, T., Ito, S., Kamura, H., Ueki, T., Okamoto, H. 2000: Experimental study on hysteretic damper with low yield strength steel under dynamic loading. In *Proceedings of the 12th WCEE*, Auckland, New Zealand.

Khatib, I., Mahin, S. 1987: Dynamic inelastic behaviour of chevron braced steel frames. In *Proceedings of the 5th Canadian Conference on Earthquake Engineering*, Balkema, Rotterdam, 211–220.

Kunnath, S.K., Malley, J.O. 2002: Advances in seismic design and evaluation of steel moment frames: Recent findings from FEMA/SAC Phase II Project. *ASCE Journal of Structural Engineering*, 128(4), 415–419.

Lee, C-H., Jeon, S-W., Kim, J-H., Kim J-H., Uang, C-M. 2004: Seismic performance of reduced beam section steel moment connections: Effects of panel zone strength and beam web connection method. In *Proceedings of the 13th World Conference on Earthquake Engineering*, Vancouver, Canada, 1–6 August, Paper 3449.

Liu, Z., Goel, S. 1987: Investigation of concrete-filled steel tubes under cyclic bending and buckling. Research Report UMCE 87–3, Department of Civil Engineering, University of Michigan, Ann Arbor, April.

Lubell, A.S., Prion, H.G.L., Ventura, C.E., Rezai, M. 2000: Unstiffened steel plate shear wall performance under cyclic loading. *ASCE Journal of Structural Engineering*, 126(4), 453–460.

Maheri, M.R., Sahebi, A. 1997: Use of steel bracing in reinforced concrete frames. *Engineering Structures*, 19(12), 1018–1024.

Malley, J.O., Popov, E.P. 1983a: Design considerations for shear links in eccentrically braced frames. EERC Report 83–24, University of California, Berkeley.

Malley, J.O., Popov, E.P. 1983b: Design of links and beam-to-column connections for eccentrically braced steel frames. EERC Report 83–03, University of California, Berkeley.

Manheim, D.N., Popov, E.P. 1983: Plastic shear hinges in steel frames. *ASCE Journal of Structural Engineering*, 109(10), 2404–2419.

Mazzolani, F.M. 2006a: *Seismic Upgrading of RC Buildings by Advanced Techniques: The ILVA-IDEM Research Project.* Polimetrica International Publisher, Monza, Italy.

Mazzolani, F.M. 2006b: Innovative metal systems for seismic upgrading of RC structures. In *Proceedings of the SDSS (Stability and Ductility of Steel Structures) Conference*, (ed. D. Camotim), Lisbon, Portugal, 6–8 September, Keynote Lecture.

Mazzolani, F.M. 2008: Innovative metal systems for seismic upgrading of RC structures. *Journal of Constructional Steel Research*, 64(7–8), 882–895.

Mazzolani, F.M. (ed.) 2009: Protection of historical buildings. In *Proceedings of the PROHITECH 09 Conference*, Rome, June 21–24. CRC Press, Taylor & Francis Group, London.

Mazzolani, F.M., Landolfo, R., De Matteis, G. 1997: Analysis of the contributing effect of building panels on steel structure resistance to seismic and aeolian phenomena, Consortium CREA Technical Final Report. European Community Commission, Executive Committee F6 "Steel Structures", Agreement no. 7210–SA/421.

Mazzolani, F.M., Piluso, V. 1996: *Theory and Design of Seismic Resistant Steel Frames*. Chapman & Hall, London.

McDaniel, C., Uang, C.-M., Seible, F. 2003: Cyclic testing of built-up steel shear links for the new bay bridge. *ASCE Journal of Structural Engineering*, 129(6).

Mimura, H., Akiyama, H. 1977: Load-deflection relationship of earthquake resistant steel shear walls with a developed diagonal tension field. *Transactions of the Arch. Inst. of Japan*, Tokyo, Japan, 109–114.

Mistakidis, E.S., De Matteis, G., Formisano, A. 2007: Low yield shear panels as an alternative for the seismic upgrading of concrete structures. *Advances in Engineering Software*, 38(8–9), 626–636.

Montuori, R., Piluso, V. 2000: Plastic design of steel frames with dog-bone beam-to-column joints. In *Proceedings the 3rd International Conference on the Behaviour of Steel Structures in Seismic Areas (STESSA 2000)*, Montreal, Canada, 627–634.

Nakagawa, S., Kihara, H., Torii, S., Nakata, Y., Matsuoka, Y., Fujisawa, K., Fukuda, K. 1996: Hysteretic behavior of low yield strength steel panel shear walls—Experimental investigation. *Proceedings of the 11th WCEE*, Acapulco, Mexico, June 23–28.

Nakamura, H., Maeda, Y., Sasaki, T., Wada, A., Takeuchi, T., Nakata, Y., Iwata, M. 2000: Fatigue properties of practical-scale unbonded braces. Nippon Steel Technical Report, No. 82.

Nakashima, M. 1995: Strain-hardening behaviour of shear panel made of low yield steel. I: Test. *ASCE Journal of Structural Engineering*, 121(12), 1742–1749.

Nakashima, M., Akazawa, T., Tsuji, B. 1995: Strain-hardening behaviour of shear panel made of low yield steel. II: model. *ASCE Journal of Structural Engineering*, 121(12), 1750–1757.

Nakashima, M., Iwai, S., Iwata, M., Takeuchi, T., Konomi, S., Akazawa, T., Sabouri, K. 1994: Energy dissipation behaviour of shear panels made of low yield steel. *Earthquake Engineering and Structural Dynamics*, 23, 1299–1313.

Okazaki, T., Engelhardt, M.D. 2007: Cyclic loading behaviour of EBF links constructed of ASTM A992 steel. *Journal of Constructional Steel Research*, 63, 751–765.

Plumier, A. 1990: New idea for safe structure in seismic zone. In *IABSE Symposium*, Brussels, Belgium.

Popov, E.P., Engelhardt, M.D. 1988: Seismic eccentrically braced frames. *Journal of Constructional Steel Research*, 10, 321–354.

Rai, D.C., Wallace, B.J. 2000: Aluminium shear-link for seismic energy dissipation. In *Proceedings of the 12th World Conference on Earthquake Engineering*, Auckland, New Zealand, Paper No. 0279.

Rezai, M. 1999: Seismic behaviour of steel plate shear walls by shake table testing. PhD dissertation, University of British Columbia, Vancouver, Canada.

Ricles, J.M., Popov, E.P. 1993: Inelastic link element for EBF seismic analysis. *ASCE Journal of Structural Engineering*, 120(2), 441–463.

Ricles, J.M., Sause, R., Garlock, M.M., Zhao, C. 2001: Posttensioned seismic-resistant connections for steel frames. *ASCE Journal of Structural Engineering*, 127(2), 113–121.

Ricles, J.M., Sause, R., Peng, S.W., Lu, L.W. 2002: Experimental evaluation of earthquake resistant posttensioned steel connections. *ASCE Journal of Structural Engineering*, 128(7), 850–859.

Roberts, T.M., Sabouri-Ghomi, S. 1992: Hysteretic characteristic of unstiffened perforated steel plate shear panels. *Thin Walled Structures* 0263-8231/92.

Roeder, C.W. 2002: Connection performance for seismic design of steel moment frames. *ASCE Journal of Structural Engineering*, 128(4), 517–525.

Roeder, C.W., Foutch, D.A., Goel, S.C. 1987: Seismic testing of full-scale steel building—Part II. *ASCE Journal of Structural Engineering*, 113(11), 2130–2145.

Roeder, C.W., Popov, E.P. 1978: Eccentrically braced steel frames for earthquakes. *ASCE Journal of Structural Engineering*, 104(3), 391–412.

Rojas, P., Ricles, J.M., Sause, R. 2005: Seismic performance of post-tensioned steel moment resisting frames with friction devices. *ASCE Journal of Structural Engineering*, 131(4), 529–540.

Sabelli, R., Aiken, I. 2004: U.S. building-code provisions for buckling-restrained braced frames: Basis and development. In *Proceedings of the 13th World Conference on Earthquake Engineering*, Vancouver, Canada, CD-ROM, Paper No. 1828.

Sabouri-Ghomi, S., Roberts, T.M. 1991: Nonlinear dynamic analysis of thin steel plate shear walls. *Computers & Structures*, 39(1/2), 121–127.

Sabouri-Ghomi, S., Ventura, C., Kharrazi, M.H.K. 2003: Shear analysis and design of ductile steel plate walls. In *Proceedings of the 4th International Conference on Behaviour of Steel Structures in Seismic Areas (STESSA 2003)*, Naples, Italy, June.

SAC 96–03. 1997: Interim guidelines. FEMA 267/A, SAC Joint Venture, California, USA.

Seilie, I.F., Hooper, J.D. 2005: Steel plate shear walls: Practical design and construction. *Modern Steel Construction, AISC*, 45(4), 37–43.

Seydel, E. 1933: The critical shear load of rectangular plates. Zeitshrift fur Flugtechnik und Motorluftschiffahrt, vol. 24, n. 3, p. 78. Verlag von Oldenbourg, Munchen und Berlin.

Skan, S.W., Southwell, R.V. 1924: On the stability under shearing forces of a flat elastic strip. *Proc. Soc. (London), Series A*, 105, 582.

Stein, M., Fralich, R.W. 1949: Critical shear stress of infinitely long, simply supported, plate with transverse stiffeners. NACA Tech. Note 1851.

Stein, M., Neff, J. 1947: Buckling stress of simply supported rectangular flat plates in shear. NACA. TN. 1222.

Stratan, A., Dubina, D. 2002: Seismic performance of eccentric braced frames with removable low yield steel link. In *Proceedings of International Conference Earthquake Loss Estimation and Risk Reduction*, 24–26 October, Bucharest, Romania.

Tamai, H. , Takenaka, H., Nakano, T., Kojima, O., Kondoh, K., Hanai, M. 1991: On hysteretic damper using low-yield stress steel plate installed in K-braced frame, Part 1. Outline of the device and its mechanism. Summaries of technical papers of annual meeting AIJ, Series C, 1447–1448.

Tanaka, K., Sasaki, Y. 2000: Hysteretic performance of shear panel dampers of ultra low yield strength steel for seismic response control of buildings. *Proceedings of the 12th WCEE*, (CD-ROM, paper no. 1248), 30 January–4 February, Auckland, New Zealand.

Tanaka, K., Torii, T., Sasaki, Y., Miyama, T., Kawai, H., Iwata. M. et al. 1998: Practical application of damage tolerant structures with seismic control panel using low yield point steel to a high-rise steel building. In *Proceedings of Structural Engineering World Wide*, Elsevier; CD-ROM Paper T190-4.

Thorburn, L.J., Kulak, G.L., Montgomery, C.J. 1983: Analysis of steel plate shear walls. University of Alberta Structural Engineering Report No. 107.

Timler, P.A., Kulak, G.L. 1983: Experimental study of steel plate shear walls. University of Alberta Structural Engineering Report No. 114.

Timoshenko, S. 1915: Stability of rectangular plates with stiffeners. *Mem. Inst. Engrs. Ways of Commun*, 89.

Timoshenko, S. 1936: *Theory of Elastic Stability*. McGraw-Hill Book Company, Inc., New York.

Tromposch, E.W., Kulak, G.L. 1987: Cyclic and static behaviour of thin panel steel plate shear walls. Struct. Eng. Rep. No. 145, Dept. of Civ. Engrg., University of Alberta, Edmonton, Alta, Canada.

Tsai, K.C., Lai, J.W., Hwang, Y.C., Lin, S.L., Weng, Y.T. 2004a: Research and application of double-core buckling restrained braces in Taiwan. In *Proceedings of the 13th World Conference on Earthquake Engineering*, Vancouver, Canada.

Tsai, K.C., Weng, Y.T., Lin, S.L., Goel, S. 2004b: Pseudo-dynamic test of a full-scale CFT/BRB frame–Part I: Performance-based specimen design. In *Proceedings of the 13th World Conference on Earthquake Engineering*, Vancouver, Canada, CD-ROM, Paper No. 750.

Uang, C.-M., Yu, Q.-S., Noel, S., Gross, J.L. 2000: Cyclic testing of steel moment connections rehabilitated with RBS or Welded haunch. *ASCE Journal of Structural Engineering*, 126(1), 57–68.

Wada, A., Nakashima, M. 2004: From infancy to maturity of buckling restrained braces research. In *Proceedings of the 13th World Conference on Earthquake Engineering*, Vancouver, Canada, CD-ROM, Paper No. 1732.

Wagner, H. 1931: Flat sheet metal girders with very thin metal webs. Part I—General theories and assumptions. National Advisory Committee for Aeronautics. Technical Memo, No. 604.

Wolski, M., Ricles, J.M., Sause, R. 2006: Seismic resistant self-centering steel moment resisting frames with bottom flange friction devices. In *Proceedings of 5th International Conference on the Behaviour of Steel Structures in Seismic Areas (STESSA 2006)*, Yokohama, Japan, 14–17 August, 481–487.

Xue, M., Lu, L.W. 1994: Interaction of infilled steel shear wall panels with surrounding frame members. In *Proceedings of the Structural Stability Research Council Annual Technical Session*, Bethlehem, 339–354.

Zekioglu, A., Mozaffarian, H., Le Chang, K., Uang C-M., Noel, S. 1997: Design after Northridge. Modern Steel Construction, March, *AISC*, 37(3), 36–42.

Chapter 5

Advances in steel beam ductility

5.1 NEW CONCEPTS ON STRUCTURAL DUCTILITY

The design of earthquake-resistant structures is characterized by many uncertainties that the structural designer has to face. The good behavior of a structure during a seismic attack is guaranteed by safety checks, which are required according to three levels of design approach: serviceability, damageability, and ultimate limit states. For the first two limit states, exceeding the design values of seismic actions cannot produce important effects. In contrast, as the ultimate limit state is referred to as the structural behavior near-collapse in case of severe earthquakes, the variability, dispersion, scatter, and, in general, the uncertainty in the evaluation of the design value of seismic actions seem to be the rule, not the exception. The result could be the total or partial structural collapse, which is not acceptable to the seismic design philosophy.

In order to consider this situation, the structure must be endowed by design with the ability to develop and maintain its bearing capacity, even when the considered seismic action exceeds the design limits. A measure of this ability is the *ductility*, that is, the capacity of the structure to sustain this overload by large deformations in plastic range without significant loss of resistance. So, the prediction of the available ductility under seismic loads is a key point in seismic design. A measure of the ductility is the *plastic rotation capacity* of the member sections. Unfortunately, in present design codification, there are no provisions for the determination of this rotation capacity. Eurocode 8 (EN 1998-1-1) proposes the classification of the structural cross-sections into four ductility classes: *ductile, compact, semicompact*, and *slender*, but it is clearly stated that only the first three classes can be considered for seismic design. The main criticism of this classification refers to the fact that ductility evaluation is just based on the use of the cross-sectional characteristics, without considering the third dimension of the member. As an alternative, the Italian code OPCM 3431 (2005) introduced a new classification criterion into three ductility classes, *ductile*, *plastic*, and *slender*, which was based on the characteristics of steel members and introduced an overstrength factor.

According to the context of performance-based design, a structure is designed so that, under a specific ground motion, the structural performance is lying within prescribed bounds. To achieve these levels of verification, seismic design must be based on the required-available formulation (Gioncu and Mazzolani, 2002, 2010):

REQUIRED CAPACITY < AVAILABLE CAPACITY

Currently, the required-available equation is applied to the triad of mechanical properties, which must be considered in seismic design: *rigidity*, *strength*, and *ductility:*

REQUIRED RIGIDITY < AVAILABLE RIGIDITY

REQUIRED STRENGTH < AVAILABLE STRENGTH

REQUIRED DUCTILITY < AVAILABLE DUCTILITY

A coherent strategy for seismic design requires that the structure must be verified for rigidity at the serviceability level, for strength at the damageability level, and for ductility at the ultimate limit.

Looking at these three verifications, one can remark that the first two are not difficult to fulfill, but the last one is often difficult to satisfy. The main reason for this is the difficulty in defining the required and available ductility, on the basis of the quite vague provisions of the codes for the determination of structural ductility. In addition, the new design philosophy states that the required ductility must be defined in function of the earthquake characteristics (Figure 5.1):

1. Reduced ductility but high strength (reduced q factor) for earthquakes with a short duration and a reduced number of cycles
2. High ductility but reduced strength (high q factor) for earthquakes with a long duration and a large number of cycles

where q is the behavior factor.

This classification is a function of the structural capacity to dissipate seismic energy in a given time, by a reduced or large number of cycles. Therefore, this methodology should require the possibility of designing the structural ductility (reduced or high) according to the earthquake characteristics, but this choice is not introduced in the existing code rules.

A solution to improve this situation is to define ductility using the rotation capacity of structural members. Therefore, analytical or numerical

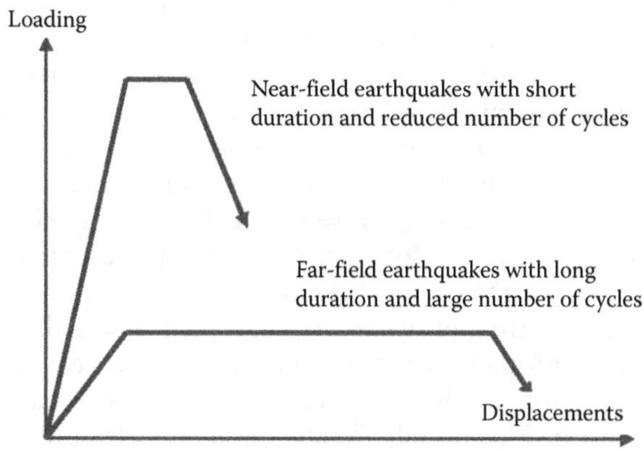

Figure 5.1 Different conceptions on strength and ductility.

methodologies for determining the available rotation capacity of steel members are a crucial design demand. Among the existing methodologies (experimental, theoretical, and empirical), it was proved that the most efficient one for design purposes is the local plastic mechanism model based on yield lines and plastic zones (Gioncu and Petcu, 1997; Gioncu and Mazzolani, 2002).

Seismic-resistant structures are usually designed relying on their ability to sustain high plastic deformations. The design philosophy considers that the earthquake input energy is dissipated through the hysteretic behavior of a member; plastic hinges are formed in predetermined positions due to a number of cycles of seismic loading. This concept is based on the condition that the plastic hinge must show a stable hysteretic behavior with a sufficient rotational ductility to allow for dissipation of this input energy. Before the 1960s, the notion of ductility was used only to characterize material behavior. After Housner's studies regarding earthquake problems and Baker's research works on plastic design, this concept has been extended to the structural level. According to this design philosophy, the structure may be designed for lower forces than those it has to resist, taking into account the inelastic reserves of the structural system. Therefore, evaluation of the available ductility is of primary importance.

The use of the *monotonic ductility* for seismic actions has provided to be a valuable concept in many earthquakes and it corresponds to the methodology included in the modern codes. For instance, Eurocode 8 specifies the use of the ductility classes prescribed in Eurocode 3, mainly determined for static design. This approach is based on the observation that in many cases the load-deformation skeleton curves (built on the cyclic curve) fit very well to the monotonic curves. The ductility of steel members under monotonic loads has been studied by Gioncu et al. (2012) and Anastasiadis et al. (2012).

Unfortunately, the 1977 Vrancea and the 1985 Mexico City earthquakes, both far-field earthquakes with relevant soft soil conditions, as well as the 1994 Northridge and 1995 Kobe earthquakes, both characterized as near-field events, caused unexpected damage, which has seriously compromised the validity of the code provisions, with respect to required and available ductility, based on monotonic ductility. In this direction, they revealed a significant gap in the knowledge of seismic behavior of steel structures in extreme conditions of seismic loading. As a consequence, after these series of devastating earthquakes, it has been recognized by the scientific society that both seismic hazard and risk treatment to be reassessed.

In accordance with these remarks, a new concept, based on the *seismic ductility,* which considers the influence of the main characteristics of the earthquake, must be introduced. This target is possible only if the impressive progress in seismology will be transferred by engineering seismology into earthquake engineering (Gioncu and Mazzolani, 2011). The basic concepts of today's seismic codes were born almost 70 years ago, when knowledge on the seismic actions and structural response were rather poor. Today, the earthquake-resistant design has grown within the new multidisciplinary fields of engineering seismology and earthquake engineering, wherein many exciting developments have taken place and are predicted in the near future. Therefore, it is very clear that any progress is impossible without considering the new amount of knowledge recently cumulated in seismology, due to the fact that the same structural system behaves in a different manner as a function of earthquake type. Lessons learned since the past 40 years of real earthquake excitations reveal that the following issues should be considered in seismic design:

1. The ground motions characteristics in function of source–site distance (far-field and near-field earthquakes)
2. The source types (interplate, intraplate, and intraslab) with very different rupture characteristics

There are such large differences among the ground motions generated by these sources at different epicenter–site distances that ignorance of these aspects can be considered to be a shortcoming in seismic design. In this context, engineering seismology is now paying more attention to the establishment of differences in the main characteristics of the sources. At the same time, the task of earthquake engineering is to take more care about the structural response to different ground motions.

Current design practice is based only on the earthquake magnitude and the corresponding spectrum; however, it does not assure a proper and safe seismic design for all earthquake types. In the past, owing to the reduced number of records during severe earthquakes, mainly obtained far from the recording stations, the codified design methodologies were developed on the consideration that the ground motions during earthquakes are characterized only by the large number of reversal cycles in accelerograms and by site soil conditions. Very recently, owing to the development of a large network of instrumentation all over the world, there is a large number of records of ground motions for different epicentral distances as well as different local site conditions. The analysis of this new information has emphasized the diversity of ground typologies. For instance, this information offers a possibility to consider, for design purposes, the great differences in the ground motions between far-source and near-source seismic regions. In spite of this situation, the effects of near-field versus far-field ground motions are not well understood. Moreover, until now, structural design codes did not recognize the principal behavioral differences among the aforementioned earthquake types.

Each event is basically unique due to the influences of many different factors. An earthquake can be considered as the result to be the behavior of a nonlinear system that is extremely sensible to a very small change in initial conditions, approached by the theory of chaos and the science of seismology (Gioncu and Mazzolani, 2011). Therefore, the determination of the actual characteristics of an earthquake is a very difficult task and, as a consequence, the seismic actions are determined with great uncertainty.

The basic ductility design criterion, which any earthquake-resistant structure must satisfy, is generalized in the following way (Gioncu and Mazzolani, 2002):

SEISMIC REQUIRED DUCTILITY < SEISMIC AVAILABLE DUCTILITY

The seismic required ductility is the effect of the earthquake on the structure, determining the maximum values of seismic demand for design purposes. The seismic available ductility is the structure's ability to resist the earthquake effects without failure.

The analysis of the available ductility must also consider the joint behavior. Steel-framed building structures with full-strength beam-to-column joints are quite standard nowadays. Buildings utilizing such framing systems are widely used in design practice (Coelho et al., 2004). However, there is a growing recognition that there are significant benefits in designing joints as partial strength and semirigid. The design of joints within this partial-strength/semirigid approach is becoming more and more popular. The main components of a beam-to-column joint are the column flange and web, as well as the connecting elements and the beam flanges and web. The evaluation of the ductility of a steel joint requires the proper characterization of each component (Da Silva et al., 2002). Therefore, the ductility of a beam occupies the first place in full-strength joints and a very important place in defining the joint ductility.

Because the main results concerning the available joint ductility are determined for monotonic actions, a very important issue is to use these results for the evaluation of the influence of seismic actions. The general aspects of seismic required and seismic available ductility, in function of earthquake types, are presented by Gioncu and Mazzolani (2000, 2011). The main characteristics of near-field versus far-field earthquakes in determining the required ductility are presented by Hall et al. (1995), Krawinkler and Alavi (1998), Gyorgyi et al. (2006), and Gioncu and Mazzolani (2011). The main conclusion of these studies is that very large differences exist in postelastic response, when the effects of far-field and near-field earthquake loading as well as the source type are considered.

In contrast, the references are very poor in case of seismic available beam ductility. For far-field earthquakes, characterized by a large number of reversal cycles and accumulation of plastic deformations, the available ductility is treated by Gioncu and Mazzolani (2002) and Lee and Stojadinovic (2008). For near-field earthquakes, where the seismic loads are characterized by velocity pulses, a reduced number of reversal cycles, and the influence of strain rate, some results are presented by Gioncu and Mazzolani (2011) and Gioncu et al. (2013).

In this context, the determination of seismic available ductility must consider the differences in ground motions, which is a very difficult task. The simplest methodology is to evaluate the ductility for monotonic loads and to correct accordingly the determined values, considering the specific characteristics of each earthquake type:

$$\text{SEISMIC DUCTILITY} = \{\text{Correction factor}\} \times \text{MONOTONIC DUCTILITY}$$

Considering the complexity of problems related to seismic available ductility in function of earthquake types, the analysis must be divided into two parts:

- The far-field earthquakes characterized by cycle loadings
- The near-field earthquakes characterized by effects of seismic wave propagation with high-velocity pulses and strain rate

5.2 DUCTROT-M COMPUTER PROGRAM

5.2.1 Investigation on local plastic mechanism models for beams

During the experimental tests on the steel wide-flange beam, one can observe that the plastic deformations are produced only in a limited zone, the remaining part of the beam remaining in an elastic field. In this plastic

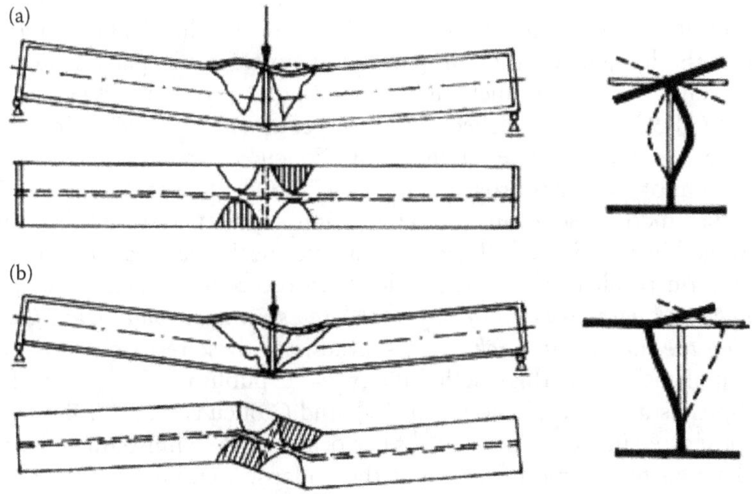

Figure 5.2 Plastic buckling types for standard beam SB1: (a) in-plane buckling; (b) out-of-plane buckling.

zone, large rotations are concentrated, working as plastic hinges. The plastic rotations are amplified if in these zones plastic buckling of the flange and web occurs. Two main buckling types were observed during the experimental tests: in-plane and out-of-plane buckling modes (Lukey and Adams, 1969) (Figure 5.2).

After these experimental results, a series of tests were performed, confirming the theoretical results (Gioncu and Petcu, 1997; Gioncu and Mazzolani, 2002). Two beam specimens were tested, symmetrically loaded by one or two vertical forces, having different moment variations in the zones where plastic hinges are formed. The first loading system models the linear moment gradient, while the second system models the constant moment. Important differences between the obtained results are observed, indicating that the moment variation is an important factor in determining the beam ductility.

After these experimental results, a series of studies were performed for finding analytical expressions of moment–rotation curves in plastic range. The first application of the plastic mechanism method was given in the paper of Climenhaga and Johnson (1972), which considers the plastic hinge formed in the flange and web from the experimental evidence. The most important aspect of this paper is the fact that both in-plane and out-of-plane behaviors are considered. An important progress in using this methodology was given by Ivanyi (1979), based on the use of the virtual work to analyze the local plastic mechanism. Kuhlmann (1985) and Feldmann (1994) proposed a plastic mechanism for in-plane plastic buckling, by considering the interaction between the flange and the web.

Intensive research works started in Timisoara in 1989 to determine the ductility of wide-flange members, by using the local plastic mechanism methodology, for both in-plane and out-of-plane plastic buckling modes. The first results, including the computer programs POSTEL and DUCTROT 93, were published by Gioncu et al. (1989, 1994). The results of the new researches were published in Gioncu and Petcu (1997) and the new variant of computer program, the DUCTROT-96, was elaborated. The elaborated methodology allows the rotation capacity and ductility to be determined in function of all geometrical and mechanical parameters.

The main result of these studies leads to the conclusion that the *plastic buckling operates as a filter against large strains in the tension flange, reducing the danger of cracking.* A synthesis of these research works and some applications of this methodology were published by Anastasiadis (1999), Anastasiadis and Gioncu (1999), and Gioncu et al. (2000).

Another step in the research works performed at the Timisoara University was devoted to the modification of the collapse mechanism, mainly for improving the dimensions of the plastic zone in the flange local mechanism and the shape of the web local mechanism. Based on this new local plastic mechanism, a new computer program, DUCTROT-M, was elaborated in 2001 (Gioncu and Petcu, 2001) and the results were published in Gioncu and Mazzolani (2002) and Petcu and Gioncu (2003).

This computer program provided a very good correspondence with the experimental results. Details concerning the new shape of plastic mechanisms for in-plane and out-of-plane mechanisms can be found in Gioncu and Mazzolani (2002). Further applications of these studies, aimed at improving the rotation capacity, were published by Gioncu (2006a,b) and Gioncu et al. (2009).

In the last period, new aspects of the use of local plastic mechanisms for evaluating rotation capacity, based on the development presented by Climenhaga and Johnson (1972), were published by Lee and Stojadinovic (2003, 2008) and Tan et al. (2007).

5.2.2 Characteristics of DUCTROT-M computer program

5.2.2.1 Modeling the member behavior

The available rotation capacity must be determined taking into account that the member belongs to a structure with a complex behavior. But this is a very difficult task, due to the great number of factors influencing the behavior of the actual member. Thus, it is important to simplify the analysis by substituting the actual member with a simple member having a very similar behavior. This substitutive member is the so-called *standard beam*, which was used for the first time by Gioncu and Petcu (1997) to evaluate the rotation capacity.

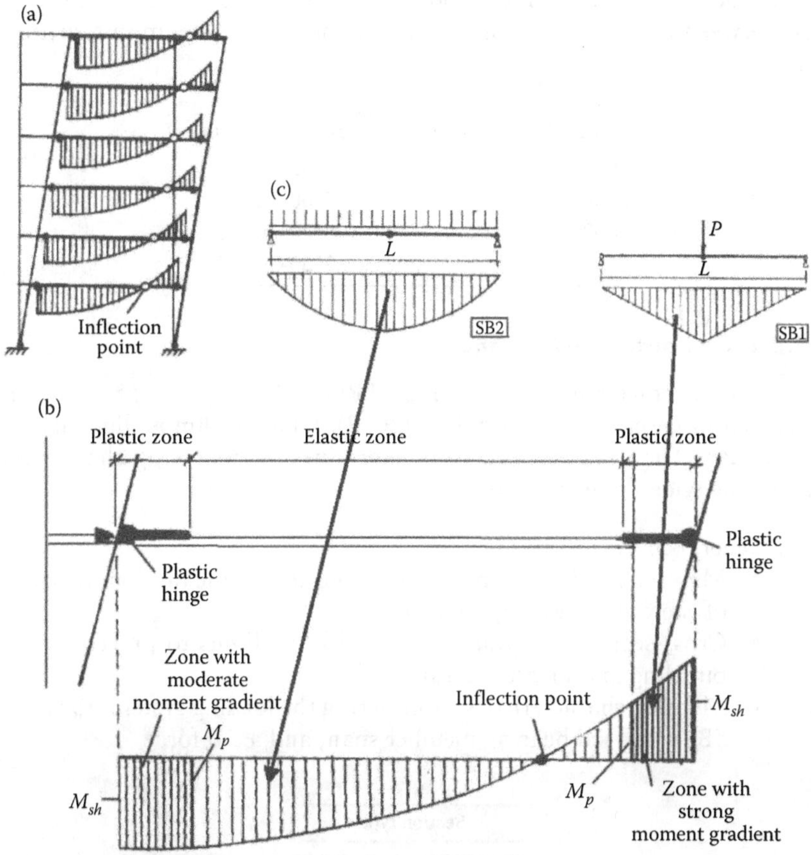

Figure 5.3 Using standard beams to determine rotation capacity: (a) structure scheme; (b) influence of moment gradient; (c) standard beam types.

Figure 5.3a shows the behavior of a framed structure, where the *inflexion points* divide the beams into two portions, with positive and negative bending moments. The rotation capacity of the beam ends must be determined under different conditions. For positive moments, the plastic hinge forms in a quasi-constant gradient zone, while, for negative moments, it forms in important moment gradient zone (Figure 5.3b). Therefore, the actual member behavior in the structure can be replaced by two standard beam types (Figure 5.3c):

- SB1, with a central concentrated load in the zone under a quasi-linear moment gradient
- SB2, with a distributed load beam in the zone with a weak moment gradient

Considering that the inflexion point is situated at (0.2–0.3) L_b, the relation between the standard beam span, L, and the real beam in structure, L_b, is

Standard beam span (mm)	*Beam span (mm)*
2000	3500–5000
3000	5000–7500
4000	6500–10,000
5000	8000–12,500

5.2.2.2 Computer performance

The main characteristics of the DUCTROT-M computer program are presented in Gioncu and Mazzolani (2002). Its algorithm is illustrated in Figure 5.4a. The computer program determines the following characteristics for the wide-flange members:

1. General characteristics:
   - Material mechanical properties, considering the random variation of the yield stress superior limits
   - Cross-sectional characteristics, with the limits to prevent elastic buckling and brittle fracture
   - Member characteristics, considering the loading systems (SB1 and SB2 standard beams), member span, and axial force

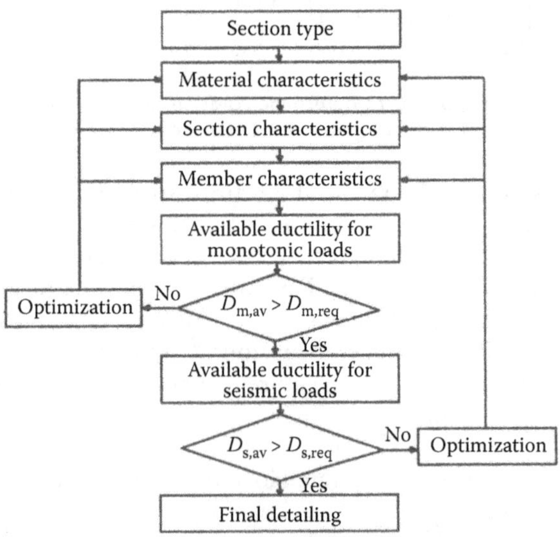

Figure 5.4 DUCTROT-M computer program: analysis algorithm.

2. Ductility characteristics for monotonic loads:
   - Two main plastic buckling loads, which produces in-plane and out-of-plane plastic mechanisms
   - Moment–rotation curves for the two mechanism types and for gradient or quasi-constant moments
   - Determination of ultimate rotation and rotation capacity for the defined level related to the fully plastic moment (1.0 or 0.9)
   - Main geometrical dimensions of plastic mechanism shape
   - Influence of the beam–column connection details
   - Selection between in-plane and out-of-plane plastic mechanisms
   - Influence of fabrication type: rolled or welded (fillet or penetrated welds)
3. Ductility for seismic actions:
   - Influence of pulse loading, in function of strain-rate level and exterior temperature
   - Influence of cyclic loading, in function of cycle type (increasing, constant, or decreasing) and cycles number

The computer program DUCTROT-M is now freely available on CD-ROM as an Appendix of the book of Gioncu and Mazzolani (2002); it is also free on the site of Petcu and Gioncu (2002).

5.2.3 Local plastic mechanism for gradient moments

5.2.3.1 In-plane local plastic mechanism

The local plastic mechanism is characterized by a deformation located in the beam plan, without lateral displacements of flanges (Figure 5.2a). The shape of collapse mechanisms observed during the experimental tests is composed by yield lines (true mechanism) or by a combination of yield lines and plastic zones (quasi-mechanism). The last mechanism type involves a larger amount of energy than the first type, due to the membrane yielding of the plastic zone (Gioncu and Mazzolani, 2002). An evolution can be noted in the attempts to transform the experimental mechanism into analytical and numerical models.

The Climenhaga and Johnson (1972) and Ivanyi (1979) collapse models are based on a perfect symmetry with respect to the vertical axis and consider the length of the flange and the web plastic zone equal to the flange width.

Kuhlmann's model (1985) is based on yield lines only, for symmetric local plastic mechanisms, formed by plastic buckling of the compressed flange and compressed part of the web.

Feldmann's model (1994) considers only the collapse mechanism formed in the compressed flange.

Table 5.1 Symmetric Collapse Plastic Mechanism Types

Author(s)	*Flange*	*Web*
Climenhaga and Johnson (1972) Ivanyi (1979)	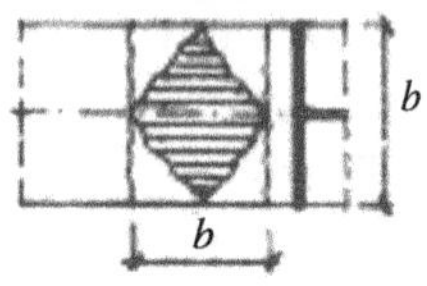	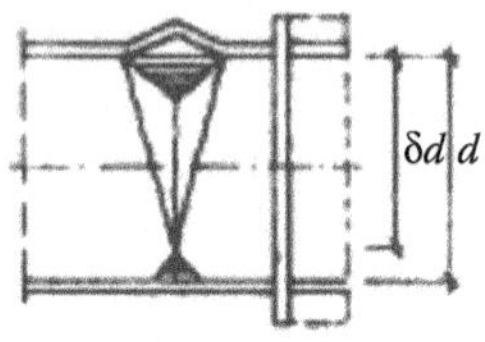
Kuhlmann (1986)	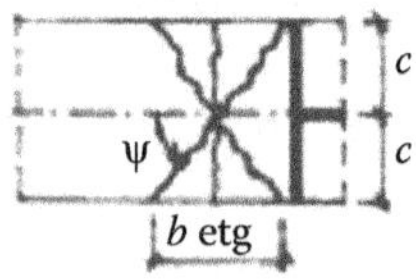	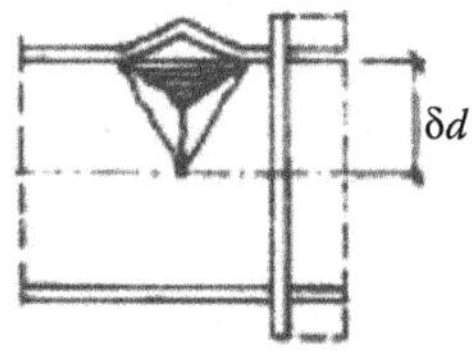
Feldmann (1994)	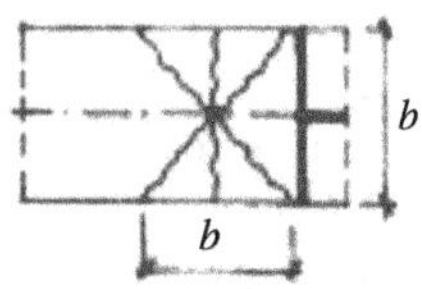	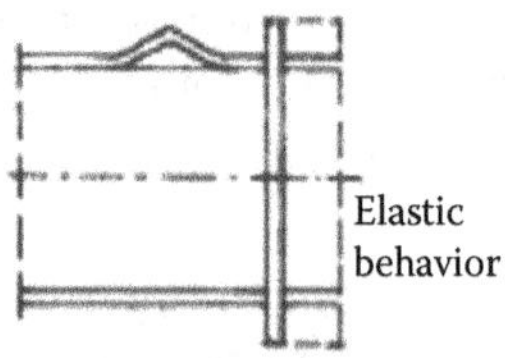

The POSTEL (Gioncu et al., 1989) and DUCTROT 93 (Gioncu et al., 1994) computer programs, elaborated at the Timisoara University, are based on the above considerations (Table 5.1).

The second stage consists in the improvement of plastic mechanism shapes, examining the experimental tests, proposing an asymmetric plastic mechanism (Table 5.2).

The new versions of the computer program, DUCTROT-96, devoted to the evaluation of the rotation capacity, were based on an improved mechanism, consisting in a more suitable shape for the flange and web plastic local deformations. By a more careful examination of the experimental tests (SAC, 1996), one can observe that the width of the web plastic zone is different from the flange width (Figure 5.5a), as was considered in the previous plastic mechanism models. Taking into account that the dimensions of plastic zones of the flange and the web must be the same, the shape of the flange plastic mechanism was improved.

The second improvement refers to the rotation point of the web plastic mechanism. Two cases are studied, considering the beam–column connection details:

- The mechanism type corresponds to the case when the shape of the mechanism is limited to the presence of the column flange or vertical ribs (Figure 5.5b). Applications of the first program version were published in Gioncu (2006a,b) and Gioncu et al. (2009).

Table 5.2 Asymmetrical Collapse Mechanism Types

Author(s)	*Flange*	*Web*
Gioneu and Petcu (1995, 1997)	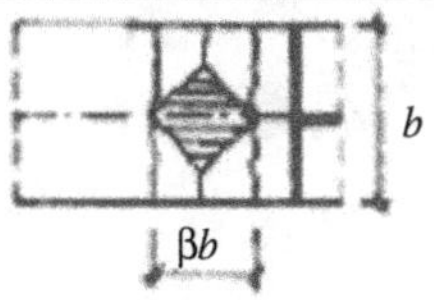	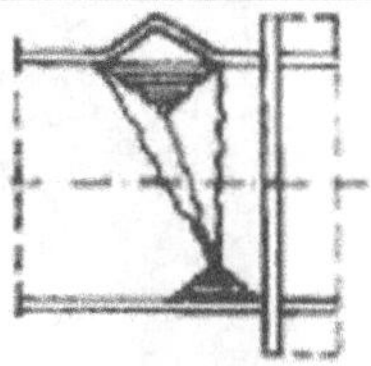
Tehami (1997)	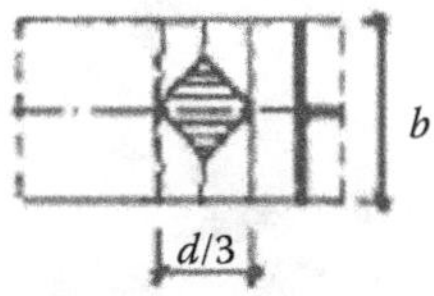	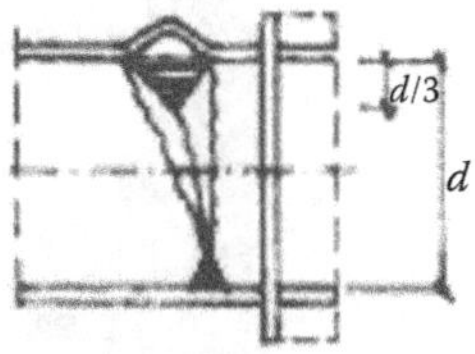
Gioneu and Petcu (2011)	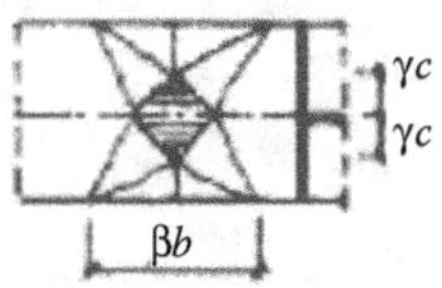	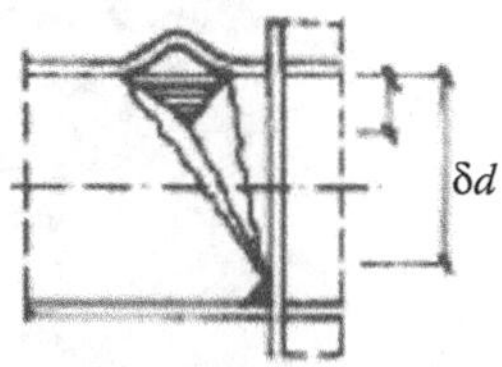

- The formation of the web shape plastic deformation is free, having no imposed rotation point (Figure 5.5c), in case of the plastic hinge formed far from the column (cover plates or dog-bone solutions for the beam–column connection).

The final version of the computer program, DUCTROT-M (Gioncu and Mazzolani, 2002; Gioncu and Petcu, 2001; Petcu and Gioncu, 2002; Gioncu, 2011; Gioncu et al., 2012) was elaborated considering both cases as shown in Figure 5.5.

For the standard beam SB1 (Figure 5.6a), the global plastic mechanism is presented in Figure 5.6b, being composed by two local plastic mechanisms. The mechanisms rotate around the rotation center O. The experimental moment–rotation curve is presented in Figure 5.6c. There are some important characteristic points, which mark some important changes in the beam behavior. The first point A refers to the reach of the flange yielding; the second one B is defined by the occurrence of the fully plastic moment. To develop the plastic hinge, the rotation must increase; a very important observation, the increase in the bending moment over the fully plastic moment, is due to the behavior in the strain-hardening range. The maximum moment value is reached at point C, when plastic buckling occurs in the yielding zone of the compression flange and web. At this step, the local plastic mechanism is formed. After point C, the bending begins to decrease with the

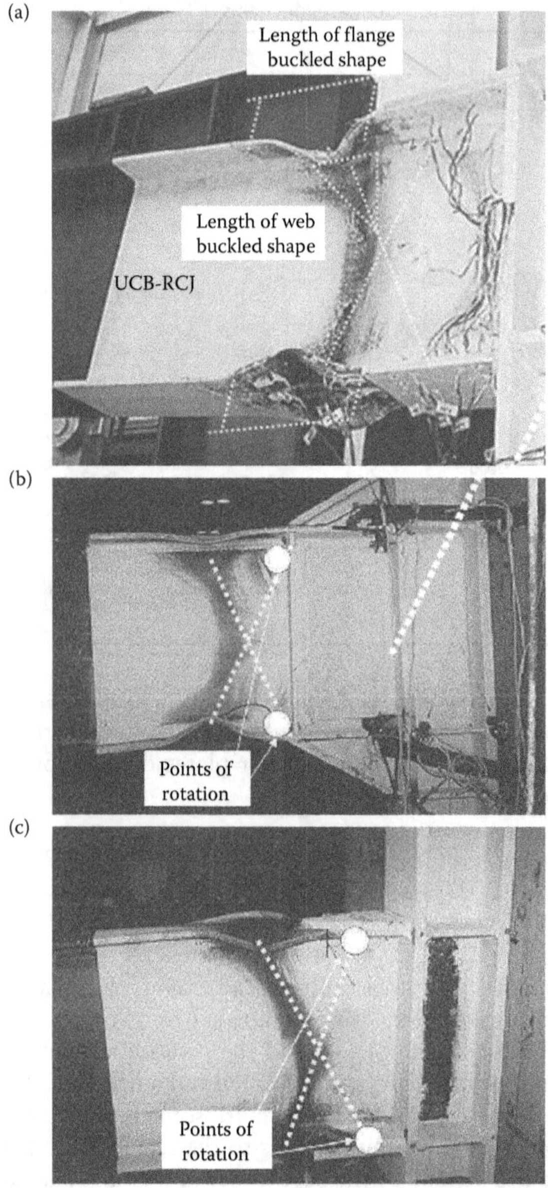

Figure 5.5 Experimental aspects of local in-plane plastic mechanisms: (a) flange and web lengths of buckled shape relation; (b) position of rotation point in case of vertical ribs; (c) position of rotation point in case of free conditions. (Modified after SAC 1996: Connection test summaries. Report No. SAC-96–02, SAC Joint Venture, Sacramento, California, http://www.sacsteel.org/design/test-summaries.html.)

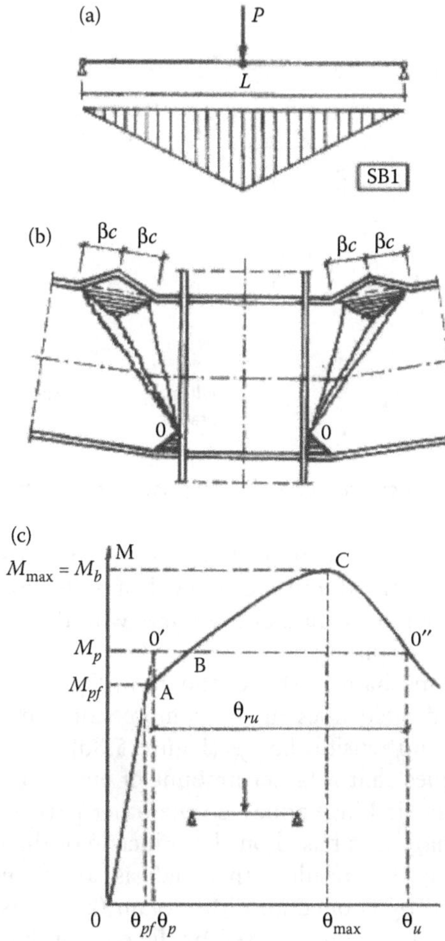

Figure 5.6 Local plastic mechanisms for gradient moment: (a) SB1 standard beam; (b) local plastic mechanism; (c) actual moment–rotation curve.

increase in rotation and the equilibrium of the beam becomes unstable. The ultimate rotation capacity is determined in the lowering postbuckling curve at the intersection with the theoretical fully plastic moment M_p C(point 0″).

The theoretical moment–rotation curve provided by the computer program reproduces the above experimental curve, as shown in Figure 5.7, where both elastic and plastic behaviors are easily described by analytical relationships. The difficulty in completing the moment–rotation curve is related to the plastic postbuckling behavior, where the degradation of the moment capacity is due to the plastic buckling. For this purpose, the use of local plastic mechanisms proved to be very operative. The maximum

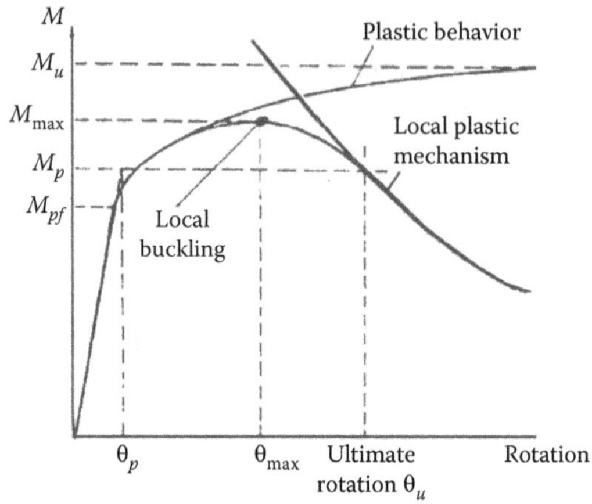

Figure 5.7 Theoretical postcritical curve of local plastic mechanism.

moment results at the intersection of the plastic and postcritical curves, eroded by local geometrical and mechanical imperfections. The ultimate rotation results at the intersection of this curve with the line corresponding to the fully plastic moment M_p.

The local plastic mechanism shape, shown in Figure 5.8a, is composed of plastic zones and yield lines in the compression flange (Figure 5.8b), web (Figure 5.8c), and tension flange (Figure 5.8d). The work of a plastic mechanism implies that a larger amount of energy is absorbed in the small area of the plastic hinge zones, so the other parts can be neglected. The rigid-plastic analysis is based on the principle of the minimum of the total potential energy. The result of this analysis, after some mathematical operations presented by Gioncu and Mazzolani (2002), is the postcritical curve of local plastic mechanism, given by the relationship

$$M/M_p = A_1 + A_2\, \theta^{-1/2} \tag{5.1}$$

The coefficients A_1 and A_2 contain the mechanical and geometrical characteristics of beams and the shape of the plastic mechanism. They are given in Gioncu and Mazzolani (2002) and Petcu and Gioncu (2003).

The postcritical curve depends on the geometrical parameters of the local plastic mechanism. The length of mechanism is studied by Gioncu and Petcu (1997), based on theoretical studies and experimental data. The new version of the computer program is based on these studies, keeping unmodified the length of the plastic mechanism. The influence of other parameters defining the shape of the plastic mechanism is presented in Figure 5.9. The conclusions of this analysis are as follows:

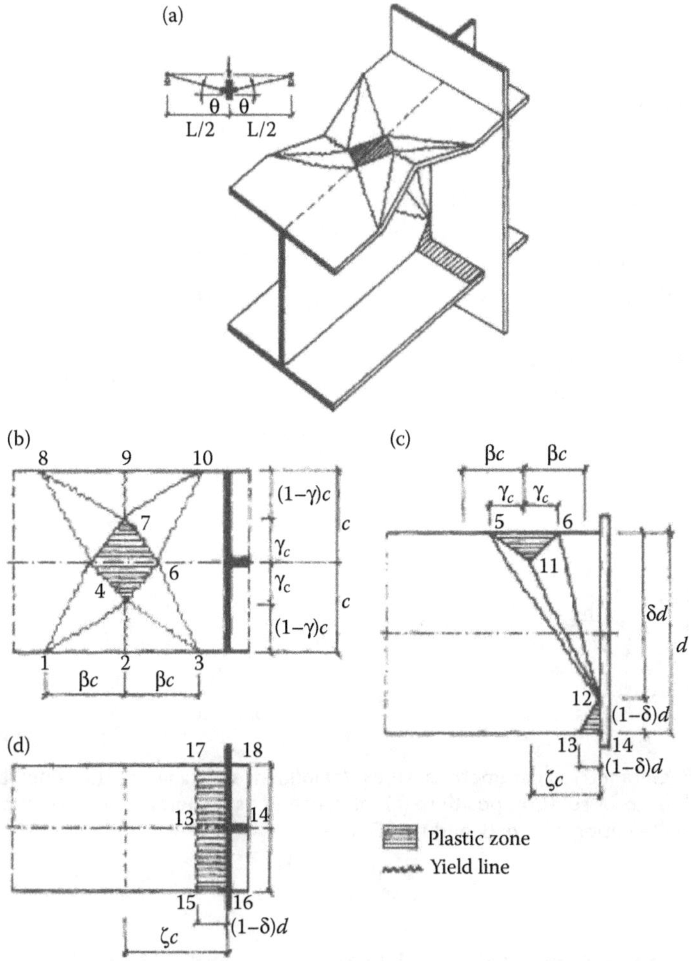

Figure 5.8 In-plane plastic mechanism: (a) general view; (b) compression flange plastic mechanism; (c) web plastic mechanism; (d) tension flange plastic mechanism. (Gioncu, V., Mazzolani, F.M. 2002: *Ductility of Seismic Resistant Steel Structures*. Spon Press, London.)

- The main dimension is the one defining the plastic zones of compression flange and web (Figure 5.9a), which presents a minimum in determining the rotation capacity.
- The parameter defining the position of the rotation point of the plastic mechanism shows a minimum for the position in the tension flange (Figure 5.9b).

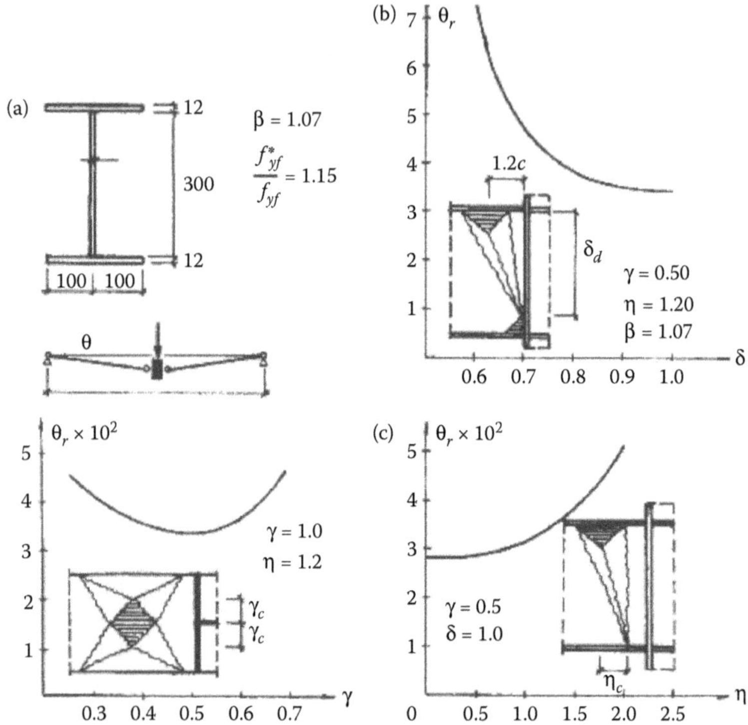

Figure 5.9 Geometrical parametrical study: (a) influence of flange plastic zone; (b) influence of rotation position; (c) influence of asymmetry of web plastic shape. Definition of γ, δ, β, and η in Figure 5.8.

- The asymmetry of the mechanism shape produces an increase in the rotation capacity, the minimum being obtained for a symmetric shape (Figure 5.9c).
- The minimization of the postcritical curve in function of the first two parameters is included in the computer program performance. For the last parameter, which depends on the position of the rotation point determined by constructional details (Figure 5.10), the characteristic of the plastic local mechanism asymmetry must be introduced in the performance data.

Figure 5.11 shows the determination of in-plane rotation capacity using the DUCTROT-M computer program.

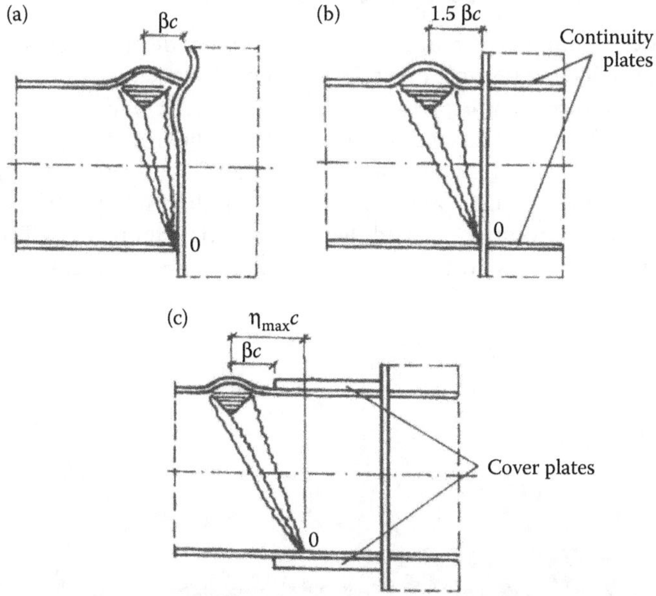

Figure 5.10 Asymmetry of plastic mechanism: (a) column without continuity plates; (b) column with continuity plates; (c) beam with cover plates.

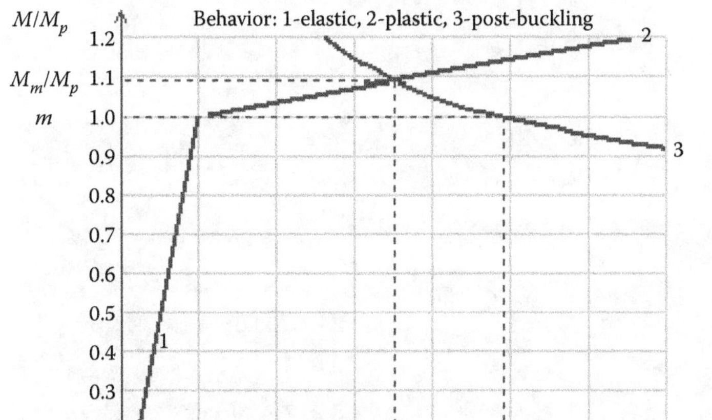

Figure 5.11 Determination of ultimate rotation for in-plane plastic mechanism using the DUCTROT computer program.

5.2.3.2 Out-of-plane local plastic mechanism

This plastic mechanism is characterized by the lateral displacements of flanges (Figure 5.2b). The experimental shapes of out-of-plane plastic mechanisms, called S-mechanisms, are presented in Figures 5.12 and 5.13 (Mateescu and Gioncu, 2000; Smith and Adams, 1968). Theoretical studies on this mechanism type are presented by Gioncu and Petcu (1997), Climenhaga and Johnson (1972), and Ivany (1979). This mechanism type is characterized by free lateral rotation around the vertical axis (Figure 5.14). Both these research works are related to joints where there are no elements to prevent

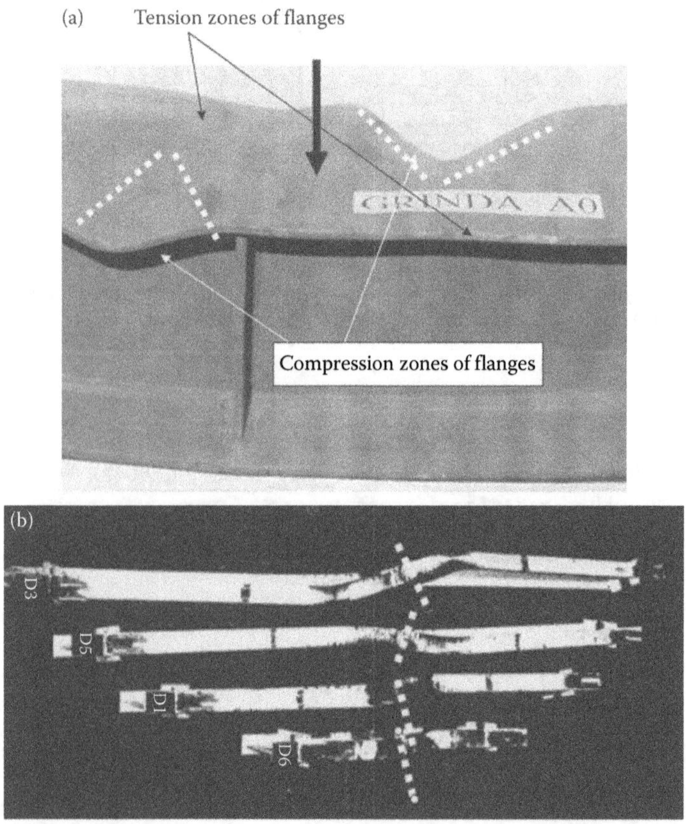

Figure 5.12 Experimental shape for out-of-plane plastic mechanism: (a) view of plastic mechanism. (Modified after Mateescu, G., Gioncu, V. 2000: *Behaviour of Steel Structures in Seismic Areas, STESSA 2000* (eds. F.M. Mazzolani, R. Tremblay), August 21–24, Montreal, Balkema, Rotterdam, 55–62.) (b) Lateral rotation of beam. (Modified after Smith, R.J., Adams, P.F. 1968: Experiments on wide-flange beams under moment gradient. Structural Engineering Report No. 13, May 1968, University of Alberta, Department of Civil Engineering.)

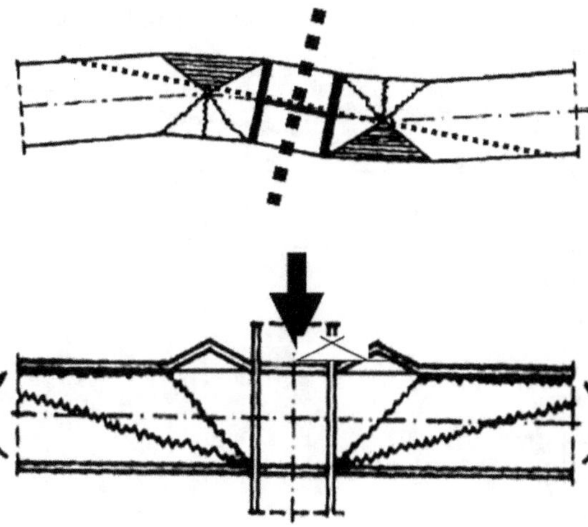

Figure 5.13 Out-of-plane plastic mechanisms.

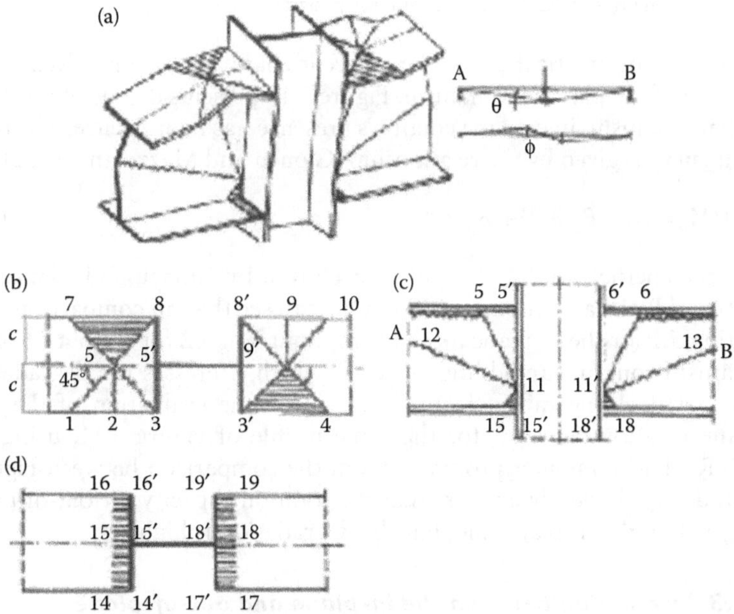

Figure 5.14 S-shaped plastic mechanism: (a) general view; (b) compression flange plastic mechanism; (c) web plastic mechanism; (d) tension flange mechanism. (Gioncu, V., Mazzolani, F.M. 2002: *Ductility of Seismic Resistant Steel Structures.* Spon Press, London.)

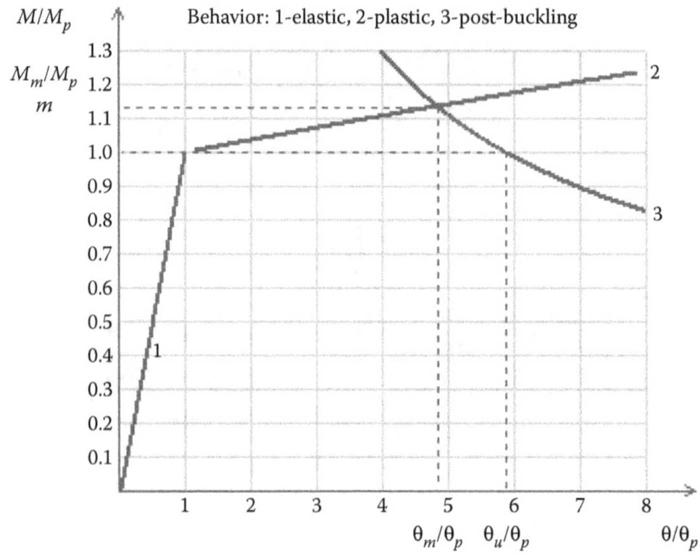

Figure 5.15 Determination of ultimate rotation capacity for out-of-plane mechanism using the DUCTROT-M computer program.

the rotation. In practical cases, the adjacent column is also involved in the formation of the plastic mechanism. Figure 5.15 presents details of this local plastic mechanism, in case the column's presence has no influence. The post-critical curve is given by the relationship (Gioncu and Mazzolani, 2002)

$$M/M_p = B_1 + B_2\,\theta^{-1/2} + B_3\,\theta^{-3/4} \tag{5.2}$$

where the coefficients B_1, B_2, and B_3 are given by Gioncu and Mazzolani (2002) and Petcu and Gioncu (2003). One can see that, in comparison with Equation 5.1 for the in-plane mechanism, Equation 5.2 for the out-of-plane mechanism contains an additional term, which expresses the degradation in the postbuckling range. Figure 5.15 shows the evaluation of the out-of-plane rotation capacity, for the same profile of Figure 5.11, using the DUCTROT-M computer program. From the comparison between Figures 5.11 and 5.15, it clearly appears that the rotation capacity for out-of-plane is larger than the in-plane one, but the degradation is higher.

5.2.3.3 Interaction between the in-plane and out-of-plane local plastic mechanisms

The experimental evidence (Lukey and Adams, 1969) shows that the first formed mechanism is the in-plane one in the majority of tests and only in the postbuckling range the beam buckles out-of-plane, due to the considerable

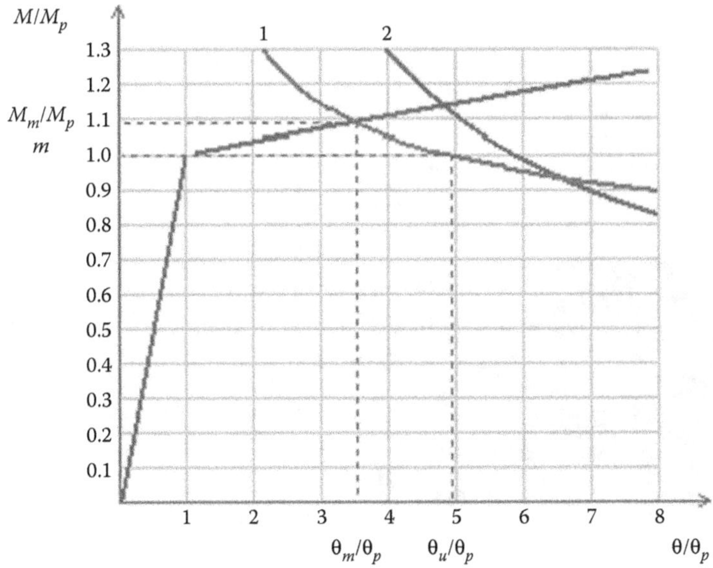

Figure 5.16 Interaction between in-plane and out-of-plane local plastic mechanisms: (1) in-plane mechanism; (2) out-of-plane mechanism.

weakening of the flange rigidity, caused by plastic deformations. There are two cases of interaction (Figure 5.16):

1. The intersection of the two postbuckling curves takes place under the line $M/M_p = 1.0$, when the rotation capacity is defined by the in-plane mechanism.
2. The intersection occurs over this line and the rotation capacity is defined by the out-of-plane mechanism, taking into account the interaction of these two buckling modes.

It is well known that the coupling of two buckling forms can increase the influence of the imperfections (Gioncu, 1998). But this form of coupling belongs to the category of weak interaction in a postcritical range, when the interaction could be neglected, being covered by the scatter caused by other factors that have a higher influence on the rotation capacity than the one produced by this interaction (Gioncu et al., 1996).

5.2.4 Local plastic mechanism for quasi-constant moments

The experimental data for the determination of the quasi-constant moment are obtained from beams loaded by two vertical forces (Ivanyi, 1979; Haaijer

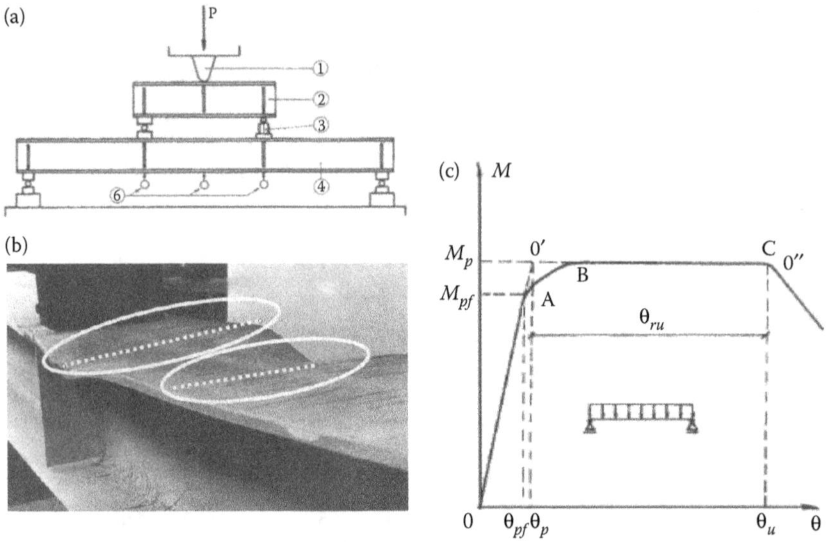

Figure 5.17 Experimental shape for plastic mechanism under constant moments: (a) Experimental test. (b) Local plastic mechanisms. (Modified after Hoglund, T., Nylander, H. 1970: Maximiforhallnde B/t for tryckt flans hos valsad I-balk vid dimensionering med graslastruetod. Technical University Stockholm, Report 83/197.) (c) Experimental moment–rotation curve.

and Thurlimann, 1958; Hoglund and Nylander, 1970) (Figure 5.17a). The local plastic mechanism is characterized by two symmetric in-plane shapes (Figure 5.17b). The corresponding experimental moment–rotation curve is given in Figure 5.17c. Figure 5.18 shows the standard beam for the evaluation of the rotation capacity (a) and the local plastic mechanism (b). The evaluation of the rotation capacity using the DUCTROT-M computer program is presented in Figure 5.19. Generally, the rotation capacity for quasi-constant moments is larger than the one for moment gradient.

5.2.5 Definition of ultimate rotation and rotation capacity

Using the DUCTROT-M computer program, it is possible to determine the ultimate rotation at the intersection of the postcritical curve with the theoretical fully plastic moment. In order to decide whether a structural member has sufficient ductility or not to provide a good response in seismic-resistant structures, the practice requires some indicators to be checked, like the ductility and/or the rotation capacity. One must recognize that, in the absence of standard definitions of these indicators, the most rational one could be related to the ultimate rotation, which is universally accepted by the specialists (Gioncu

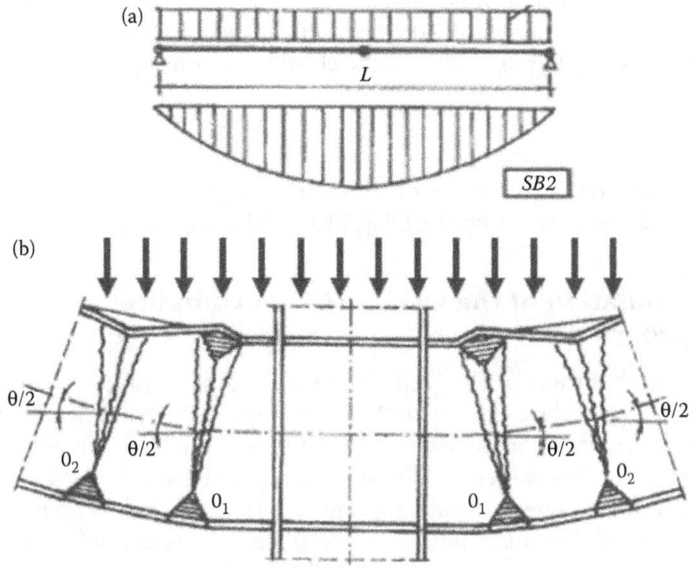

Figure 5.18 Local plastic mechanism for quasi-constant moments: (a) SB2 standard beam; (b) local plastic mechanism.

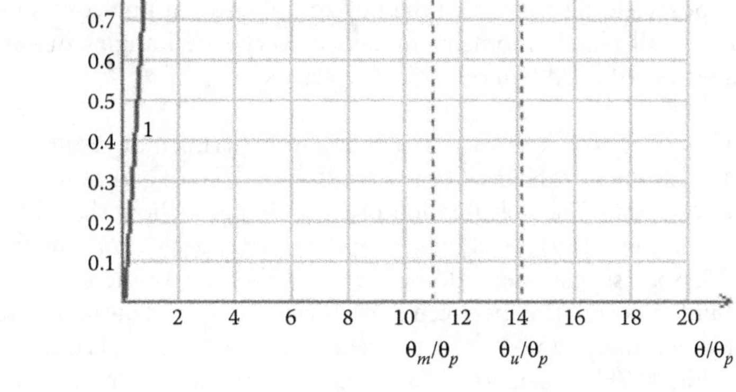

Figure 5.19 Determination of ultimate rotation using the DUCTROT-M computer program.

and Petcu, 1997). The member ductility is based on the determination of the rotation capacity parameter R, defined as the ratio between the plastic rotation at collapse θ_p ($\theta_p = \theta_u - \theta_y$) and the elastic limit one θ_y:

$$R = \theta_p/\theta_y \tag{5.3}$$

This definition requires the determination of the ultimate rotation θ_u, which can be done by using the DUCTROT-M computer program.

5.2.6 Validation of the DUCTROT-M computer program

After the elaboration of the DUCTROT-M computer program, the most important question was to evaluate how realistic the results were. First of all, in order to obtain an answer to this question, an extensive comparison between the theoretical results obtained using the DUCTROT-M computer program and the experimental tests reported in the technical literature has been performed. The most important testing results presented in the literature are selected in Gioncu and Petcu (1997), Lukey and Adams (1969), Spangemacher (1991), Boerave et al. (1993), Kemp and Dekker (1991), and Susuki (1997), numbering 81 specimens. A very careful examination of the related experimental values was performed, considering the possibility of identifying some errors or at least some imperfections connected with the following aspects:

1. Evaluation of the material properties in plastic and hardening ranges, which are less supervised than those in the preyielding field.
2. Measurement of rotations for elastic and plastic deformations, which might be affected by some errors due to testing arrangements.
3. Determination of the ultimate rotation, as the intersection between the postyielding curve and the horizontal straight line corresponding to the full plastic moment, in case of very small angles in-between, might be affected by nonnegligible errors.

Generally, it is well known that scatters in experimental results are very large in the stability problems. For instance, the studies of Nakashima (1994), referring to the statistical evaluation of steel strength, show that the coefficient of variation is 0.01–0.13, while for ductility capacity, this coefficient is 0.54–1.35. So, a special methodology for the validation of theoretical results must provide a correlation between experimental data and numerical results. A step-by-step methodology can be used to process the experimental data (Rumshiskii, 1978), considering the rough, systematic, or aleatory errors of measurements. Based on this methodology and using the 81 selected results coming from the technical literature (Gioncu and Tirca, 1996), the first step consisted in eliminating the six experimental values obviously wrong because they are out of the rational field (their values were very different from the other ones in the same group of specimens with the same characteristics). Within the remaining 75 data (variation coefficient 0.379), some errors are identified as the result of the above difficulties in measurements. Some of these errors can be discovered and eliminated, using probabilistic considerations, based on an adopted confidence level. Therefore, the mean value, standard deviation, and coefficient of variation are determined. The elimination of the wrong values is performed in two steps (15 in the first and 5 in the second), until a good confidence level $P = 0.95$ and coefficient of variation 0.202 are obtained. For the remaining 55 data, a good correspondence between experimental and theoretical results is achieved.

By using the new version DUCTROT-M, the eliminated values have remained the same, but a better coefficient of variation was obtained (0.185). The correlation between the existing experimental results (RE) and the calculated rotation capacity (RC) is given in Figure 5.20. One can see that, despite a quite large scatter area typical for the stability experimental data, the correlation is good, creating confidence in the developed local plastic mechanism as a methodology for determining the available rotation capacity.

A very important observation refers to the type of local plastic mechanism: the large majority of experiments (especially for rolled sections) show that the plastic buckling is due to the out-of-plane mechanism. This fact is confirmed by the numerical FEM analyses performed by Espiga (1997), showing that the dominant type of plastic buckling for rolled standard beams is the out-of-plan local plastic mechanism.

5.3 MONOTONIC AVAILABLE DUCTILITY

5.3.1 Applications of the DUCTROT-M computer program

A methodology based on the local plastic mechanism and the development of a computer program DUCTROT-M for determining the rotation capacity of wide-flange beams has been presented in Section 5.2. Based on the experimental evidences, two main plastic mechanisms are developed, in-plane and out-of-plane, for gradient and quasi-constant moment loading cases. The validation of this methodology using the experimental data from the technical literature shows that the results obtained by means of DUCTROT-M can be used in practical design. Some applications of this computer program are presented as follows, by analyzing in particular the practical aspects of structural design.

The first analyzed problem refers to the disputed cross-section ductility versus member ductility: the first methodology is commonly adopted

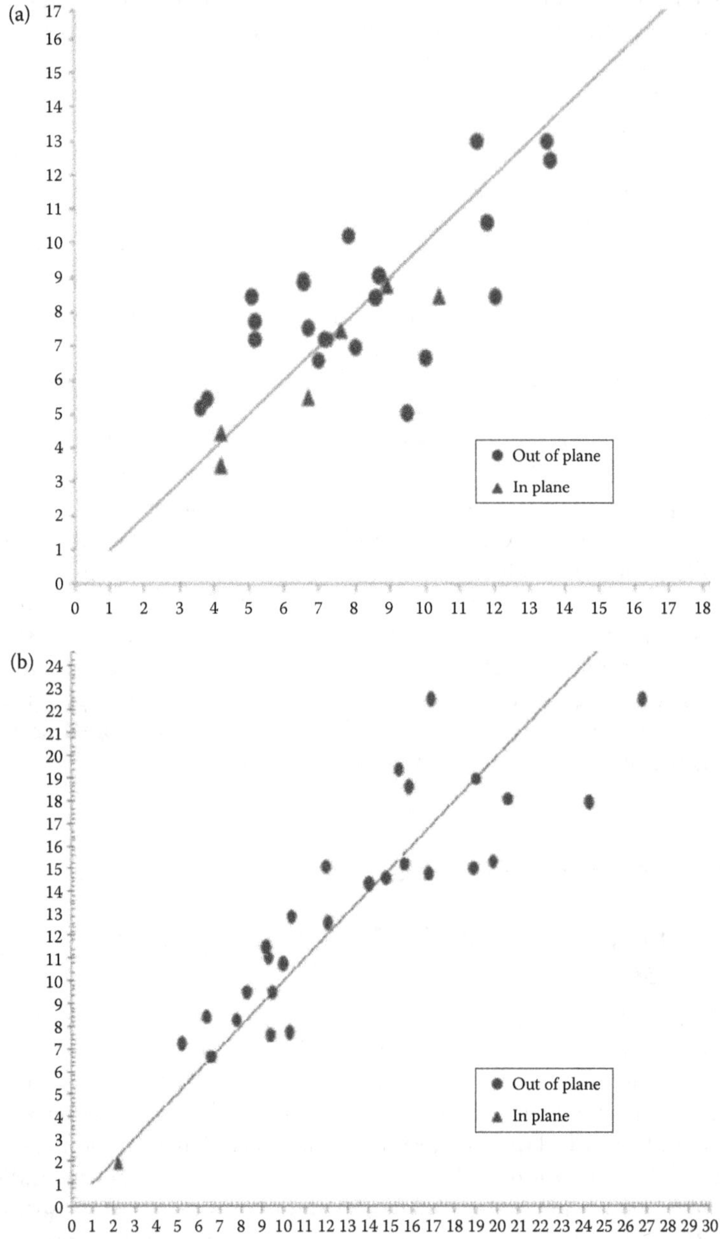

Figure 5.20 Correlation of theoretical-experimental values for rotation capacity; (a) Lukey-Adams (1969) and Kulhmann (1989) experimental values; (b) Spangemacher (1991), Boerave et al. (1993), Kemp and Dekker (1991), and Suzuki et al. (1994) experimental values.

in the codes, but it has been contested in many research works (Gioncu and Mazzolani, 1994). The possibility to compare the results of the two approaches represents the best way to clarify the terms of this dispute.

By examining the theoretical and experimental results obtained on standard beams and relating them to the practical aspects, two main criticisms can be mentioned:

- It is not realistic to neglect the presence of steel or concrete floors in the evaluation of the rotation capacity, the formation of the plastic local mechanism being impeded by the presence of floors. Therefore, in many cases, the quasi-constant moment has no meaning to be considered in design practice.
- All the theoretical and experimental research works neglect the fact that the beam ends are connected to columns and the out-of-plane rotation of the nodes is not free (as in the case of standard beams), due to the presence of transversal beams connected to the node. In these conditions, the main local plastic mechanism is the in-plane one.

5.3.2 Cross-section ductility versus member ductility

There are two ductility types widely used in the literature to characterize a beam inside of a structure (Gioncu and Mazzolani, 2002) (Figure 5.21):

1. *Cross-section ductility*, or curvature ductility, which refers to the plastic deformations of the cross-section, considering the independent behavior of the parts composing the cross-section itself.

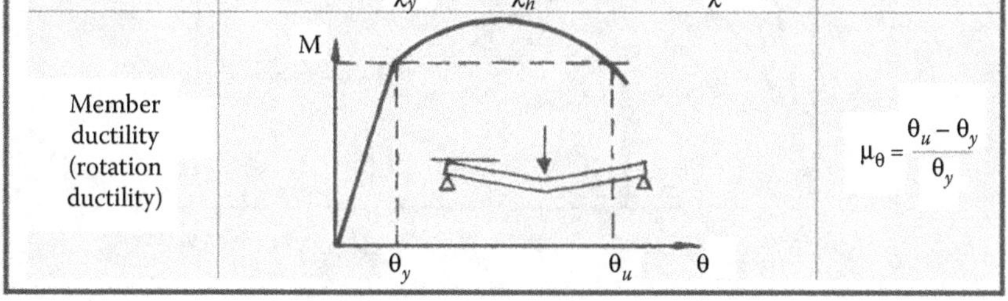

Figure 5.21 Ductility types: cross-section and member ductility.

2. *Member ductility*, or rotation ductility, when the properties of members (interaction between cross-section parts, influence of the beam span and loading system) are considered.

The characterization of the structural ductility by means of one of these two types gives rise to many discussions among the specialists.

The first definition is mainly used in the code provisions, based on the cross-section behavior classes (Figure 5.22c):

- *C1, class 1, plastic sections:* sections are characterized by the possibility to develop a plastic hinge with high rotation capacity.

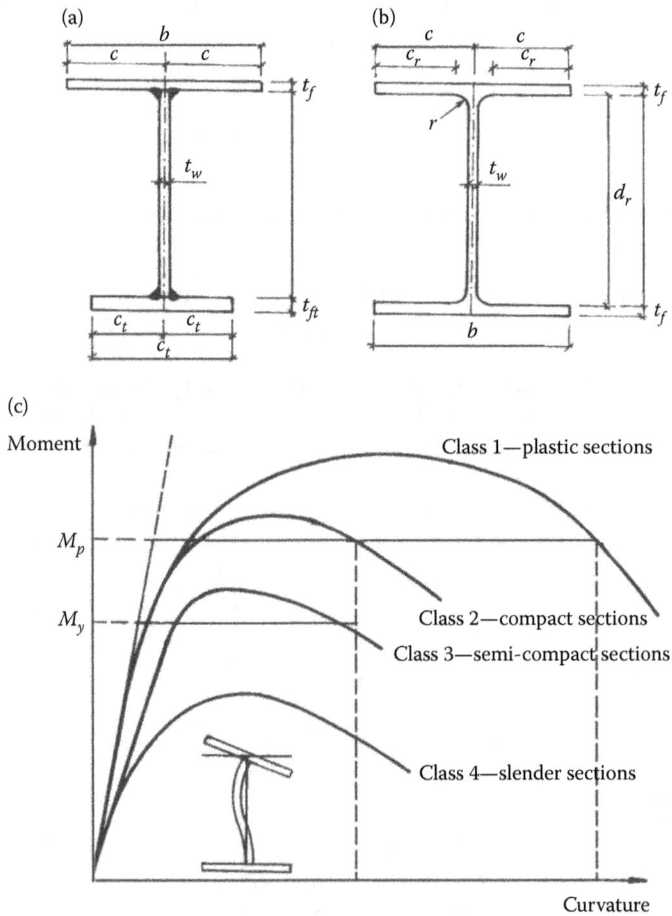

Figure 5.22 Cross-section classes: (a) welded sections; (b) rolled sections; (c) moment–curvature curve.

- *C2, class 2, compact sections:* sections are able to provide their maximum plastic flexural strength, but they have a limited rotation capacity due to some local effects.
- *C3, class 3, semicompact sections:* the bending moment capacity can be attained just for the first yielding, without reaching the plastic moment.
- *C4, class 4, slender sections:* sections are not able to develop their total flexural resistance due to the premature occurrence of local buckling in their compression parts.

For the wide-flange sections, the flange and web slenderness are defined by the ratios (Eurocode 3) (Figure 5.22a and b) (for steel S235):

Section types	*Flange c/t_f*	*Web d/t_w*
Rolled	C1 < 10	<72
	C2 < 11	<83
	C3 < 15	<124
Welded	C1 < 9	<72
	C2 < 10	<83
	C3 < 14	<124

This classification, being limited at the cross-section level only, has many deficiencies (Gioncu and Mazzolani, 2002):

1. Independent limitations between the flanges and web ratios are unreasonable because, obviously, the flange is restrained by the web, and the web by the flange.
2. The local ductility depends not only on the cross-section dimensions but also on the ratio between the width of the flange and web, and member span and loading type.
3. The subdivision in different classes does not correspond to the actual behavior of beams, which is continuous, and the given discrete values of slenderness to define different classes seem to be very arbitrary.

Despite these recognized deficiencies, this classification is commonly used in the codification, due to its simplicity in design practice.

As an alternative, a more effective classification at the level of member ductility has been proposed by Mazzolani and Piluso (1993) and Gioncu and Mazzolani (2002) (Figure 5.23):

- *HD, high ductility,* corresponding to the members designed, dimensioned, and detailed such that they ensure the development of large plastic rotations.

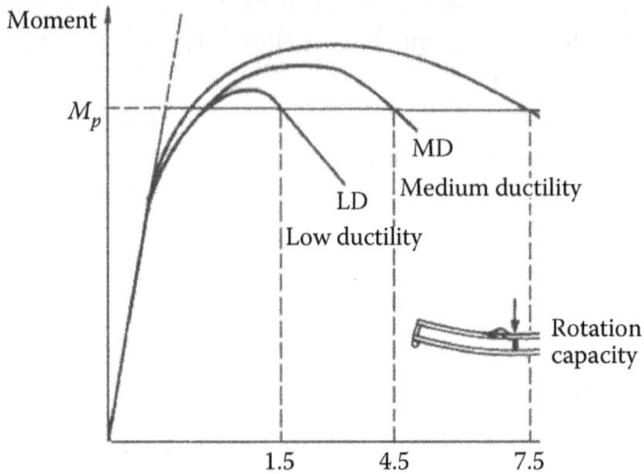

Figure 5.23 Member ductility classes, moment–rotation capacity curve.

- *MD, medium ductility,* corresponding to the members designed, dimensioned, and detailed such that they ensure the development of moderate plastic rotations.
- *LD, low ductility,* corresponding to the members designed, dimensioned, and detailed such that they ensure the development of low plastic rotation only.

Considering the definition of ultimate rotation capacity, given by Equation 5.3, the quantification of the above classification is based on the following criteria:

- HD, $R > 7.5$
- MD, $4.5 < R < 7.5$
- LD, $1.5 < R < 4.5$

Members having $R < 1.5$ are considered nonductile.

The above limits of the ultimate rotation capacity have been selected based on the results obtained by using the DUCTROT-M computer program, which has been calibrated on the comparison with experimental results.

The comparison between the results obtained using the DUCTROT-M computer program for member ductility and the code provisions for cross-section ductility is presented in Figures 5.24 and 5.25. Only in-plane plastic mechanisms are considered. In these figures, the hachured areas show the cases when the two classifications are coincident, but one can see that this coincidence is very poor.

Figure 5.24 refers to a section with web slenderness belonging to the C1 class. Assuming different values of the flange thickness $t_f = 8$, 10, and

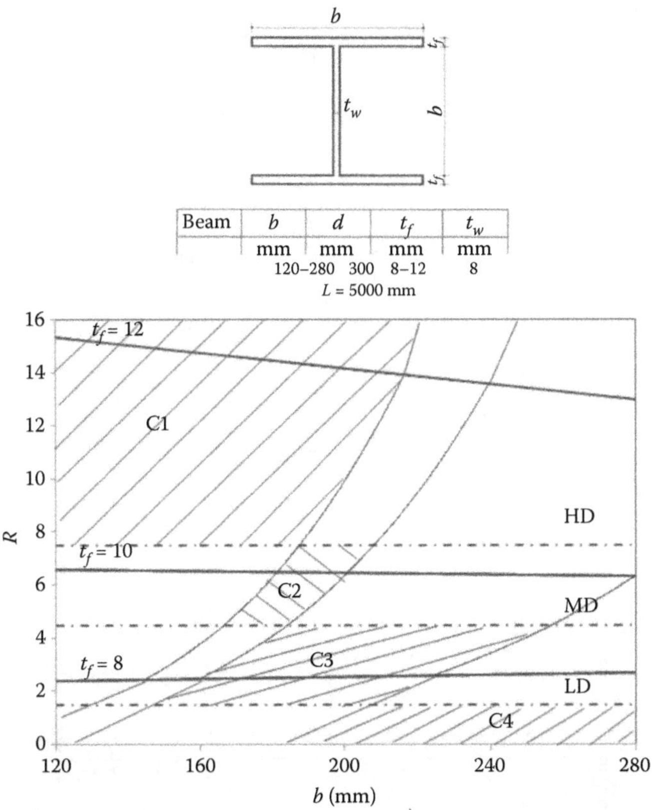

Figure 5.24 Cross-section versus member classes for a profile with C1 class for web.

12 mm, the member ductility of the profile belongs to the LD, MD, and HD classes, respectively, independent of the flange width. In contrast, the cross-section ductility classes change in function of flange width. For instance, the profiles with $t_f = 12$ mm are framed in the HD class for the entire field of the flange width $b = 120–280$ mm, while for $b > 216$ mm, the profile is framed in the C2 class, and for $b > 240$ mm, in the C3 class. So, it seems that the framing in different classes in function of the flange slenderness is questionable.

Figure 5.25 considers a profile with web slenderness belonging to the C2 class. Despite this classification, having $t_f = 12$ mm, the profile can be classified in the HD class. So, even the framing in different classes in function of web slenderness is questionable.

Therefore, examining these figures, one can observe that the two classifications according to cross-section ductility and member ductility lead to very different results. The incongruence of the codified classification based

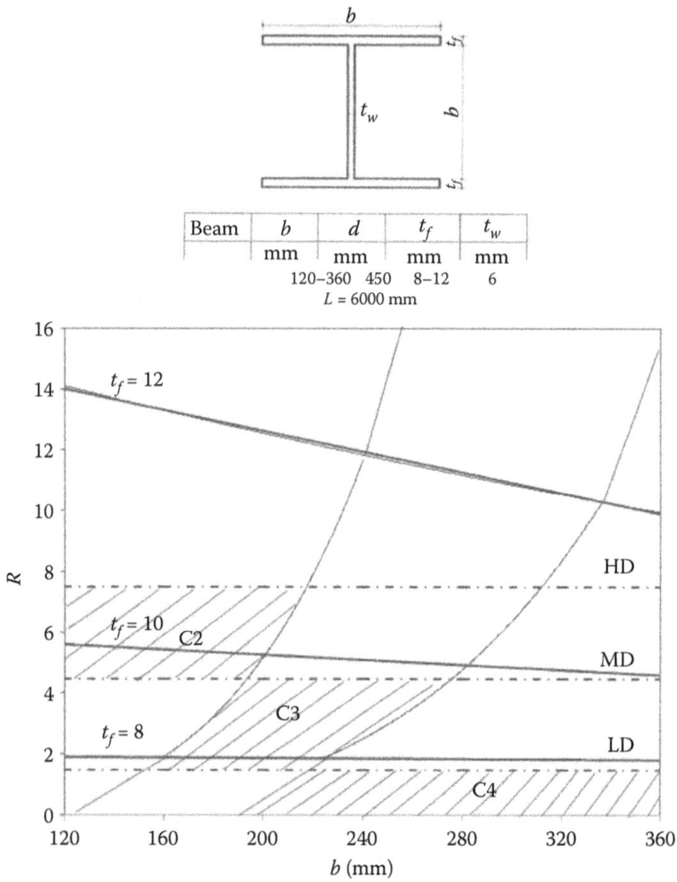

Figure 5.25 Cross-section versus member classes for a profile with C2 class for web.

on cross-section ductility requires serious consideration to improve it in light of the above consistent criticisms.

A first tentative classification to codify the member ductility has been done in Italian seismic code OPCM 3431 (2005), based on the research results given by Mazzolani and Piluso (1993), Formisano et al. (2006), and D'Aniello et al. (2012), considering, as criterion for the member ductility classification, the overstrength factor s (Figure 5.26), instead of the rotation capacity factor **R** used in the application of DUCTROT-M computer program. The s–**R** relation is the following:

- Ductile sections s > 1.2, **R** = 4.11–8.00
- Plastic sections 1.15 < s < 1.20, **R** = 2.6–4.11

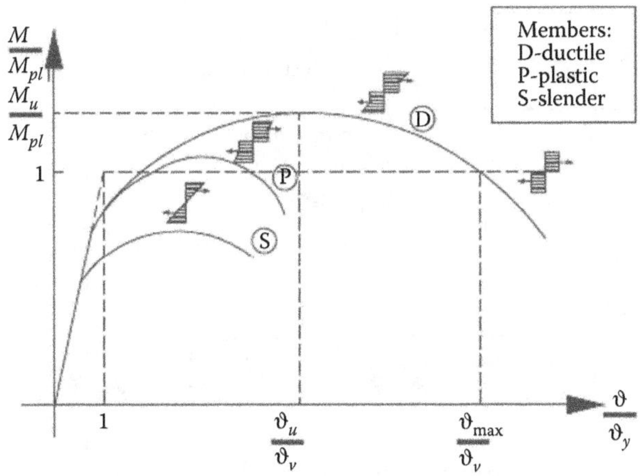

Member categories	s
Ductile	s > 1.2
Plastic	1 < s < 1.2
Slender	s ≤ 1

Figure 5.26 Overstrength classes. (After Formisano, A. et al. 2006: *Behaviour of Steel Structure in Seismic Areas, STESSA 2006* (eds. F.M. Mazzolani, A. Wada), Yokohama, August 14–17, Taylor & Francis, London, 225–232.)

5.3.3 Gradient versus quasi-constant moments

The methodology presented in Section 5.2.2.1 for the evaluation of the beam ductility considers the use of two standard beams: SB1, loaded by a gradient moment in a short span beam and SB2, loaded by a quasi-constant moment in a long span beam (Figure 5.27).

The ductility of the real beam is given by the minimum value evaluated for the two standard beams:

$$R = \min\,(RSB1, RSB2)$$

This methodology neglects the presence of floors, considering that the formation of local plastic mechanisms is free. In reality, the beam–floor interaction does not allow the plastic buckling of the beam flange and the formation of the plastic mechanism in the compression zone of the steel profile of the SB2 standard beam (Figure 5.28). The plastic hinge in this zone with sagging moment could be formed only producing the concrete crushing. As a consequence, the first plastic hinge forms for a hogging moment in the SB1 standard beam, where the plastic mechanism in the steel beam is free to develop. Therefore, the evaluation of the beam ductility

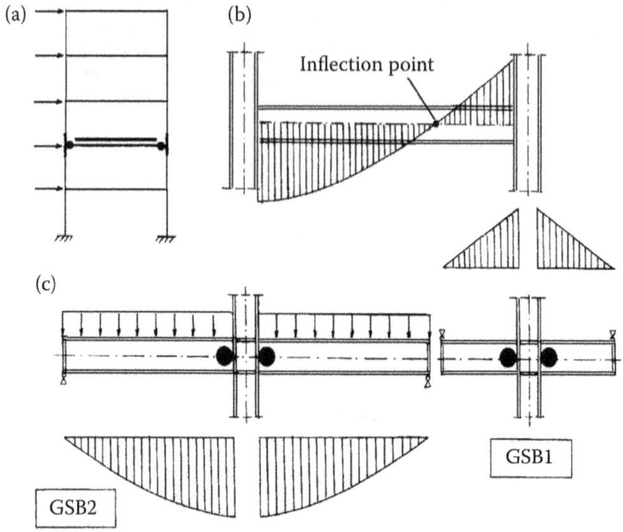

Figure 5.27 Standard beam types: (a) structure scheme; (b) substructure configuration; (c) SB1 and SB2 standard beams.

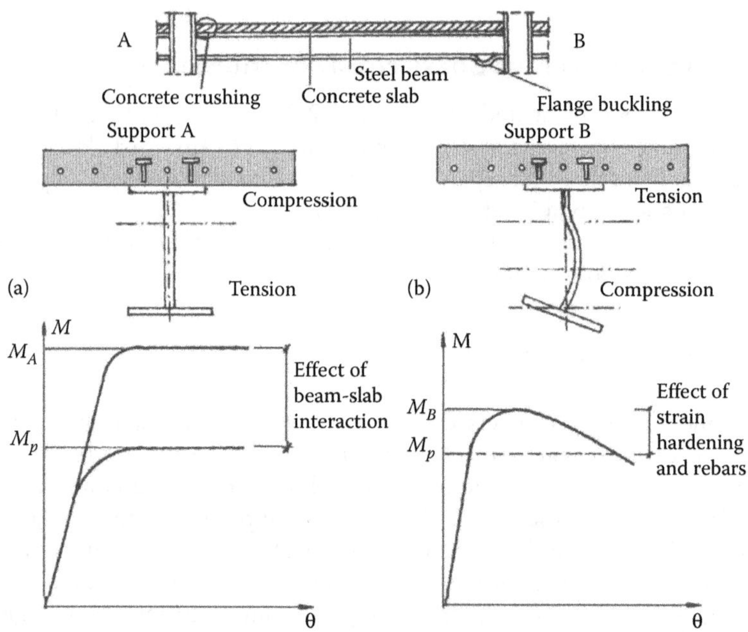

Figure 5.28 Beam–slab interaction: (a) sagging moments; (b) hogging moments.

in practical design just requires to consider the SB1 standard beam, which is characterized by a gradient moment:

R = RSB1

5.3.4 In-plane versus out-of-plane plastic mechanisms

The plastic buckling of the SB1 standard beam occurs for a hogging moment in the compressed part by the in-plane or out-of-plane plastic mechanism (Figure 5.29).

The tests considered for the validation of the DUCTROT-M computer program did not avoid the lateral rotation of the beam during plastic buckling. The analysis of theoretical and experimental results has shown that the great majority of tested beams (especially rolled profiles) lost their carrying capacity for an out-of-plane plastic mechanism. But the beam ends in practical situations belong to a complex node, composed by a column with horizontal reinforcements and transversal beams that impede the free lateral rotation (Figure 5.30). Therefore, the results on the plastic mechanism type, obtained during the experimental tests on standard beams, must be carefully examined.

In addition, it is essential to mention that the experimental test developed during the SAC (1996) and RECOS experimental programs (Mazzolani, 2000) on beam-to-column connections never did show collapse modes of out-of-plane type.

Only in the case of the reduced beam section, that is, RBS (dog-bone), were the out-of-plane plastic mechanisms observed, associated with an

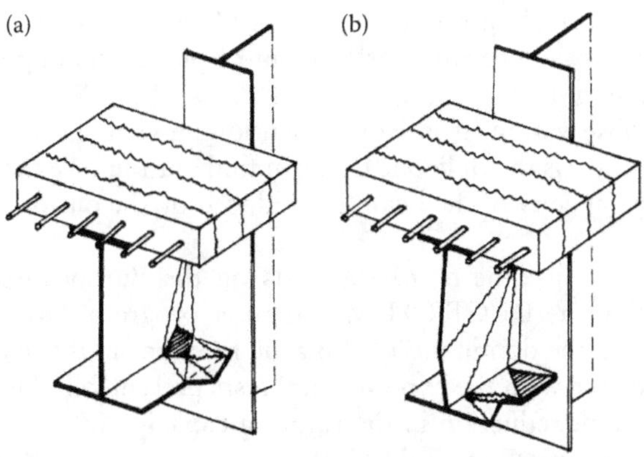

Figure 5.29 Local plastic mechanisms for hogging moments: (a) in-plane mechanism; (b) out-of-plane mechanism.

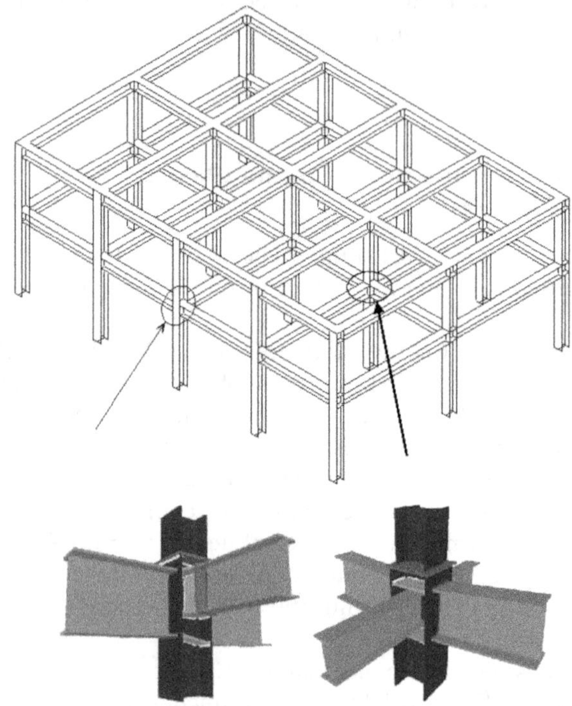

Figure 5.30 Beams in complex structures.

accelerated degradation of connection rigidity, in the experimental tests (Chi and Uang, 2002), performed after the Northridge earthquake in order to protect the column–beam connections. Such degradation was due to the lateral-torsional buckling producing out-of-plane force components. This is an additional reason to eliminate the out-of-plane buckling mechanisms in practical design.

After this observation, it is worth mentioning that the DUCTROT-M computer program is the only one that can find out the cases when the ultimate carrying capacity of the beam is produced by the out-of-plane plastic buckling.

For rolled sections, the numerical tests on the European sections IPE and HEA, using the DUCTROT-M computer program, have shown that for these profiles the dominant local plastic mechanism is the out-of-plane one (in accordance with the experimental tests presented in Section 5.2.6), with an important reduction in the rotation capacity, in comparison with the in-plane mechanism (see Table 5.3).

Therefore, a very important remark is that a consistent increase in local ductility for European rolled sections can be obtained by studying the

Table 5.3 Rotation capacity for European Sections

	L = 5000			*L* = 6000		
Rolled sections	IPE300	IPE400	IPE500	HEA400	HEA500	HEA600
In-plane	9.03	13.41	15.00	21.13	36.12	40.59
Out-of-plane	6.27	8.12	8.83	9.75	10.29	10.51

details of the beam ends in nodes in order to impede the out-of-plane plastic buckling; it can be achieved by increasing the node twisting rigidity to be greater than the lateral rigidity of the beam (Figure 5.30). For this purpose, transversal stiffening plates were inserted between the column flanges and rigid transversal beams in the joints. In addition, it is important to also consider the effects of RC floors in preventing the local twisting of columns (Figure 5.29b). So, considering these aspects, the out-of-plane local plastic mechanism can be eliminated and the in-plane plastic mechanism only (Figure 5.29a) will govern the ductility, producing an important increase in the rotation capacity of beams.

For welded sections, there are two ways to eliminate the out-of-plane plastic mechanism:

1. To use the constructional measures recommended for rolled sections.
2. To choose the section dimensions, so that the out-of-plane mechanism is directly eliminated.

When the dimensions of cross-sections are chosen in order to obtain the elimination of the out-of-plane mechanism, the solutions are presented by Gioncu et al. (2009) (Figure 5.31). One can observe that for an increase in the flange thickness, an increase in the ultimate rotation is obtained for the in-plane plastic mechanism. But from a given value of the flange thickness, the in-plane mechanism is transformed into the out-of-plane one and after the ultimate rotation remains practically constant as far as the flange thickness increases. In order to eliminate the out-of plane plastic mechanism, the ratio between the web and flange thickness must be chosen in the range of 0.7–0.8. As the European profiles do not respect these conditions, this is the reason why they have a tendency to lose the carrying capacity by out-of-plane plastic mechanisms.

5.3.5 Available rotation capacity of rolled beams

5.3.5.1 Influence of junction

Hot-rolled I-section members, widely used for beams and columns in multistory buildings, have different values of the available rotation capacity

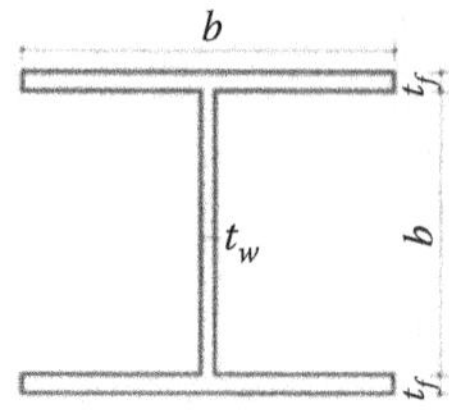

Beam	b (mm)	d (mm)	t_f (mm)	t_w (mm)
A1	160	240	8÷14	8
A2	200	300	10÷16	10
A3	300	400	12÷18	12
A4	400	500	14÷20	14

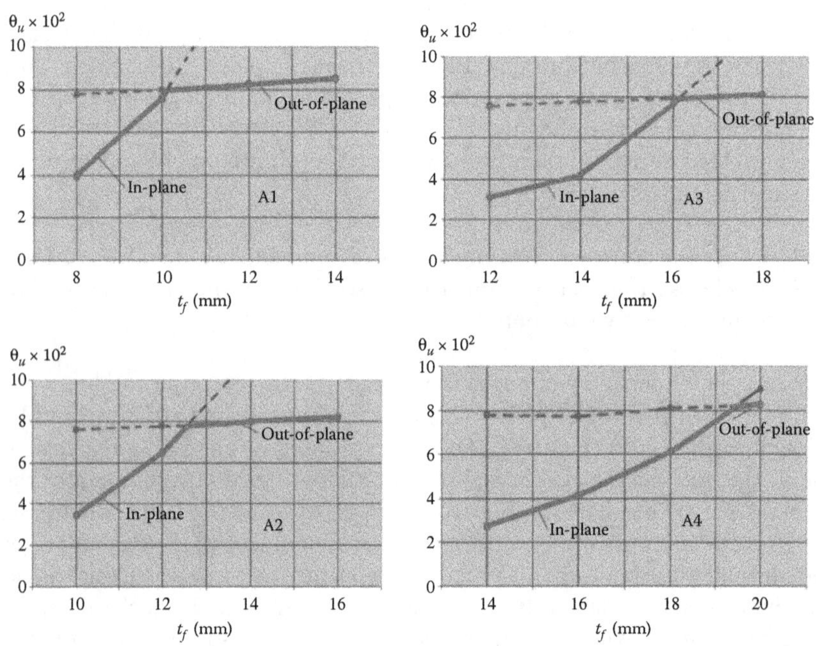

Figure 5.31 Ultimate rotations for in-plane and out-of-plane mechanisms for different double *T* sections.

than the welded sections due to a rigid zone, named junction, which connects the web to the flanges. This effect is introduced in the provisions of Eurocode 3, by changing the limits between ductility classes (see Section 5.3). The junction zone with radius r creates a condition under which the flange buckles around the rigid zone (Figure 5.32a), thus reducing the flange width and, as a consequence, increasing the rotation capacity of

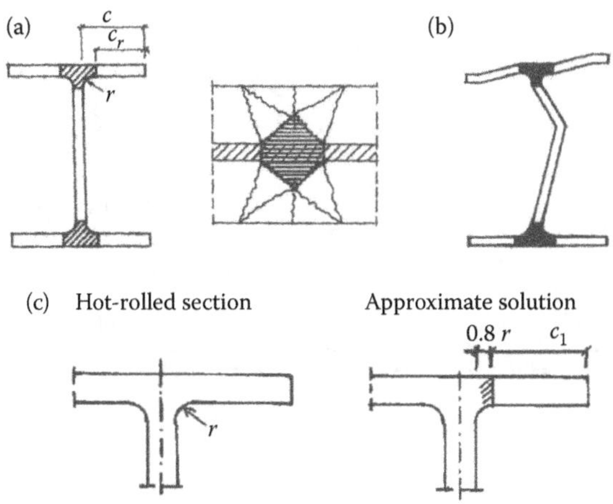

Figure 5.32 Flange–web junction for rolled profiles: (a) junction characteristics; (b) modified plastic mechanism; (c) simplified approach.

the element. In order to evaluate the available ductility under real constructional circumstances, an improved plastic collapse mechanism was proposed (Gioncu and Mazzolani, 2002; Anastasiadis, 1999) (Figure 5.32). In this mechanism, the increased plastic zone due to the junction is introduced in local plastic zone.

By using the DUCTROT-M computer program, which implements the usual constructional details of I-wide-flange sections, a parametrical analysis took place. The main target is to demonstrate the contribution of the junction on the plastic rotation capacity. Concerning the contribution of the junction to the plastic rotation capacity, one can remark that there is a very significant increase in available ductility attaining approximately 50% and 85% for the HEA and IPE sections, respectively, as compared with the same sections without junctions (Figure 5.33).

A simplified method to consider the effect of this junction consists in correcting the value of the rotation capacity determined without junction (Anastasiadis, 1999):

$$R_j = r_j\, R \tag{5.4}$$

$$r_j = (c/c_r)^2;\; c_r = c - 0.5\text{tw} - 0.8\text{r} \tag{5.5a,b}$$

The correlation between the values obtained using the modified collapse mechanism and the one obtained using Equation 5.4 is presented in Figure 5.34, showing that this simple procedure allows good values of the rotation capacity of rolled profiles to be obtained.

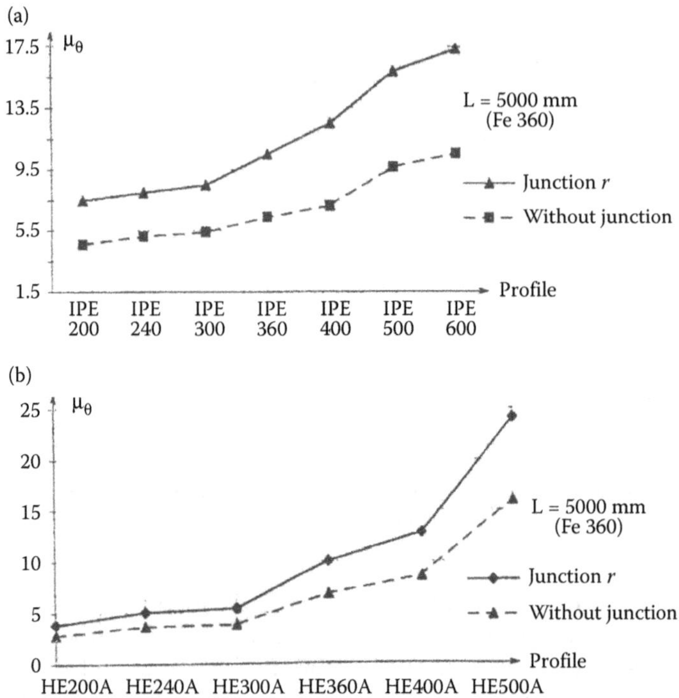

Figure 5.33 Influence of flange–web junction: (a) IPE profiles; (b) HEA profiles.

Figure 5.34 Correlation between theoretical values and simplified approach.

One must mention that the influence of junctions is greater than that given by the Eurocode 3 approach, which just changes the limits between the ductility classes.

5.3.5.2 Ductility of the RBS

Among the solutions for safeguard of brittle joints, the idea of the weakening of flanges at the beam ends has proven to be very effective. By cutting the beam flange near the joint in a specific zone (dog-bone), the formation of plastic hinges away from the column-to-beam flange welds, allowing the stable yielding of the beam, is assured by the reduced moment capacity (Figure 5.35). The geometrical characteristics of the RBS are presented by Anastasiadis et al. (2005).

Figure 5.36 presents the example of the IPE profile, for which the rotation capacity is determined in function of the flange reduction for the in-plane and out-of-plane plastic mechanisms. One can see that an important increase in the in-plane rotation capacity is obtained, provided that the measure to impede the node twisting is assured. The experimental tests performed on standard beams have shown that plastic buckling always starts with an in-plane mechanism, followed by an out-of-plane mechanism, due to the weakening of the buckled flange. Experimental tests performed by Chi and Uang (2002) have shown that this phenomenon is accelerated, due to weakening of the flange and consequent reduction of lateral rigidity.

Because the plastic buckling occurs in the RBS, away from the node, the measures for increasing the node rigidity have no effect. So, the collapse practically occurs by means of the out-of-plane plastic mechanism and, as a consequence, a reduction of rotation capacity must be considered. In addition, owing to the presence of out-of-plane forces (Figure 5.37), the degradation of lateral rigidity during cyclic loading is accelerated. Therefore, the presence of extra lateral bracing near the RBS region is recommended to impede the out-of-plane buckling.

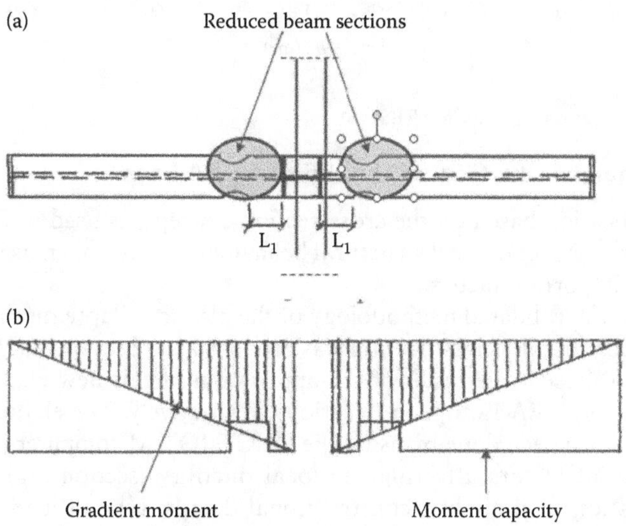

Figure 5.35 Reduced beam section: (a) beam scheme; (b) positions of RBS.

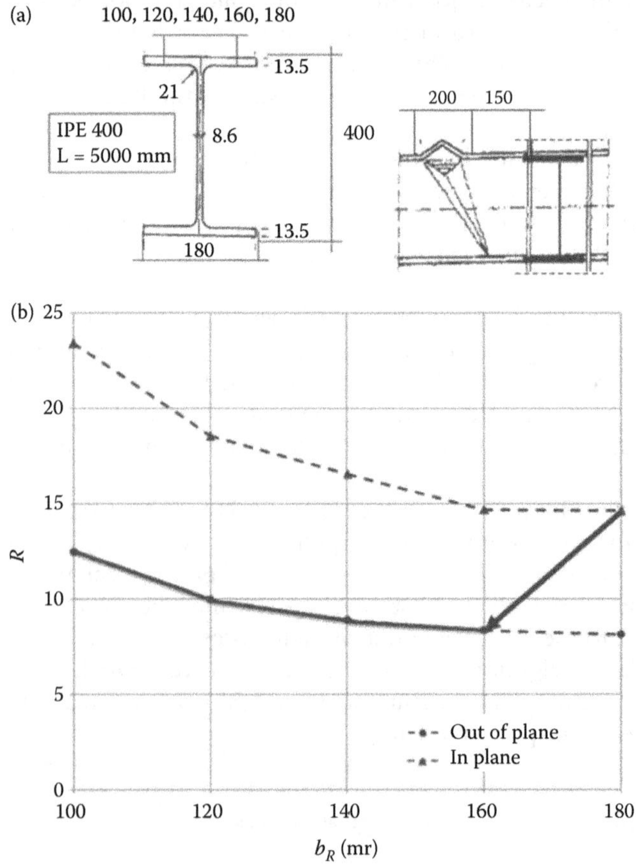

Figure 5.36 Rotation capacity for RBS.

5.3.5.3 Member ductility of the IPE and HEA beams

The classification based on the cross-section concept, as used in Eurocode 3, does not interpret the real structural behavior of the member, as it ignores some very important factors.

Based on the validated methodology of the plastic collapse mechanism, a new member ductility classification was proposed (see Section 5.3.2). Tables 5.4 and 5.5 present the result of the application of this new classification to the IPE and HEA European I-sections, respectively. The elaboration of these tables has been done by using the DUCTROT-M computer program; all the crucial factors affecting the local ductility (section and member characteristics, steel quality, constructional details, type of loading) were taken into account. These tables refer to the in-plane plastic mechanism, under the hypothesis that the out-of plane plastic mechanism is prevented

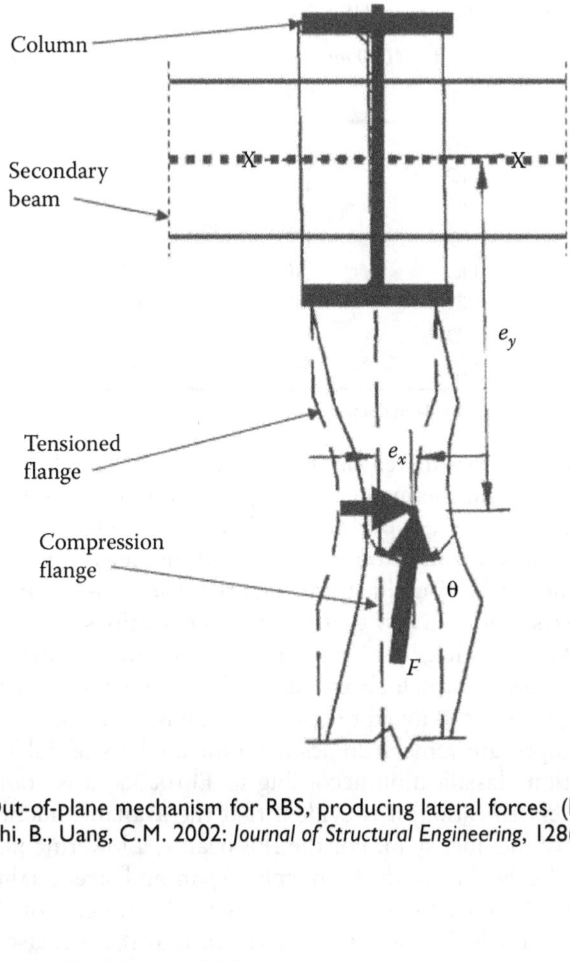

Figure 5.37 Out-of-plane mechanism for RBS, producing lateral forces. (Modified after Chi, B., Uang, C.M. 2002: *Journal of Structural Engineering*, 128(4), 464–473.)

Table 5.4 Member Classification of IPE

	L = 2000 mm		*L = 3000 mm*		*L = 4000 mm*		*L = 5000 mm*	
Steel section	*S235*	*S355*	*S235*	*S355*	*S235*	*S355*	*S235*	*S355*
IPE 300	HD	HD	HD	MD	—	—	—	—
IPE 330	HD	HD	HD	MD	HD	MD	—	—
IPE 360	HD	HD	HD	HD	HD	MD	MD	LD
IPE 400	HD	HD	HD	HD	HD	MD	MD	LD
IPE 450	—	—	HD	HD	HD	MD	MD	MD
IPE 500	—	—	HD	HD	HD	MD	MD	MD
IPE 550	—	—	—	—	HD	HD	HD	MD
IPE 600	—	—	—	—	—	—	HD	MD

Note: HD, high ductility; MD, medium ductility; LD, low ductility.

Table 5.5 Member Classification of HEA Profiles

	L = *4000 mm*		*L* = *5000 mm*	
Steel section	*S235*	*S355*	*S235*	*S355*
HEA 320	HD	MD	HD	MD
HEA 340	HD	HD	HD	MD
HEA 360	HD	HD	HD	HD
HEA 400	HD	HD	HD	HD
HEA 450	HD	HD	HD	HD
HEA 500	HD	HD	HD	HD
HEA 550	HD	HD	HD	HD
HEA 600	HD	HD	HD	HD

Note: HD, high ductility; MD, medium ductility.

by special measures for increasing the torsional rigidity of the nodes. The empty cells mean that the member sizing does not depend on the ductility limit state, but that the serviceability limit state could be predominant.

As a first remark, it can be observed that all the European profiles reveal an out-of-plane mechanism due to the fact that the web-to-flange thickness ratio t_w/t_f varies from 0.63 to 0.66 for the IPE sections and from 0.52 to 0.63 for the HEA sections. This is a confirmation of the numerical results (Gioncu et al., 2009), which clearly showed that this ratio must be greater than 0.7–0.8, in order to avoid the out-of plane mechanism.

A second important remark comes from the analysis of Table 5.6, where the cross-section classification according to Eurocode 3 is compared with the proposed classification; it is evident that the available ductility changes as a function of member span. For the IPE beams, a lowering plastic capacity could be observed as both the member span and steel quality increase. Generally, the same conclusions could be observed in the case of HEA beams.

As a general conclusion, it can be observed that the increase in rotation capacity has two opposite effects:

1. One favorable for the structural behavior, due to the large capacity to dissipate seismic energy during the earthquakes.
2. One unfavorable, due to the fact that the filter against large strains in the tension flanges disappears, increasing the danger of brittle cracking in case of ground motions, which produce high strain rates in the structures. Owing to this fact, all the connections tested within the frame of SAC activity (1996) collapsed for the brittle fracture.

Current codes do not prescribe direct criteria for the evaluation of local available ductility. An important advancement of the Eurocodes should be the implementation of the ductility-based design together with the strength-based design. Exploiting the current knowledge, such an achievement

Table 5.6 Comparison between Cross-Section and Member Classifications

		IPE 300		*IPE 330*		*IPE 360*		*IPE 400*		*IPE 450*		*IPE 500*		*IPE 550*		*IPE 600*	
Code/proposal	*Steel grade(S)*	*235*	*355*	*235*	*355*	*235*	*355*	*235*	*355*	*235*	*355*	*235*	*355*	*235*	*355*	*235*	*355*
Classification according to EN 1993 Part 1-1		I	I	I	I	I	I	I	I	I	I	I	I	I	I	I	I
Classification according to plastic collapse mechanism	L = 20 m	HD	HD	HD	HD	HD	HD	HD	HD	—	—	—	—	—	—	—	—
	L = 30 m	HD	MD	HD	MD	HD	HD	HD	HD	HD	HD	HD	HD	—	—	—	—
	L = 40 m	—	—	MD	MD	HD	MD	HD	MD	HD	MD	HD	MD	HD	HD	—	—
	L = 50 m	—	—	—	—	MD	LD	MD	MD	MD	MD	MD	MD	HD	MD	HD	MD

Note: ▒, Worse behavior than predicted in EN 1993; —, Sizing of the member does not depend on the ductility limit state.

could be obtained, considering that the basic components of a frame (beam, column, connection, and slabs) interact between them, always taking into account the constructional details of the structural system.

5.3.6 Available rotation capacity for welded beams

5.3.6.1 *Influence of welding type*

The definition of the flange width as a function of welding type in welded beams is necessary for the ductility evaluation in practical design as for rolled sections. As a first step, the rotation capacity can be determined by using the same method based on a correction (see Section 5.3.5a). The rotation capacity is obtained by using Equations 5.4 and 5.5a, where for the effective widths the following values are suggested:

- For filled welds (Figure 5.38a):

$$c_r = c - 0.5\ t_w - 1.1a \tag{5.6a}$$

- For penetrated welds (Figure 5.38b):

$$c_r = c - 0.5\ (t_w + a) \tag{5.6b}$$

where a is the weld thickness.

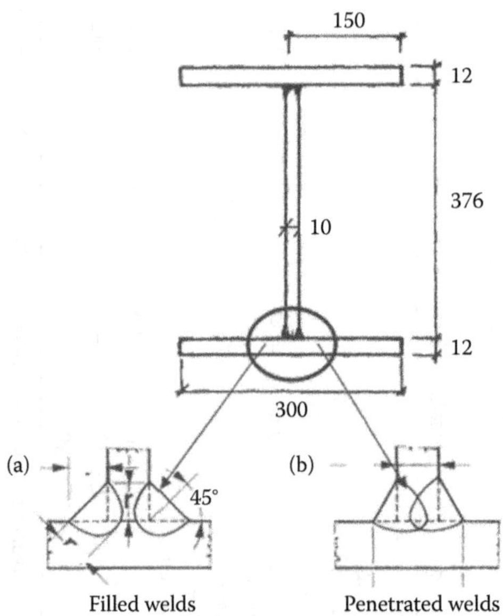

Figure 5.38 Welded profile types: (a) Filled welds; (b) penetrated welds.

Table 5.7 Rotation Capacity of Welded Profiles

	a (mm)				
Welding type	*0*	*4*	*5*	*6*	*7*
Without welding		3.14			
Filled welds		3.71	3.78	3.86	3.91
Penetrated welds		3.55	3.58	3.62	3.65

The rotation capacity of the welded profile of Figure 5.38 (with L = 5000 mm), by using the DUCTROT-M computer program, is influenced by the welding type (see Table 5.7).

One can notice that the influence of the welding thickness and the welding type are not very important.

5.3.6.2 *Influence of steel grade and yield stress random variability*

Figure 5.39 presents the influence of steel grade on rotation capacity. One can see that the ductility decreases with the increase in yield stress. The

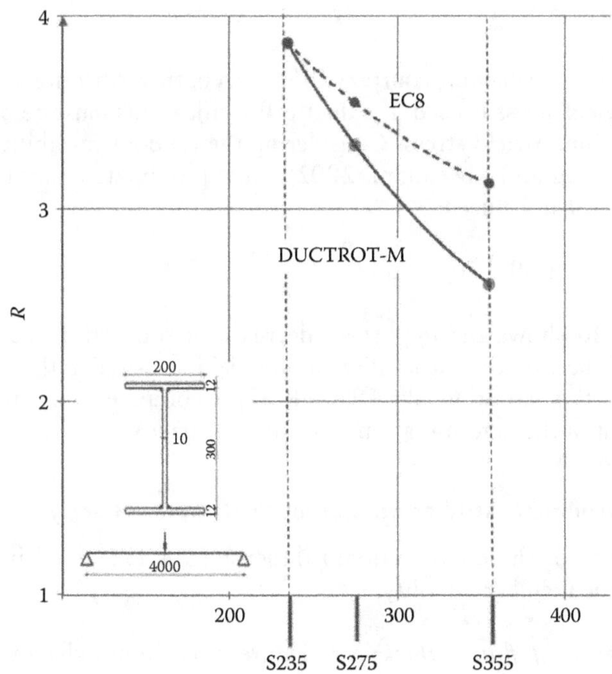

Figure 5.39 Influence of steel grade on rotation capacity.

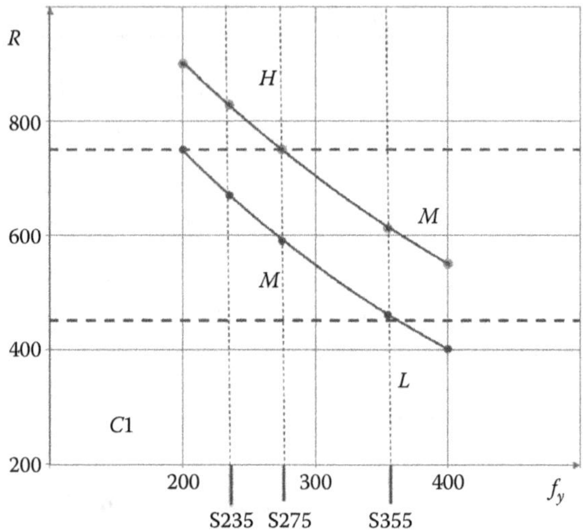

Figure 5.40 Influence of random variability of yield stress on rotation capacity.

decreasing trend evaluated with DUCTROT-M is more important than the one proposed in Eurocode 3, where the reduction depends on the factor $(235/f_y)^{1/2}$.

Owing to this behavior, contrary to the strength verifications where the minimum yield stress is used, the ductility verifications must be performed for the maximum yield stress. Considering the random variability of yield stresses (Gioncu and Mazzolani, 2002), an approximate relation between f_{ymax} and f_{ymin} can be used:

$$f_{ymax} = f_{ymin} + 50\ldots70 \text{ (N/mm}^2\text{)} \tag{5.7}$$

Figure 5.40 shows the important decrease in rotation capacity taking account of the random variability of the yield stress. For the important influence of this variation, the DUCTROT-M computer program takes it into account in the determination of rotation capacity.

5.3.6.3 Parametric studies on member rotation capacity

The influence of the cross-sectional dimensions is examined in order to maximize the member ductility.

1. *Influence of flange thickness.* Figure 5.41 shows the influence of the flange thickness on the member ductility, the increase in rotation capacity being very high with the increase in flange thickness.

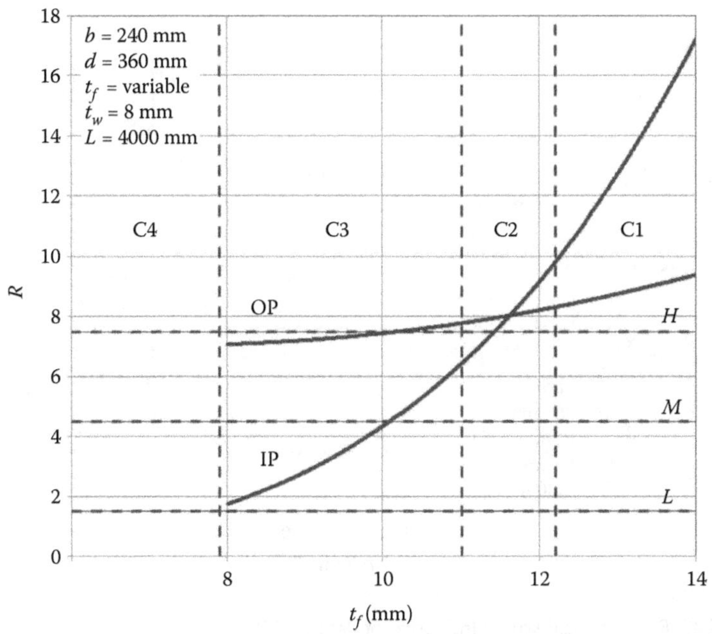

Figure 5.41 Parametrical study: Influence of flange thickness.

A comparison between cross-section classes (C1, C2, C3, C4) and member classes (high H, medium M, low L) is presented in Figure 5.41. One can observe that the criticism on the discrete subdivision in classes is confirmed, the member class M being over the cross-section class C3. The out-of-plane mechanism limits the increase in ductility resulting from the in-plane mechanism.

2. *Influence of flange width.* The profile belongs to class C1 according to the cross-section classification, but the same profile is in class M (medium) for the member classification, the variation of flange width having no significant influence (Figure 5.42).
3. *Influence of web thickness.* In contrast to the influence of flange thickness, the decrease in rotation capacity with the increase in web thickness is very important (Figure 5.43). The profile of class C1 can belong to M (medium) or L (low) ductility classes for member classification.The out-of-plane mechanism limits the increase in rotation capacity with the decrease in web thickness.
4. *Influence of web height.* Like for the flange width, the web height also does not play a very important effect in the increase in rotation capacity (Figure 5.44). One can see that the profile of class C1 has an M (medium) rotation capacity.

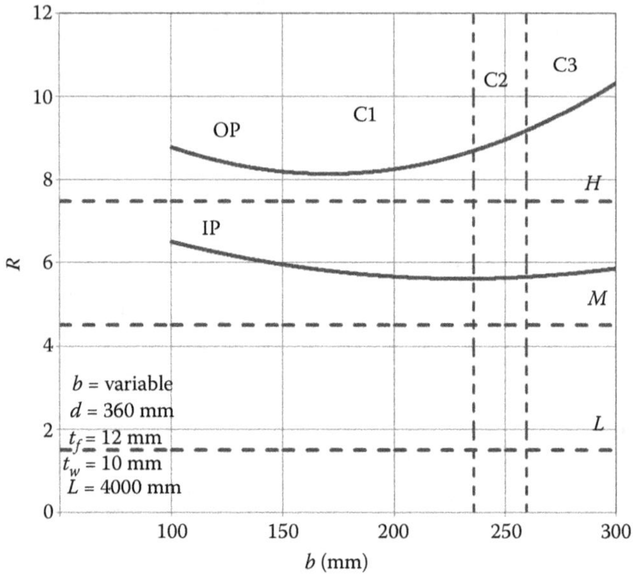

Figure 5.42 Parametrical study: Influence of flange width.

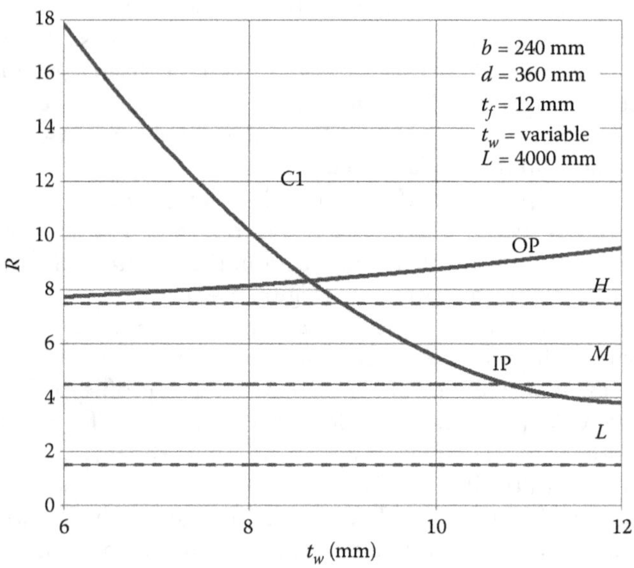

Figure 5.43 Parametrical study: Influence of web thickness.

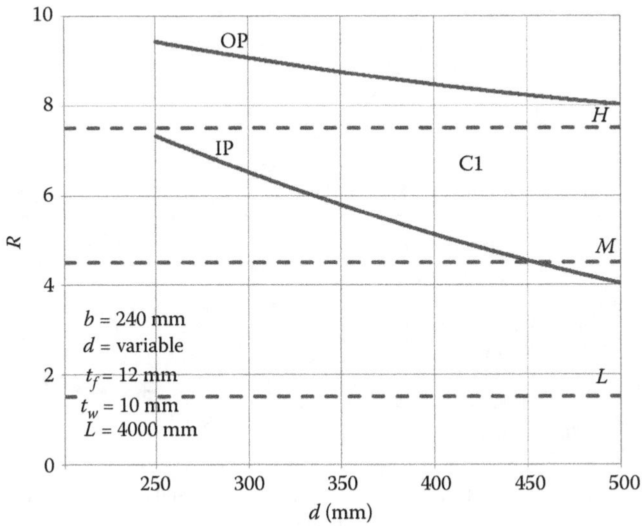

Figure 5.44 Parametrical study: Influence of web height.

5. *Influence of beam span*. Figure 5.45 shows that the rotation capacity of a profile of class C1 can move from M (medium) to L (low), in function of the beam span. This is another demonstration on the limits of the cross-section classification for providing proper information on the ductility of profiles.

5.3.6.4 Suggestions for a proper selection of profile dimensions

Because the beam span is not a parameter that can be modified, a suitable value of rotation capacity can be reached just selecting the optimum dimensions of the cross-section, that is, thicknesses and width of flanges, and thickness and height of web. The parametrical studies have shown that the flange width and web height have no important effects on increasing the rotation capacity and that the increase in flanges and web thicknesses has contrary effects. The reason for these results is related to the wavelength of the plastic mechanism. For the in-plane mechanism, the rotation capacity is a function of this length, the capacity of dissipating energy being multiplied by the increase in yielding line lengths. The length of the plastic mechanism, based on theoretical researches and experimental data, is given in Gioncu and Petcu (1997), as

$$L_m = 0.6(t_f/t_w)^{3/4}(d/b)^{1/4}b \tag{5.8}$$

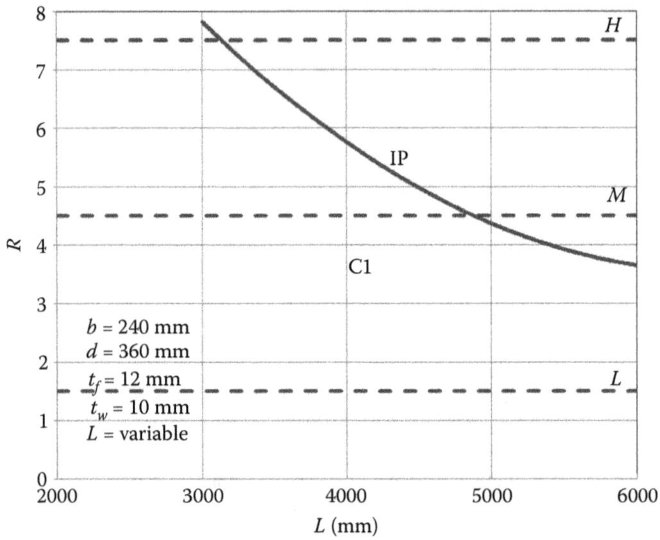

Figure 5.45 Parametrical study: Influence of beam span.

One can notice that the ratio t_f/t_w has a great influence on the plastic mechanism length, while the ratio d/b has a reduced influence. These observations are confirmed by the numerical tests on rotation capacity, the main factor for the increase in rotation capacity being the flange thickness and the factor producing the most important decrease in rotation capacity being the web thickness.

As is shown in Section 5.3.4, an important attention must be paid to eliminate the possibility of occurrence of the out-of-plane mechanism, due to the lowering effect on the rotation capacity and the acceleration of degradation during cycle loads. Figure 5.46 presents the influence of the flange width (b), flange thickness (t_f), and web height (d), and keeping constant the web thickness $t_w = 8$ mm. One can see that for $t_f = 10$ mm (Figure 5.46a) for all the section parameters, the dominant mechanism is the in-plane one. For $t_f = 12$ mm (Figure 5.46b), both the plastic mechanisms are present, in function of the flange width and web height. For $t_f = 14$ mm (Figure 5.46c), only the out-of-plane mechanism is present.

5.3.7 Other applications of DUCTROT-M computer program

The DUCTROT-M computer program gives the possibility to study the influence of some beam–column details on rotation capacity, taking into

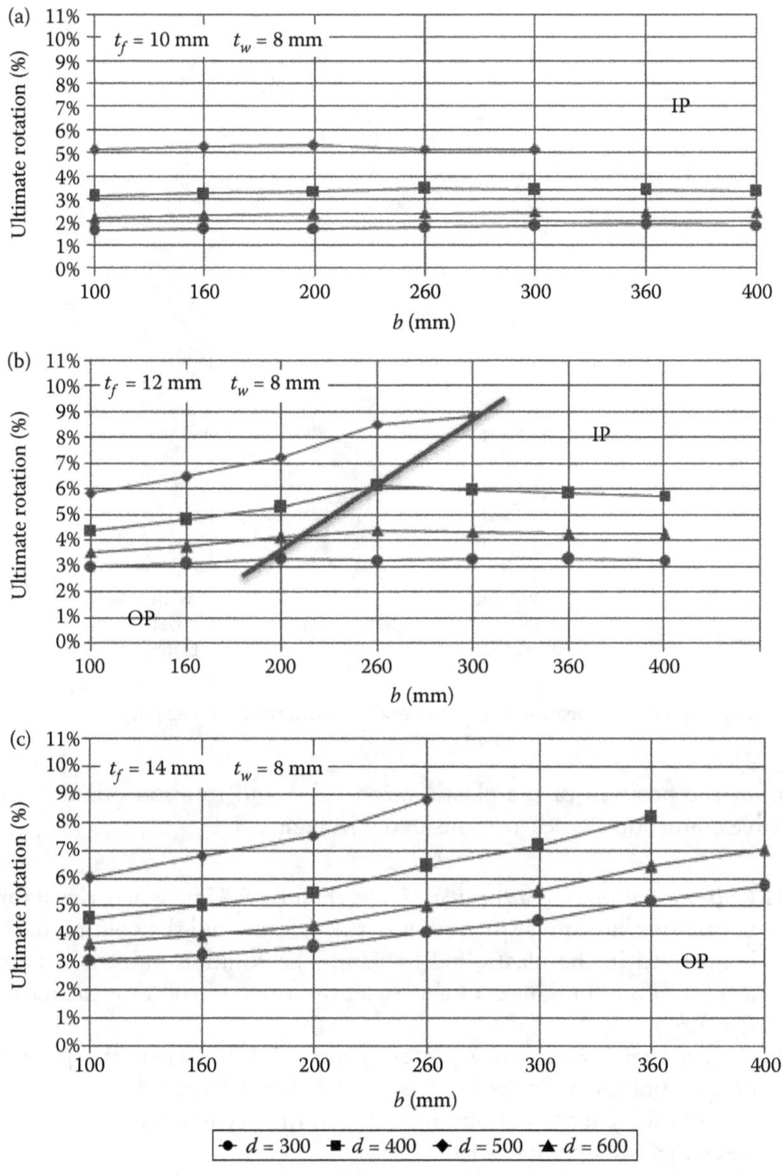

Figure 5.46 Optimum adjustment of the shape of welded profiles: (a) t_f = 10 mm; (b) t_f = 12 mm; (c) t_f = 14 mm.

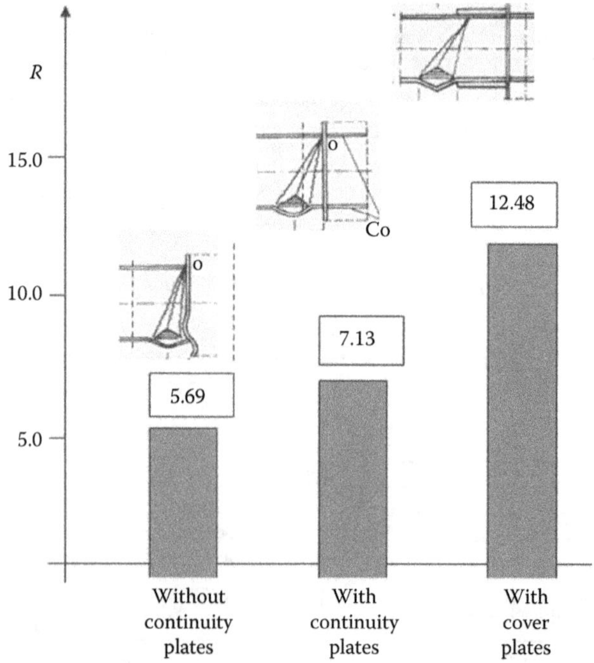

Figure 5.47 Influence of beam-to-column details on the rotation capacity.

account the position of the plastic mechanism and rotation point. Three possible connection types are considered in Figure 5.47.

1. *Column without continuity plates* (Figure 5.47a), when the flange plastic mechanism is situated near the column and the column flange is involved in the plastic mechanism. The rotation point is situated near the column flange. In this case, the value of rotation capacity is $R = 5.69$.
2. *Column with continuity plates* (Figure 5.47b), when the position of the rotation point is far from the column flange, an increase in the rotation capacity is obtained due to the asymmetry of the plastic mechanism shape: $R = 7.13$.
3. *Connections with cover plates* (Figure 5.47c), used in order to protect the welded connections and to force the development of plastic hinges in a given position far from the column; in this case, the plastic mechanism is free to develop an optimum shape by choosing a favorable rotation point. Owing to this fact, the increase in the rotation capacity is very important: $R = 12.48$.

5.4 LOCAL DUCTILITY UNDER FAR-FIELD EARTHQUAKES

5.4.1 Characteristics of far-field earthquakes

Usually, for engineering purposes, earthquake ground motions are characterized by their acceleration amplitudes. A more complete description of the ground shaking also requires the introduction of additional indicators, such as the shaking duration and the number of cycles of the motion. Only in this way is it possible to evaluate the structural damage to which the effect of the available ductility is referred.

5.4.1.1 Crustal earthquakes (subduction or strike-slip types)

These occur in shallow sources (Figure 5.48). The earthquake characteristics rely on the propagation path of body waves (P and S), surface waves (L and R), and the local soil conditions.

As a first remark, the propagation path effect depends on the percentage of the path travel through soft sediments. Deviations from a uniform horizontal-layered crust model, along the path of the waves propagating from the epicenter to the site, occur. These deviations are produced by a collection of sedimentary basins with alluviums, separated by irregular basement rock, forming mountains as well as geological and topographical irregularities. Therefore, two opposing phenomena occur: the first one is the attenuation of ground motions compared to the distance from the epicenter, and the second one is the amplification of these motions due to the presence of the sedimentary deposits. Owing to these effects, the recorded acceleration, velocities, and displacements are very different in the near-field and far-field sites. Generally, the near-field records are related to the velocity pulse with a reduced number of cycles and a very short earthquake duration (produced by the P and S waves), while the far-field records are characterized by many acceleration cycles and a long duration of earthquakes (produced by the L and R waves). The earthquakes in intermediate sites have the characteristics of both the far-field and near-field earthquakes. Even though it is very well known that the body waves (P and S) are characterized by high frequencies, and the surfaces waves (L and R) by low frequencies, it must be considered that the higher frequencies attenuate more rapidly than the lower ones as the distance increases. Consequently, the earthquakes in the intermediate sites are dominated by the far-field earthquake characteristics. It should be mentioned that for this earthquake type the site soil conditions play a leading role, in many cases being more important than the earthquake type itself or the traveled path.

As a second remark, the soil profile is composed by multilayers with different mechanical properties and thickness. The alternate of layers is a very

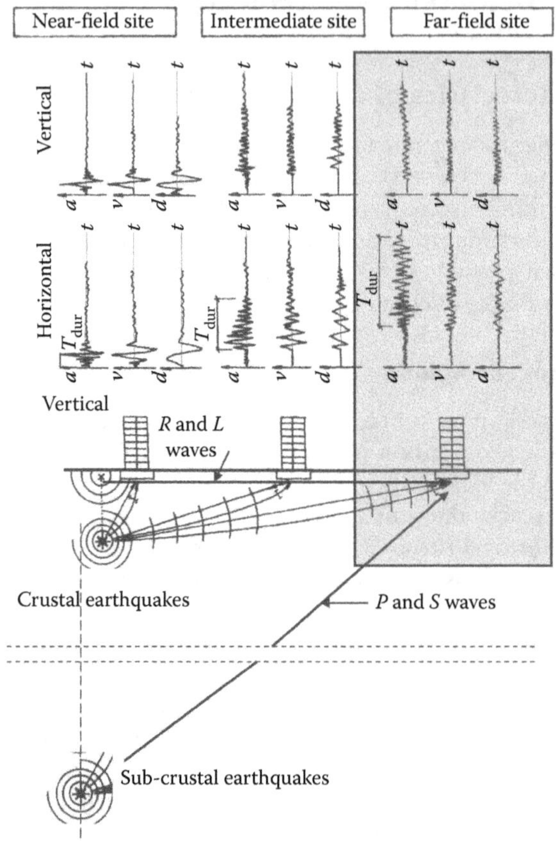

Figure 5.48 Far-field earthquakes.

important factor, which has been largely recognized; the nature of layers changes the amplitude and frequency with a direct effect on the amount of the structural damage.

5.4.1.2 Subcrustal earthquakes

These are produced by deep sources (Figure 5.48). In this case, the earthquake characteristics are given as a function of both the L and R surface waves, P and S body waves, and also the reflected and refracted waves. In contrast to crustal earthquakes, the earthquake characteristics depend on the succession of vertical layers and not on the horizontal travel path. In addition, the effects of near-field source disappear, the earthquake characteristics being practically the same over an extended surface. Therefore, for this type of earthquake, it is difficult to make a clear distinction between near-field or far-field earthquakes.

In any case, owing to the great depth of the source, the main characteristics of subcrustal earthquakes correspond very closely to the ones of far-field earthquakes, as the duration, number of cycles, influence of soil site conditions, and so on.

5.4.2 Ductility under cycle loading produced by far-field earthquakes

5.4.2.1 Shaking duration

Ground motion duration is an important parameter for the seismic design of a structure situated in far-field sites. However, this parameter is not yet taken into account in the seismic codes, due to the fact that the evaluation of the effective duration of an earthquake and the influence of this duration of the structural behavior are a very complex task. There are several hundred papers in the technical literature related to the influence of duration of strong ground motions on structural damage, but with different conclusions. Some of them suggest that the duration should be incorporated into the specification of seismic design loads. But, at the same time, other research works support the policy of the current seismic design codes, which neglect the duration (Hancock and Bommer, 2006). From the engineering point of view, it is very clear that, considering two earthquakes having the same magnitude but different durations, the damage is more important for the one with greater duration.

Recently, the increase in the network of recording stations offers the possibility to overcome this problem. The duration is defined by *braked duration* (the time interval between the first and the last time when the acceleration exceeds the level of 0.05*g*), *uniform duration* (the time interval during which the ground motion acceleration exceeds a fixed threshold value), and *significant duration* (the time interval between the two time points when the Arias intensity exceeds two separate fixed threshold values) (Snaebjornsson and Sigbjornsson, 2008). The seismic design must include the effects of duration on the number of cycles, because long duration has been recognized as a potential source of structural damage. It is well known that the structure's strength is not related to the ground motion duration; however, the earthquake duration, as a result of the cumulative plastic deformations, has a significant effect on the available ductility.

In addition, the earthquake duration depends on the duration of the rupture process at the source, prolongation of the duration due to the propagation path effects, and prolongation effects caused by the local soil layers at the recording site (Lindt and Goh, 2004; Lee, 2002). Figure 5.49 presents the increase in duration in function of earthquake magnitude and source-to-site distance (Gioncu and Mazzolani, 2002). It can be observed that the

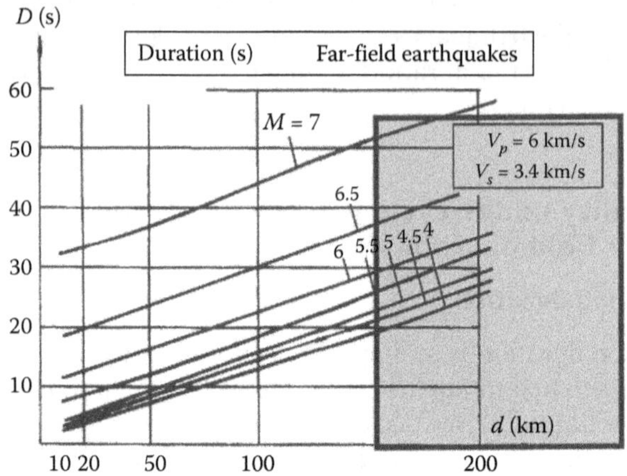

Figure 5.49 Duration effect in function of the epicentral distance.

increasing duration is proportional to the epicentral distance and, further, that the duration is greater for large magnitudes.

Generally, the duration in far-field sites can be framed in the interval of 30–50 s. During this interval, the structure is subjected to a large number of cyclic loadings, but only a reduced number of them have important effects in damaging the structure.

5.4.2.2 *Effective number of cycles of earthquake ground motions*

The cycle number of ground motions is widely used in geotechnical earthquake engineering (liquefaction and landslide problems), but not for the structural analysis. Despite all specialists recognizing that the number of cycles is a robust indicator of the structural destructive capacity of an earthquake, the approach to evaluate this damaging indicator is not universally accepted (Stafford and Bommer, 2009). There are two main methodologies: the first one by directly counting the dominating cycle using ground motion time history and the second by indirectly counting it, using the structural response time history. Figure 5.50 very clearly shows that the number of cycles is related not only to the ground motions, but also for the structural response. The number of cycles is larger for rigid structures than the flexible ones. Hence, the second methodology is more reliable. The effective number of cycles of an earthquake ground motion can be obtained by dividing the total hysteretic energy by the equivalent energy, which would be absorbed by a section of the structure (Figure 5.51a) (Hancock and Bommer, 2004). Figure 5.51b, as an application of the above-mentioned methodology, shows the equivalent number of cycles for an interplate action, like in the case of

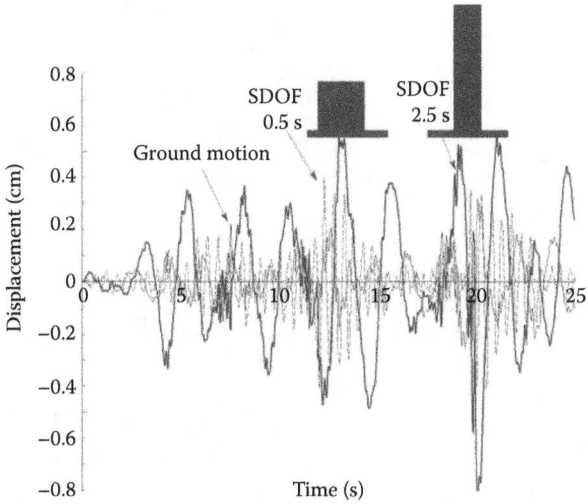

Figure 5.50 Displacement time history in function of ground motion and structural type. (After Stafford, P.J., Bommer, J.J. 2009: *Soil Dynamics and Earthquake Engineering*, 29, 1425–1436.)

the Northridge earthquake. The main observation from this figure is that the number of cycles for far-field sites is about 10–12 cycles.

Unfortunately, owing to the complexity of the problem, there are no systematic research works dealing with the main factors influencing the number of cycles, which are source type, earthquake magnitude, structural rigidity (rigid structures, T < 1 s, flexible structures, T > 1 s). Taking into account the main characteristics of different earthquake types (Gioncu and Mazzolani, 2000), the number of effective cycles given in Table 5.8 can be used just for structural design purposes in far-field regions.

It must be recognized that the above proposal is only a qualitative attempt without any pretension to be very accurate; it presents some contradictions with the ECCS Recommendation for experimental tests (1985), which determines the number of cycles for collapse of the tested specimen. This number does not correspond to the actual number of the structural members involved in the earthquake. In any case, the problem of the evaluation of the number of effective cycles during far-field and near-field earthquakes is far from being solved and it remains an open task that is still under debate.

5.4.2.3 Typology of cycle loading in function of earthquake type

The evaluation of the effect of cyclic actions on ductility, in case of structures situated in far-field areas, is based on the reduction of the rotation capacity due to the accumulation of plastic deformations.

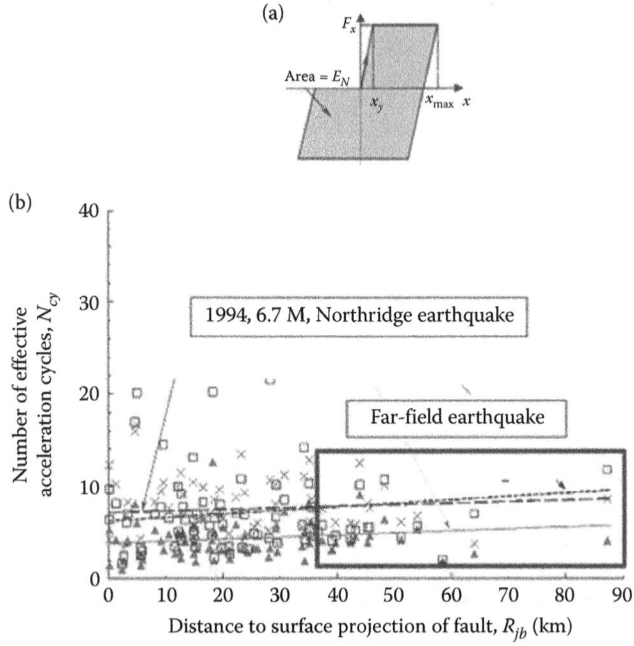

Figure 5.51 Relationship between number of effective cycles and source–site distance: (a) input energy dissipated by a section of the structure; (b) effective cycles in function of the epicentral distance. (After Stafford, P.J., Bommer, J. J. 2009: *Soil Dynamics and Earthquake Engineering*, 29, 1425–1436.)

Table 5.8 Number of Effective Cycles

Earthquake type	*Rigid structure*	*Flexible structure*
Intraplate earthquakes ($M < 6.0$)	Max 5 cycles	—
Interplate earthquakes		
Subduction ($M < 9.0$)	Max 10 cycles	Max 6 cycles
Strike-slip ($M < 7.0$)	Max 12 cycles	Max 8 cycles
Intraslab earthquakes ($M < 8.0$)	Max 15 cycles	Max 10 cycles

The seismic actions are characterized by a first period with a slow increase in accelerations. After the culminating phase, a decrease in ground motion occurs until the movement completely stops. The effect of this process is the formation of a collapse mechanism, produced by a sufficient number of plastic hinges (Figure 5.52a). The development of such a mechanism is composed in two phases: the first phase with the accumulation of plastic rotation, followed by the second phase of structural

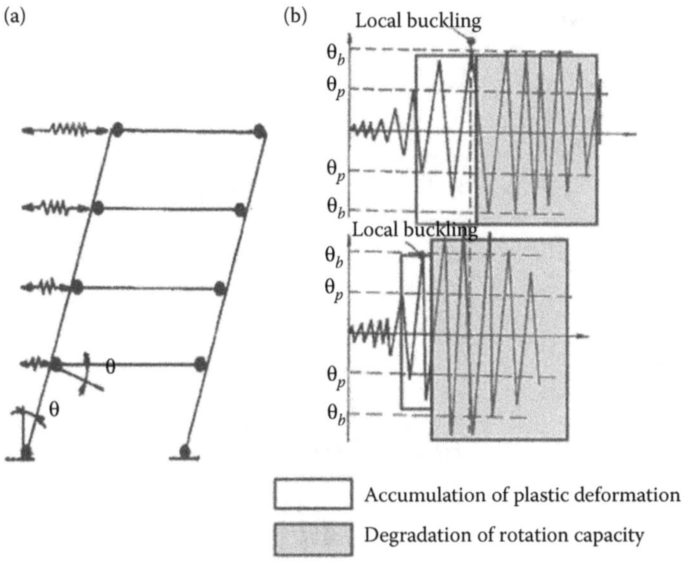

Figure 5.52 Conceptual behavior of a plastic hinge in case of far-field earthquakes: (a) structural scheme; (b) plastic hinge rotation.

degradation, which can be characterized by a decrease, constancy, or increase in seismic actions (Figure 5.52b). Generally, the decrease in seismic actions has no influence on structural degradation. Therefore, in order to study the influence of cyclic loading on the ductility reduction, only two types of cyclic actions should be considered, namely, under constant (Figure 5.53a) or increased amplitudes (Figure 5.53b). Generally, for far-field earthquakes, the type of cyclic action is strictly connected to the effects of soil conditions. Therefore, for the prediction of the ductility reduction, the following rules, corresponding to ones of Eurocode 8, are recommended:

- For normal soil conditions (soils Class A, B, and C), cyclic actions under constant amplitude
- For soft soil conditions (soils Class D, E, $S_{1,2}$), cyclic actions under increasing amplitude

Different loading protocols are proposed trying to simulate the earthquake excitation; for instance, the ECCS Recommendations (1985) for increasing actions suggest a repetition of three times each cycle until collapse (Figure 5.53c). This procedure is very good to interpret the member behavior during laboratory tests and to make their results comparable at an international level, but it is very far from the real behavior when the structure is subjected to an actual earthquake.

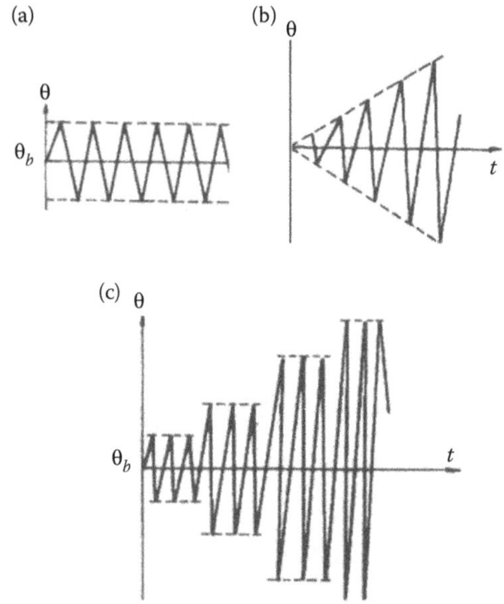

Figure 5.53 Cycle loading types: (a) constant amplitude; (b) gradually increasing amplitude; (c) ECCS experimental proposal.

5.4.2.4 Main effects of cycle loadings

The seismic design philosophy accepts the approach that significant excursions into inelastic range will occur during severe earthquakes, but the current methodologies, adopting an elastoplastic model, do not take into account the effects of accumulation of plastic deformations due to cycle loading. There are two main effects: the first referring to the increase in required ductility; the second to the degradation of the available ductility. Figure 5.54a shows the effect of accumulation of plastic rotation on the required rotation capacity, while Figure 5.54b presents the erosion of the available rotation capacity, due to the multiple cycles (Gioncu and Mazzolani, 2002). The erosion refers both to the moment and rotation capacities. The main effect to the accumulation of plastic deformations during cycle loading can be observed in the moment–rotation curve, which shows the degradation of ductility with respect to the one determined for monotonic loading (Figure 5.54).

5.4.3 Ultralow-cycle fatigue: A new opportunity to solve the dispute on cycle fatigue-accumulation of plastic deformations?

As is well known, the fatigue phenomenon corresponds to the progressive and localized structural damage that occurs when a material is subjected to cyclic loading.

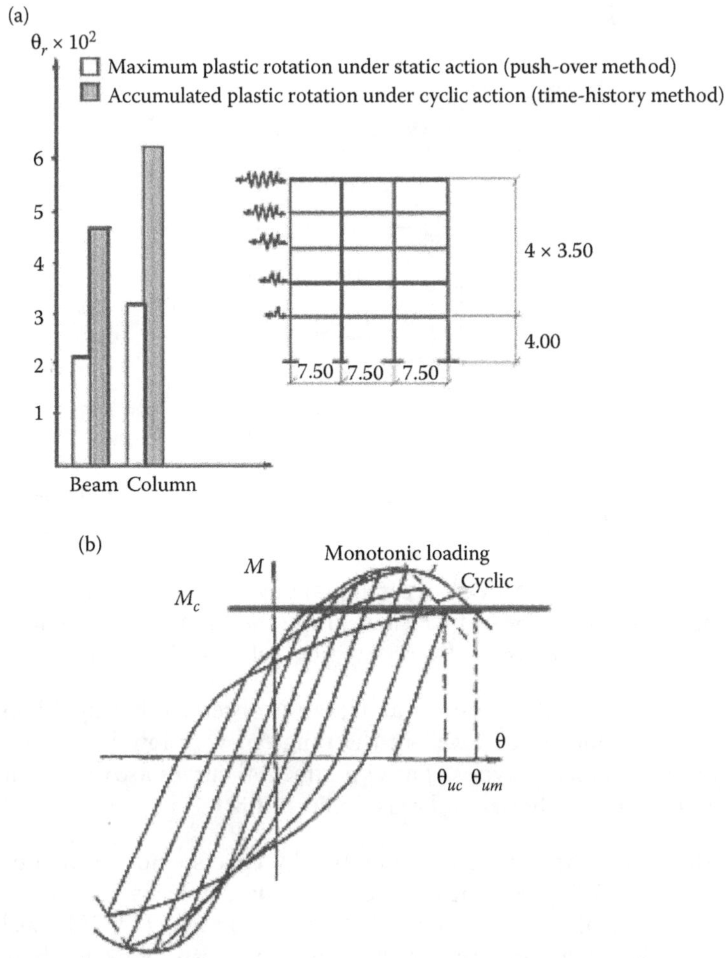

Figure 5.54 Effects of cyclic loadings: (a) at the required level; (b) at the available level.

There are two fatigue types: *high-cycle fatigue* (HCF) and *low-cycle fatigue* (LCF) (Figure 5.55) (Feldmann et al., 2011). Traditionally, fracture mechanics, with damage mechanics for some special cases, was used for the investigation of HCF problems in order to predict the fracture in elastic range as well as the initiation of cracking. The HCF occurs for more than 10^4 cycles and the minimum values of stress for fatigue verification are obtained for 10^6 cycles, the fatigue behavior being characterized by the Woehler curve. The LCF, instead, occurs close to (or at) the yield limit in elastoplastic range under maximum 10^4 cycles, characterized by the Manson–Coffin law.

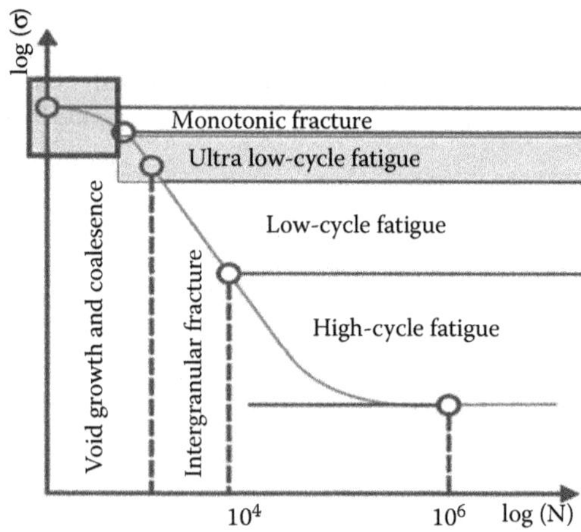

Figure 5.55 Classification of fatigue. (After Feldmann M. et al. 2011: *Steel Construction*, 4(2), 94–113.)

In the last decades, the effects of earthquake cycle loadings were attempted to be studied in the frame of LCF rules, using the Manson–Coffin law. But it is not reliable in the range of very low life cycles, as declared below:

> From the recorded experimental data, it is observed that the Manson–Coffin law does not fit well in the range of very low life cycles, i.e. about less than 100 cycles. The experimental curves are often concave smoothly toward horizontal axis. (Xue, 2008)

So, the experimental curves practically correspond to monotonic actions (Figure 5.56). Recently, a new category, namely, *ultralow-cycle fatigue* (ULCF), (also called *extremely low-cycle fatigue* [ELCF]) has been introduced. For civil engineering structures, and especially for building structures, some members may be stressed in plastic range when they are cyclically loaded, in many cases the damage being caused by plastic buckling. Considering the maximum effective number of cycles (less than 20), it is very clear that the effect of cycles produced by seismic excitation belongs to the inferior limit of ULCF. As a consequence, the damage curves for earthquake effects are similar to the one of monotonic loads (Xue, 2008). The LCF is characterized by a slope, corresponding to the lowering effect of cyclic loading. In contrast, the ULCF shows a curve tangent to the horizontal line, corresponding to monotonic loading. This observation strengthens the basic concept and supports the proposed methodology to use the results obtained for monotonic ductility, with the proper implementation of correction factors taking into account the specific characteristics of earthquake type (see Section 5.1). At the same time, considering the

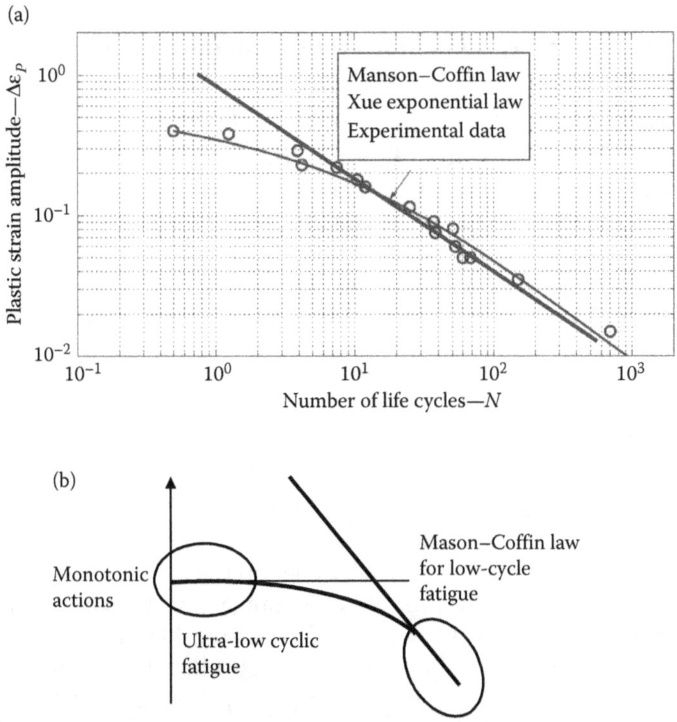

Figure 5.56 Recorded experimental data for ultralow-cycle fatigue and low-cycle fatigue: (a) Manson–Coffin and Xue laws versus experimental data. (After Xue, L., 2008: *International Journal of Fatigue*, 30, 1691–1698.) (b) Ultralow-cycle fatigue and monotonic actions.

reduced number of effective cycles during earthquakes, a basic question arises: namely, should these effects be framed in the category of fatigue or is it more valuable to consider them just as a result of accumulation of plastic deformations? LCF is connected to crack initiation–propagation and mainly to brittle fracture of a tensile part of a section. However, in case of strengthening (e.g., using ribs and cover plates) or weakening (e.g., RBS) of the connection, where the plastic hinge is moved away from the column face, the flange plastic buckling works as a fuse dissipating energy and avoiding brittle fractures (SAC, 1996; Anastasiadis et al., 1999). This was demonstrated from experimental evidence (SAC, 1996). In fact, the conceptual approach of this topic is strictly related to the detailing of the potential region of the formation of the plastic hinge. Moreover, the concept of plastic accumulation looks more open, can be combined well with the established buckling theories, and finally overcomes the experimental constraints of the fatigue approach.

5.4.4 Cyclic actions on steel I-shaped beams

5.4.4.1 Review on experimental studies and theoretical approaches

The I-shaped cross-section is the one most used for steel beams. Therefore, the effects of cyclic actions are studied for these sections, but the results are also valuably applicable for other sections.

For the sake of completeness, a concise review of the experimental and theoretical works from the available literature will be given in this section together with a detailed discussion in this so complex field, which still remains open. A selection of some existing research results on experimental tests is presented in Table 5.8.

5.4.4.2 Experimental testing

A comprehensive overview of the main cyclic tests developed in the last decades is given in Table 5.9. The experimental tests conducted by Bertero and Popov (1965) on rolled steel sections showed three important aspects: (i) the control of deformations at each cycle; (ii) the influence of the number of cycles on fracture; and (iii) the effect of flange local buckling, which can cause a rapid reduction in the number of cycles until fracture.

The influence of lateral buckling on steel-rolled beams subjected to repeated and reversal loading was examined by Takanashi (1974), revealing a great reduction in the rotation capacity as the lateral slenderness ratio increases (LS from 51.20 to 81.70 for the examined cases), obtaining values of cyclic rotation capacity from 1.0 to 2.50 for low-carbon steel. Furthermore, as the lateral instability effect differs between monotonic and cyclic loading, the ductility is drastically reduced in the second loading type. Guruparan and Walpole (1990) also performed tests on hot-rolled and welded I-shaped beams, showing that the strength degradation was still acceptable for lateral slenderness LS = 37, the deterioration increased for a value of LS = 57, while, for the larger value of LS = 97, a twisting was observed and the reduction was no more acceptable.

Tests by Van et al. (1974), Mitani et al. (1977), and Suzuki and Ono (1977) indicated that the deterioration is severe only when local flange buckling is accompanied by web buckling or lateral-torsional buckling. Web buckling or flange buckling produces strength degradation, while the lateral-torsional buckling produces a loss of stiffness or a deterioration of the load-carrying capacity. In addition, in many cases the moment capacity under repeated cyclic action is a little higher than the monotonic one, due to strain-hardening effect, but always a drastic deterioration after local buckling is remarked, when the beam was subjected to cyclic actions. Suzuki and Ono (1977) recorded from their tests the values of cyclic rotation capacity between 3.0 and 9.80, as a function of the lateral slenderness

Table 5.9 Selective Cyclic Experimental Tests on Steel Beams

						Slenderness			
No.	*Author*	*N. tests*	*Section type*	*Test scheme*	*Loading history*	*FS*	*WS*	*LS*	*Main objective*
1	Bertero and Popov	11	RS	MG	CD	6.35	1.24	—	Revealed the main parameters of cyclic loading (controlled deformations, number of cycles to failure)
2	Van et al.	50	RS	MG	CD	—	—	30–60	Local flange buckling associated with web buckling of lateral buckling produce strength-stiffness degradation
3	Takanashi	16	RS	MG	CD	—	—	51.20–80.90	
4	Mitani et al.	90	WS	MG	CD	6.2–16.0	158–640	25.0–40.0	
5	Suzuki and Ono	—	WS	CM	—	—	—	—	
6	Takanashi and Udegawa	10	WS	MG	CD	15	35	61	Slab effect reducing the deformation capacity and increasing the stress level
7	Ballio and Calado	4	WS	MG	CD	10–25	38.3–60.5	—	Once the flange has buckled the maximum load cannot be reached
8	Castiglioni	14	RS	MG	ID	7–10	19.8–33.2	—	
9	Yoshuzimi and Matsui	10	WS	MG	ID	7.9–16.2	31.3–42.2	—	Interaction between local and lateral buckling
10	Gioncu et al.	6	WS	CM	ID	9.0–13.0	60.0–71.0	—	Superposition of plastic mechanism in cyclic loading

continued

Table 5.9 (continued) Selective Cyclic Experimental Tests on Steel Beams

No.	Author	N. tests	Section type	Test scheme	Loading history	Slenderness FS	Slenderness WS	Slenderness LS	Main objective
11	Guruparan and Walpole	6	WS/RS	MG	ID	8.2–8.5	53.7–80.0	—	A lateral slenderness greater than 60–70 reduces seriously the deformation capacity
12	Ballio and Castiglioni	45	RS	MG	CD/RD	6.9–10.0	19.8–39.2	—	Web slenderness is of crucial importance for a stable cyclic deformational capacity
13	Lee and Lee	8	RS	CM	CD	9.6	30.4	—	60% deterioration in case of cyclic action
14	Green et al.	13	WS	MG	ID	5.9–9.0	27.9–29.3	—	Influence of yielding ration of more 0.85
15	Mateescu and Gioncu	11	WS	MG	ID/PD	10.9–12.3	24.0–26.3	—	Early fractures in impulsive loading
16	Valente and Castiglioni								Effect of slab
17	D'Aniello et al.	4	RS	MG	ID	14.28–16.52	22.33–27.46	—	$E_{monotonic}$ ~0.20 E_{cyclic} of the dissipated energy
18	Jiao et al.	4	WS	MG	ID	7.69	50	—	Effect of Baushinger effect on plastic capacity

Type of section: WS, welded sections; RS, hot-rolled sections.

Slenderness ratio: FS, flange slenderness; WS, web slenderness; LS, lateral slenderness.

Moment distribution: MG, moment gradient; CM, constant moment.

Loading conditions: CD, constant displacement; ID, increasing displacement; RD, random displacement; PD, pulse displacement.

with ratios of about 65, 55, and 45, respectively. Unfortunately, their paper does not provide detailed information about the geometrical and mechanical characteristics of the tested specimens. Lee and Lee (1994) also noticed the strength degradation after local buckling, showing that the deformation capacity of cyclically loaded beams is reduced about 60% in comparison with monotonic loading.

In Europe, an important experimental program was developed at the Politecnico di Milano (Ballio and Calado, 1986a,b; Ballio and Castiglioni, 1993, 1994, 1995), investigating the flange local buckling of I-shaped steel welded and hot-rolled sections under predominantly cyclic bending moments; different increasing displacement histories as well as the ECCS Recommendations ones (1985) were applied (Figure 5.57). Among the interesting results, the following are noted:

1. The indication that once the flange has buckled, the maximum load cannot be reached in the subsequent cycles.
2. The amplification of local flange buckling gives rise to a significant decrease in the load-carrying capacity.
3. Cracks appearing at the flange-web welded junction tend to extend up to the flange tip.

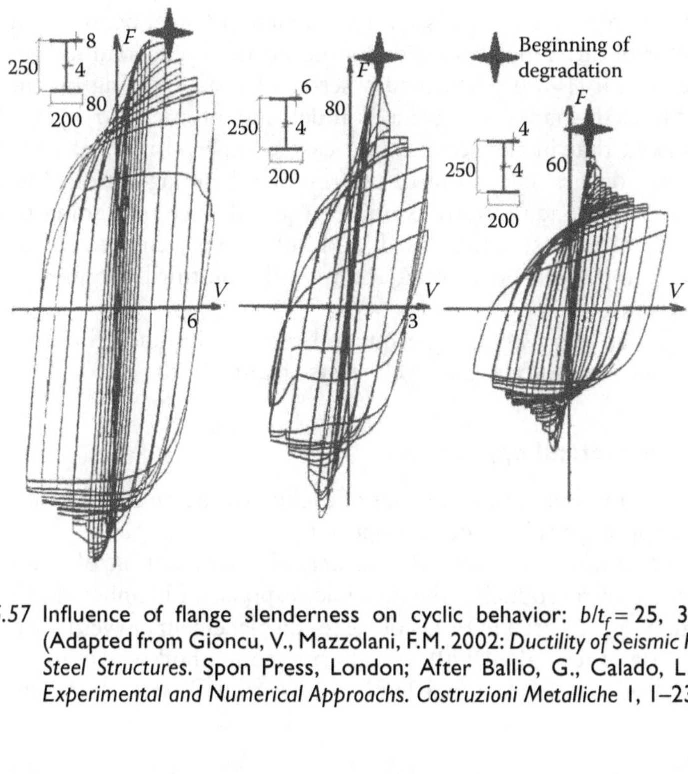

Figure 5.57 Influence of flange slenderness on cyclic behavior: $b/t_f = 25$, 33.3, 50. (Adapted from Gioncu, V., Mazzolani, F.M. 2002: *Ductility of Seismic Resistant Steel Structures*. Spon Press, London; After Ballio, G., Calado, L. 1986a: *Experimental and Numerical Approachs. Costruzioni Metalliche* 1, 1–23.)

Another testing program at the same university (Castiglioni and Losa, 1992; Ballio and Castiglioni, 1995; Castiglioni, 1987), focusing on the local ductility of European I-shaped sections (IPE, HEA, HEB) subjected to different cyclic loading histories proposed by ECCS, was carried out, revealing that IPE sections, having a comparable flange slenderness ratio with that of HEB, have shown a reduced ductility due to the increased web slenderness. Reanalysis of the aforementioned experimental programs by Castiglioni (2005) evidenced that a premature brittle failure may occur, if the cycle amplitude is not large enough to produce flange local buckling. Gioncu and Petcu (1997) had already mentioned that local buckling works as a fuse preventing early undesirable failure mechanisms.

Tests by Takanashi and Udegawa (1989), Castiglioni et al. (2000), and Valente and Castiglioni (2003) on steel beams and composite members indicated that the presence of the slab reduces the cyclic inelastic capacity, due to the fact that the upper flange local buckling is prevented. As a consequence, in highly stressed locations, early fractures can appear, as was confirmed by real earthquake events (Northridge, 1994; Kobe, 1995).

The influence of yielding strength on the cyclic ductility of steel beams was experimentally examined, among others, by Takanashi (1974) and Green et al. (2002, 2007). The tests showed that the increase in yield strength or in yield ratio produces a decrease in the cyclic inelastic capacity of members.

Jiao et al. (2011) revealed the important contribution of the Bauschinger effect on the plastic energy dissipation capacity of steel beams, which is a basic factor of difference between monotonic and cyclic loading.

Gioncu et al. (1989) performed a series of tests focusing on the study of the buckled shapes as formed under monotonic and cyclic loads. They showed that in the second load case a superposition of two plastic mechanisms occurs as a result of the reversed bending action. Moreover, Mateescu and Gioncu (2000) examined the influence of loading types, as defined by an increased, decreased, and pulse displacement. The first two actions develop a plastic accumulation, while the third one just produces early fractures.

D'Aniello et al.'s (2012) experimental test results underline the differences between monotonic and cyclic tests (Figure 5.58).

5.4.4.3 Theoretical approaches

For the examination of the influence of cyclic loading on steel members, five principal approaches have been proposed.

The *constitutive law method* considers the formulation of constitutive laws capable of reproducing the inelastic response of members in the presence of damage, plotted in terms of force–displacement curves (Castiglioni, 1987; Castiglioni et al., 1990). An alternative method was used by Vayas (1997), based on the material law, and associated with an extension to

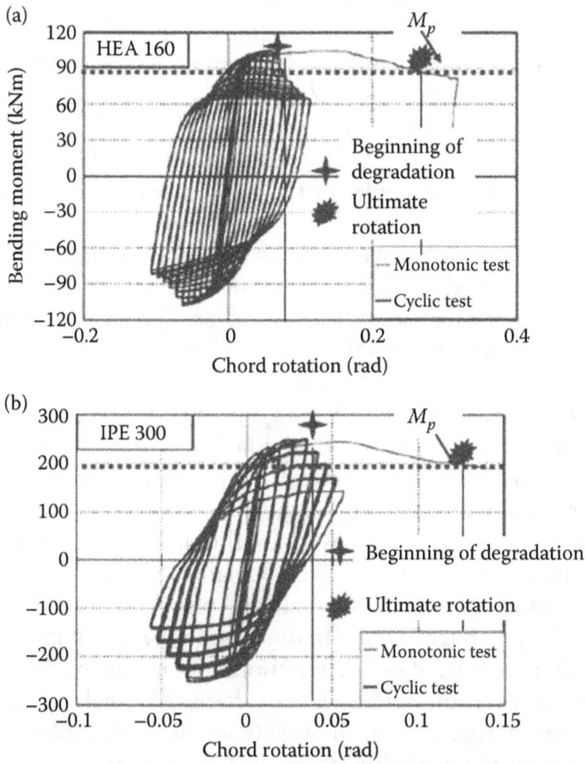

Figure 5.58 Monotonic versus cyclic tests on HEA (a) and IPE (b) profiles. (After D'Aniello, M., et al. 2012: *Journal of Constructional Steel Research*, 78, 144–158.)

cyclic loading of the methodology of effective width, already examined for monotonic loads, in order to predict the strength and stiffness degradation. Zambrano et al. (1999) proposes a constitutive law with degrading strength and stiffness characteristics, taking into account the damage accumulation and the maximum excursion in the inelastic range.

The *strip method* divides the cross-section of the member into a finite number of strips, each strip being characterized by area, distance from the centroid, residual stress, yield stress, and ultimate strain (Figure 5.57) (Ballio et al., 1986). To account for the member damage, due to local buckling, fracture, and gradual degradation, the strip area is assumed to be reduced according to a certain law depending on the type of structural damage. FEM analysis is also used in many research activities (Butterworth and Beamish, 1993; Liu et al., 2002; Okazaki et al., 2006).

LCF approaches. Yamada et al. (1988) and Yamada (1992) proposed low cyclic fatigue fracture limits for all the element components in order to describe the structural ductility. Using the classical rule of Mason–Coffin, Ballio and Castiglioni (1995) and Castiglioni et al. (1997) developed a procedure that unifies both high and low cyclic fatigue for the design and damage assessment of steel members. Recently, Lee and Stojadinovic (2008), by making use of the plastic collapse mechanisms, developed a method that considers the strength degradation induced by local buckling and LCF fracture, exploiting the principles of the above-mentioned Manson–Coffin rule.

Energy-based approaches were mainly used by the Japanese structural community, in which the collapse of a member is limited by an energy criterion. Akiyama proposed a criterion in energy terms, defining that a failure can possibly occur when the cumulated plastic ductility, given by the skeleton curve, is greater than the monotonic one (Akiyama, 1988).

Plastic deformation accumulation approaches. Generally, there are two directions accounting for plastic accumulation. The first one is developed based on the theory of damage models, defining the failure as a limiting value of deterioration and further using Miner's assumption of linear damage accumulation, properly adjusted, for the attainment of collapse condition (Krawinkler and Zohrei, 1986; Calado and Azevedo, 1989). The second one is based on the plastic collapse mechanism model, where the flange and web local buckling is obtained after the accumulation of residual displacements produced by cyclic actions and its result consists in developing a moment–rotation curve with a softening branch, which describes the gradual degradation after each cycle (Gioncu et al., 2000).

Commenting on the available theoretical approaches, it could be noted that methodologies based on LCF or damage models are generally computationally cumbersome and are associated with fatigue curves, which should be developed for a series of sections and details. FEM analysis, when properly calibrated, provides accurate results, but it is not suitable for practical design, due to the large computing time as well as the difficulties and uncertainties regarding discretization. Energy-based approaches are physically compatible with seismic action; however, it is difficult to handle and, for this reason, is also relatively impractical for design purposes. Finally, the plastic collapse mechanism methodology seems to be simple and accurate, based on classical theories of plastic and buckling analysis, and will be further analyzed.

5.4.4.4 Comments about experimental and theoretical results

In order to develop a comprehensive methodology to evaluate the effects of cyclic loadings on local ductility, the experimental and theoretical results presented in the literature are examined, giving rise to the following conclusions:

1. There are not enough experimental data to be used for developing a comprehensive approach for the evaluation of the effects of cycle loading in the frame of the new concept of ULCF. So, the methodology developed must be based mainly on theoretical results and some conclusions obtained during experimental tests.
2. The start of degradation due to cyclic loadings is yielded by the plastic buckling of compression flanges, corresponding to the buckling produced during monotonic loading.
3. Experimental results concerning the I-shaped sections have shown a high sensibility to the cyclic actions, for both the strength and rotation capacities (Figure 5.58). But the loading tests were governed by a large number of cycles, which generally exceeds the maximum number of cycles expected during an earthquake. This can give wrong information about the actual influence of cyclic loading on the beam ductility.
4. Experimental and theoretical results have shown that among the total number of cycles, there are some which do not produce any degradation of strength and ductility and others giving important contribution to this degradation. As shown at point (ii), the stable cycles can be determined in function of plastic buckling produced during a monotonic test. The compact sections present a large number of nondegradation cycles, while this number is small for slender sections (Figure 5.57).
5. Experimental evidence has revealed that it is possible to make use of the monotonic ductility, with proper reduction, in order to predict the cyclic ductility. Therefore, the experimental and theoretical works, previously examined, provide the basic data for the correction of the monotonic ductility.
6. Commenting on the available theoretical approaches, it could be noted that methodologies based on LCF (or for ULCF) or damage models are generally computationally cumbersome and are associated with fatigue curves that should be developed for a series of sections and details. FEM analysis, when properly calibrated, provides accurate results, but it is not suitable for practical design, due to large time consumption as well as difficulties and uncertainties regarding discretization. Energy-based approaches are physically compatible with seismic action; however, they are difficult to handle and therefore relatively impractical for design purposes.
7. Finally, the plastic collapse mechanism methodology seems to be simple and accurate, based on classical theories of plastic and buckling analysis, and can be used to determine the degradation of monotonic ductility due to cyclic actions. The use of the DUCTROT-M computer program, based on the accumulation of plastic deformations, offers the possibility to analyze this degradation in function of beam geometrical characteristics, total number of important cycles, and number of nondamaging and damaging cycles.

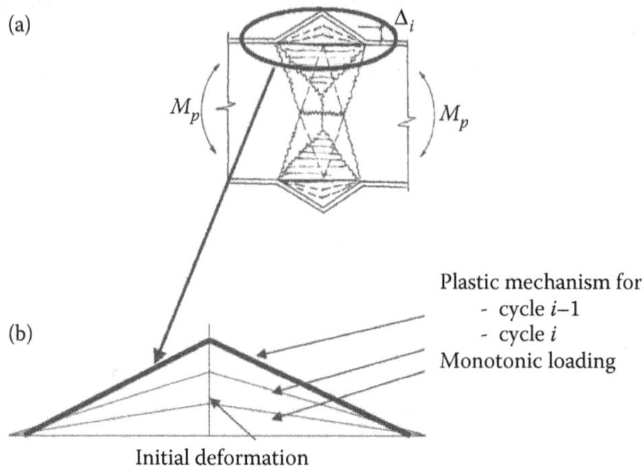

Figure 5.59 Accumulation of plastic deformations: (a) shape of a plastic-buckled section during an experimental test; (b) plastic mechanism for the compressed flange.

5.4.5 Erosion of monotonic ductility due to accumulation of plastic deformations

5.4.5.1 Accumulation of plastic deformations

During cyclic loading, a specific plastic mechanism is observed, being characterized by the superimposed plastic effect due to two opposite bending moments. This plastic mechanism is characterized by buckling in the plastic range of the compressed flanges (cycle i–1). A clearly deformed local plastic shape remains after this cycle (Figure 5.59a). During the following cycle i, a new plastic deformation, in the form of a local mechanism, superimposes over the previously deformed shape corresponding to cycle i–1. Like in stability problems, the deformed shape of the i–1 cycle can be considered as a geometrical imperfection (initial geometrical deformation), over which the new plastic deformation, corresponding to cycle i, is superposed t (Figure 5.59b). So, the effects of a specific number of effective cycles, corresponding to the earthquake type, can be approached as a continuum accumulation of plastic deformations; in this way, the reduction in the strength and rotation capacity of a section can be determined.

5.4.5.2 Example for bended plate under cyclic loading

A bended plate forms a plastic mechanism composed of three yield lines (Figure 5.60a) (Gioncu and Mazzolani, 2002; Park and Lee, 1996; Feldmann,

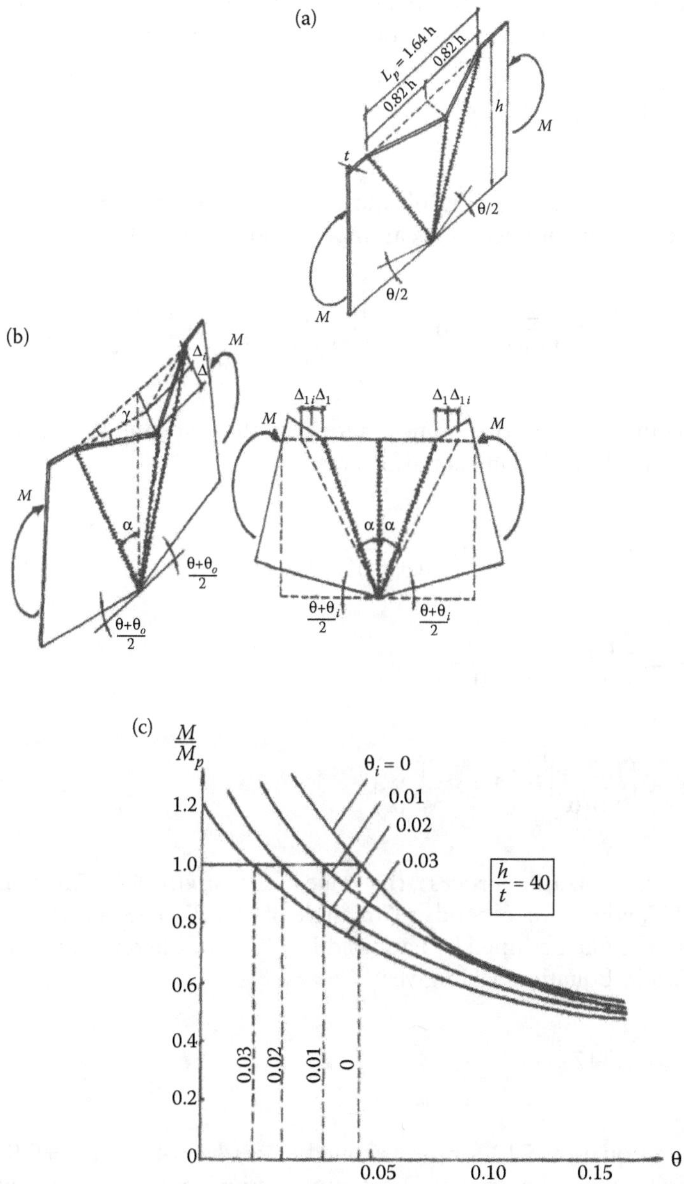

Figure 5.60 Yielding mechanism of a bended plate: (a) mechanism configuration; (b) mechanism with initial deformations; (c) moment–rotation curve relationship, influence of initial deformations.

1994). Considering the plate of Figure 5.60b with an initial geometrical imperfection, the following relationships can be written:

$$\Delta_{1i} = h\frac{\theta_i}{2}; \Delta_i = h\frac{\theta}{2} \tag{5.9a,b}$$

Using the principles of rigid-plastic analysis, the internal potential energy for a mechanism with geometrical imperfection is given by

$$U = 2f_y\frac{t^2}{2}h\left(\frac{2}{\tan\alpha}\right)^{1/2}\left(1+\frac{1}{\cos^2\alpha}\right)\left[\left(\frac{\theta+\theta_i}{2}\right)^{1/2} - \left(\frac{\theta_i}{2}\right)^{1/2}\right] \tag{5.10}$$

Taking into account the moment acting on the plate and the corresponding rotation, the following equations result:

$$M = \frac{dU}{d\theta} - 2f_y h\frac{t^2}{2}k(\alpha)\frac{1}{((\theta+\theta_i)/2)^{1/2}} \tag{5.11a}$$

$$\frac{M}{M_p} = 2\frac{t}{h}k(\alpha)\frac{1}{((\theta+\theta_i)/2)^{1/2}} \tag{5.11b}$$

$$k(\alpha) = \left(\frac{2}{\tan\alpha}\right)^{1/2}\left(1+\frac{1}{\cos^2\alpha}\right) \tag{5.11c}$$

After a minimization process, the value of factor $k(\alpha) = 4.1736$ results for $\alpha = 0.6847$, which corresponds to an angle of 39.29°. The obtained value is similar to the one obtained by Park and Lee (1996). Therefore, substituting $k = 4.1736$ in Equation 5.11a gives

$$\frac{M}{M_p} = 8.347\frac{t}{h}\frac{1}{(\theta+\theta_i)^{1/2}} \tag{5.12}$$

Plotting Equation 5.12 for the assumed values $h/t = 40$ and $\theta_i = 0.0–0.03$, Figure 5.60c shows how geometrical imperfections progressively reduce the rotation capacity at the level of $M = M_p$. The rotation capacity results from the intersection of $M/M_p = 1$ with the curves given by Equation 5.12, thus providing a rotation capacity equal to $\theta = 0.04355$ rad.

In order to study the influence of pulsatory loading (Figure 5.61), the evaluation of the rotation limit of a bended plate can be done by means of

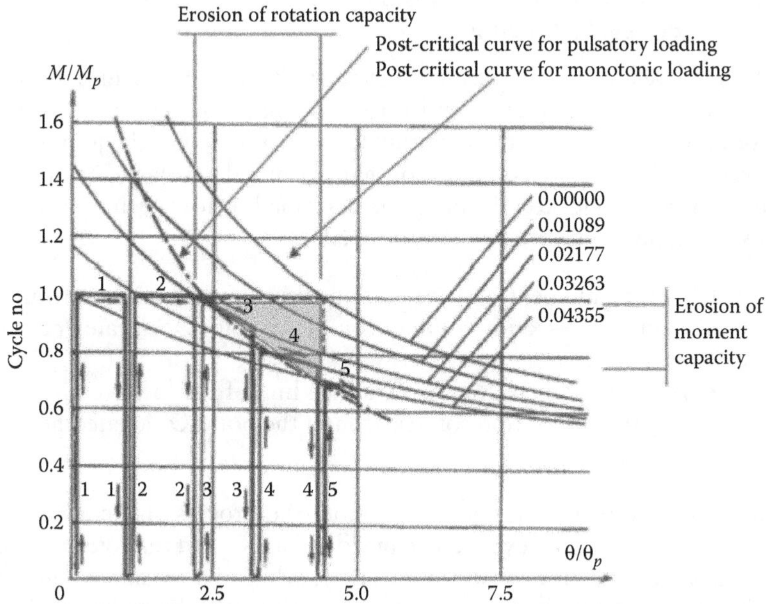

Figure 5.61 Moment–rotation curve of a cyclically bended plate with initial imperfection.

four steps for a plastic rotation, $\theta_p = 0.04355/4 = 0.01089$ rad, each being composed by loading, unloading, and reloading of a rigid-ideal plastic material. After the first step, the remaining rotation is equal with θ_p. The second step starts with this initial rotation and at the final stage of this cycle, the remaining rotation is equal to $2\theta_p$. After the first two steps, the curve corresponds to an initial plastic rotation of $2\theta_p$, achieving the value of 0.02177 rad, while a decrease in the bending moment occurs. At the following step, with an initial rotation of $3\theta_p = 0.03263$ rad, a new decrease is produced due to the increase in initial rotation. Therefore, a new postcritical curve results, due to pulsatory loading, different from the one obtained for monotonic loading.

Hence, the concept of using geometrical imperfections (initial geometrical deformation), in order to approach the softening behavior of a plate under repetitive actions, seems to be compatible with the real behavior.

The following conclusions yield from this bended plate model:

- The first two pulses produced in the range of $M = M_p$ have no contribution to the reduction of rotation capacity.
- The rotation capacity for pulsatory loading is reduced to about half of the one corresponding to monotonic loadings.

5.4.5.3 *Plastic collapse mechanism of I-shaped steel beams under cyclic loading*

At the member level, the plastic mechanism is introduced in a position where the plastic hinges are intended to be formed (adjacent to the column face or at a distance from the column face as a function of the joint detailing). Hence, this mechanism is a tool that permits the inelastic behavior of the beam that belongs to a frame to be described. Under cyclic action, the following response is observed (Figure 5.62a):

- The positive moment, M^+, causes buckling of the compressed upper flange, while the section rotates around the point O^+ situated near the tension flange.
- The negative moment, M^-, causes buckling of the lower compressed flange, while the section rotates around the point O^- located near the opposite flange.

Under the reversals of the seismic action, the process also continues in the same way in the next cycles. The buckled flanges and the rotation point are always situated in a different position, while the tension strains are not able to straighten the buckled flanges. As a result, the collapse mechanism is developed by the superposition of two local plastic mechanisms. Moreover, during each next cycle, the element works with an initial geometrical imperfection as resulted from the previous cycle. Consequently, an accumulation of plastic deformations in the buckled flanges, associated with the number of cycles, could be observed (Figure 5.62b). Physically, the aforementioned process produces a gradual degradation, which creates the condition for a differentiation between the monotonic and cyclic loading effects.

For monotonic loading, the local plastic mechanism methodology was used by many researchers for the prediction of the local ductility of steel members. According to the systematic studies developed at the "Politehnica" University of Timisoara, the local plastic mechanism is composed by yielding lines and plastic zones, which primarily dissipate the input energy. The proposed plastic collapse mechanism, developed under monotonic loading, was further extended by introducing the effect of cyclic bending, which produces the accumulation of plastic deformations with the main consequence of a gradual strength and deformation deterioration. In case of cyclic actions, a steel beam works under positive and negative bending moments, which, due to the repetitive character, cause accumulation of plastic deformations. From experimental observations and analytical works, the following issues could be remarked:

- A superposition of two plastic mechanisms was observed during experimental tests under cyclic loading (Figure 5.63a).

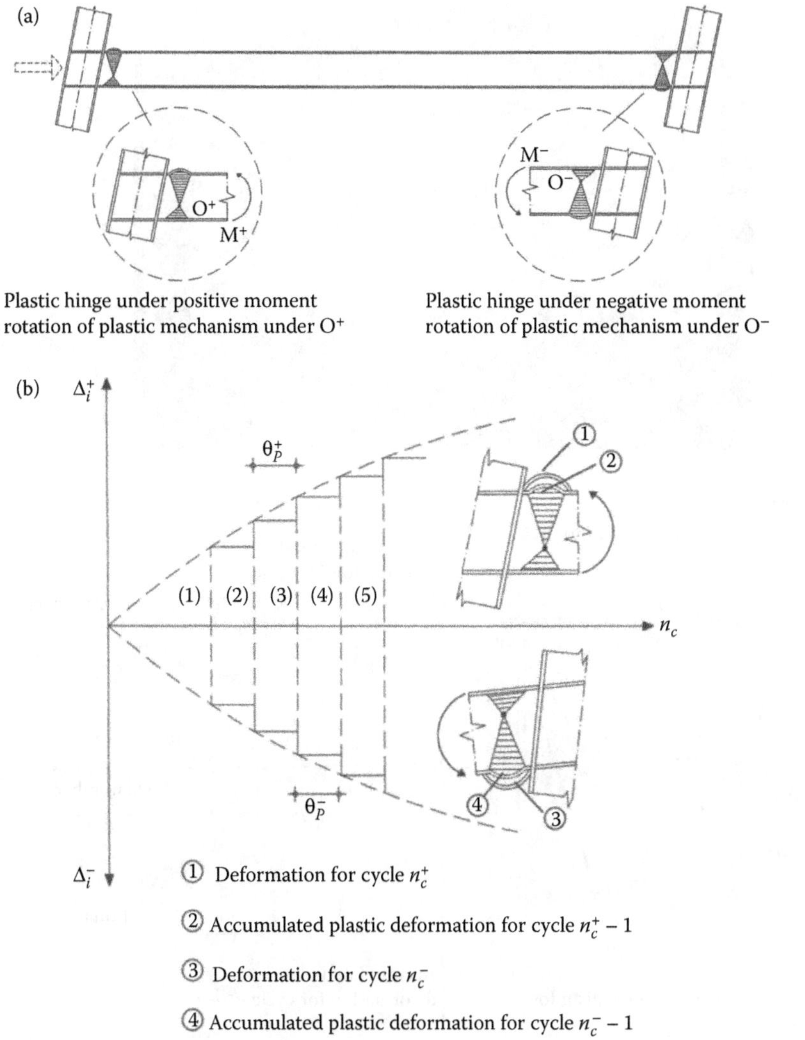

Figure 5.62 Plastic collapse mechanism at the member level: (a) positions of plastic hinges; (b) accumulated plastic deformation of a plastic collapse mechanism in function of the number of cycles.

- Buckling in the upper compressed flange occurs at the first semicycle and the section rotates around a point in or near the opposite flange, where the tensile force is very small (Figure 5.63b).
- At the reversal semicycle, the bottom compressed flange buckles too, but the reversal action is not able to straighten the upper buckled flange due to the fact that the tensile force is small.

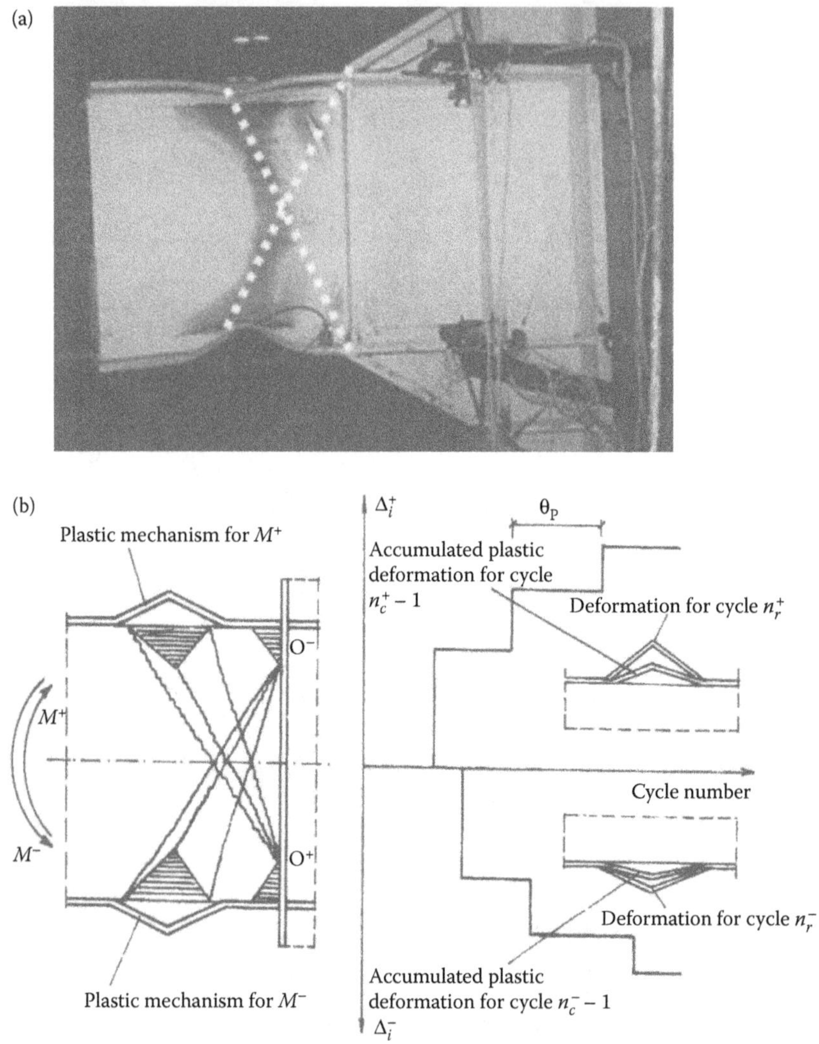

Figure 5.63 Plastic local mechanism for cycle loading: (a) experimental evidence; (b) accumulation of plastic deformations for buckled flanges.

- During the next cycles $\mathbf{n}_r^+$ (for the upper flange) and n_r^- (for the bottom flange), the section works with an initial geometrical deformation, Δ_i, resulting from the previous cycles, $n_r^+ - 1$ and $n_r - 1$, respectively. Therefore, after each cycle, an additional deformation is superimposed on the previous one and in this way an accumulation of plastic deformations is achieved.
- The initial deformation increases with the increase in cycles.

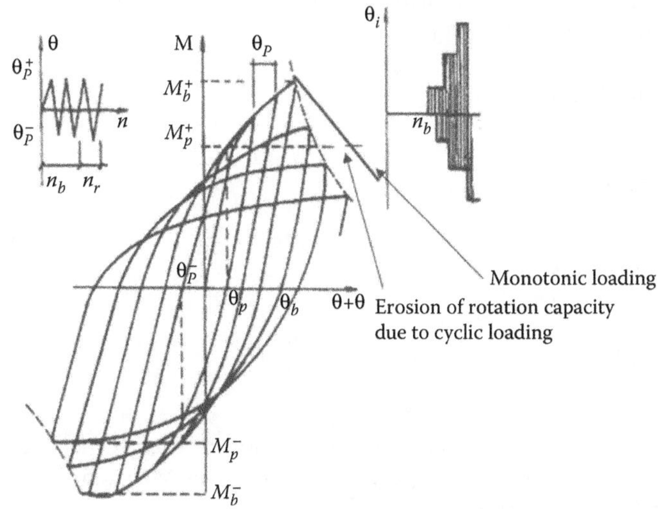

Figure 5.64 Moment–rotation curve eroded by cyclic loading.

After the occurrence of the flange local buckling, a deviation from the characteristic monotonic behavior is generally observed (Figure 5.64), due to the fact that the yielding lines and plastic zones produce the lowering of the ultimate available rotation capacity. There are no differences between monotonic and cyclic loading until the buckling of the compressed flange. This observation allows for an extension of the stable part of the moment rotation curve to its unstable part. The difference in behavior begins to be significant only after the plastic buckling occurs at the cycle n_b. It is evident that the level of degradation is directly associated with a number of cycles. Thus, the accumulation of residual deformations in the flange and web deteriorates the load-carrying capacity and reduces the rotation capacity of the steel section.

Hence, the proposed model for cyclic loadings is based on the concept of accumulated initial deformations, described by the shape of the local plastic mechanism, in the same way that the initial geometrical imperfections are considered in the models for the analysis of stability problems.

5.4.5.4 Local member plastic mechanism

Owing to the uncertainties and inherent variability of the earthquake ground motions, it is difficult to simulate the structural response without making a series of assumptions regarding the local inelastic behavior, which is defined as an envelope. By using the methodology of the plastic collapse mechanism (Gioncu and Mazzolani, 2002; Gioncu et al., 2012), the moment–rotation relationship, with the introduction of the accumulated plastic rotation θ_{ac}, is given by

- For monotonic loading

$$\frac{M}{M_p} = \alpha_{1(M)} + \alpha_{2(M)} \frac{1}{\theta^{1/2}} \tag{5.13a}$$

- For cyclic loading

$$\frac{M}{M_p} = \alpha_{1(M)} + \alpha_{2(M)} \frac{1}{\theta^{1/2}} \tag{5.13b}$$

where $\alpha_{1(M)}$ and $\alpha_{2(M)}$ are constants in function of the geometrical dimensions of the cross-section, as given by Gioncu and Mazzolani (2002) and Petcu and Gioncu (2003).

It is clearly evidenced that the erosion of the rotation capacity is related to the number of cycles. As was shown, not all cycles produce an erosion of member ductility. Therefore, if the maximum number of pulses used in Mateescu and Gioncu (2000) is considered, the total number of cycles n can be defined as:

$$n = n_b + n_c \tag{5.14a}$$

where

- n_b is the number of cycles produced in the stable elastoplastic field, before the reduction of moment capacity due to local buckling.
- n_c is the number of decisive strong pulses producing the erosion of rotation capacity being

$$n_c = n - n_b \tag{5.14b}$$

Hence, the cyclic ultimate rotation θ_{uc} can result from the monotonic one θ_{um}:

$$\theta_{uc} = \theta_{um} - \theta_{ac} \tag{5.15}$$

where

- For constant loading:

$$\theta_{ac} = (n_c - 1)\, \theta_p \tag{5.16a}$$

- For increasing loading:

$$\theta_{ac} = n_c\,(n_c - 1)\,\theta_p/2 \tag{5.16b}$$

In terms of the rotation capacity, considering that the step for each cycle is θ_p, according to the ECCS Recommendations (1985) for experimental tests, the following relationships can be set up:

- Cyclic rotation capacity under constant amplitude:

$$R_{cyclic.c} = R_{mon} - (n_c - 1), \quad \text{with } n_c \geq 2 \tag{5.17a}$$

- Cyclic rotation capacity under increasing amplitude:

$$R_{cyclic.in} = R_{mon} - n_c(n_c - 1)/2, \text{ with } n_c \geq 2 \tag{5.17b}$$

where R_{mon} is the monotonic rotation capacity given by

$$R_{mon} = (\theta_u/\theta_p) - 1 \tag{5.18}$$

As a result of the conventional step for cyclic loading proposed in ECCS Recommendation (1985), the number of cycles corresponding to the stable elastoplastic field, considering that the moment M_p and the corresponding rotation θ_p occurs at the value 0.8–09 θ_u (Figure 5.65), is given by

$$n_b = (0.8 - 0.9)\,R_{mon} \tag{5.19}$$

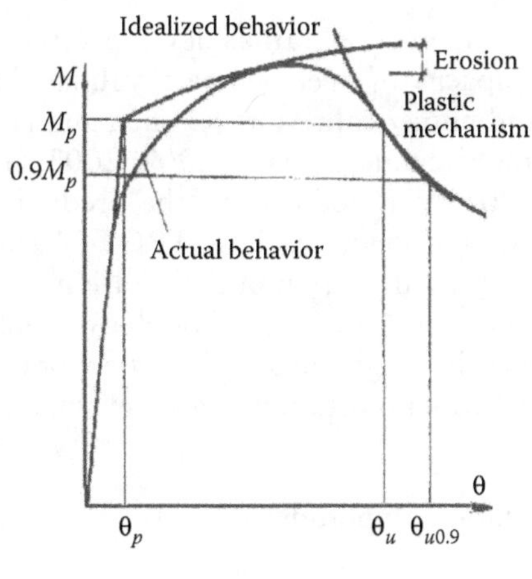

Figure 5.65 Determining the rotation capacity for monotonic loading.

From the above relationship (5.19), two different cases can be identified:

- If $n < n_b$, $n_c < 0$, all cycles occur in a stable field, before the plastic buckling of the compression flange, and therefore cyclic loading has no effect on the monotonic rotation capacity.
- If $n > n_b$, $n_c > 0$, a reduction of rotation capacity occurs, and therefore a correction of the monotonic rotation capacity due to the effect of cyclic loading is necessary.

The relationship (5.19) estimates the cycle number in the stable field, providing a prediction on the maximum number of cycles for an expected earthquake and also offering the possibility to establish the required R_{mon}, in order to eliminate or to minimize the damaging effects of cyclic loading. For instance, if the expected effective cycle number for a given earthquake is 6 (flexible structure, soil class B), it is possible to choose a beam section with $R_{mon} = 6/(0.8–0.9) \simeq 6.3–6.5$, for which the cycle loading does not produce any reduction.

The monotonic rotation capacity definition is conventionally related to the theoretical fully plastic moment at the intersection of its value with the lowering postcritical curve. The use of the value 0.9 M_p for this intersection is suggested by Gioncu and Mazzolani (2002) for the reduced rotation capacity in case of cyclic loading.

5.4.6 Available beam ductility for far-field earthquakes

5.4.6.1 The DUCTROT-M computer program for the prediction of available cyclic ductility by considering the affecting factors

The DUCTROT-M computer program is developed with the aim of determining the rotation capacity of steel members, validated by the verification with experimental and numerical results (Gioncu and Petcu, 1997; Gioncu and Mazzolani, 2002; Petcu and Gioncu, 2002, 2003; Gioncu et al., 2012; Anastasiadis et al., 2013). In addition to the prediction of the rotation capacity under monotonic load, the DUCTROT-M software contains a series of functions for the calculation of the rotation capacity under cyclic load. Based on the above-mentioned methodology, some further relationships are presented as follows giving a holistic view of this program.

The erosion of the ultimate rotation can be evaluated by using the following equations:

- For constant rotation amplitude:

$$\Delta\theta = i\theta_p \tag{5.20}$$

where i is equal to n_c or less than this value (e.g., the greatest value for which $\theta_{uj} - \Delta\theta_u \geq 0$).

- For increasing rotation amplitudes:

$$\Delta\theta_u = \frac{i(i+1)}{2}\theta_p \tag{5.21}$$

where i is equal to n_c or less than this value (e.g., the greatest value for which $\theta_{uj} - \Delta\theta_u \geq 0$).

Finally, the ultimate or the rotation capacity is

$$\theta_{ujc} = \theta_{uj} - \Delta\theta_{un} \tag{5.22a}$$

$$R_{\text{cyclic}} = \frac{\theta_{ujc}}{\theta_p} - 1 \tag{5.22b}$$

Regarding fracture rotation, the software also facilitates its calculation, taking into account the yield ratio, the form of the plastic mechanism as well as the sectional and member characteristics.

With the aid of the DUCTROT-M computer program, an analysis was performed focusing on the main influence parameters, which affect the cyclic available rotation capacity. In this study, the European I-sections (IPE and HEA) were considered.

5.4.6.2 Influence of loading type

The effect of cyclic loading conditions, with constant or increased amplitude, on the rotation capacity is presented in Figure 5.66. For the case of increasing amplitude, taking into account the number of cycles that produces a gradual deterioration, it is pointed out that after several cycles the rotation capacity is exhausted as compared with the monotonic behavior. Also for constant amplitude, one can observe a dramatic erosion of the available rotation capacity, which reaches more than 50% of the monotonic one. Owing to great uncertainties with regard to the ground motion, the available ductility could be determined as an envelope that is defined by an upper bound given by the increasing amplitude and a lower bound described by the constant amplitude.

Looking from a different point of view and assuming that the increasing loading type is directly related to soft soil conditions, and the constant one to stiff soil conditions, it could be remarked that the same member has a different available inelastic rotation capacity with respect to a different

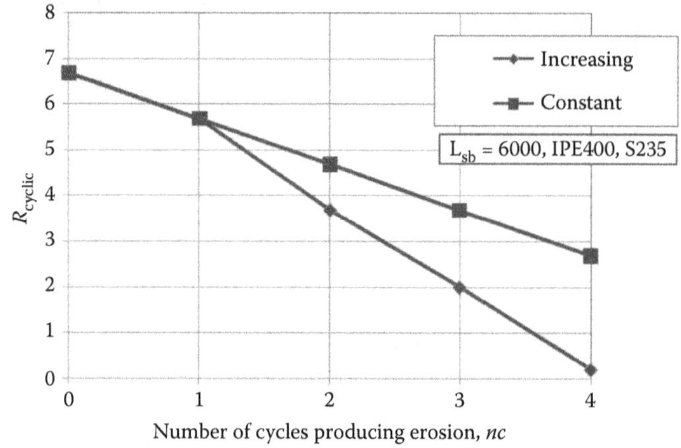

Figure 5.66 Influence of loading type on the available cyclic member ductility.

process of the accumulation of plastic deformations. Following this concept, the effect of local soil conditions on the member available ductility is a very important issue. Therefore, it is not only the required ductility, but also the available one, that is influenced by the earthquake type (for instance, far-source seismic action) and by local soil conditions.

5.4.6.3 Influence of cross-section shape

The influence of section type on the number of cycles producing erosion after the flange buckling is plotted in Figure 5.67. Comparing two different types of hot-rolled sections with approximately the same load-carrying capacity and slenderness ratios, one can observe that the cyclic rotation

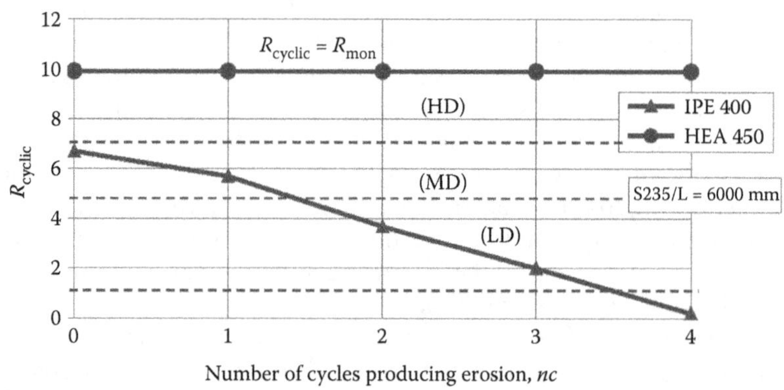

Figure 5.67 Influence of cross-section shape on the available cyclic member ductility.

capacity of HEA 450 section is constant, and it is the same for the monotonic behavior, while the IPE section undergoes a gradual degradation of the cyclic rotation capacity. Using the DUCTROT-M computer program, an extensive analysis for the whole range of the HEA and IPE sections was carried out, revealing the same conclusion (Figure 5.68).

The HEA sections have a better behavior, due to the greater web thickness and higher buckling length. Moreover, identical remarks were pointed out from the experimental work published by Ballio and Calado (1986a,b), Castiglioni and Losa (1992), and Castiglioni (2005).

Therefore, for design purposes, the following indications can be given:

- The use of HEA sections is recommended in the conformation of moment-resisting frames supported on soft soils.
- A structural detail can improve the inelastic capacity of a plastic hinge as presented in Figure 5.69. A combination of the HEA and IPE sections can be used according to the "column tree" concept: the stub made of an HEA section is shop-welded from, and the remaining part of the beam is made of, the IPE section. In this way, both the economical and behavioral requirements could be achieved. In addition, the HEA section could be fabricated as an RBS, which is more advantageous than in the case of an IPE section, due to better adjustments of flange reduction, as a function of plastic capacity, as well as easier fabrication, further enhancing the rotational capacity of the steel beam.

5.4.6.4 Influence of yield and ultimate stress ratio

The investigation on the influence of the yielding ratio, $\rho_y = f_y/f_u$, on the cyclic available ductility is directly related to the number of cycles, n_c, producing erosion of the local available ductility. Figure 5.70a and b, refer to the IPE sections, shows the accumulation effect, introduced by a certain number of cycles, comparatively to the variation of yielding ratio; however, one can observe the following three main issues. First, the effect of the cyclic cumulative action has a more pronounced character than the increase in the yield to ultimate stress ratio. Second, the effect of the aforementioned ratio becomes noticeable for values greater than 0.85 where the reduction is approximately 20–30% as compared with a ratio equal to 0.65 and as a function of loading type (increasing and constant amplitude). Finally, the reduction is more severe in case of increasing amplitude than in case of a constant one, but in any case the dramatic reduction rotation capacity is due to the number of cycles producing ductility degradation. The same representation for the HEA sections (Figure 5.71a and b) shows that low values of ρ_y equal to 0.65 do not produce any effect for cyclic actions, but the increase of ρ_y in the range

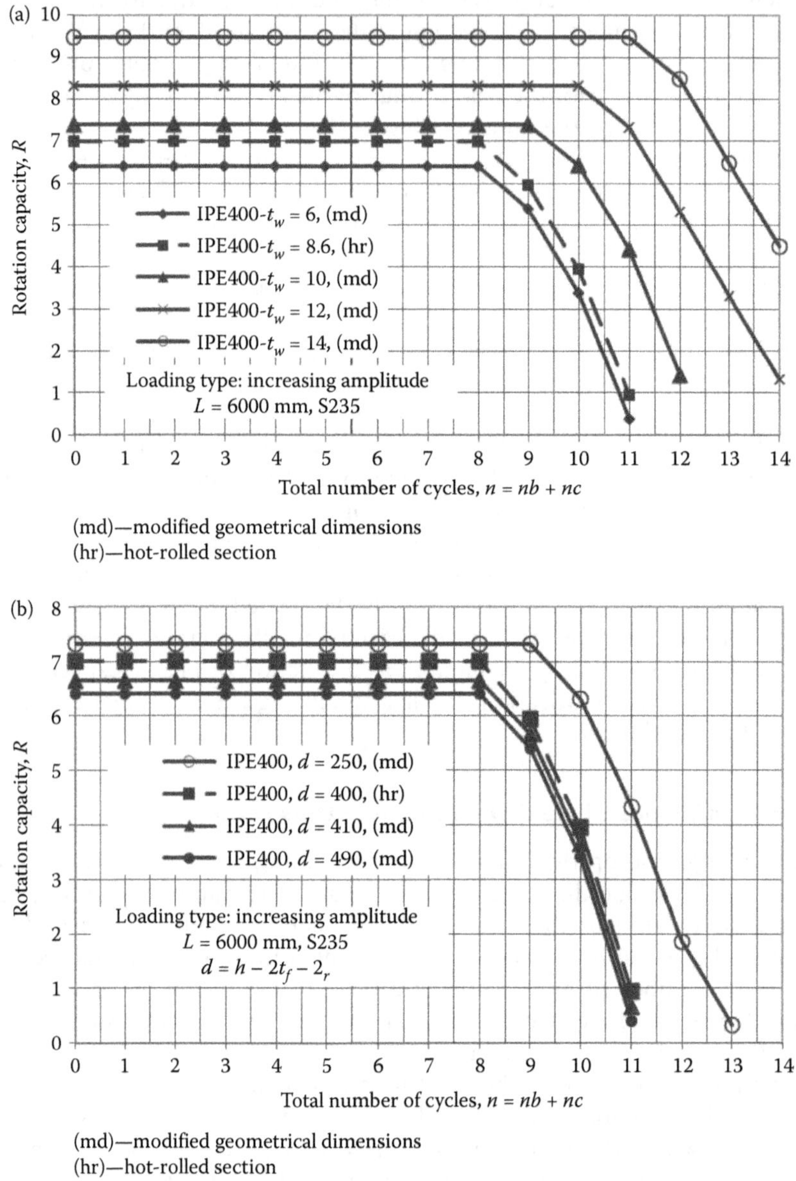

Figure 5.68 Influence of web slenderness on the available ductility: (a) effect of web thickness; (b) effect of web height.

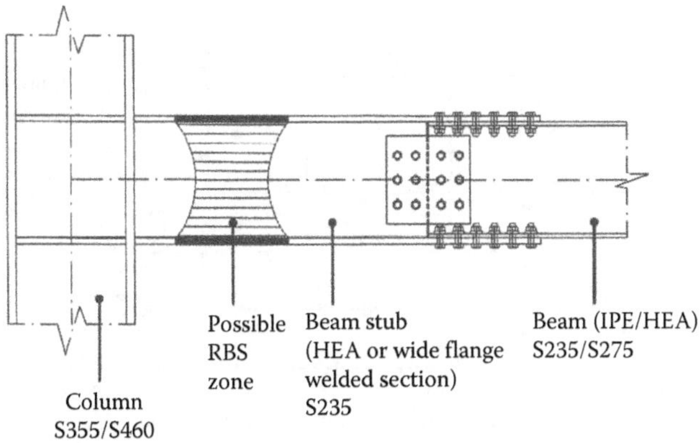

Figure 5.69 Structural details enhancing the ductile capacity of a beam in a frame.

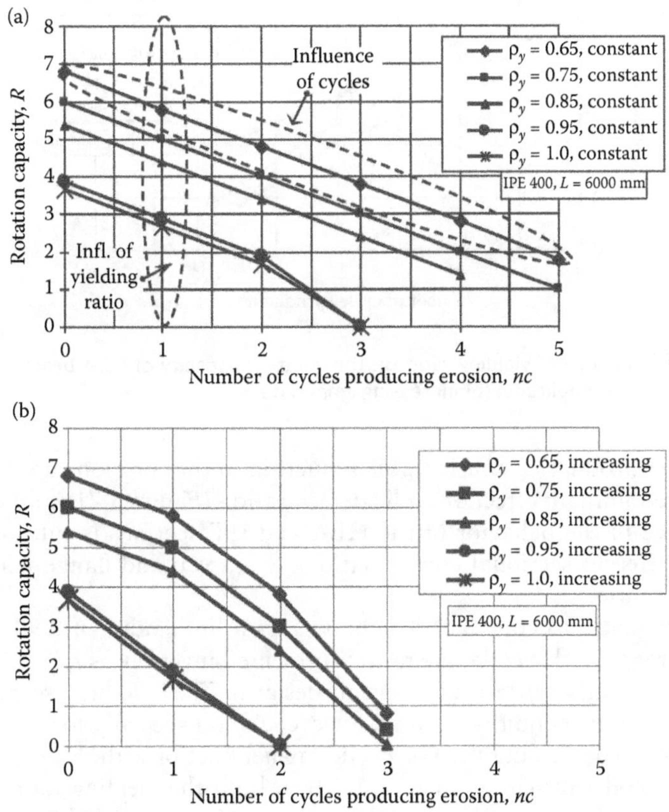

Figure 5.70 Influence of yielding ratio on the rotation capacity of IPE beams: (a) constant amplitude; (b) increasing amplitude.

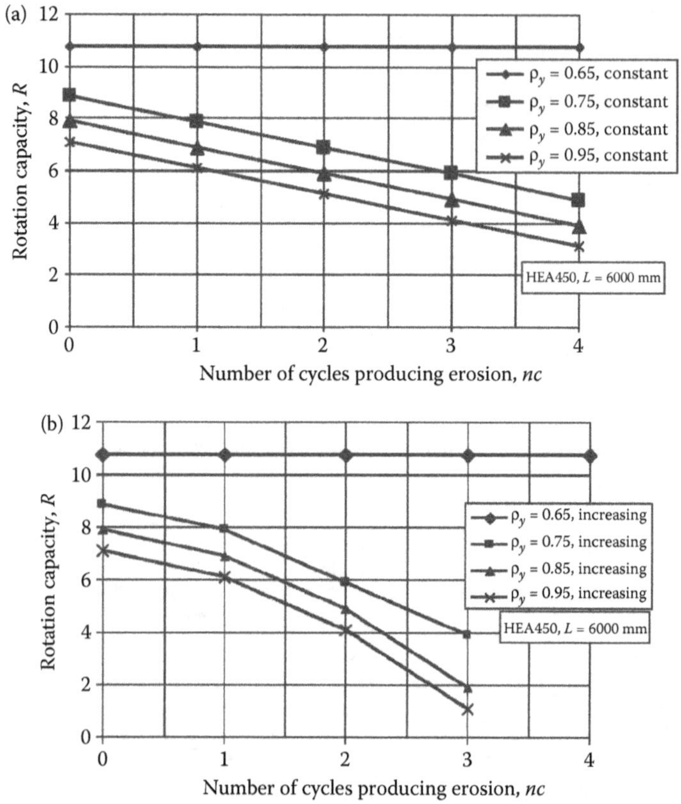

Figure 5.71 Influence of yielding ratio on the rotation capacity of HEA beams: (a) constant amplitude; (b) increasing amplitude.

0.75–0.95 produces a considerable reduction in the rotation capacity due to the predominant effect of cyclic deterioration (Figure 5.71a and b). The divergence in the behavior of the HEA and IPE sections is substantially due to different sectional conformations (e.g., web and flange width-to-thickness ratio).

The current Eurocode 3 (2005) prescribes a limit value of 0.90 for the yielding ratio (in this code, the ratio is presented inversely as $f_u/f_y > 1.1$, but Eurocode 8 states $f_u/f_y > 1.2$ for seismic designs). The 0.90 limit seems to be proper for static conditions, while for cyclic and especially for strain-rate conditions, it is unsuitable. For cyclic far-field actions, the value of 0.85 (or in the code format, a value of 1.20) to limit the yielding ratio is also considered to be convenient for taking into account the reliability of steel producers. Eurocode 8 specifies measures regarding the assurance of the yielding strength f_y (clauses 6.2 and 6.11).

Accordingly, the experimental and analytical works presented in SAC (1996), El-Tawil et al. (2000), and Mao et al. (2001) conclude that for a yielding ratio of 0.85–0.95 the rotation capacity is reduced under the value of 0.03 rad, which is considered as a benchmark value. However, owing to a great number of factors affecting the local ductility (e.g., detailing, loading, and workmanship), the difficulty, and the differentiation in the experimental conditions to capture the critical deformations, a proper level of conservatism should be considered.

After an extensive parametrical study, it is observed that the conclusions mentioned previously are valid for the whole range of IPE and HEA sections. It is a very important remark when one tries to associate the monotonic and the cyclic local ductility with the ductility classification. Hence, for design conditions within the range of material variability between $\rho_y = 0.65–0.80$, the influence of the yielding ratio does not affect the cyclic available rotation capacity, the number of cycles being of primary importance. For values larger than 0.85, it is necessary for the designer to proceed in the prediction of the cyclic rotation capacity, further reducing the rotation due to the material variability.

5.4.6.5 Influence of yield strength

At first glance, the influence of yield strength appears to be decisive regarding the rotation capacity under cyclic action, but a distinction is necessary between the behavior of the HEA and IPE sections. In fact, Figure 5.72 shows that the ductility reduction associated with the number of cycle, as far as the yielding strength increases, is particularly severe for the IPE sections; the same is true only for the HEA sections made of S355 steel, but not

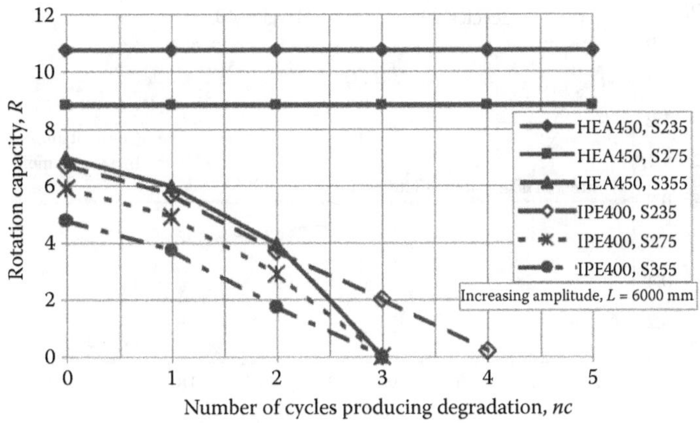

Figure 5.72 Influence of steel quality on the available cyclic member ductility.

for low steel qualities, for which the effect of cycles seems to be ineffective. As observed previously, the stable behavior of this type of profiles HEA derives from the cross-section shape (Figure 5.67).

As a conclusion from both paragraphs (d) and (e), it could be pointed out that the influence of material variability and steel quality is related to the effective action of a given number of cycles.

5.4.6.6 *Influence of strength degradation*

In order to study the strength degradation of hot-rolled profiles, the loading history proposed from ECCS was applied in combination with the concept of the cumulated initial deformation. Figure 5.73a and b shows the

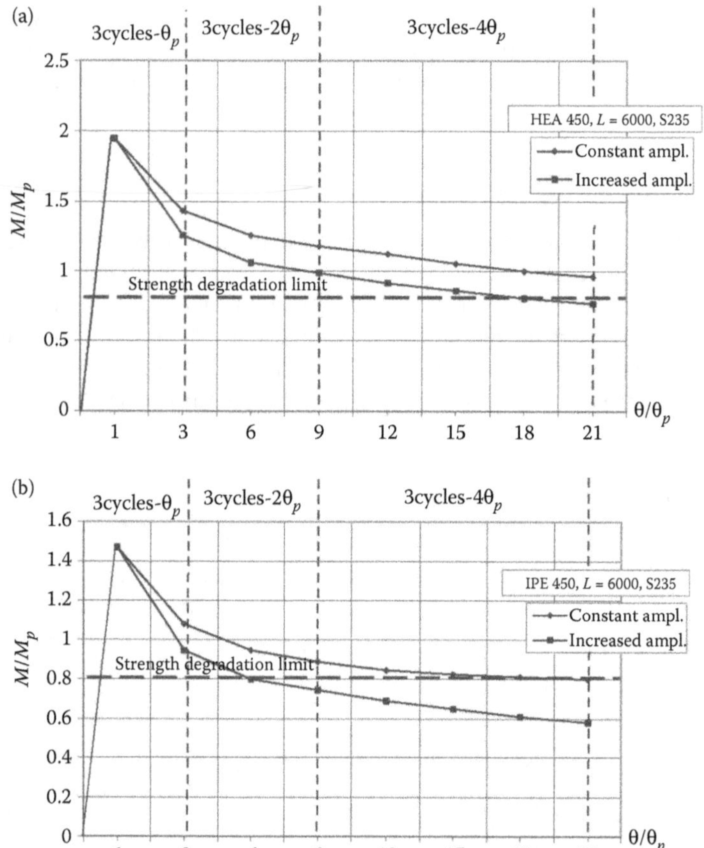

Figure 5.73 Moment–rotation curves for HEA (a) and IPE (b) profiles (loading history according to the ECCS Recommendations).

comparative gradual strength degradation of I-shaped beams conformed as HEA(a) and IPE(b) sections.

As expected, the HEA sections have a better strength capacity than the IPE sections with about 30–35% as shown from a parametrical analysis. The superiority of the HEA sections was also proved in the experimental works presented by Ballio and Calado (1986), Castiglioni and Losa (1992), and Calado and Azevedo (1989). Furthermore, the experimental findings also validated the concept of the initial cumulated deformation.

5.4.6.7 *Classification of the cyclic available member ductility*

In case of cyclic action, the classification could be associated with one of the monotonic behaviors, taking into account the main effect, which corresponds to the number of cycles producing erosion of the monotonic ductility, in particular, the local buckling producing strength and ductility degradation. Moreover, Gioncu et al. (2012) and Anastasiadis et al. (2013) pointed out the unsatisfactory framework for the prediction of the available ductility under monotonic loading. In this direction, a new classification was proposed, based on the member level of inelastic behavior. The numerical analysis revealed that a reduction of 2–4 of the rotation capacity, as compared with the monotonically determined ductility, is the effect of cyclic action on the IPE and HEA steel beams; in particular, 3–4 for the IPE sections and 2–3 for the HEA sections. Consequently, it results that $R_{\text{cyclic}} = R_{\text{monotonic}} - n_c$, where the factor $n_c = 2–4$ is given as a function of the cross-section member conformation. Accordingly, exploiting the limits proposed by Gioncu et al. (2012), making use of the basic relation for the rotation capacity, $R = (\theta_u/\theta_p) - 1$, and dividing it by a mean factor reduction of $n_c = 3$, the following member cyclic classification is proposed:

High cyclic ductility, HCD ≥ 2.50

Medium cyclic ductility, $1.50 \leq$ MCD < 2.50

Low cyclic ductility, LCD ≤ 0.50

Similar values were obtained from experimental studies (Takanashi, 1974; Suzuki and Ono, 1977). However, considering, for instance, the target value of 0.035 rad, specified in EUROCODE 8 for high ductility class performance, applied to the IPE and HEA sections, values of the same order were obtained (Table 5.10).

Therefore, the previously mentioned limits appear to be reasonable for practical design. In this manner, taking the results of Anastasiadis et al. (2013) and applying a reduction factor of 4 for the IPE sections and 3 for

Table 5.10 Calculation of Rotation Capacity for the Target Value Given in EC8

Section[a]	*Span (mm)*	*Target ultimate rotation, θ_u (rad)*	*Plastic rotation, θ_p (rad)*	*Rotation capacity, R*	*Classification*
IPE 300	6000	0.035	0.015	1.33	MCD
IPE 400			0.012	1.92	MCD
HEA 300			0.012	1.76	MCD
HEA 450			0.008	3.37	HCD

Note: $R = (\theta_u/\theta_p) - 1; \theta_p = M_pL/4EI$.

[a] Steel quality S235.

the HEA sections, the evaluation of cyclic member ductility is presented in Tables 5.11 and 5.12 for the IPE and HEA sections, respectively. Ductility of the HEA sections remains unchanged, while for the IPE sections, a degradation of one class is observed. In any case, the section classification specified in Eurocodes 8 and 3 seems to be also ineffective for cyclic conditions, as was demonstrated for monotonic conditions too (Anastasiadis, 2012).

Table 5.11 Comparative Presentation of Member Classification for IPE under Monotonic and Cyclic Loading

		L = 2000 mm		*L = 3000 mm*		*L = 4000 mm*		*L = 5000 mm*	
Steel shape	*Ductility type*[a]	*S235*	*S355*	*S235*	*S355*	*S235*	*S355*	*S235*	*S355*
IPE 300	Monotonic	HD	HD	HD	MD	—	—	—	—
	Cyclic	HCD	MCD	MCD	MCD	—	—	—	—
IPE 330	Monotonic	HD	HD	HD	MD	HD	MD	—	—
	Cyclic	HCD	MCD	MCD	MCD	MCD	MCD	—	—
IPE 360	Monotonic	HD	HD	HD	HD	HD	MD	MD	LD
	Cyclic	HCD	MCD	MCD	MCD	MCD	MCD	MCD	LCD
IPE 400	Monotonic	HD	HD	HD	HD	HD	MD	MD	LD
	Cyclic	HCD	HCD	MCD	MCD	MCD	MCD	MCD	LCD
IPE 450	Monotonic	—	—	HD	HD	HD	MD	MD	MD
	Cyclic	—	—	HCD	MCD	MCD	MCD	MCD	LCD
IPE 500	Monotonic	—	—	HD	HD	HD	MD	MD	MD
	Cyclic	—	—	HCD	MCD	MCD	MCD	MCD	LCD
IPE 550	Monotonic	—	—	—	—	HD	HD	HD	MD
	Cyclic	—	—	—	—	MCD	MCD	MCD	MCD
IPE 600	Monotonic	—	—	—	—	—	—	HD	MD
	Cyclic	—	—	—	—	—	—	LCD	LCD

Note: —, Sizing of the member would be other than for the ductility limit state. For instance, serviceability limit state would be the predominant criterion for member sizing. L, standard beam span; HD, high ductility; HCD, high cyclic ductility; MD, medium ductility; MCD, medium cyclic ductility; LD, low ductility; LCD, low cyclic ductility.

[a] Rotation capacity for an in-plane postelastic buckling mechanism.

Table 5.12 Comparative Presentation of Member Classification for HEA under Monotonic and Cyclic Loading

		L = *4000 mm*		*L* = *5000 mm*	
Steel section	*Ductility type*[a]	*S235*	*S355*	*S235*	*S355*
HEA 320	Monotonic = cyclic	HD/HCD	MD/MCD	HD/HDC	MD/MCD
HEA 340	Monotonic = cyclic	HD/HCD	HD/HDC	HD/HDC	MD/MCD
HEA 360	Monotonic = cyclic	HD/HCD	HD/HCD	HD/HCD	HD/HCD
HEA 400	Monotonic = cyclic	HD/HCD	HD/HCD	HD/HCD	HD/HCD
HEA 450	Monotonic = cyclic	HD/HCD	HD/HCD	HD/HCD	HD/HCD
HEA 500	Monotonic = cyclic	HD/HCD	HD/HCD	HD/HCD	HD/HCD
HEA 550	Monotonic = cyclic	HD/HCD	HD/HCD	HD/HCD	HD/HCD
HEA 600	Monotonic = cyclic	HD/HCD	HD/HCD	HD/HCD	HD/HCD

Note: L, standard beam span; HD, high ductility; HCD, high cyclic ductility; MD, medium ductility; MCD, medium cyclic ductility; LD, low ductility; LCD, low cyclic ductility.

[a] Rotation capacity for an in-plane postelastic buckling mechanism.

5.5 NEAR-FIELD EARTHQUAKE EFFECTS ON THE AVAILABLE DUCTILITY OF STEEL BEAMS

5.5.1 Ductility problems for near-field earthquakes

In contrast to far-field earthquakes, where the ground motions are mainly produced by the surface waves *R* and *L* (the body waves having negligible effects), the near-field earthquakes are directly influenced by the first arrived body waves *P* and *S*, the surface waves *R* and *L* only rounding up the damage produced by these waves.

Therefore, all characteristics of these earthquakes are different. But the main difference is in the transmission of seismic forces to the structure. If for the far-field earthquakes the transmission is characterized by inducing vibrations in the structure due to horizontal seismic actions, for the near-field earthquakes the transmission is made by seismic wave propagation caused by vertical seismic actions. Situated right over the source, the seismic action is dominated by the direct seismic waves *P* and *S* propagating with very high velocity along the structure and violence due to the "last ball effects" (Newton's cradle). This difference demands different approaches in structure design, the actual modern design codes neglecting this aspect.

The near-field earthquakes are characterized by very high-velocity pulse loading, a reduced number of pulses, short duration, and important vertical components. Owing to the reduced number of pulses and the short duration, the amount of dissipated energy is reduced and the problem of ductility must be reevaluated for near-field earthquakes in comparison with far-field earthquakes. Also owing to the very high velocity of long pulse seismic action, the effect of strain rate is a crucial problem, the fracture of members and connections being the real danger for the structure collapse. Therefore, the design approaches for structures in near-field areas must consider, besides the fulfilling of the required ductility, the strength needs to avoid the fracture collapse.

An analysis of the existing research works and code rules reveals that the studies are mainly concentrated on the structural requirements in near-field areas (especially on the design forces) and not in the available structural response. The research works considering the effects of the near-field earthquakes on the code requirements started in the last decades, especially after the Northridge and Kobe earthquakes (Hall et al., 1995; Whittaker, 2000; Alavi and Krawinkler, 2001; Huang 2003; Park et al., 2004; Mollaioli et al., 2006; Kalkan and Kunnath, 2006). Just the American code UBC 97 (1997) and the Chinese code GB50011-2001 (2001) consider the characteristic values of near-field earthquakes, but, unfortunately, very few other design codes consider these effects. The studies concerning the effects of near-field earthquakes on the available capacity are very few as well.

First, the effects of strain rate on steel properties are experimentally examined on small specimens and, due to the testing machine capacity, in the strain-rate field out of an interesting range for structures in the near-field area. Gioncu et al. (2000), Gioncu (2000), and Mateescu and Gioncu (2000) made some theoretical investigations on the structural response of steel members subjected to loading producing important strain rates. Experimental tests on welded connections in dynamic conditions, with strain rates in the field of 10^{-2} s^{-1} and 5×10^{-2} s^{-1}, were performed by El Hassouni et al. (2011). This domain corresponds to the far-field earthquakes as well as the near-field low and moderate earthquakes. Though these values were reduced in comparison with the ones produced during strong near-field earthquakes, the cumulative plastic rotation and the dissipated energy decreased by about 20%, due to the strain-rate effects. In conclusions, the authors underline the possible effects in the actual domain (which cannot be studied in the laboratory due to the limitation of testing machines) in the case of structures subjected to near-field actions:

> *A veritable increase of strength and decrease of ductility can be expected at near-source earthquake where the strain-rate rate can vary from 10% (10^{-1})* to 1000% (10^{1}) sec^{-1} in the beam near weld, which is much greater than strain rate recorder during the tests. From this high strain-rate results an increasing the yield ratio of about 0.75–0.95, impeding the development of plastic rotation at the beam end near weld. This situation induces a dramatic reduction of the available ductility; the fracture rotation of the beam is reduced by about 80%–96% when the strain rate varies from 10% (10^{-1}) to 1000% (10^{1}) sec^{-1}. The ductile plastic collapse of beam is replaced by the fracture collapse after the first few cycles. This was a common phenomenon

> during the Kobe earthquake. The section with thick beam flanges is more dangerous for fracture than thinner ones. Thus, the effect of the strain rate can be reduced by a good design conception where beam dimensions can be selected so that a small to medium plastic buckling occurs during strong earthquake. Finally, the use of static performance of the members cannot be applied in the case of seismic loading with high strain-rate effect in the near-field areas. (El Hassouni et al., 2011)

From this conclusion, it very clearly appears that a consistent research program has to consider the following aspects of phase 1 (action of *P* and *S* seismic waves):

- Examining the decrease in rotation ductility due to the effect of strain rate
- Determining the fracture rotation influenced by the strain rate
- Proposing some constructive provision to reduce the danger of fracture rotation

The determination of available ductility for monotonic loads was presented by Gioncu et al. (2012), Anastasiadis et al. (2013), and Gioncu and Mazzolani (2002). The correction factor is estimated on the same local plastic mechanism, using the modified mechanical steel properties due to the strain rate. This can be done by using the DUCTROT-M computer program (Gioncu and Mazzolani, 2002; Petcu and Gioncu, 2002).

For the last two approaches, a simple approximate methodology to determine the fracture rotation is proposed based on the reaching of the ultimate strength in a point of plastic mechanism. The results of the new analysis methods for near-field earthquakes based on near-surface wave propagation in the structure are used, together with the DUCTROT-M computer program for determining the fracture rotation.

5.5.2 Near-field earthquakes

5.5.2.1 General

Near-field (also named near-fault or near-source) earthquakes are produced by ground motions situated right over the source (Figure 5.74). The seismic behavior of a structure can be divided into two phases:

- *Phase 1*, due to the wave propagation of the *P* and *S* body waves with great velocity and high strain rate, producing early earthquake damage of connections in some zones of structure.
- *Phase 2*, due to arrival at the structure site of the surface waves *R* and *L*; the vibrations of damaged structure occur, resulting in the resonance between the damaged structure and soil ground motions, amplifying the structural damage produced during phase 1.

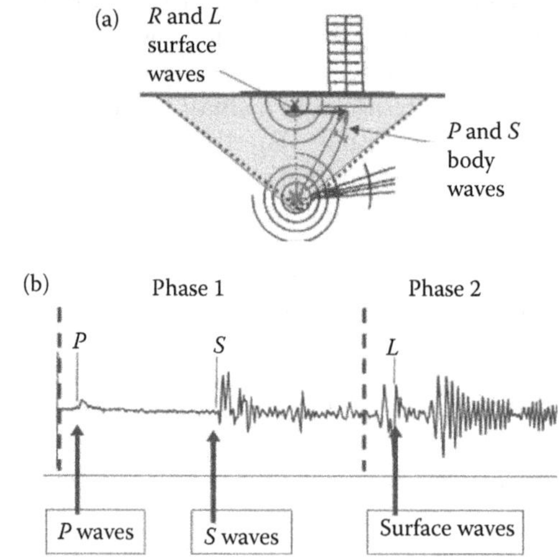

Figure 5.74 Near-field earthquakes: (a) body and surface waves; (b) the two phases of structural response.

Therefore, one can understand that the near-field earthquakes perform many characteristics different from the far-field earthquakes regarding acceleration, velocities, number of cycles, and duration. These differences may be attributed to the ray paths of the direct P and S waves.

Owing to the source position near the earth's surface, the near-field earthquakes are always the result of some rupture in the crustal plate. Two types of crustal earthquakes should be considered: the first type includes those generated by interplate faults and the second by intraplate faults, with very different characteristics.

In contrast to crustal far-field earthquakes, where the effects of the traveling path and site soil conditions lead to the disappearance of the effects of the different source type characteristics, for near-field earthquakes, these differences in the source type characteristics play a crucial role.

5.5.2.2 Interplate crustal earthquakes

There are three different interplate types, with very different characteristics:

1. *Thrust-subduction earthquake.* These earthquakes are characterized by the highest magnitudes, the records in magnitude being the result of the tectonic plate thrust-subductions (1960, Chile, M 9.5; 1964, Alaska, M 9.2; 1957, Aleutian Islands, M 9.1; 2004, Sumatra, M 9.0; 2011, Tohoku, M 9.0). The effects of near-field earthquakes on the structures are important and cannot be neglected in design, especially

due to the characteristics of seismic actions, very different from the ones recorded for far-field sites (Gioncu and Mazzolani, 2002 and 2011). The main features are as follows:

- The main characteristics of the rupture mechanism are large velocity pulses and short duration of the forward directivity, lower amplitudes, and long duration in backward directivity.
- Few effective cycles in acceleration.
- Long-velocity pulses with very high values. During the 1999 Taiwan earthquake, the peak of velocity recorded on a free site was about 4 m/s, while the maximum recorded velocity during the 1994 Northridge (Rinaldi Station) was about 2 m/s. But these values are inferior to that determined in case of the presence on site of structures when the values are 50–300 m/s (recorded during the Californian earthquakes).
- Reduced number of velocity cycles, with a maximum of 3–4 pulses. The number of significant pulses is likely related to slip distributions in the fault and, consequently, it is difficult to be predicted. Generally, for faults, normal components are significant only for one full cycle of pulse.
- The pulse period is influenced by the earthquake magnitude and local site conditions. In function of magnitude, the pulse period can vary between 0.6 and 6 s. A longer period occurs at soil sites than at rock sites for events with a magnitude lower than M = 7.0.
- Significant vertical components, larger than the ones in the horizontal direction, can occur.

2. Strike-slip earthquakes. These earthquakes are characterized by moderate magnitudes (1989, Loma Prieta, M 7.1; 1995, Kobe, M 6.9; 1999, Izmit, M 7.4), but they are very damaging due to the long pulse and long duration. Many characteristics are similar to the thrust-subduction earthquakes, but with very important differences (Gioncu and Mazzolani, 2011):
   - The directivity plays a very important role, but the components normal to directivity are more severe than the fault parallel components.
   - The dominant velocities occur in a plan which is normal to the rupture surface.
   - The numbers of acceleration and velocity cycles are higher than the ones for the thrust-subduction case.
   - The pulse and duration of strike-slip earthquakes are longer than those for thrust-subduction earthquakes.
3. *Thrust-collision earthquakes* produced by some microplates, which are divisions of the major tectonic plates. The faults are the result of rupture due to compression forces, or of the borders of some ancient tectonic plates (Gioncu and Mazzolani, 2011). The most important

collision earthquakes in Europe are the ones produced along the Pyrenees and Alps mountains, caused by the tectonic forces of the Iberian and Italian microplates in active contact with the Euro-Asian tectonic plate.

5.5.2.3 Intraplate crustal earthquakes

The occurrence of these earthquakes in areas that are tectonically relatively stable violates the plate tectonic model, which considers that the earthquakes are concentrated on the narrow bands of plate boundaries. While interplate earthquakes are well understood, as the mechanisms causing intraplate earthquakes are subject to considerable debate, crustal-fracture earthquakes occur in the so-named diffused zones, where the presence of faults is due to the fracture of weak zones in crustal plates, caused by a plume rising from the mantle. In Europe, the most well-known zones with such crust fractures are the Rhine rift, and the networks of active faults in the Balkans and Greece. In the United States, the eastern zones (i.e., New Madrid, Missouri) are the major seismic ones. In Asia, the North China rift basin hosted several large earthquakes in the last decades. Although only 5% of the global seismic energy is released by intraplate earthquakes and less than 10% of all earthquakes are of intraplate type, some of the most damaging earthquakes have occurred in the supposedly stable zones. In 1963, a relative moderate earthquake, M 6.2, occurred under the Skopje city (Macedonia), producing large building damage and claiming 1070 victims. It was the first event that can be considered as the manifestation of a near-field earthquake, but, due to the lack of sufficient records and knowledge at that time, this aspect was not recognized. However, the maximum credible earthquakes produced by these faults are generally, in normal conditions, smaller in magnitude, in frequency of occurrence, and in area of influence as compared to the interplate earthquakes. Generally, the positions of interplate faults are known due to their manifestations during the historical time and the corresponding zones can be supervised by a network of seismic stations, having the possibility to obtain valuable basic data. In contrast, the positions of intraplate faults are more difficult to detect, in many cases being buried. Under these circumstances, it is very difficult to determine the main characteristics of earthquakes. Anyway, it is possible, even from this poor database, to characterize the intraplate earthquakes by

- Very long recurrence intervals for a particular location, so many people consider this event only as an accidental one.
- The majority of intraplate earthquakes do not exceed the magnitudes 5–6, with the exception of the ones in the China rift, where the magnitudes are over 7, and the case of the U.S. New Madrid earthquakes, where the magnitudes can exceed 8, due to the liquefiable soil.

- The velocities are moderate on the free sites, under 50 cm/s, the pulse period is very short, and the vertical components have reduced values. There is no information about the velocities in case of the presence of some structures.

Considering these aspects, the effects of intraplate near-field earthquakes on structures are moderate, with the exception of the case where an earthquake occurs under a densely constructed area, especially with liquefiable soil.

5.5.3 Phase 1: Wave propagation for *P* and *S* body waves

5.5.3.1 Ground motions on free sites or in the presence of buildings

A very important question for near-field earthquakes: can the ground motions recorded on free sites be used or not for designing structures against the actions of the *P* and *S* body waves?

During an earthquake, the rupture of a fault initiates seismic waves *P* and *S* toward the earth's surface (Figure 5.75), with a tendency to assume a vertical direction, as far as the surface layers are concerned. Then, the *P* wave motions are thought to be dominant in the vertical direction, and the *S* wave motions in the horizontal direction. *P* waves are always characterized by higher frequencies, but they attenuate more rapidly with the horizontal distance, compared with *S* waves. Consequently, the near-field ground motions are characterized by important vertical components and high-frequency energy.

The wave propagation phenomenon gives rise to an important aspect, which is not relevant for the seismic records of far-field ground motions: the difference in records obtained on free sites or on sites with buildings.

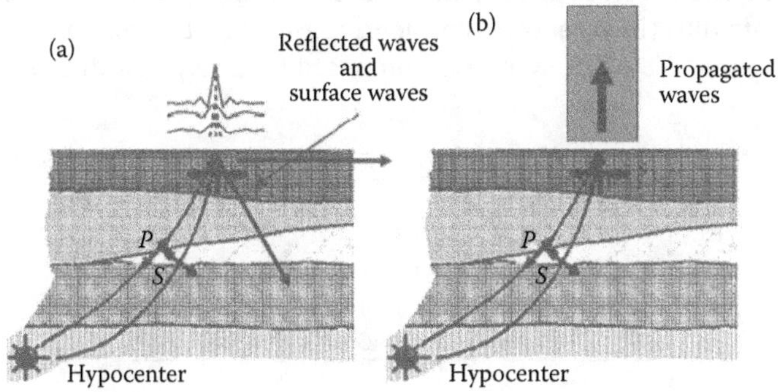

Figure 5.75 Ground motion types: (a) free site; (b) site with structure.

When the *surface is free* (Figure 5.75a), the recorded ground motions represent only a part of the site movement, because, when the waves reach the surface, they are reflected back to the earth or to the earth's surface, changing into the surface's waves, *R* and *L*, respectively. However, if there is a *building on the surface*, the waves continue to propagate into the structure until its top (Figure 5.75b).

Examining the amplifications of waves at the surface layer in a free site and in a built site, it is possible to consider the soil surface layer or the structure as representing the last object in the way of the wave propagation, drawing a parallel with the very well-known Newton's cradle (Figure 5.76).

Newton's cradle is a device that demonstrates the conservation of kinetic energy via a series of swinging balls. If one lateral ball is pulled away and left to fall, it strikes the first ball in the series and comes to a near dead stop. The impact produces a shock wave that propagates through the intermediate balls. The ball on the opposite side acquires most of the velocity and almost instantly swings in an arc almost as high as the released height of the first ball. This shows that the final ball receives most of the energy that was in the first ball.

For a free site, it is very important to observe that a similar amplification of accelerations along depth is observed for the last layer in the case of multilayer soil. During the Kobe earthquake, a variation of peak accelerations with depth was recorded on the free surface (Figure 5.77) (Adam et al., 2000). A deamplification is observed for the horizontal component: from a value of 0.3*g* at the surface, a value of 0.6*g* was recorded at a given depth. In contrast, a very important amplification is observed for the vertical component, the surface motion of 0.55*g* being recorded in comparison with the one recorded at depth, 0.2*g*. Measurements of the vertical variation of the acceleration components have shown that the amplification of vertical accelerations is due to the passing of the main shock in the superficial soil deposits, which constitute the last layer. But this amplification disappears after a short time, having pulse characteristics.

In contrast to the free surface, on a *site with buildings*, the waves continue to propagate into the structure, considering the soil-structure as a common system (Figure 5.78). The system is composed by a series of different layers,

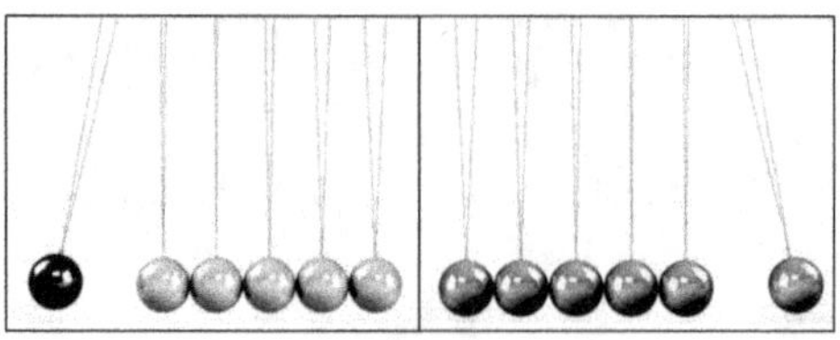

Figure 5.76 Newton's cradle. (From Wikipedia (nd): Newton's cradle. http://en.wikipedia.org/wiki/Newton's_cradle.)

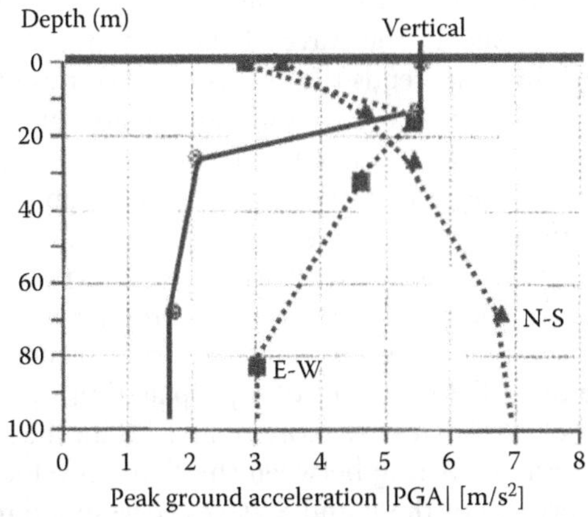

Figure 5.77 Amplification of vertical peak acceleration at the free site. (After Adam, M., Schmid, G., Chouw, N. 2000: *12th World Conference on Earthquake Engineering*, Auckland, 30 January–4 February 2000, Paper No. 1313.)

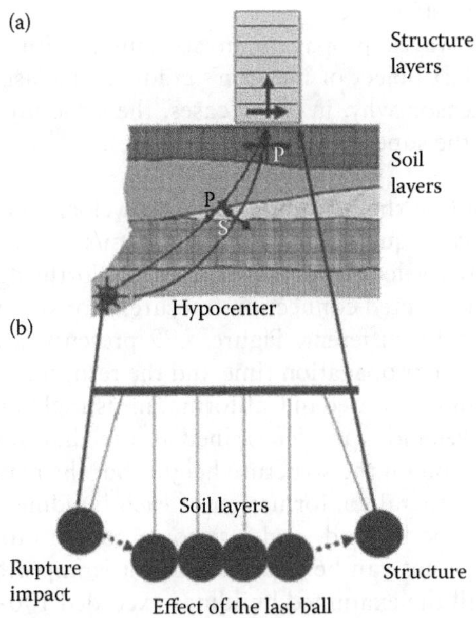

Figure 5.78 Waves propagation in soil and in site with structure: (a) soil-structure system; (b) Newton's cradle—effect of last ball.

playing the role of balls, the first being the source and the last the structure. Owing to the rupture shock, the wave of the first layer strikes that of the second layer, which in turns sends the energy to the series of layers, causing the last layer, the structure, to receive the total kinetic energy of the seismic wave. Therefore, the structure situated in a near-field site receives an important part of the kinetic energy developed by the source rupture process and its behavior is very different from the one of structures situated in far-field sites.

The structure is reached by two different wave types, *P* and *S*, at the same time interval. Unfortunately, the only recorded wave velocities related in the literature correspond to the *P* waves. So, there is not sufficient information about the recorded velocities of *S* propagation waves. It is rational to consider that the *S* velocities are more reduced than the *P* velocity, being $V_s \sim 0.6\ V_P$, the relation existing between the *P* and *S* velocities in the soil. The main characteristics of the *P* and *S* waves correspond to pulse actions.

In conclusion, there are great differences in seismic waves measured at the soil surface:

- If the site is free, the wave must accommodate to this situation and the last object in Newton's cradle is the last layer, and an amplification of accelerations is observed there.
- If the site is occupied by some structures, they represent the last object in Newton's cradle and a great part of the seismic energy is concentrated in these structures.
- The seismic waves propagate in structures with great velocities, because the last object of Newton's cradle is the last structure story. This is the reason why, in many cases, the structure damage is concentrated in the superior levels of structures.

It is well known that the maximum recorded velocities on free sites during the near-field earthquakes do not exceed 4 m/s (as during the Taiwan earthquake) and the velocities recorded during Northridge and Kobe do not explain the unexpected connection fractures. For sites with structures, the situation is totally different. Figure 5.79 presents (after Snieder and Safak 2006) the wave propagation time and the resulting velocities for five instrumented buildings located in California, measured during some recent earthquakes. The velocities are determined as a medium value considering a uniform distribution on the structure height. But the material, structure, irregularities, and general conformation of each building are different, so no exact rules can be learned, and just some concentration of velocities in some structure zones can be expected. But it is important to note that the velocities for all the examined buildings exceeded 100–200 m/s, many times more than the velocities obtained on free sites.

Therefore, the recorded ground motions on free sites cannot represent the actual actions due to the *P* and *S* waves for designing structures situated

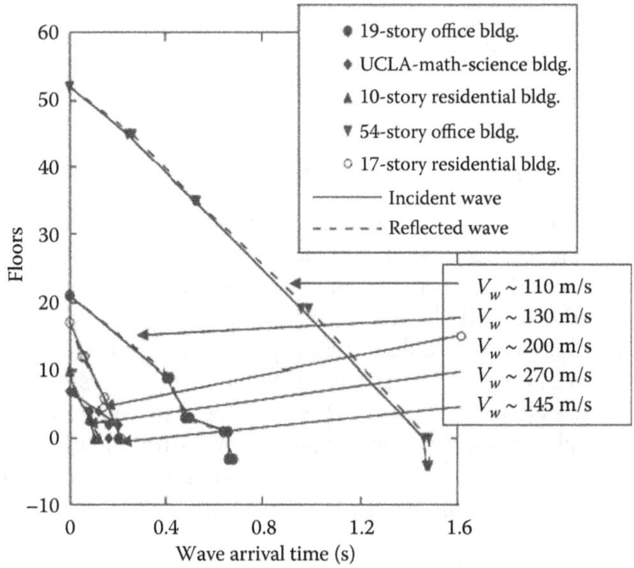

Figure 5.79 Measured wave arrival time and calculated corresponding *P* velocities. (Modified after Snieder, R., Safak, E. 2006: *Bulletin of the Seismological Society of America*, 96, 586–598.)

on near-field areas. At the same time, the design methodologies used for structures situated in far-field areas must be replaced by new approaches, considering the seismic wave propagation along the structure.

This observation is important because it gives rise to a debate on the following problems:

- The possibility to use the accelerograms recorded on a free site for the design of a structure, despite the knowledge that the building presence drastically changes the feature of the ground motion.
- One must be aware of the fact that the recorded acceleration and the corresponding spectra can be used only to analyze the structure in phase 2 (see Figure 5.74), after the actions of the *P* and *S* waves have produced significant damage.

These aspects complicate very much the structural design against near-field earthquakes and warn the designer about the uncertainties of the present approaches of the code provisions. Considering the great velocities of wave propagation in design procedures, the following aspects must be taken into account in the evaluation of structural ductility:

- Changing the material characteristics (especially the nominal yield stress) due to the important strain-rate effect

- Reducing the ductility determined for monotonic loading, due to the decrease in the yielding plateau
- Replacing the rotation ductility with the local fracture, due to the important strain rate

5.5.3.2 Wave propagation approach

Today, the vibration approach is a methodology accepted by all the code provisions for the seismic design of structures situated in earthquake-prone areas, basically developed for structures situated in far-field areas. The new methodology, the so-called wave propagation approach, which is mainly used by seismologists to calculate the propagation in the soil of waves generated by an earthquake, also starts to be used for analysis of structures situated in near-field areas (Safak, 1999). At the same time, the wave propagation approach is used for monitoring the structural health following an earthquake, in order to rapidly detect the damage suffered (Snieder and Safak, 2006). Two different types of approach are followed:

1. *Vibration approach* (Figure 5.80). This approach is based on the recorded accelerations measured at the structure site and uses these records to analyze the structure, which is modeled as a multi-degree-of-freedom system, whose response induced by earthquakes is evaluated by means of modal analysis. This approach has proved to be a very adequate methodology in practice, due to the fact that it is developed mainly for far-field records, where the first mode of vibration is the dominant one. However, in the last period, it was proved that this simplified methodology has limitations for the important case of strong near-field earthquakes, because it does not provide accurate information on the large local shape, internal important deformation, and local damage in the structures, as was emphasized during the Northridge and Kobe earthquakes. Therefore, the

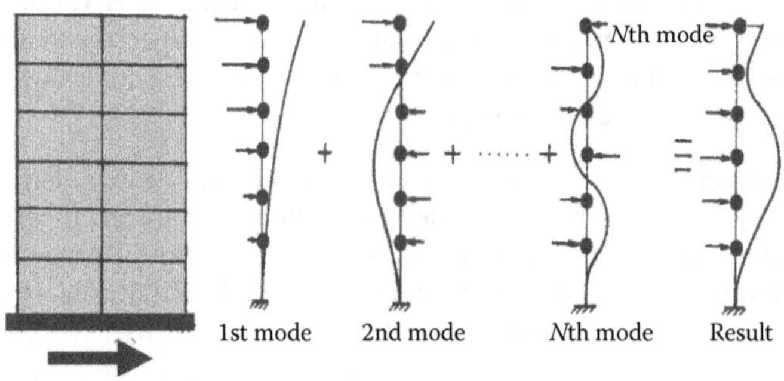

Figure 5.80 Vibration approach.

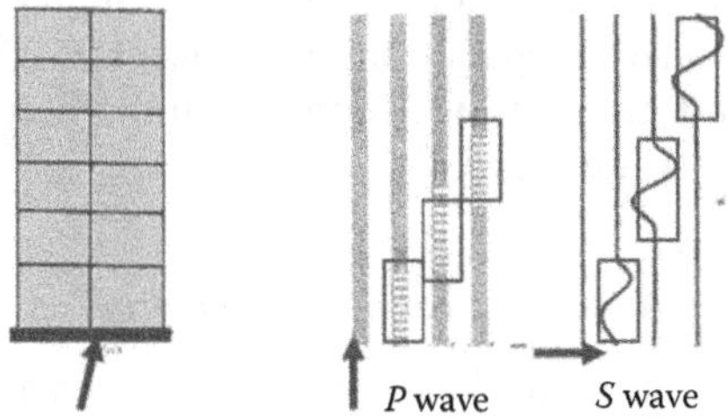

Figure 5.81 Wave propagation approach.

response spectrum alone is not a sufficient representation of near-field earthquakes (Chichowicz, 2011).

2. *Wave propagation approach* (Figure 5.81). This concept has been developed in the last years to solve the special problems of the structural response for near-field earthquakes, being based on the wave velocities (not the velocities obtained from acceleration records on free sites) measured on the structure site by the geologists. There are some methodologies to study the wave propagation in structures:
   - The use of the ideal shear beam model (Hall et al., 1995; Iwan, 1997; Zhang et al., 2011), which allows the localization of large interstory drifts. Iwan's drift spectrum (1997) is based on the evaluation of the shear wave propagation along a beam and computes the drift by analogy with the shear strain of a beam. This method does not consider the phenomenon in the right way, being based on the recorded ground motions on a free site.
   - Considering each story as a layer where the wave propagates as in soil layers (Safak, 1995, 1998, 1999), the wave propagation approach has been mainly used to investigate the layer ground motions due to earthquakes, but this method can also be applied in structural engineering, considering a story as a layer of the soil-structure system.
   - Developing a special method, named the normalized input–output minimization (NIOM) analysis (Kawakami and Oyungchimeg, 2004, 2005; Oyungchimeg and Kawakami, 2003), based on a statistical correlation between earthquake motions observed at different points.
   - Using finite-element analysis (Kohler et al., 2007; Trifunac and Todorowska, 1998) and a computer program calibrated to determine the structural response for base impulsive wave impulsive excitation.
   - Owing to the fact that this approach has been developed in the last years, the wave propagation through a building has not been investigated enough. However, the obtained partial results can

be used for the development of a preliminary picture about the effects of wave propagation in structures. This approach can describe and analyze local damage and fractures during strong near-field earthquakes.

5.5.3.3 *Applications in earthquake engineering*

For some structures situated over a source, the wave propagation into structure is examined both instrumentally and theoretically. Applications have been selected where the wave propagation time and the propagation velocity are measured.

A steel moment-resisting frame building, the UCLA Doris and Luis Factor in Los Angeles (Figure 5.82a), with 17 stories (a superstructure of 15 stories with 61.3 m height plus two underground stories embedded at 8.3 m depth) was densely and permanently instrumented (Kohler et al., 2007).

Its behavior was examined during the December 16, 2004 Santa Monica earthquake (M 3.6). Using the ETABS computer program, an FEM dynamic analysis was performed for the structure subjected to a horizontal pulse loading at the base and the wave propagation was determined (Figure 5.82b).

The resulting impulse response shows that the wave is transmitted up by incident waves and down by reflected waves. In the figure, the wave represents the lateral displacement of the corresponding level during the wave propagation along the structure. Therefore, considering this lateral displacement together with the lower and upper lateral displacements, the wave can be regarded as a story drift and, therefore, the wave propagation as the story drift propagation. Examining Figure 5.82b, one can see that the wave propagation time is very short (practically instantaneous) in the two underground levels, both of them being much stiffer than the upper floors. The wave travel velocity of seismic waves between the ground floor and the roof, considering it as uniform along the structure height, is given by $V_w = H/T_w$, where $H = 64$ m is the building height and $T_w = 0.4$ s is the travel wave time. Therefore, the transversal wave velocity for this structure results:

$$V_w = 64/0.4 = 160 \text{ m s}^{-1}$$

This value of propagation velocity is also confirmed by the values obtained during the small and moderate February 14, 2004 (M 4.3), February 21, 2004 (M 2.7), December 16, 2004 (M 3.6) and January 6, 2005 (M 4.4) earthquakes, measured in some densely instrumented buildings of the University of California, Los Angeles. One can see that the propagation velocity is the same for all earthquakes, being independent of the earthquake magnitude. Only the wave amplitude, not the velocity, is a function of the earthquake magnitude.

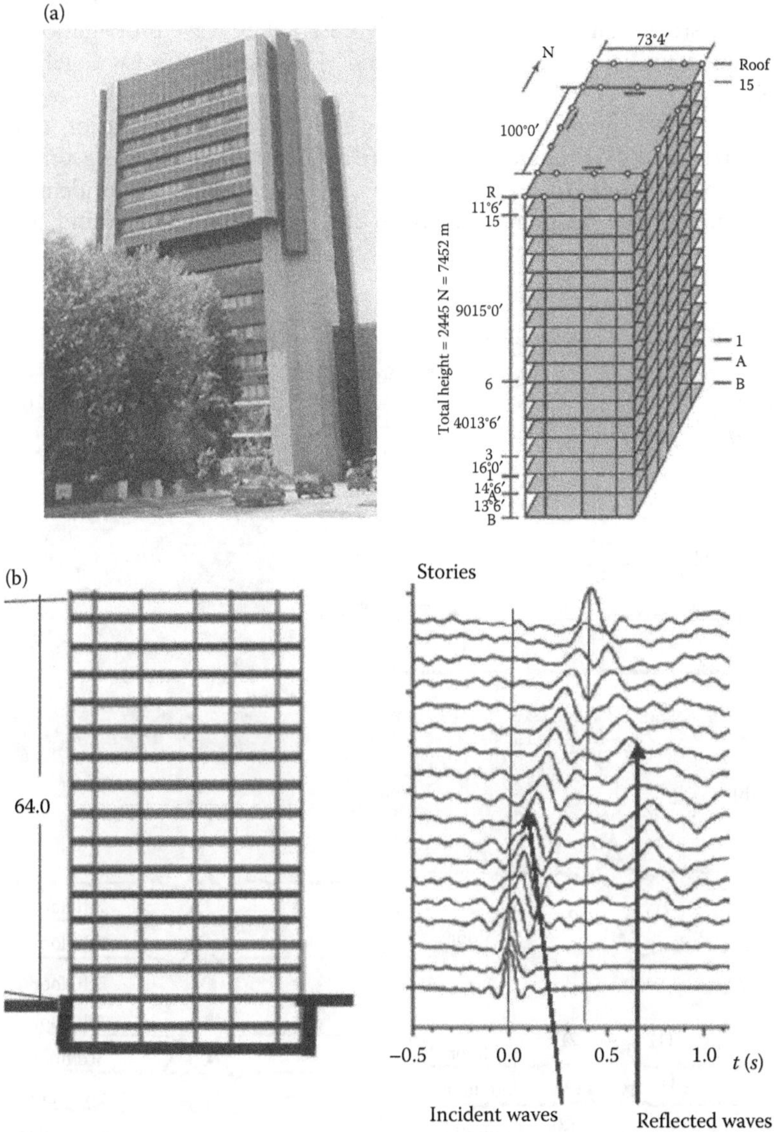

Figure 5.82 UCLA Doris and Luis factor: (a) view and structure; (b) wave propagation. (Modified after Kohler, M.D., Heaton, T.H., Bradford, S.C. 2007: *Bulletin of the Seismological Society of America*, 97(4) 1334–1345.)

A very damaged RC structure during the Northridge earthquake was the seven-story Van Nuys hotel (Figure 5.83a). The wave propagation was determined by using the NIOM method (Kawakami and Oyungchimeg, 2004).

Figure 5.83b shows the results using the recorded acceleration during the Northridge earthquake, and Figure 5.83c the results using only the first vibration mode. In both the figures, two clear peaks are evident, corresponding to the incident wave from the ground to the top and to the

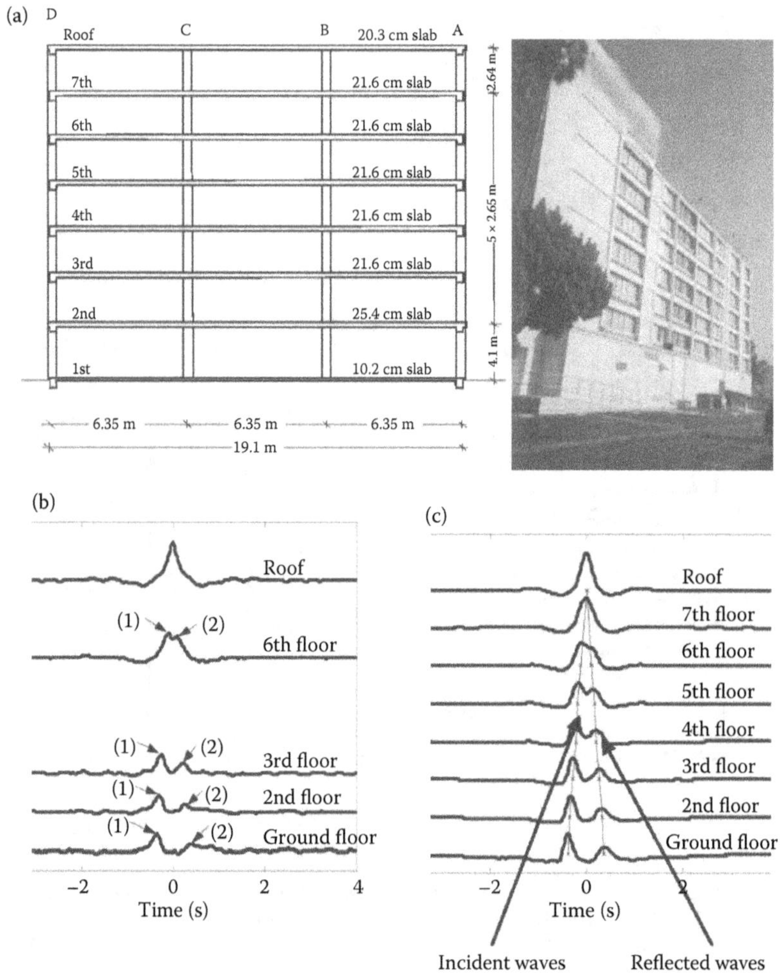

Figure 5.83 Van Nuyts hotel: (a) structure and view; (b and c) wave propagation. (Modified after Kawakami, H., Oyungchimeg, M. 2004: *13th World Conference on Earthquake Engineering*, Vancouver, 2004, 1–6 August, Paper No. 317.)

reflected wave from the roof to the ground, respectively. This observation means that the recorded accelerogram on a free site, used in seismic design, is changed in the presence of a building due to the reflected waves. Examining Figure 5.83c, one can see that the amplitude of the roof wave is larger than the induced wave. The same observation can be made from Figure 5.82b. The incident-wave travel time from the ground floor to the roof is $T_w = 0.36$ s and the corresponding wave velocity:

$$V_w = 20/0.36 = 55.6 \text{ m s}^{-1}$$

In Todorowska et al. (1988), for an RC six-story building frame, the wave travel velocity was found to be equal to 420 m s^{-1}, with a mention that during similar research works, for another structure, the wave velocities exceeded 300 m s^{-1}.

For the Millikan Library Building in Pasadena (California), a very well-instrumented and reinforced 10-story building, the measured shear wave velocity was 322 m s^{-1} (Safak, 1995; Snieder and Safak, 2006).

Other studies about the propagation velocity were performed by Todorowska and Rahmani (2012). For a Los Angeles 54-story steel office building excited by the 1994 Northridge earthquake, the measured velocities were about 140 m s^{-1}. For the Millikan Library Building, excited by the Californian 2002 Yorba Linda earthquake, using the direct algorithm, the propagation velocity was 405 m s^{-1}. But values of 700–800 m s^{-1} occurred, by using some more refined methodologies, for which the velocity distribution is not uniform.

In conclusion, the variation of the propagation velocities, in function of earthquake magnitude and structure type, is very high. The variability for steel structures is smaller than that for the RC ones.

5.5.4 Phase 2: Effects of surface wave

5.5.4.1 Damage produced during the first phase due to the body P and S waves

The above wave velocities were determined using the direct algorithm, representing average values, and considering a uniform distribution of velocity along the structure height. But it is shown that the actual distribution is not uniform, due to the superposing of different wave types: direct, reflected, and refracted waves (Figure 5.84a). Owing to this effect, a concentration of some velocities for some stories occurs and the member damage is concentrated only in some zones of the structure (Figure 5.84b), even if the structure is regular. In addition, some structural irregularities such as

- Different structural types (structural walls or bracing only in the first story)

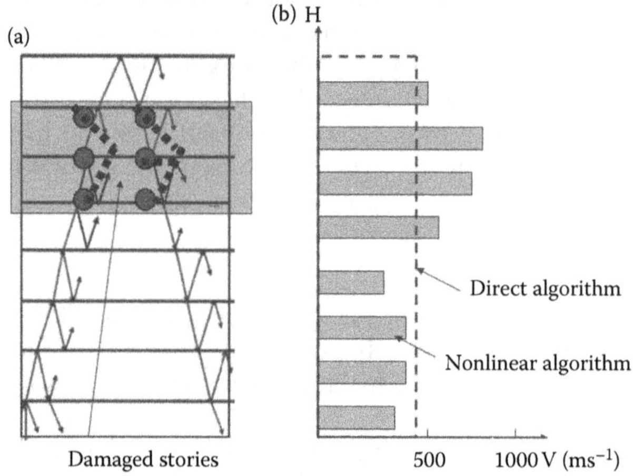

Figure 5.84 Effects of body waves: (a) direct, reflected, and refracted waves; (b) actual variation of velocities.

- Different materials (composite, reinforced concrete, steel at different stories)
- Different occupancy (office and residential parts, with different utilities)
- Setback or soft story

can produce a rigidity change along the structure height and, consequently, an increase in wave velocities in some stories and in the sensibility to local damage.

Therefore, owing to the strain rate, some fracture of connection components occurs in these zones. The main characteristic of these fractures, due to the high velocity of waves, is that they produce a local partial damage (start of fracture), preventing the total fracture of connections.

5.5.4.2 Damage produced during the second phase due to the surface R and L waves

When the surface waves reach the damaged structure, the local initial fracture could develop in a total connection fracture, or in a general collapse of a structure story, depending on both the structure and earthquake characteristics. During the Northridge earthquake (thrust earthquake), only the total collapse of connections was observed (Figure 5.85a), without having any effect on the story collapse. This behavior is due to the almost uniform distribution of velocities and damage on the structure height, with a consequent reduction of the uniform rigidity. In contrast, during the Kobe

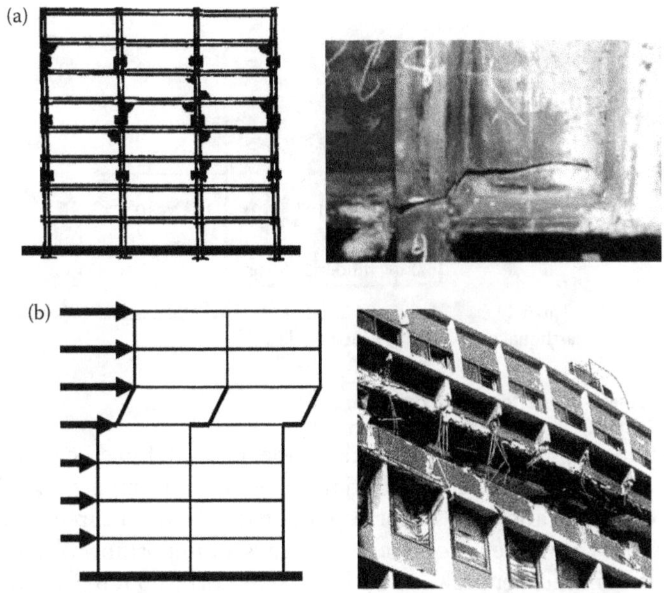

Figure 5.85 Effects of surface waves: (a) connection fracture in Northridge; (b) story collapse in Kobe.

earthquake (strike-slip earthquake), the collapse modes, due to both the local fracture and story collapse, were evident (Figure 5.85b). Concentration of damage on the intermediate stories was a general phenomenon during the Kobe earthquake for both steel and RC structures.

5.5.5 Influence of strain rate on available rotation ductility

5.5.5.1 *Effects of strain rate on steel characteristics*

There are two main factors influencing the available ductility.

1. *Strain rate* $\dot{\epsilon}$ is the rate of change in strain with respect to time and it is given by

$$\dot{\epsilon} = d\varepsilon/dt = d\{(1 - 1_0)/1_0\}/dt = dl/dt \times 1/1_0 = V/1_0(s^{-1}) \qquad (5.23)$$

where l is the length under applied stress, l_0 is the original length, and V is the speed of deformation (Wikipedia, nd). Therefore, the strain rate is directly proportional with the loading velocity.

Referring to the influence of loading velocity, Figure 5.86 shows a classification of the strain-rate effects. For strain rates smaller than 10^{-2} s^{-1},

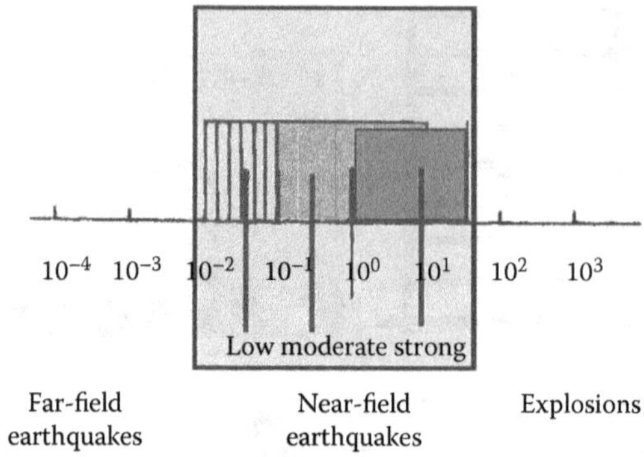

Figure 5.86 Strain-rate effects and dynamic actions.

corresponding to far-field earthquakes, the inertial forces can be disregarded. For greater values, the inertial forces play an important role. The range between 10^{-2} and $10^{1.5}$ s^{-1} corresponds to the field of near-source earthquakes, where the influence of strain rate plays an important role. There are three levels of influence: between 10^{-2} and 10^{-1}, corresponding to low earthquake, the influence of strain rate also being low; between 10^{-1} and 10^{0}, corresponding to moderate earthquakes, the effects also being moderate; from 10^{0} to $10^{1.5}$, corresponding to strong earthquakes, the influence also being very strong. Unfortunately, due to the limits of laboratory machines, there are very few experimental results in the range of near-field and only on small specimens. Strain rates larger than $10^{1.5}$ correspond to explosion actions.

2. *Yield ratio* ρ_y is the ratio between yield stress f_y and tensile strength f_u, denoted as

$$\rho_y = f_y/f_u \tag{5.24}$$

Yield ratio is considered as a measure of material ductility: it is considered to be low when $\rho_y < 0.75$ and high when $\rho_y > 0.75$. The yield ratio increases with an increase in yield stress. As a consequence, the material ductility is substantially impaired for elevated yield ratios. A yield ratio that assures a good ductility is within 0.5 and 0.7. The used European steels S235, S275, and S355 frames in this range, but the tendency exists to realize high-strength steels with yield ratio out of this range.

A crucial problem is to clarify how the strain rate affects the material ductility of structures situated in near-field zones. One of the first research works dealing with the effect of strain rate on the behavior of steels was performed by Manjoine in 1944 (Gioncu and Mazzolani, 2002). The results of these experimental tests are presented in Figure 5.87, indicating a very important increase in the yield stress with the strain rate increasing, especially for a

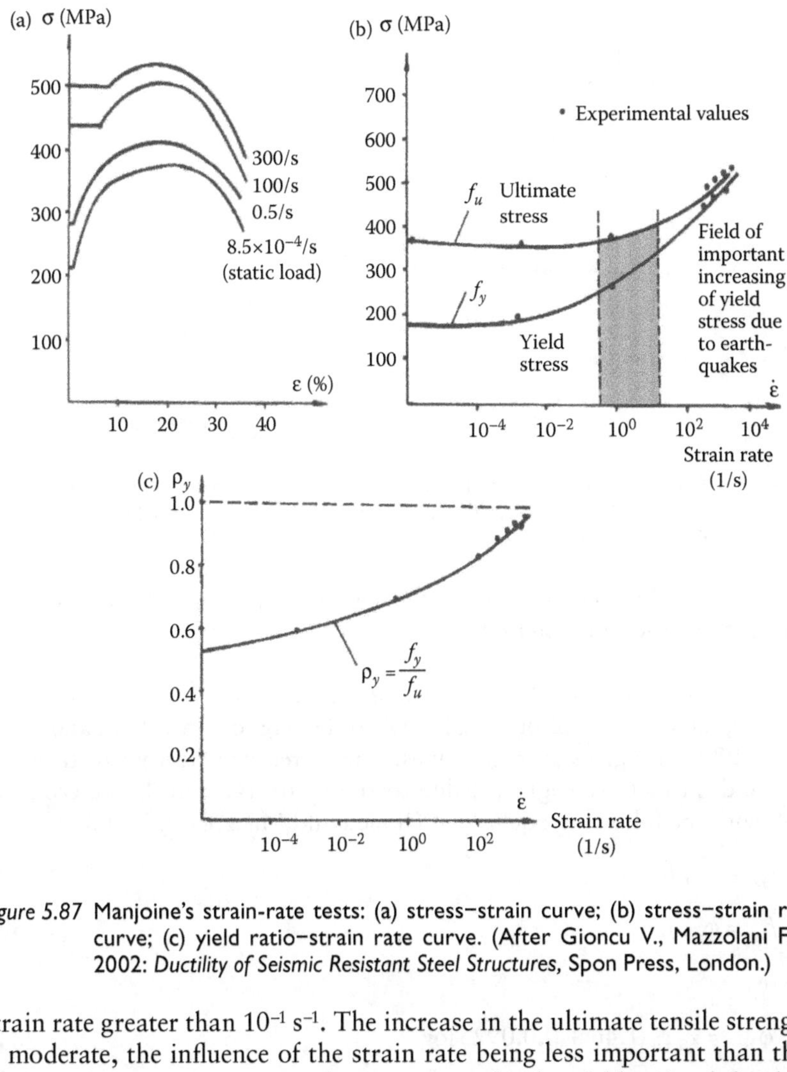

Figure 5.87 Manjoine's strain-rate tests: (a) stress–strain curve; (b) stress–strain rate curve; (c) yield ratio–strain rate curve. (After Gioncu V., Mazzolani F.M. 2002: *Ductility of Seismic Resistant Steel Structures*, Spon Press, London.)

strain rate greater than 10^{-1} s^{-1}. The increase in the ultimate tensile strength is moderate, the influence of the strain rate being less important than that of the yield stress (Figure 5.87b). Consequently, the yield ratio, defined by Equation 5.24, increases as far as the strain rate increases, with a tendency to reach the value 1 (Figure 5.87c). So, a reduction in material ductility occurs, especially for a strain rate greater than $10^{-1.5}$ s^{-1}, corresponding to structures situated in near-field moderate and strong earthquake areas.

Different constitutive laws modeling the influence of strain rates were proposed by Gioncu and Mazzolani (2002). A comparison among these different laws is presented in Figure 5.88. One can see that in the field of interest, near-field earthquakes, there is a very large scatter of values, from very high to small. Operating from the safe side, two proposals to use these

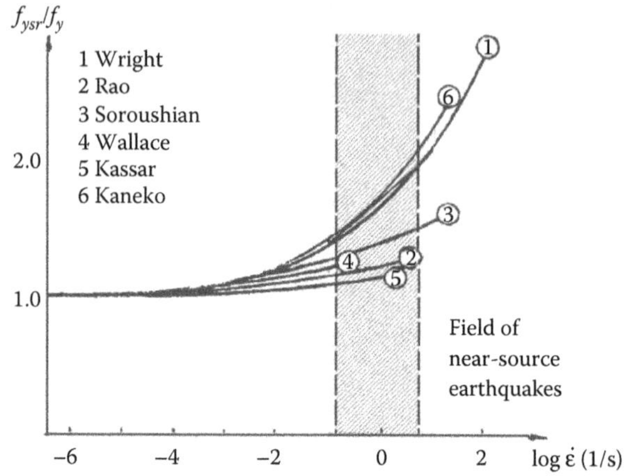

Figure 5.88 Comparison between the different strain-rate laws. (After Gioncu, V., Mazzolani, F.M. 2002: *Ductility of Seismic Resistant Steel Structures*. Spon Press, London.)

laws are available, one to evaluate the rotation capacity and the second to analyze the structural collapse.

1. For the analysis of the *influence of the strain rate on rotation capacity*, the most useful proposal seems to belong to Soroushian and Choi (1987) as it gives average values. The increase in steel yield stress f_{ysr} and ultimate strength f_{usr} due to the strain rate can be determined with the following equations (Gioncu and Mazzolani, 2002):

$$f_{ysr} = \varphi_{ysr} f_y \tag{5.25a}$$

$$f_{usr} = \varphi_{usr} f_u \tag{5.25b}$$

where

$$\varphi_{ysr} = c_T c_w (1.46 + 0.0.0925 \log \dot{\epsilon}) \tag{5.26a}$$

$$\varphi_{usr} = 1.15 + 0.00496 \log \dot{\epsilon} \tag{5.26b}$$

The influence of temperature on yield stress is given by the correction factor c_T, which, in the range +20°C to −20°C, is not high. In this range of temperature, the following values of c_T are proposed (Gioncu and Mazzolani, 2002; Wright and Hall, 1964):

$$c_T = 1.0 \text{ for room temperature} \tag{5.27a}$$

$$c_T = 1.05\text{–}1.1 \text{ for low temperature}$$

where the room temperature corresponds to protected structures and the low temperature to unprotected structures.

The influence of material behavior in rolled and welded sections is considered by the correction factor c_w (Gioncu and Mazzolani, 2002; Kaneta et al., 1986; Kohzu and Suita, 1996), whose proposed values are the following:

$$c_w = 1.0 \text{ for rolled sections} \tag{5.27b}$$

$$c_w = 1.10\text{–}1.15 \text{ for welded sections}$$

Therefore, the influence of strain rate on the yield ratio may be determined with the equation:

$$\rho_{ysr} = \{\varphi_{ysr}/\varphi_{usr}\}\rho_y = \{c_T c_w(1.46 + 0.0925\log\dot{\epsilon})/(1.15 + 0.00496\log\dot{\epsilon})\}\rho_y \tag{5.28}$$

The increase in yield ratio due to the high strain rate has a very bad influence on the member ductility, significantly reducing the capacity of seismic energy dissipation, especially in the strain rate range from 10^{-1} to 10^{1} s^{-1}, corresponding to velocities of near-field earthquakes.

Figures 5.89a and 5.89b present the application of Equations 5.26 for two steel quality, S235 and S275, rolled profiled, welded profiles, in room and low temperature conditions, for strain-rate values corresponding to near-field ratios. The yielding ratios ρ_{ysr} are also given for each studied case. The main conclusions learned from these figures are:

- There are situations in which the fracture can occur during seismic actions for given values of the strain rate.
- In conditions of high strain rate, high steel quality S275 has a more favorable behavior than low steel quality S235.
- Rolled profiles have a better behavior than the welded ones.
- The worst behavior during strong earthquakes in near-field areas is in the case of welded profiles under low-temperature conditions.

2. For the analysis of *structural collapse* due to the effects of strain rate, when a law with the largest values in the field of interest is required (Figure 5.88), Kaneko's law (Gioncu and Mazzolani, 2002) is proposed:

$$\begin{aligned} \frac{f_{ysr}}{f_y} &= 1 + \frac{21}{f_y}\log\frac{\dot{\varepsilon}}{\dot{\varepsilon}_0} \quad \dot{\varepsilon}_0 = 10^{-4}/\text{s} \\ \frac{f_{usr}}{f_u} &= 1 + \frac{7.4}{f_u}\log\frac{\dot{\varepsilon}}{\dot{\varepsilon}_0}, \ (\text{N/mm}^2) \end{aligned} \tag{5.29}$$

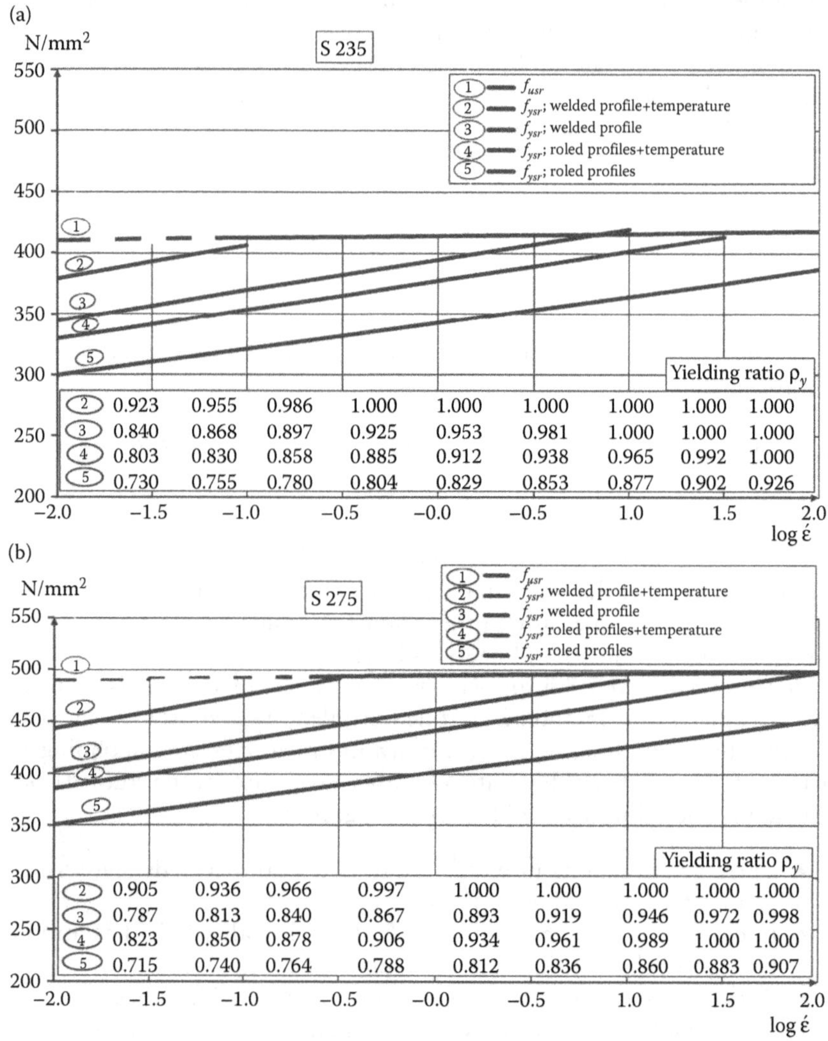

Figure 5.89 Effects of strain rate: (a) steel quality S235; (b) steel quality S275.

5.5.5.2 Effects of strain rate on local ductility

As generally accepted, the monotonic local ductility is defined at the intersection between the values of the rotation capacity, θ_u, determined in the lowering postbuckling curve and the theoretical fully plastic moment M_p (Figure 5.90) (Gioncu, 2000; Mateescu and Gioncu, 2000; El Hassouni et al., 2011). This definition is very useful for practical purposes, because a bilinear moment–rotation curve, with horizontal postyielding behavior,

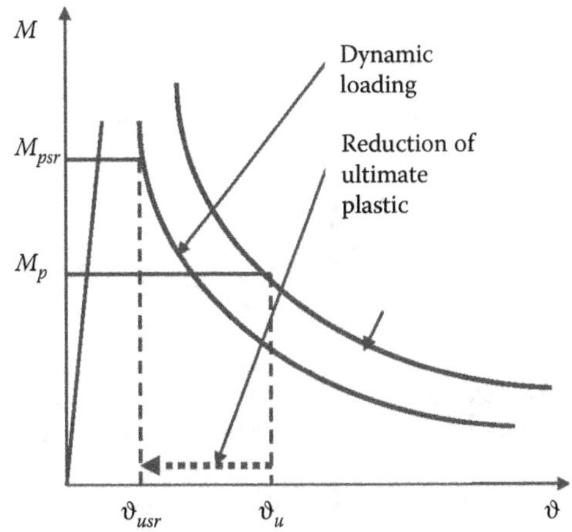

Figure 5.90 Effects of strain rate on ultimate plastic rotation.

is used in many computer programs for structural analyses under seismic actions. In case of important strain rates, the fully plastic moment increases:

$$M_{psr} = \varphi_{ysr} M_p \tag{5.30}$$

where φ_{ysr} is given by relation (5.26a). The rotation capacity φ_{usr}, which considers the effect of strain rates, is determined at the intersection with the postbuckling curve, resulting in a reduction in rotation capacity (Figure 5.90).

The effects of strain rates on a rolled section (HEA 400) and a welded section with the same shape as IPE 400 are presented in Figure 5.91a and b for room and low temperatures. The DUCTROT-M computer program was used for preparing these diagrams. An important reduction is in ultimate rotation of about 30–40% in case of near-field earthquakes, which is more severe for low temperatures. The worst situation is for welded sections under low temperature.

5.5.6 Influence of strain rates on local fracture

5.5.6.1 Replacement of ductile rotation by local fracture due to strain rate

In case of a near-field earthquake, the failure of the steel member, due to the propagation wave action may occur by local fracture, when the carrying capacity is exceeded by one or two excursions in the plastic range. In this case, the strain rate, produced by a very high velocity of seismic action, can transform the ductile rotation capacity into local fracture. From the

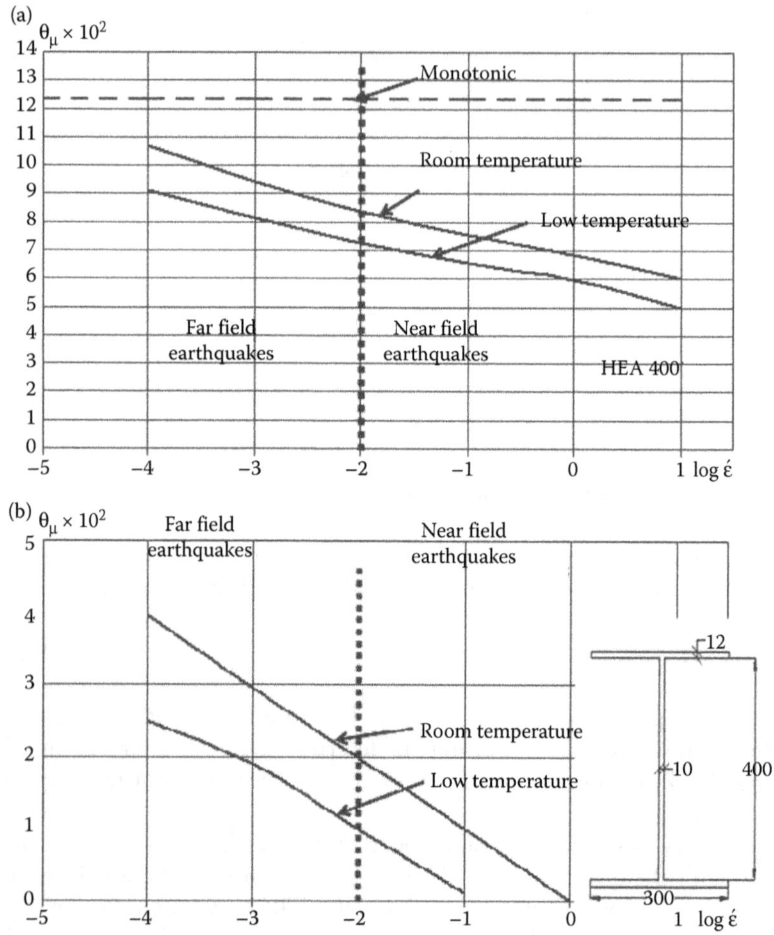

Figure 5.91 Influence of strain rate on rotation capacity: (a) rolled profile HEA400; (b) welded profile.

ductility point of view, the local fracture due to the strain rate may be classified as (Figure 5.92)

1. *High ductile fracture*, with an almost unlimited increase in the deformation capacity; a ductile fracture causes collapse, when the ultimate strain is reached.
2. *Limited ductile fracture*, when the moment–rotation curve presents a maximum value and the collapse is produced due to local plastic buckling.
3. Brittle fracture, produced after limited amount of strain.

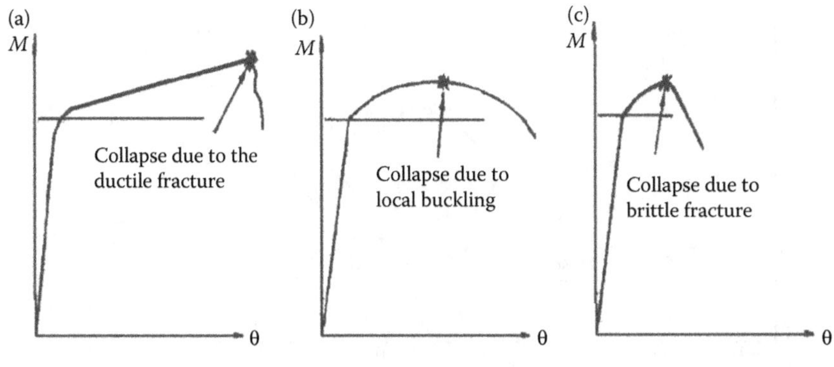

Figure 5.92 Collapse types: (a) ductile fracture; (b) local buckling; (c) brittle fracture.

The main problem of seismic design is to eliminate the brittle fracture by constructional measures. For a simply supported beam, Figure 5.93 shows the moment–rotation curve for a welded profile under different values of strain rates. The fracture rotation is also identified on these curves. One can see that this parameter has a great influence; for strain rates higher than 10^{-1} s^{-1}, the fracture occurs before plastic ductility is reached.

Figure 5.94 shows the influence of strain rates on rotation capacity as a function of flange thickness. In the figure, the domains of plastic and fracture collapse are marked. It is clearly evident that the influence of strain

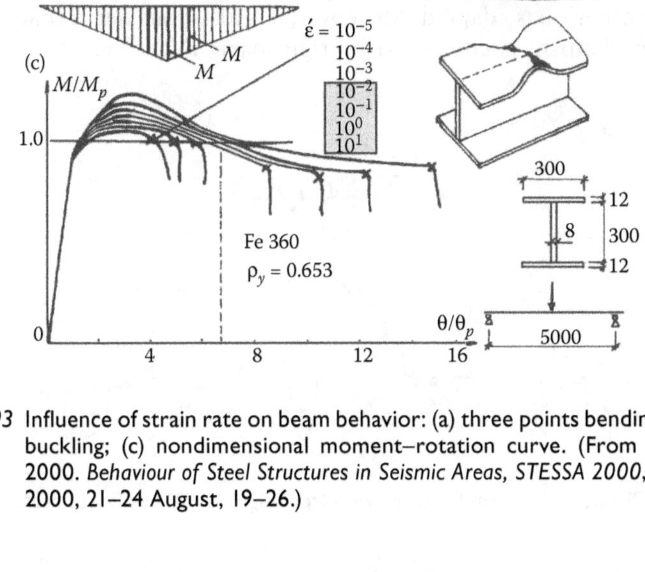

Figure 5.93 Influence of strain rate on beam behavior: (a) three points bending; (b) local buckling; (c) nondimensional moment–rotation curve. (From Gioncu V. 2000. *Behaviour of Steel Structures in Seismic Areas, STESSA 2000*, Montreal, 2000, 21–24 August, 19–26.)

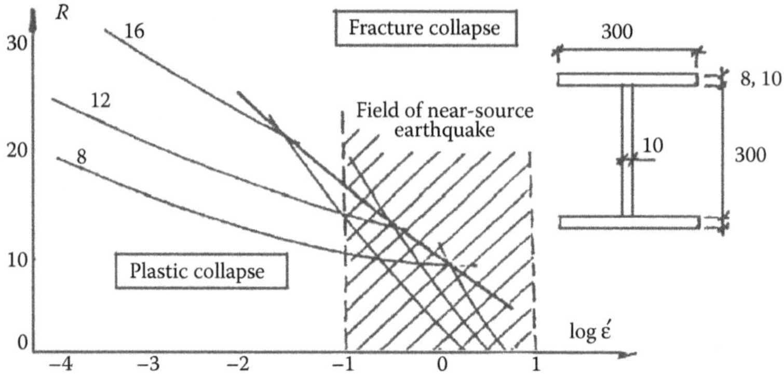

Figure 5.94 Plastic and fracture collapses related to strain rate. (From Gioncu, V. 2000. *Behaviour of Steel Structures in Seismic Areas, STESSA 2000*, Montreal, 2000, 21–24 August, 19–26.)

rates is higher for thick flanges. So, some constructional rules can be given in order to reduce the danger of fracture.

5.5.6.2 Fracture rotation of yield lines

For a plate with a free edge, the plastic mechanism is formed by a yield line situated at the middle of the plate span (Figure 5.95). Two approaches are valuable for evaluating the fracture rotation:

1. *Plastic hinge approach* (Figure 5.96). One assumes that the plate element is composed by rigid perfectly plastic material, where Tresca's yield criterion is adapted. Moreover, it is also considered that the plastic mechanism is composed by rigid parts and concentrated plastic

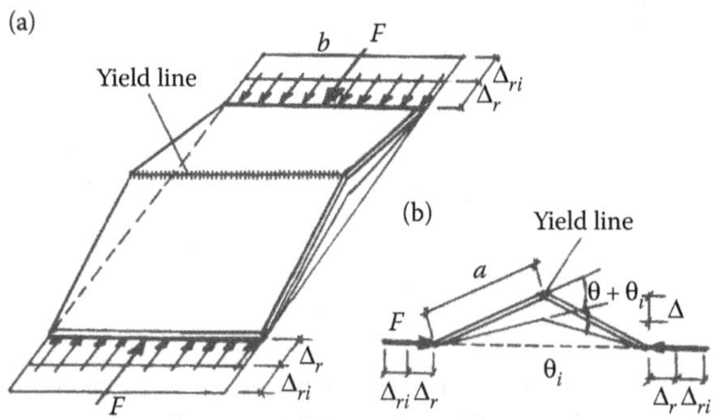

Figure 5.95 Plastic mechanism for plate with free edges: (a) 3D view; (b) section.

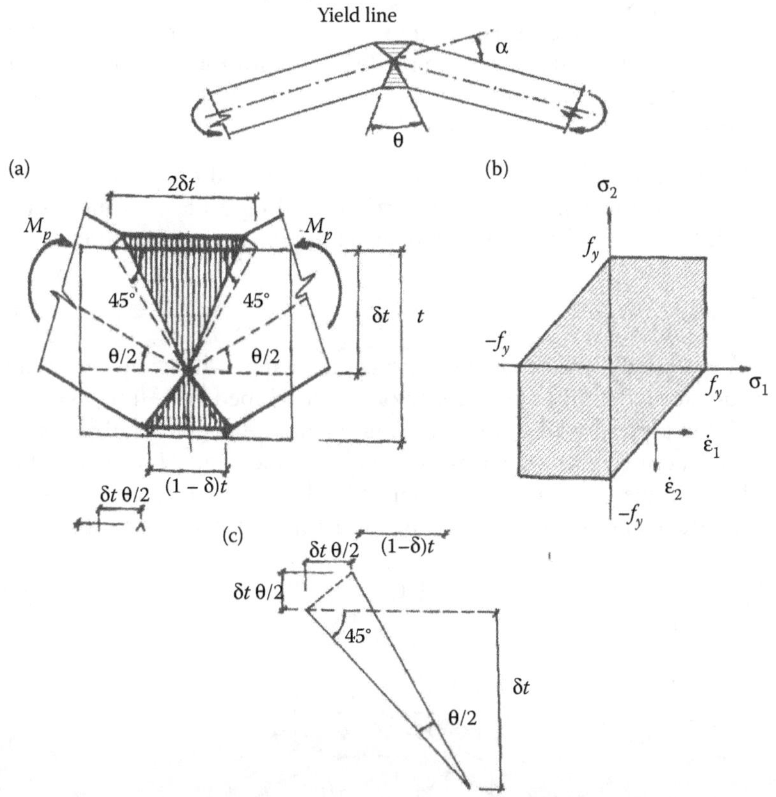

Figure 5.96 Concentrated plastic yield line: (a) plastic hinge; (b) yield criterion; (c) plastic rotation.

lines, whose rigid boundaries form a 45° angle, with respect to the longitudinal axis. A simple relationship between strain and rotation can be obtained:

$$\varepsilon_1 = \vartheta/2; \ \varepsilon_2 = \vartheta/2 \tag{5.31a,b}$$

Thus, the fracture rotation of a yield line α_f, in case of a concentrated plastic hinge, is given by the following equation:

$$\alpha_f = 2\varepsilon_{uf} \tag{5.32a}$$

$$\varepsilon_{uf} = (\varepsilon_u + \varepsilon_t)/2 \sim 1.5\ \varepsilon_u \tag{5.32b}$$

where ε_u is the uniform strain corresponding to f_u, ultimate tensile stress; and ε_t is the maximum strain attained in the stress–strain curve immediately before fracture.

So, the adopted fracture criterion considers that the yield line failure is produced between the uniform strain ε_u and maximum strain ε_t. Compared to the very complex criteria used in the literature for the fracture definition, this approach seems to be very primitive, but it may be satisfactory from the engineering point of view.

2. *Plastic zone approach* (Figure 5.97). It is based on the remark that, in reality, the plastic hinge is distributed along a plastic zone. Kotelko (1996) proposed a simplified relation for the fracture rotation:

$$\alpha_f = 2k\varepsilon_{uf} \quad (5.33)$$

where k is a coefficient determined either experimentally or by a minimization procedure, defining the number of the developed yield lines. Values of $k = 3–4$ correspond well to experimental results. A more detailed determination of fracture rotation was presented by Gioncu and Mazzolani (2002). The plastic hinge is defined by the flange buckling length (Figure 5.94b). In this way, the length of the plastic zone under axial force is given by

$$L_p = (M_{uN}/M_{pN} - 1)L = (1/\rho_y - 1)L \quad (5.34)$$

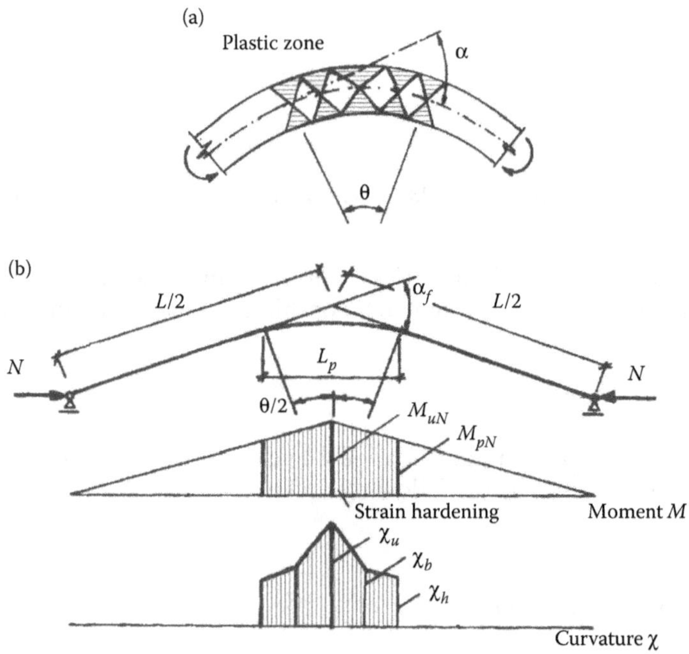

Figure 5.97 Distributed plastic yield line: (a) plastic zone; (b) moment–curvature distribution.

where L_p is the plastic zone, L is the buckling length defined by the distance between the two inflexion points of the plastic mechanism, M_{uN} and M_{pN} are the ultimate and plastic moments affected by the axial force, and ρ_y is the yielding ratio. From an ideal stress–strain diagram, the curvature of the plastic zone, χ, and the corresponding strain, ε, are

$$\chi = \vartheta / L_p \; ; \; \varepsilon = t\chi/2 \tag{5.35}$$

Consequently, for the ultimate limit state, a simple relation for the prediction of the fracture rotation, α_f, of a compressed plate, which takes into account the main factors affecting the fracture behavior as the yield ratio, plate thickness, ultimate strain, and buckling wave, is given by (see also Gioncu and Mazzolani, 2002)

$$\alpha_f = \zeta\,(1/\rho_y - 1)(L/t)\varepsilon_{uf} \tag{5.36}$$

where

$$\xi = 1 + 2\frac{\varepsilon_b}{\varepsilon_u} + \frac{\varepsilon_h}{\varepsilon_u}$$

$$\varepsilon_b = \frac{\chi_b^t}{2}; \; \varepsilon_h = \frac{\chi_h^t}{2}; \; \varepsilon_u = \frac{\chi_u^t}{2}; \tag{5.37}$$

For practical purposes, the fracture rotation can be evaluated by using a simplified value $\zeta = 1.5$. For ST235 and $L/t = 6$, results $\alpha_f = 6.38\ \varepsilon_{uf}$, which corresponds very well to the values proposed by Kotelko (1996). One can see that the fracture rotation depends on the yield ratio, the length of the plastic mechanism, and the steel quality. Figure 5.98 shows the influence of the European, U.S., and Japanese steel qualities on the fracture rotations (Gioncu and Mazzolani, 2002). A very important reduction in fracture rotation may be noticed when the yield stress and yield ratio increase, showing the importance of steel quality in the protection against premature collapse.

5.5.6.3 Fracture of beam flanges

During both strong earthquakes and experimental tests, some cracks may be observed both in the compression flange and web (Figure 5.99). The main characteristic of seismic loading in near-field zones is the velocity of the pulse, while at the level of structure it is the strain rate for producing plastic hinges. In case of buckling of compressed flanges (Figure 5.100a), the rotation of section occurs around a point situated at the tension flange. Owing to the increase in plastic moment as the effect of strain rate, the fracture of a flange can occur before reaching the ultimate plastic rotation (Figure 5.100b).

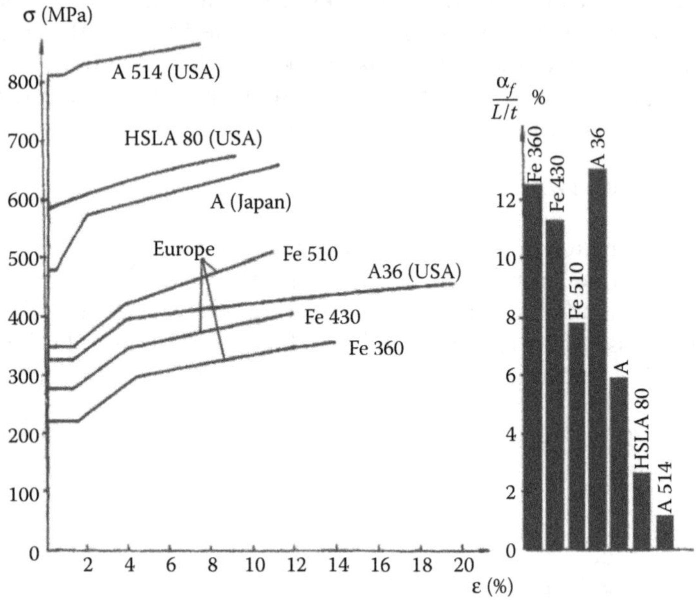

Figure 5.98 Influence of steel quality on plate collapse.

According to the method of plastic collapse mechanism proposed for the evaluation of the ultimate rotation (Gioncu and Mazzolani, 2002), the mechanism is formed by a series of rigid plates separated by yield lines, where the plastic deformations are concentrated. All the yield lines work in the plastic range with the plastic moment M_p.

But when the load increases until the collapse, the moments along the yield lines have different values, as function of rotations: some of them work in the plastic range, others in strain hardening. In contrast to the rigid-plastic analysis, the moment values are different, depending on the rotation amount: ultimate moment for fracture lines, moment in strain-hardening range for some lines and plastic moment for yield lines.

The image of the fracture mechanism of a beam in Figure 5.96 shows that it is dominated by the fracture of the yield line of the compression flange. The relationship between the rotation fracture of the yielding line, α_f, and the member rotation, ϑ_f, results from the compatibility of the plastic mechanism (Gioncu and Mazzolani, 2002):

$$\vartheta_f = (c/4d)\,\alpha_f^2 \tag{5.38}$$

For the beam, the fracture occurs for a couple of values (Gioncu and Mazzolani, 2002):

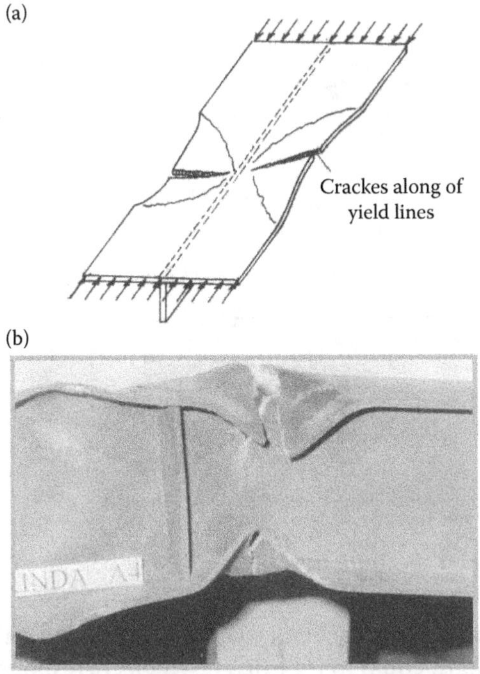

Figure 5.99 Cracking of yield lines: (a) cracking of flanges; (b) cracking of flanges due to cyclic loading with high velocity. (From Mateescu, G., Gioncu, V. 2000: In *Behaviour of Steel Structures in Seismic Areas, STESSA 2000* (eds. F.M. Mazzolani, R. Tremblay), 21–24 August, Montreal, Balkema, Rotterdam, 55–62.)

1. Fracture rotation

$$\vartheta_{uf} = 2\varepsilon_{uf} \tag{5.39a}$$

The collapse by fracture of the tension flange occurs for the ultimate strain (Gioncu and Mazzolani, 2002):

$$\varepsilon_{uf} = 1.5\varepsilon_u \tag{5.39b}$$

where the uniform strain, ε_u, is the strain attained at the maximum point of the engineering stress–strain curve, which corresponds to the beginning of necking, and total strain, ε_t, the maximum strain attained in the stress–strain curve immediately before fracture occurs.

2. Fracture moment

$$M_u/M_p = (1 + 3/\rho_y)/4 \tag{5.40}$$

Because it is unreasonable to expect that a plate section in bending reaches the large strains corresponding to the attainment of yield stress

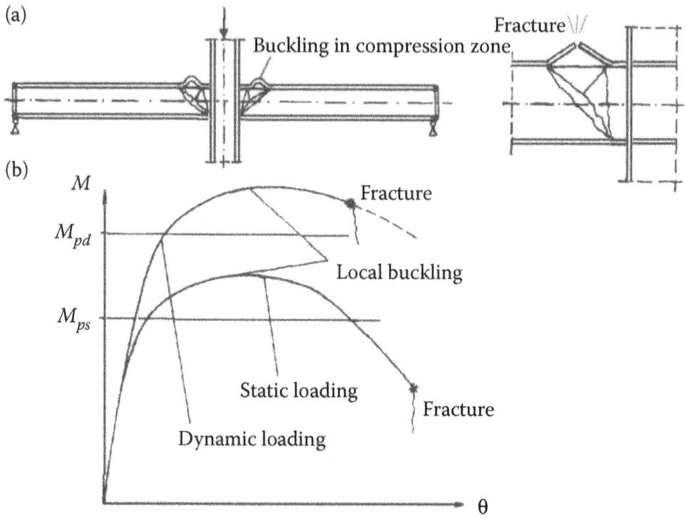

Figure 5.100 Strain-rate effects for beams.

in a tension test, it is more rational to take for the bending a reduced ultimate tensile strength equal to $(f_u + 3f_y)4$, which corresponds to the experimental tests of the ratio $M_u/M_p = 1.30–1.40$ (Gioncu and Mazzolani, 2002).

The fracture rotation of the yield line of compressed flange, α_f, is given by the following relation:

$$\alpha_f = 1.5(1/\rho_y - 1)(L_f/t_f)\varepsilon_{uf} \tag{5.41}$$

After some algebra (Gioncu and Mazzolani, 2002, Chapter 8; Petcu and Gioncu, 2003) results

$$\vartheta_f = 0.2(b^3/dt_f^2)\phi_y\, \varepsilon_{uf}^2 \tag{5.42}$$

with (see Figure 5.101)

$$\phi_y = (1/\rho_y - 1)^2/(\rho_y + 1)^3 \tag{5.43}$$

Figure 5.102 shows the moment–rotation relationship for the fracture of the compression flange in a double T section. The fracture rotation corresponds to the ductile rotation of the plastic hinge for $\rho_y = 0.78$. The ductility corresponds to the ultimate plastic rotation for $\rho_y < 0.78$, and to the flange fracture for $\rho_y > 0.78$.

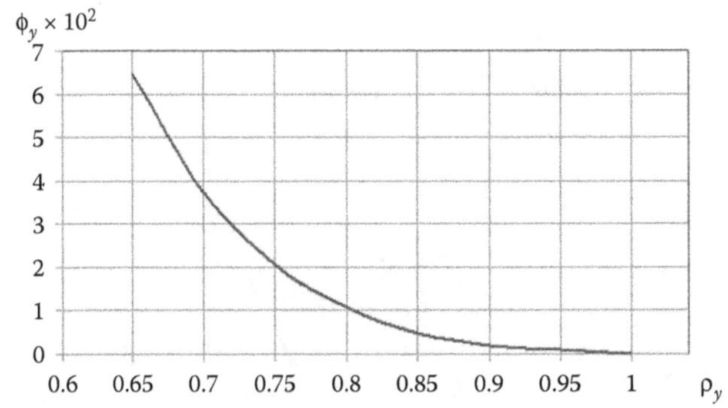

Figure 5.101 The function ϕ_y versus ρ_y.

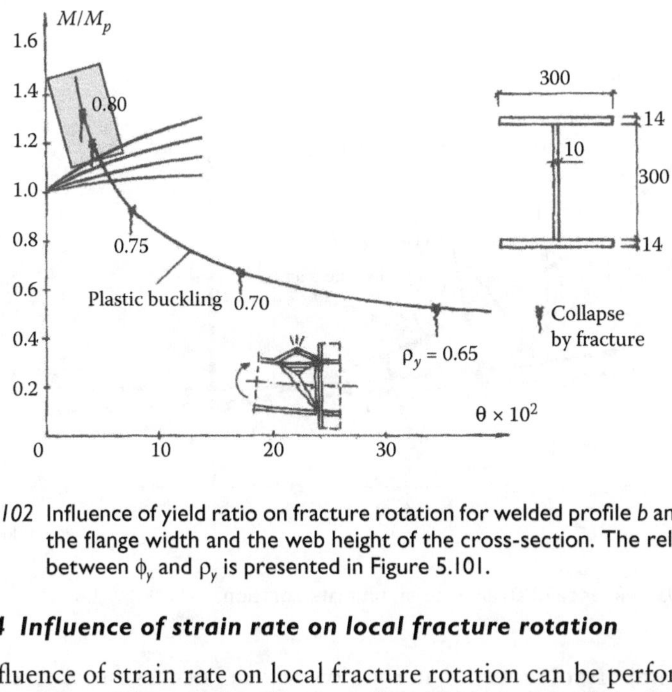

Figure 5.102 Influence of yield ratio on fracture rotation for welded profile *b* and *d* being the flange width and the web height of the cross-section. The relationship between ϕ_y and ρ_y is presented in Figure 5.101.

5.5.6.4 Influence of strain rate on local fracture rotation

The influence of strain rate on local fracture rotation can be performed by using Equations 5.25 and 5.26 between the yield ratio ρ_y and the strain rate $\dot{\epsilon} > 10^2$. Using the DUCTROT-M computer program, the curves of Figure 5.103 are given for rolled (a) and welded (b) profiles, in conditions of room and low temperatures. Rotation capacity and fracture rotation are examined in function of strain rate, for far-field and near-field earthquakes. Comparing these diagrams, the following observations result:

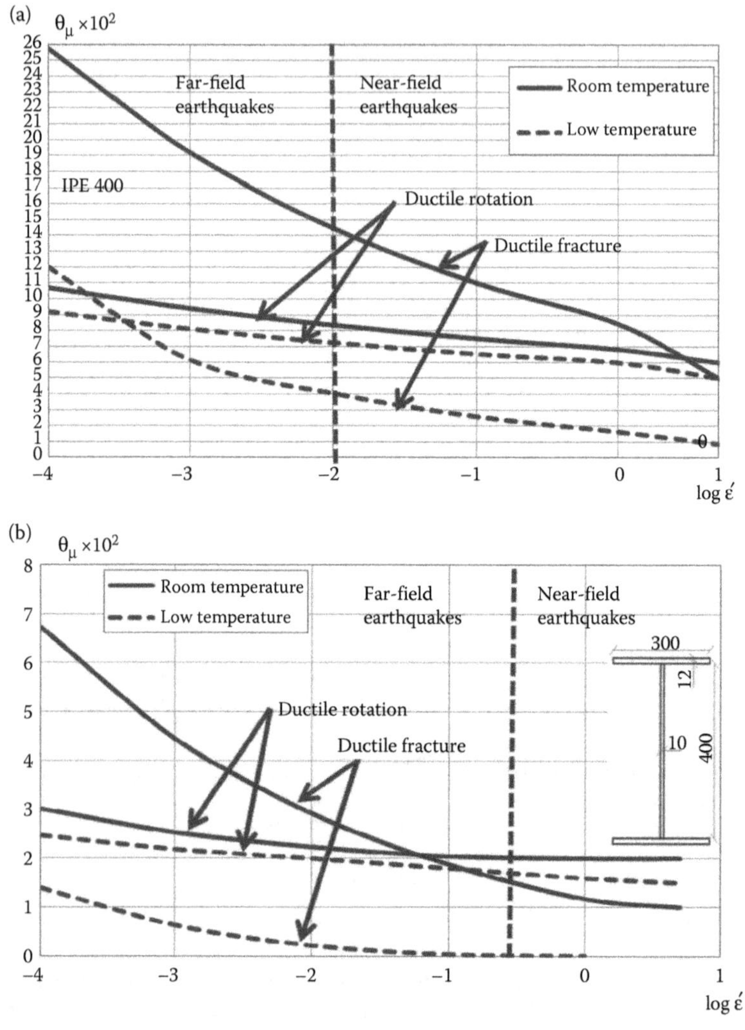

Figure 5.103 Influence of strain rate on ultimate rotation.

- An important reduction in rotation capacity occurs due to strain rates both for far-field and near-field earthquakes, such a reduction being pointed out for the first time in the experimental tests performed by El Hassouni et al. (2011).
- In conditions of room temperature (protected structure), for rolled sections (IPE 400) (Figure 5.103a), the ultimate section capacity results from the ductile rotation, in both cases of far-field and near-field earthquakes. In contrast, for welded sections in case of near-field

earthquakes (Figure 5.103b), the ultimate section capacity is given by the fracture rotation.

- The influence of low temperature (unprotected structure) is not relevant in both cases of far-field and near-field earthquakes and the section capacity is always given by the fracture rotation.

5.5.7 Another vision about the Northridge and Kobe damage

The investigations concerning the causes of damage in steel buildings during the Northridge and Kobe earthquakes have led to a wide discussion within the international scientific community. It is generally admitted that the near-field effects are the main reason for this damage, but there are many opinions about the detailed causes.

The main question under dispute is why the steel structures were not able to dissipate seismic energy by plastic deformations, as is prescribed by the codes, and why this energy was concentrated to produce fracture in the frame joints. The specialists are divided into two different opinions: one considering that only the details and execution are guilty; the other that the code prescriptions are incomplete.

On the one hand, it can be ascribed to the use of field welding, so that the poor workmanship is solely to blame, and, therefore, it is necessary to increase the on-site supervision and controls, as well as to improve the welding details and procedures. Because the damage was concentrated in the vicinity of the bottom welds, it is reasonable to believe that the technology of welding is to blame in this respect. In addition, the presence of a backing bar produces an artificial crack and initiates the rupture. One must mention that the U.S. connection types were tested in laboratories and none of these problems were observed. But the loading frames in the category of semistatic conditions were without significant values for strain rate.

On the other hand, damage causes can be attributed to an excess of seismic loading and to defective design guidance, leading to the available characteristics of near-source earthquakes being lower than the earthquake required ones. The main deficiencies are considered to be the neglect of velocity pulse characteristics of seismic loading, introduction of important effects of strain rate, and the vertical components of seismic loads. On the other side, the measured velocities on free field, with rare exceptions, are not so large as to justify the damage produced.

In the last decades, a new approach, the so-called seismic wave propagation in structure, allows a new vision to be developed about the damage produced during the Northridge and Kobe earthquakes. The velocities measured during some U.S. earthquakes (over 100 m s^{-1}) were able to produce large values of strain rate (extend from 10^{-1} and $10^{1.5}$), giving rise to important connection fractures.

A methodology to calculate the seismic response of structures, founded on layered soil and based on wave propagation, interprets the structure as an extension of the layered soil medium by considering each story as another layer in the wave propagation path (Safak, 1999). The earthquake generates seismic waves propagating through the soil to the earth's surface, which, after meeting a structure, will continue to propagate into the structure itself. Owing to the fact that the structure is the last object of the soil-structure system, the impulse loading increases due to Newton's cradle effect. One must recognize that the scientists who work in this field, the majority being geo-technicians and not structural engineers, are mostly interested in determining the wave propagation in structure and the interaction soil-structure, rather than in evaluating the damage produced by this propagation. It is important to underline that this approach gives rise to a new interpretation to explain the important damage produced during the Northridge and Kobe earthquakes. But it must be recognized that this approach is only at the beginning of its exploitation and this subject is still open to very interesting research works in the future.

The first attempt to understand the response of a structure under seismic loading considers the action of horizontal ground motions only, the vertical ones being considered as insignificant. But this approach is true only for far-field earthquakes, where the attenuation of the *P* and *S* body waves is very high.

In contrast, in the case of near-field earthquakes, the seismic actions can be divided into two phases: phase 1 produced by the body *P* and *S* waves, followed later by the surface *R* and *S* waves in phase 2. The first waves acted on the structure in phase 1 with great velocity vertical propagation, producing some damage mainly concentrated in the connections. The second act in phase 2 with horizontal inertial forces applied to a damaged structure.

Both the Northridge (M 6.7) and Kobe (M 7.2) earthquakes belong to the category of near-field ground motion, but have some important differences, belonging to different source types, thrust and strike-slip, respectively.

During the Northridge earthquake, the *P* and *S* waves produced in the first phase very important connection fractures, but the surface *R* and *S* waves did not have important effects on the steel structures. Owing to this fact, the first reports declared that steel structures were untouched. Titles of papers like "Northridge Earthquake Confirms the Steel Superiority" were very optimistic, comparing steel structures with the RC ones (Gioncu and Mazzolani, 2002). Unfortunately, more refined inspections have shown relevant damage in connections, which were produced during phase 1 and not pointed out in phase 2. Generally, the damage is uniformly distributed on the structure height.

In contrast, during the Kobe earthquake, the same damage occurred for the connections, but, due to the more damaging characteristics of the strike-slip earthquake (especially the duration and cycle number), the actions of the surface waves *R* and *S* were more visible, mainly producing collapse of the intermediate stories, already damaged in the first phase, by transforming them in the plastic mechanisms.

Therefore, examining the cases of the Northridge and Kobe earthquakes, it is possible to conclude that

- The body waves *P* and *S* produced the fracture of connections at the levels with maximum velocity propagation.
- The surface waves *R* and *L*, in function of earthquake characteristics, can or cannot transform the structure in a global or local plastic mechanism, formed because of the damaged connections.

5.5.8 Fracture of welded MRF structures due to near-field earthquakes

The wave propagation must be examined for the two distinct waves, axial *P* waves and shear *S* waves, each having very distinct characteristics.

5.5.8.1 P wave propagation

The *P* wave propagation (Figure 5.104) can be separated into the up-going waves until structure top (*incident waves*) and the down-going waves to the structure base (*reflected waves*). Owing to the incident waves, high tensile forces occur in columns, while for reflected waves the axial forces are in compression. Owing to the fact that tension effects of strain rate are more

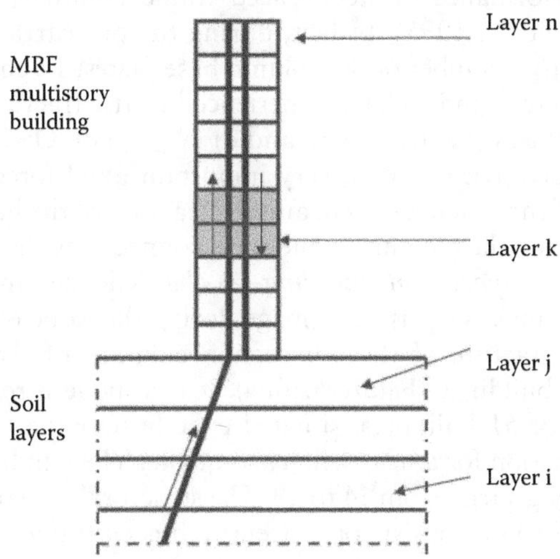

Figure 5.104 Propagation of *P* waves in structure.

important than the compression ones (Sierakowski, 1997), very high tension forces occur in structure columns for incident waves.

Examining Figure 5.79, the P wave velocities exceed 100 m s^{-1}, which means that, using the relation (5.23), the strain rate is over $\dot{\epsilon}$. For these very high values of strain rate and the yield ratio $\rho_y = 1$, the collapse of columns occurs directly by fracture without any plastic deformations. The fracture axial force is given by

$$N_{\text{fracture}} = A\, f_{usr} \tag{5.44}$$

where A is the area of the cross-section of columns and f_{usr} is the ultimate stress, also considering the effect of strain rate (see Equation 5.25b). The structures with concentrically bracing systems, with important axial forces, are very exposed to these axial fractures.

Two very interesting cases are presented as follows:

1. *Oviatt Library, California State University—Northridge* (Trembaly et al., 1995). The building, located in the epicentral region of the Northridge earthquake, is composed of three structural parts: the original reinforced concrete core building, built in the early 1970s, and two wings that were opened in 1991. The Oviatt core suffered reparable structural damage during the Northridge earthquake and it was reopened in August 1994. The east and west wings were built as perimetral four-storied steel-braced frame structures (Figure 5.105a) to be more earthquake resistant than the existing RC part. However, contrary to expectations, the earthquake gave the engineers a lesson on the performance of steel-braced frame buildings under quakes (Trembaly et al., 1995). In fact, during the postearthquake damage survey, a large number of the column base plates, 12 cm thick (Figure 5.105b), were found to have experienced brittle fracture through the entire thickness (Figure 5.105c and d) (WJE, nd). Owing to the high velocities of vertical waves, very important axial forces occurred in columns of the bracing system and the fracture of the base plates took place, making the column–foundation connections the weakest part.
2. *Ashiyahama apartment building—Kobe* (Gioncu and Mazzolani, 2002). The most surprising damage during the Kobe earthquake was the brittle fracture of the box-section columns of the Ashiyahama apartment buildings (Figure 5.106a). It is a modern residential town consisting of 51 buildings, situated exactly over the fracture line, a typical situation for a near-field earthquake. The number of stories of each building varies from 14 to 29. The structural system, as presented in Figure 5.106b, consists of concentric-braced frames, without bracings at the bottom story. The trussed columns have welded box sections, which are made of two parallel channels welded longitudinally.

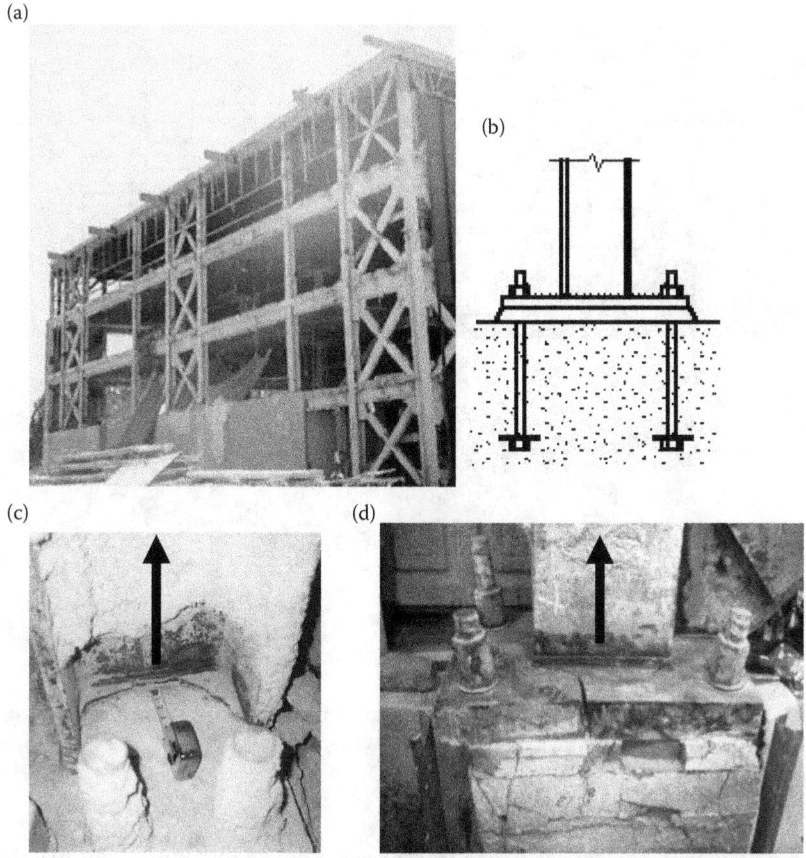

Figure 5.105 Fracture of column–foundation connections due to *P* waves during the Northridge earthquake: (a) structure of the Oviatt Library; (b) column–foundation plates; (c,d) fracture of the base plate. (From USGS-FEMA 2007: Oviatt Library at CSU Northridge fractured in 1994, Earthquake Engineering Research Institute.)

Only the lower portions of the first story columns are made from four very thick plates welded together. The brittle fracture occurred in the chord of trussed columns, but never in the girders. Figure 5.106c shows the view of the fracture in the first unprotected story column. Examining the type of fracture, the results were found to have been produced by very high tension forces, as a consequence of very important *P* wave propagation in the structure. Owing to the high velocity of the waves, also considering the condition of low temperature (below 0°C), the effects of the strain rate were so strong as to cancel the ductile properties of steel and to transform it into a brittle material.

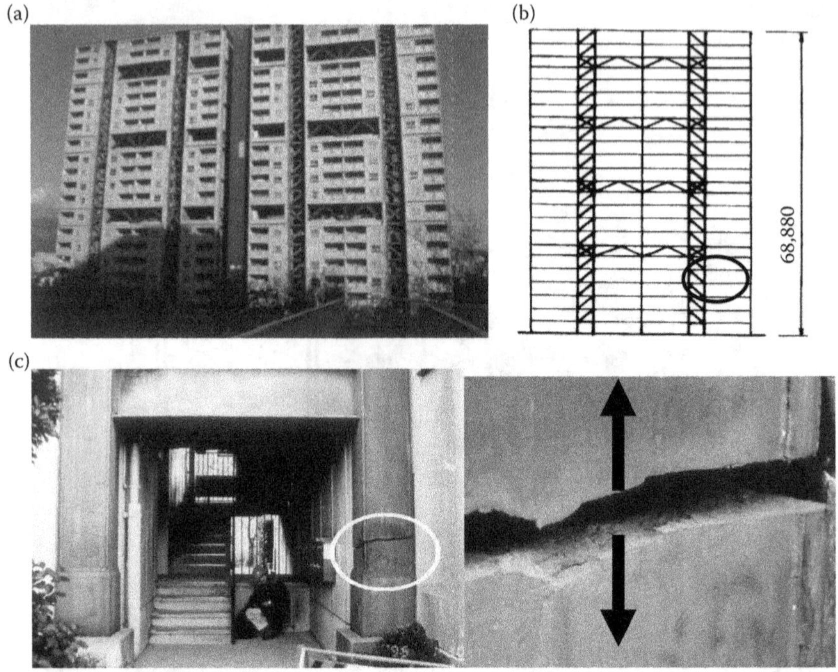

Figure 5.106 Column fracture for Ashiyahama apartment buildings: (a) building view; (b) building structure; (c) column fractures.

5.5.8.2 S wave propagation

In a new vision, the causes of the fractures produced in connections are the result of the very high velocity of the *S* wave propagation along the structure height. In case of MRF, due to the discontinuous medium composed by the story layers, the seismic waves produce the story drifts (Figure 5.107). The seismic loading acts as an impulsive force, the incident and reflected waves producing cyclic story drifts. The velocity of story drift propagation being very high, with a corresponding high strain rate, undermines the plastic behavior of structures and reduces the quantity of dissipated seismic energy. The result is the rupture of connections (Figure 5.108a) by fracture in tension in heat-affected zones (HAZs) (Figure 5.108b).

During the Northridge earthquake, approximately 140 buildings, with a welded moment-resisting frame structure, were shown to suffer unexpected fractures in or near beam-to-column-welded joints. A typical distribution of fractures in a perimetral frame is presented in Figure 5.109a. The practically uniform fracture distribution in the frame is a clear confirmation of the effect of the wave propagation over the entire structure height. The typical beam-to-column moment connection, according to the U.S. system,

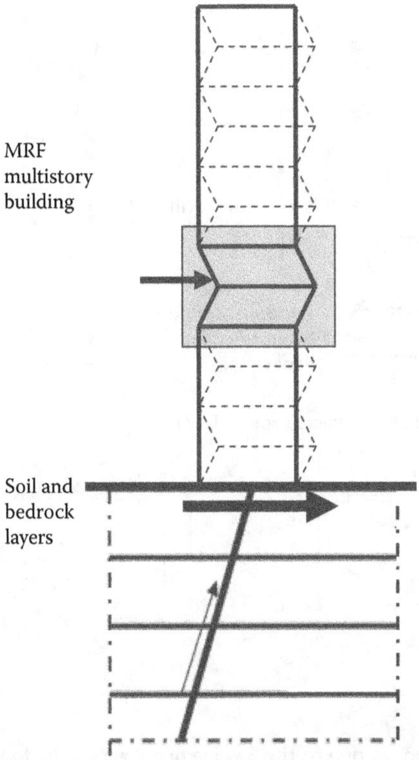

Figure 5.107 Propagation of S waves in a structure.

is presented in Figure 5.109b. Some typical damage, consisting of cracks developed in the welds of beam flanges, column flanges, and column webs, is shown in Figure 5.111c (Gioncu and Mazzolani, 2002). It must be mentioned that no fracture of buckled compression flanges was observed.

5.5.8.3 Fracture of welded connections due to S wave propagation

There are two main problems in evaluating the influence of strain rate:

- Determination of section strain rate as a function of wave velocity and local behavior of plastic hinge
- Determination of fracture rotation as a function of section strain rate

The model of story drift subjected to an impulse loading is presented in Figure 5.110a and b. The joint rotation is

$$\theta(t) = \Delta(t)/h \tag{5.45}$$

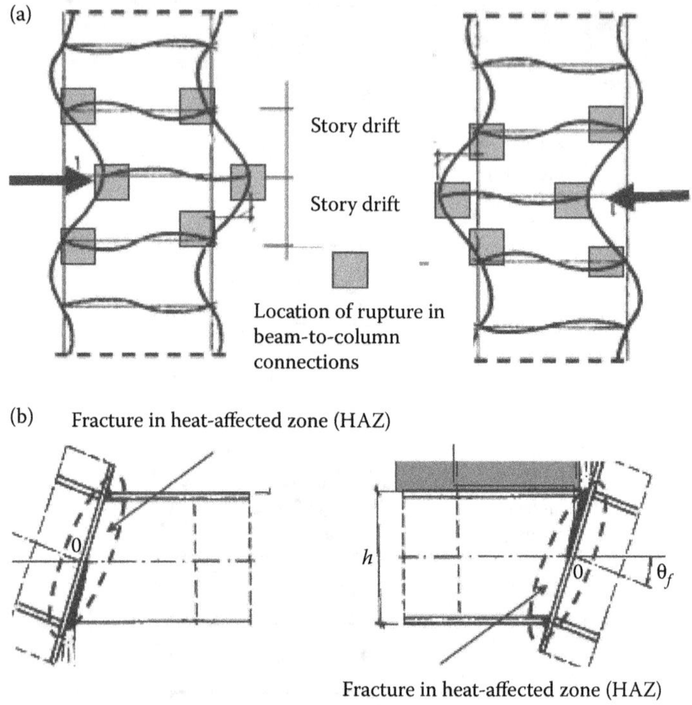

Figure 5.108 Fracture zones due to the S wave propagation: (a) location of ruptures; (b) fracture types.

and

$$\dot{\theta}(t) = d\theta(t)/dt = d\Delta(t)/dt\ 1/h = V_s(t)/h \tag{5.46}$$

Considering the relation (5.39a), it gives the following result:

$$\dot{\epsilon}(t) = V_s(t)/2h \tag{5.47}$$

In the above relations, V_s is the velocity of the S wave and h is the story height.

Unfortunately, there is no information about the S wave velocities along the structure, because usually the measurements refer to the P wave velocities only. But considering that the relation for propagation in the solid body is

$$V_S \sim 0.6\ V_P \tag{5.48}$$

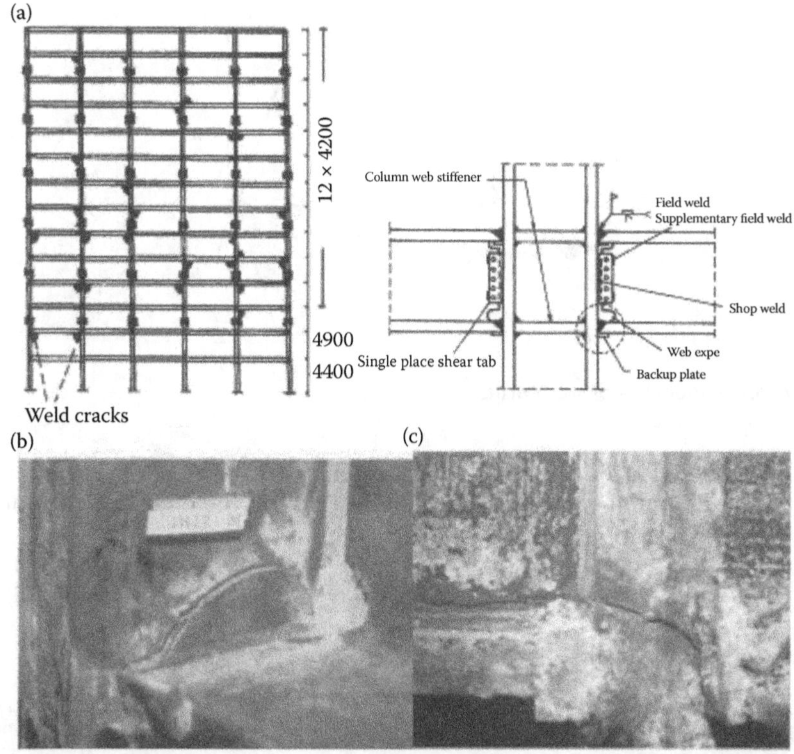

Figure 5.109 **Fractures in welded connections during the Northridge earthquake: (a) distribution of cracks; (b) connection type; (c) connection fractures.**

one can adopt the same relation even for framed structures. So, taking into account the maximum current value of $V_P \sim 350\ \mathrm{ms}^{-1}$, the result is that the maximum current value is $V_S \sim 200\ \mathrm{ms}^{-1}$.

There are two main effects of strain rate on welded connections:

1. The increase in strain rate due to high wave velocity also increases the yield ratio to the limit value $\rho_{ysr} = 1$, indicating the danger of connection fractures. Taking into account the great variability of yield ratio strain rate for high values of strain rate (see Figure 5.85), it is rational to consider that the values $\rho_{ysr} > 0.9$ are related to the danger produced by these fractures (Figure 5.111).
2. The effect of strain rate on the moment–rotation curve (Figure 5.112) increases the plastic moment and reduces the yielding plateau range. At the same time, the characteristics of strain hardening are practically unchanged. In this condition, the plastic behavior of beams can be obtained by translation of the $M - \theta$ plastic curve and the fracture

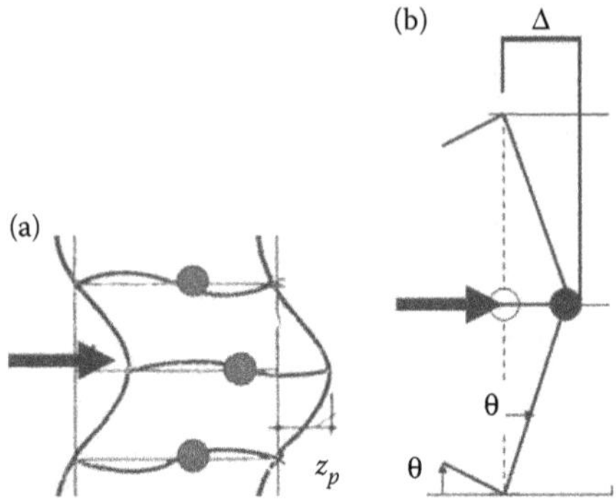

Figure 5.110 Modeling of story drift.

rotation can be determined. If θ_{rusr} and θ_{ru} are the plastic rotations corresponding to the strain-rate effect and to the monotonic action, respectively, their ratio is given by

$$\theta_{rusr}/\theta_{ru} = (M_u - M_{psr})/(M_u - M_p) \tag{5.49}$$

resulting in the yielding plateau index

$$\omega_y = \theta_{rusr}/\theta_{ru} = (1 - \rho_{ysr})/(1 - \rho_y) \tag{5.50}$$

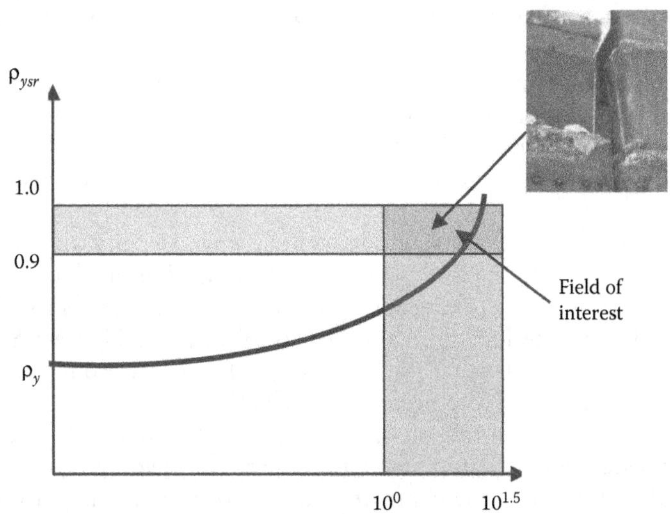

Figure 5.111 Influence of strain rate on yield ratio.

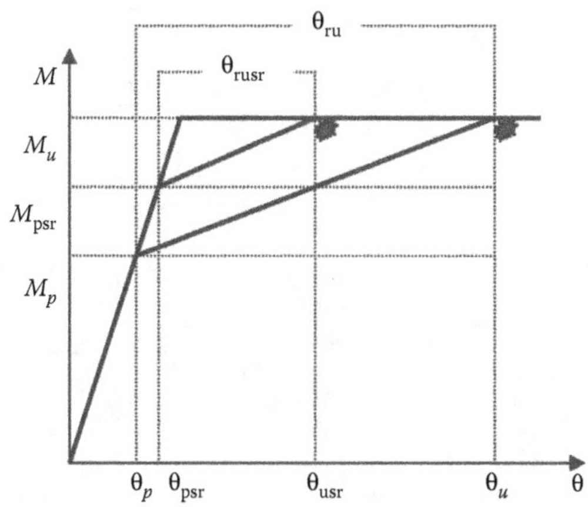

Figure 5.112 Influence of strain rate on ultimate fracture rotation.

In case $\rho_{ysr} = 1$, $\theta_{rusr} = 0$, the section fracture is produced as the effect of strain rate.

Disappearing the yielding plateau, the fracture is brittle and the HAZ is evidently the most exposed part to this rupture mode.

5.5.8.4 Applications

1. *First phase: Damage produced by the seismic wave propagation.* The analyzed-welded MRF, with story height $H = 4.0$ m, is presented in Figure 5.113a. Four cases of near-field earthquakes, corresponding to the intraplate and interplate (subduction or strike-slip) earthquakes, are examined:

• Low earthquakes	(intraplate)	$V = 10$ to 50 ms^{-1}
• Moderate earthquakes	(subduction)	$V = 50$ to 100 ms^{-1}
• Strong earthquakes	(subduction)	$V = 100$ to 150 ms^{-1}
• Very strong earthquakes	(strike-slip)	$V = 150$ to 200 ms^{-1}

As the analyses on the influence of strain rate on the rotation capacity and local fracture have shown a great influence of the type of the connected profiles (rolled or welded) and of the temperature (room or low), the studies are also extended to the case of connection fracture.

The following *S* wave velocities are considered for near-field earthquakes, in case of welded section, room temperature (protected structure), using Kaneko's law for strain rate:

• $V = 10$ ms^{-1}	$\dot{\epsilon} = 1.4375$	$\rho_{ysr} = 0.891$	$\omega_y = 0.311$
• $V = 50$ ms^{-1}	$\dot{\epsilon} = 7.1875$	$\rho_{ysr} = 0.933$	$\omega_y = 0.191$

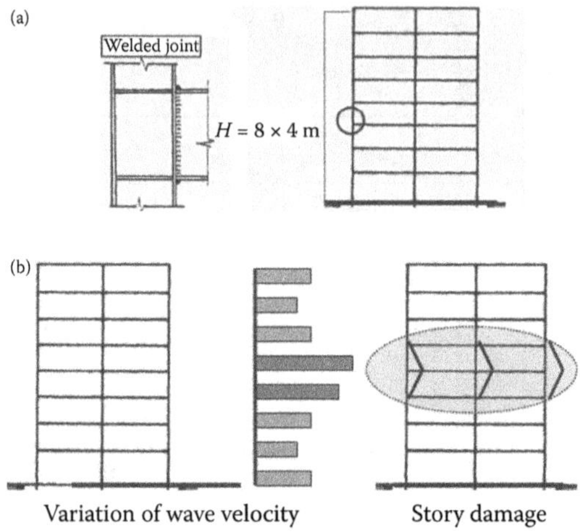

Figure 5.113 Welded MRF under S wave propagation: (a) structure conformation; (b) produced collapse of an intermediate story.

- $V = 100$ ms^{-1} $\dot{\epsilon} = 14.375$ $\rho_{ysr} = 0.950$ $\omega_y = 0.143$
- $V = 150$ ms^{-1} $\dot{\epsilon} = 21.560$ $\rho_{ysr} = 0.960$ $\omega_y = 0.114$
- $V = 200$ ms^{-1} $\dot{\epsilon} = 14.375$ $\rho_{ysr} = 0.967$ $\omega_y = 0.094$

Examining these results, the following observations are valuable for the first step:

- For velocities exceeding 50 ms^{-1}, the resulting yield ratios correspond to a field where the danger of connection fracture is very high.
- The extension of yielding plateau is much reduced, so the structure capacity to dissipate seismic energy is also much reduced.
- The design problems to assure an adequate ductility to the structure, as in the case of far-field earthquakes, are changed into design problems to assure enough strength to connections against fracture.
- A crucial solution in the connection design is to choose good details to avoid the fracture in the HAZs.

2. *Second phase: Damage produced by surface waves.* The influence of strain rate is evaluated by considering the propagation velocity to be uniform on the structure height. But, due to the variation of structure rigidity, some increase in velocity can occur in some stories (Figure 5.113b).

Therefore, the MRF damage is concentrated at an intermediate level. In the second step, the structure is damaged by surface waves, which increase

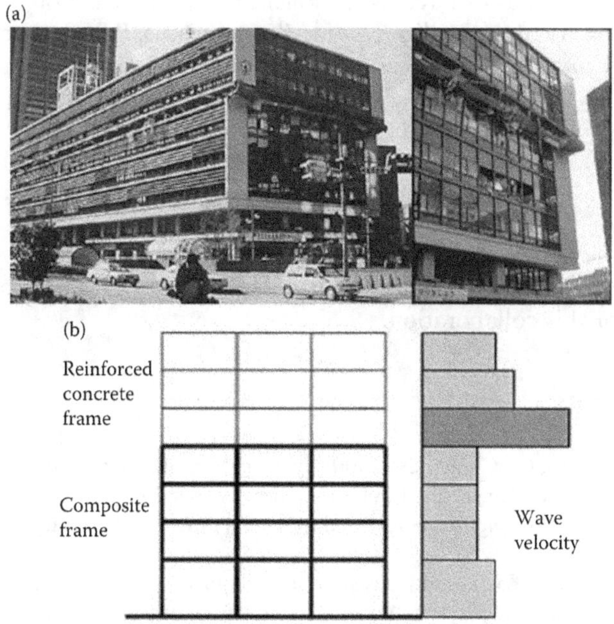

Figure 5.114 The Kobe municipal office building: (a) Collapse of the intermediate story. (After Fischinger, M. 1998: *Electronic Journal of Informatic Technology in Construction*. CDRom.) (b) Structure conformation and velocity concentration.

the structural lateral displacements, due to the impulse characteristic of seismic loads in near-field areas.

During the Kobe earthquake, many MRFs experienced the collapse at some intermediate story. Figure 5.114a presents the case of the Kobe main municipal office, in which an intermediate story collapsed, because of the change in rigidity. In fact, the first four stories are made of composite members (steel and reinforced concrete) and the upper part is made of reinforced concrete (Figure 5.114b). The resultant increase in velocity of wave propagation, which was produced at the interstory contact between the two different structural systems, had the result of creating the fracture of column sections during the first phase. Furthermore, the surface waves, which occurred in the second phase-completed the building failure.

ACKNOWLEDGMENTS

This chapter is mainly based on the following papers:

Gioncu, V., Mosoarca, M., Anastasiadis, A. 2012: Prediction of available rotation capacity and ductility of wide-flange beams: Part 1: DUCTROT-M computer program. *Journal of Constructional Steel Research*, 69, 8–19.

Anastasiadis, A., Mosoarca, M., Gioncu, V. 2012: Prediction of available rotation capacity and ductility of wide-flange beams: Part 2: Applications. *Journal of Constructional Steel Research*, 68, 176–192.

Anastasiadis, A., Mosoarca, M., Gioncu, V. Influence of earthquake type on local ductility of steel beams. Part 1: Far-field earthquakes (manuscript).

Gioncu, V., Mosoarca, M., Anastasiadis, A. Influence of earthquake type on local ductility of steel beams. Part 2: Near-field earthquakes (manuscript).

These papers were elaborated with the collaboration of Dr. Anthimos Anastasiadis and Dr. Marius Mosoarca. The authors want to express their gratitude for this collaboration.

REFERENCES

Adam, M., Schmid, G., Chouw, N. 2000: Investigation of ground motion and structural responses in near field due to incident waves. In *12th World Conference on Earthquake Engineering*, Auckland, 30 January–4 February 2000, Paper No. 1313.

Akiyama, H. 1988: Earthquake resistant design based on the energy approach. In *Proceedings of the 9th World Conference of Earthquake Engineering*, 5, 905–910.

Alavi, B., Krawinkler, H. 2001: Effects of near-faults ground motions on frame structures. Blume Earthquake Engineering Centre, Department of Civil and Environmental Engineering, Stanford University, 2001, Report No. 18.

Anastasiadis, A. 1999: Ductility problems of steel moment resisting frames (in Romanian), 1999, PhD Thesis, Politehnica University Timisoara.

Anastasiadis, A., Gioncu, V. 1999: Ductility of IPE and HEA beams and beam-columns. In *Stability and Ductility of Steel Structures SDSS 99* (eds. D. Dubina, M. Ivanyi), Timisoara, 9–11 September 1999, Elsevier, Amsterdam, 249.

Anastasiadis, A., Gioncu, V., Mazzolani, F.M. 2000: New trends in the evaluation of available ductility of steel members. In *Behaviour of Steel Structures in Seismic Areas, STESSA 2000*, (eds. F.M. Mazzolani and R. Tremblay), Montreal, 21–24 August 2000, Balkema, Rotterdam, 3–10.

Anastasiadis, A., Mateescu, G., Gioncu, V., Mazzolani, F.M. 1999: Reliability of joint systems for improving the ductility of MR-frames. In *Proceedings of the 6th International Colloquium on Stability &Ductility*, Timisoara, 259–268.

Anastasiadis, A., Mosoarca, M., Gioncu, V. 2005: Design aspects of reduced beam sections for IPE and HEA European profiles. In *8th National Conference on Metal Structures*, Xanthi, 29 September–2 October, 010/1–010/1.

Anastasiadis, A., Mosoarca, M., Gioncu, V. 2013: Influence of earthquake type on local ductility of steel beams. Part 1: Far-field earthquakes (manuscript).

Ballio, G., Calado, L. 1986a: Steel bent sections under cyclic loads. *Experimental and Numerical Approachs. Costruzioni Metalliche*, 1, 1–23.

Ballio, G., Calado, L. 1986b: Local buckling of steel sections under cyclic loads. In *Proceedings of the 8th European Conference on Earthquake Engineering*, 4, 57–64.

Ballio, G., Calado, L., Leoni, F., Perroti, F. 1986: Numerical simulation of the cyclic behavior of steel structural systems. *Costruzioni Metalliche*, 5, 269–294.

Ballio, G., Castiglioni, C.A. 1993: Le construzioni metalliche in zona sismica: Un criterio di progetto basato sull; accumulazione del danno. In *Proceedings of XVI Congresso CTA*, 99–109.

Ballio, G., Castiglioni, C.A. 1994: Damage assessment in steel members under seismic loading. In *Proceedings of Behaviour of Steel Structures in Seismic Areas*, STESSA 1994, Timisoara, Romania, 26 June–1 July 1994, 63–76.

Ballio, G., Castiglioni, C.A. 1995: A unified approach for the design of steel structures under low and/or high cycle fatigue. *Journal of Constructional Steel Research*, 34, 75–101.

Bertero, V.V., Popov, E.P. 1965: Effect of large alternating strains of steel beams. *Journal of the Structural Division*, 91(ST1), 1–12.

Boerave, Ph., Lognard, B., Janss, J., Gerardy, J.C., Schleich, J.B. 1993: Elastoplastic behaviour of steel frameworks. *Journal of Constructional Steel Research*, 27, 3–21.

Butterworth, J.M., Beamish, M.J. 1993: Inelastic buckling in three dimensional steel structures. In *Proceedings of Space Structures*, 4, 51–61.

Calado, L., Azevedo, J. 1989: A model for predicting failure of structural steel elements. *Journal of Constructional Steel Research*, 14, 41–64.

Castiglioni, C.A. 1987: Numerical simulation of steel shapes under cyclic bending effect of the constitutive material law of the material. *Costruzioni Metalliche*, 3, 154–175.

Castiglioni, C.A. 2005: Effects of the loading history on the local buckling behavior and failure mode of welded beam-to-column joints in moment resisting steel frames. *Journal of Engineering Mechanics*, 131(6), 568–585.

Castiglioni, C.A., Losa, P.L. 1992: Local buckling and structural damage in steel members under cyclic loading. In *10th World Conference on Earthquake Engineering*, Madrid, 19–24 July 1992, 2891–2896.

Castiglioni, C.A., Bernuzzi, C., Agatino, M.R. 1997: Low-cycle fatigue: A design approach. In *Proceedings of XVII Congresso CTA*, 167–175.

Castiglioni, C., Bernuzzi, C., Calado, L. 2000: Cyclic behavior of steel beam-to-column joints with concrete slab. In *Proceedings of Behaviour of Steel Structures in Seismic Areas, STESSA 2000* (eds. F.M. Mazzolani, R. Tremblay), Montreal, 21–24 August.

Castiglioni, C.A., Di Palma, N., Moretta, E. 1990: A trilinear constitutive model for the seismic steel structures. *Costruzioni Metalliche*, 2, 80–96.

Chi, B., Uang, C.M. 2002: Cyclic response and design recommendations of reduced beam section moment connections with deep columns. *Journal of Structural Engineering*, 128(4), 464–473.

Chichowicz, A. 2011: Near-field ground motion modal versus wave propagation analysis. *Shock and Vibration*, 17(4–5), 611–617.

Climenhaga, J.J., Johnson, R.P. 1972: Moment-rotation curves for locally buckling beams. *Journal of Structural Division*, 50(ST 9), 1239–1254.

Coelho, T., Almeida Souza, PeireiraCalado, P., Vieira Souza, L., Ribeiro-Neto, B., Muntz, R. 2004: Image retrieval using multiple evidence ranking, *IEEE Transactions on Knowledge and Data Engineering*, 16(4), 408–417.

D'Aniello, M., Landolfo, R., Piluso, V., Rizzano, G. 2012: Ultimate behaviour of steel beams under non-uniform bending. *Journal of Constructional Steel Research*, 78, 144–158.

Da Silva, S.L., Santiago, A., Real, P.V. 2002: Post-limit stiffness and ductility of end-plate beam-to-column steel joints. *Computer and Structures*, 80, 515–531.

ECCS-TC 13. 1985: Recommended testing procedure for assessing the behaviour of structural elements under cyclic loads. DOC. 45/86.

El Hassouni, A., Plumier, A., Cherrabi, A. 2011: Experimental and numerical analysis of the strain-rate effect on fully welded connections. *Journal of Constructional Steel Research*, 67(3), 533–546.

El-Tawil, S, Mikesell, T., Kunnath, S.K. 2000: Effect of local details and yield ratio behaviour of FR steel connections. *Journal of Structural Engineering*, 126(1), 79–87.

Espiga, F. 1997: Numerical simulation on RWTH tests. Labein Technological Research Center Report, 1997.

EUROCODE 8. EN 1998-1-1: Design indications for the seismic resistance of structures.

Feldmann, M. 1994: Zur Rotationskapazitat von I-Profilen statisch and dynamisch belastung Trager. PhD thesis, RWTH University, Aachen.

Feldmann, M., Eichler, B., Schafer, D., Sedlacek, G., Vayas, I., Karlos, V., Spiliopoulos, A. 2011: Toughness requirements for plastic design with structural steel. *Steel Construction*, 4(2), 94–113.

Fischinger, M. 1998: EASY. Earthquake engineering slides information system. *Electronic Journal of Informatic Technology in Construction*. CDRom.

Formisano, A., Faggiano, B., Landolfo, R., Mazzolani, F.M. 2006: Ductile behavior classes of steel members for seismic design. In *Behaviour of Steel Structure in Seismic Areas, STESSA 2006* (eds. F.M. Mazzolani, A. Wada), Yokohama, 14–17 August, Taylor & Francis, London, 225–232.

GB50011-2001 2001: *Code for Seismic Design for Buildings*. China Building Industry Press, China, 336 pages.

Gioncu, V. 1998: Phenomenological and mathematical modeling of coupled instabilities. In *Coupled Instabilities in Steel Structures* (ed. J. Rondal), CIMS Courses Udine, 1998, Spinger-Verlag, Wien, 85–149.

Gioncu, V. 2000: Influence of strain-rate on the behavior of steel members. In *Behaviour of Steel Structures in Seismic Areas, STESSA 2000* (eds. F.M. Mazzolani, R. Tremblay), Montreal, 21–24 August, 19–26.

Gioncu, V. 2006a: Influence of constructional details on plastic buckling and ductility of steel members. In *Stability and Ductility of Steel Structures* (eds. D. Camotim, N. Silvestre, P.D. Dinis), Lisbon, 6–8 September, IST Press, Lisbon, 895–901.

Gioncu, V. 2006b: Parametrical studies on steel member ductility. In *Steel: A New and Traditional Material for Buildings* (eds. D. Dubina, V. Ungureanu), Poiana Brasov, 20–22 September, Taylor & Francis, London, 511–517.

Gioncu, V. 2011: Characteristics of DUCTROT-M computer program for determining the available rotation capacity of steel wide-flange beams. *Pollack Periodica*, 6(1), 17–36.

Gioncu, V., Mateescu, G., Iuhas, A. 1994: Contributions to the study of plastic rotational capacity of I-steel sections. In *Behaviour of Steel Structures in Seismic Areas, STESSA 94* (eds. F.M. Mazzolani, V. Gioncu), Timisoara, 26 June–1 July, E&FN Spon, London, 169–181.

Gioncu, V., Mateescu, G., Orasteanu, S. 1989: Theoretical and experimental research regarding the ductility of welded I-sections subjected to bending. In *Stability of Metal Structures, SSRC Conference*, Beijing, 10–12 October, 289–298.

Gioncu, V., Mateescu, G., Petcu, D., Anastasiadis, A. 2000: Prediction of available ductility by means of local plastic mechanism method: DUCTROT computer program. In *Moment Resistant Connections of Steel Frames in Seismic Areas* (ed. F.M. Mazzolani), E and FN Spon, London, 95–146.

Gioncu, V., Mazzolani, F.M. 1994: Alternative methods for assessing local ductility. In *Behaviour of Steel Structures in Seismic Areas, STESSA 94* (eds. F.M. Mazzolani, V. Gioncu), Timisoara, 26 June–1 July, E&FN Spon, London, 182–190.

Gioncu, V., Mazzolani, F.M. 2002: *Ductility of Seismic Resistant Steel Structures*, Spon Press, London.

Gioncu, V., Mazzolani, F.M. 2010: Progress and challenges in the ductility control of steel structures. In *To Construct with Steel, Cluj-Napoca Seminar* (eds. Zsolt Nagy et al.), Cluj Napoca, Romania, 93–102.

Gioncu, V., Mazzolani, F.M. 2011: *Earthquake Engineering for Structural Design*, Spon Press, London, 2011.

Gioncu, V., Mosoarca, M., Anastasiadis, A. 2009: Proposal for increasing the ductility of steel structures. In *Behaviour of Steel Structures in Seismic Areas, STESSA 2009* (eds. F.M. Mazzolani, J.M. Ricles, R. Sause), Philadelphia, 16–20 August, CRC Press, Boca Raton, 679–684.

Gioncu, V., Mosoarca, M., Anastasiadis, A. 2012: Prediction of available rotation capacity and ductility of wide-flange beams: Part 1: DUCTROT-M computer program. *Journal of Constructional Steel Research*, 69, 8–19.

Gioncu, V., Mosoarca, M., Anastasiadis, A. 2013: Influence of earthquake type on local ductility of steel beams. Part 2: Near-field earthquakes (manuscript).

Gioncu, V., Petcu, D. 1997: Available rotation capacity of wide-flange beams and beam-columns. Part 1, Theoretical approaches, Part 2, Experimental and numerical tests. *Journal of Constructional Steel Research,* 43(1–3), 161–218, 219–244.

Gioncu, V., Petcu, D. 2001: Improvement of shape of local plastic mechanism (in Romanian), INCERC Report.

Gioncu, V., Tirca, L. 1996: Correlation between experimental data and numerical results in ductility of steel members. In *Conference on Material Engineering*, Lecce, 1, 93–100.

Gioncu, V., Tirca, L., Petcu, D. 1996: Interaction between in-plane and out-of-plane plastic buckling of wide-flange section members. In *Coupled Instabilities in Metal Structures, CIMS 96*, (eds. J. Rondal, D. Dubina, V. Gioncu), Liege, 5–7 September, Imperial College Press, London, 273–228.

Green, P.S., Ricle, J.M., Sause, R. 2002: Strength and ductility of HPS flexural members. *Journal of Constructional Steel Research*, 58, 907–941.

Green, P.S., Ricle, J.M., Sause, R. 2007: Response of high performance steel flexural members to inelastic cyclic loading. In *Proceedings of Behaviour of Steel Structures in Seismic Areas, STESSA 1997*, 160–167.

Guruparan, N.I., Walpole, W.R. 1990: The effect of lateral slenderness on the cyclic strength of beams. In *Proceedings of the 4th US National Conference on Earthquake Engineering*, Palm Springs, California, Vol. 2, 585–594.

Gyorgyi, J., Gioncu, V., Mosoarca, M. 2006: Behaviour of steel MRFs subjected to near-fault ground motions. In *Proceedings of Behaviour of Steel Structures in Seismic Areas, STESSA 2006*, 129–136.

Haaijer, G., Thurlimann, B. 1958: On inelastic buckling in steel. *Journal of Engineering Mechanics Division*, 83, EM 2.

Hall, J.F., Heaton, T.H., Halling, M.M., Wald, D.J. 1995: Near-source ground motion and its effects on flexible buildings. *Earthquake Spectra*, 11(4), 569–605.

Hancock, J., Bommer, J.J. 2004: Predicting the number of cycles of ground motion. In *Proceedings of the 13th World Conference on Earthquake Engineering*, Vancouver, BC, Canada, August 1–6, 2004, Paper No. 1989.

Hancock, J., Bommer, J.J. 2006: A state-of-knowledge review on the influence of strong-motion duration on structural damage. *Earthquake Spectra*, 22(3), 827–845.

Hoglund, T., Nylander, H. 1970: Maximiforhallnde B/t for tryckt flans hos valsad I-balk vid dimensionering med graslastruetod. Technical University Stockholm, Report 83/197.

Huang, C.T. 2003: Considerations of multimode structural response for near-field earthquakes. *Journal of Engineering Mechanics*, 129(4), 458–467.

Ivanyi, M. 1979: Moment rotation characteristics of locally buckling beams. *Periodica Polytechnica, Civil Engineering*, 23(3/4), 217–230.

Iwan, W.D. 1997: Drift spectrum: Measure of demand for earthquake ground motions. *Journal of Structural Engineering*, 123, 397–404.

Jiao, Y., Yamada, S., Kishiki, S., Shimada, Y. 2011: Evaluation of plastic energy dissipation capacity of steel beams suffering ductile fracture under various loading histories. *Earthquake Engineering and Structural Dynamics*, 40(14), 1553–1570.

Kalkan, E., Kunnath, S.K. 2006: Effects of fling step and forward directivity on seismic response of buildings. *Earthquake Spectra*, 22(2), 367–390.

Kaneta, K., Kohzu, I., Fujimura, K. 1986: On the strength and ductility of steel structural joints subjected to high speed monotonic tensile loading. In *The 8th European Conference on Earthquake Engineering*, Lisbon, Vol. 4.

Kawakami, H., Oyungchimeg, M. 2004: Normalized input-output minimization analysis of wave propagation in damaged and undamaged buildings. In *13th World Conference on Earthquake Engineering*, Vancouver, 1–6 August, Paper No. 317.

Kawakami, H., Oyungchimeg, M., Tingatinga, A.A.J. 2005: Analysis of earthquake wave propagation in buildings. In *Earthquake Resistant Engineering Structures V* (eds. Brebbia C.A., Beskos D.E., Manos G.D., Spyrakos C.C.), WIT Press, Southampton.

Kemp, A.R., Dekker, N.W. 1991: Available rotation capacity in steel and composite beams. *The Structural Engineer*, 69(5), 88–97.

Kohler, M.D., Heaton, T.H., Bradford, S.C. 2007: Propagation waves in the steel moment frame factor building recorded during earthquakes. *Bulletin of the Seismological Society of America*, 97(4), 1334–1345.

Kohzu, I, Suita, K. 1996: Single or few excursion failure of steel structural joints due to impulsive shocks in the 1995 Hyogoken Nanbu earthquake. In *11th World Conference on Earthquake Engineering*, Acapulco, 23–28 June, Paper No. 412.

Kotelko, M. 1996: Selected problems of collapse behaviour analysis of structural members built from strain behaviour material. *Conference on Thin-walled Structures*, 2–4 December, Glasgow.

Krawinkler, H., Alavi, B. 1998: Development of improved design procedures for near fault ground motions. In *Proceedings of SMIP98 Seminar on Utilization of Strong-Motion Data*, 21–42.

Krawinkler, H., Zohrei, M. 1986: Cumulative damage in steel structures subjected to earthquake ground motions. *Computers and Structures*, 16, 531–541.

Kuhlmann, U. 1985: Rotationskapazitat biegebeanspruchter I-Profile unter Berucksichtigung des plastischen Beulens. Technical Report, Bochum University, Mitteilung Nr. 85–5.

Kuhlmann, U. 1989: Definition of flange slenderness limits on the basis of rotation capacity values. *Journal of Constructional Steel Research*, 14, 21–40.

Lee, G.C., Lee, E.T. 1994: Local buckling of steel sections under cyclic loading. *Journal of Constructional Steel Research*, 29, 55–70.

Lee, K., Stojadinovic, B. 2003: Seismic rotation capacity and lateral bracing for US steel moment connections. In *Behaviour of Steel Structures in Seismic Areas, STESSA 2003* (ed. F.M. Mazzolani), Naples, 9–12 June, Balkema, 335–342.

Lee, K., Stojadinovic, B. 2008: A plastic collapse method for evaluating rotation capacity of full-restrained steel moment connections. *Theoretical and Applied Mechanics*, 35(1–3): 191–214.

Lee, V.W. 2002: Empirical scaling of strong earthquake ground motion—Part II; Duration of strong motion. ISET. *Journal of Earthquake Technology*, 39(4), 255–271.

Lindt van de, J.W., Goh, G.H. 2004: Effect of earthquake duration on structural reliability. *Engineering Structures*, 26, 1585–1597.

Liu, D., Nakashima, M., Kanao, I. 2002: Behaviour of complete failure of steel beams subjected to cyclic loading. *Engineering Structures*, 25, 525–535.

Lukey, A.F., Adams, P.F. 1969: Rotation capacity of beams under moment gradient. *Journal of Structural Division,* 95(ST 6), 173–1188.

Mao, C., Ricles, J.M., Lu, L, Fischer, J.W. 2001: Effect of local details on ductility of welded moment connections. *Journal of Structural Engineering*, 127(9), 1036–1044.

Mateescu, G., Gioncu, V. 2000: Member response to strong pulse seismic loading. In *Behaviour of Steel Structures in Seismic Areas, STESSA 2000* (eds. F.M. Mazzolani, R. Tremblay), 21–24 August, Montreal, Balkema, Rotterdam, 55–62.

Mazzolani, F.M. (ed) 2000: *Moment Resistant Connections of Steel Frames in Seismic Areas. Design and Reliability*, E&FN Spon, London.

Mazzolani, F.M., Piluso, V. 1993: Member behavioural classes of steel beams and beam-columns. In *XIV CTA Congress*, 24–27 October, Viareggio, 405–416.

Mitani, I., Makino, M., Matsui, Ch. 1977: Influence of local buckling on cyclic behavior of steel beam-columns. In *Proceedings of the 6th World Conference of Earthquake Engineering*, 3(6), 3175–3180.

Mollaioli, F., Bruno, S., Decanini, L.D., Panza, G.F. 2006: Characterization of the dynamic response of structures to damaging pulse-type near-fault ground motions. *Meccanica*, 41, 23–46.

Nakashima, M. 1994: Variation of ductility capacity of steel beam-columns. *Journal of Structural Engineering*, 120(7), 1941–1959.

Okazaki, T., Liu, D., Nakashima, M., Engelhardt, M. 2006: Stability requirements for beams in seismic steel moment resisting frames. *Journal of Structural Engineering*, 132(9), 1334–1342.

OPCM 3431 2005: First elements in the matter of general criteria for seismic classification of the national territory and of technical codes for structures in seismic zones. Official Gazette of the Italian Republic, 2005.

Oyungchimeg, M., Kawakami, H. 2003: A new method for propagation analysis of earthquake waves in damaged buildings: Evolutionary normalized input-output minimization (NIOM). *Journal of Asian Architecture and Building Engineering*, 16, 9–16.

Park, M.S., Lee, B.C. 1996: Prediction of bending collapse behaviour of thin-walled open section beams. *Thin-Walled Structures*, 25(3), 185–206.

Park, J., Fenves, G.L., Stojadinovic, B. 2004: Spatial distribution of response of multi-story structures for simulated ground motions. In *13th World Conference on Earthquake Engineering*, 1–6 August, Paper No. 1545.

Petcu, D., Gioncu, V. 2002: DUCTROT-M. Rotation capacity of steel members. INCERC Timisoara, Politechnica University Timisoara, West University Timisoara, http://web.info.uvt.ro/petcu/software.html#ductrot.

Petcu, D., Gioncu, V. 2003: Computer program for available ductility analysis of steel structures. *Computers and Structures,* 2003, 81, 2149–2164.

Rumshiskii, L.I. 1978: Mathematical Processing of Experimental Data (in Romanian), Editura Tehnica Bucuresti, 1978.

SAC 1996: Connection test summaries. Report No. SAC-96-02, SAC Joint Venture, Sacramento, California, http://www.sacsteel.org/design/test-summaries.html.

Safak, E. 1995: Detection and identification of soil-structure interaction in buildings from vibration recordings. *Journal of Structural Engineering*, 121, 899–906.

Safak, E. 1998: New approach to analyzing soil-building system. *Soil Dynamics and Earthquake Engineering*, 17(7/8), 509–517.

Safak, E. 1999: Wave-propagation formulation on seismic response of multistory buildings. *Journal of Structural Engineering*, 1999, 125(4), 426–437.

Sierakowski, R.L. 1997: Strain rate behavior of metals and composites. In *IGF Conference*, Cassino, 27–28 May.

Smith, R.J., Adams, P.F. 1968: Experiments on wide-flange beams under moment gradient. Structural Engineering Report No. 13, May 1968, University of Alberta, Department of Civil Engineering.

Snaebjornsson, J.Th., Sigbjornsson, R. 2008: The duration characteristics of earthquake ground motions. In *14th World Conference on Earthquake Engineering*, 2008, Beijing, China, paper no. 0102.

Snieder, R., Safak, E. 2006: Extracting the building response using seismic interferometry: Theory and application to the Millikan Library in Pasadena, California, *Bulletin of the Seismological Society of America*, 96, 586–598.

Soroushian, P., Choi, K.B. 1987: Steel mechanical proprieties at different strain rate. *Journal of Structural Engineering*, 4, 863–872.

Spangemacher, R. 1991: Zum rotationsnachweis von stahlkonstruktion, die nach dem traglastverfahren berechnet werden, PhD Thesis, Aachen University.

Stafford, P.J., Bommer, J.J. 2009: Empirical equations for the prediction of equivalent number of cycles of earthquake ground motion. *Soil Dynamics and Earthquake Engineering*, 29, 1425–1436.

Suzuki, T., Ogawa, T., Ikaraski, K. 1994: A study on local buckling behaviour of hybrid beams. *Thin-Walled Structures*, 19(1–2), 337–351.

Suzuki, T., Ogawa, T., Ikarashi, K. 1997: Evaluation of the plastic deformation capacity modified by the effect of ductile fracture. In *Behaviour of Steel Structures in Seismic Areas, STESSA 97*, (eds. F.M. Mazzolani and H. Akiyama), Kyoto, 3–8 August 1997, 10/17 Salerno, 326–333.

Suzuki, T., Ono, T. 1977: An experimental study on inelastic behavior of steel members subjected to repeated loading. In *6th World Conference of Earthquake Engineering*, 3(6), 3163–3168.

Takanashi, K. 1974: Inelastic lateral buckling of steel beams subjected to repeated and reversed loadings. In *5th World Conference of Earthquake Engineering*, 1(5), 795–798.

Takanashi, K., Udegawa, K. 1989: Behaviour of steel and composite beams at various displacement rates. *Journal of Structural Engineering*, 115(8), 2067–2081.

Tan, K.H., Huang, Z.F., Dharma, R.B. 2007: Behaviour and modelling of composite columns and beams under fire conditions. *Urban Habitat Constructions under Catastrophic Events, Workshop*, Prague, 30–31 March, 17–24.

Todorowska, M.I., Lee, V.W., Trifunac, M.D. 1988: Investigation of earthquake response of long buildings. University of Southern California, Report No. CE 88-2, 1988.

Todorowska, M.I., Rahmani, M. 2012: Recent advances in wave travel time based methodology for structural health monitoring and early earthquake damage detection in buildings. In *15th World Conference of Earthquake Engineering*, Lisbon, 24–28 September, Paper No. 5514.

Trifunac, M.D., Todorowska, M.I. 1998: Nonlinear soil response as a natural passive isolation mechanism. The 1994 Northridge, California, earthquake. *Soil Dynamic and Earthquake Engineering*, 17, 41–51.

Trembaly, R., Timler, P., Bruneau, M., Filiatrault, A. 1995: Performance of steel structures during the 1994 Northridge earthquake. *Canadian Journal of Civil Engineering*, 22, 338–360.

UBC 97 1997: Uniform Building Code. International Code Council.

USGS-FEMA 2007: Oviatt Library at CSU Northridge fractured in 1994, Earthquake Engineering Research Institute.

Valente, M., Castiglioni, C. 2003: Effects of concrete slab on the behavior of steel beam-to column joints: Experimental and numerical study. In *Behaviour of Steel Structures in Seismic Areas, STESSA 2003—Behaviour of Steel Structures in Seismic Areas: Proceedings of the 4th International Specialty Conference,* Naples, Italy, (ed. F.M. Mazzolani), 9–12 June 2003, Taylor & Francis, 369–377.

Van, W.P., Thompson, L.E., Whalley, L.E., Ozier, L.D. 1974: Cyclic behaviour of rolled steel members. In *5th World Conference of Earthquake Engineering*, 1(5), 1187–1193.

Vayas, I. 1997: Stability and ductility of steel elements. *Journal of Constructional Steel Research*, 44(1–2), 23–50.

Whittaker, A.S. (ed) 2000: *Effects of Near-Field Earthquake Shaking*. US–Japan Workshop, 20–21 March, San Francisco, California.

WJE (nd): Oviatt Library, California State University.

Wright, R.N., Hall, W.J. 1964: Loading rate effects in structural steel design. *Journal of the Structural Division*, 90(ST 5), 11–37.

Xue, L. 2008: A unified expression for low cycle fatigue and extremely low cycle fatigue and implication for monotonic loading. *International Journal of Fatigue*, 30, 1691–1698.

Yamada, M. 1992: Low fatigue fracture limits of structural materials and structural elements. In *Proceedings of Testing of Metal Structures*, 184–192

Yamada, M., Kawamura, H., Tani, A., Iwanaga, K., Sakai, Y., Nishikawa, H., Masui, A. 1988: Fracture ductility of structural elements and structures. In *9th World Conference of Earthquake Engineering*, 4, 219–241.

Zambrano, A., Mulas, M.G., Castiglioni, C.A. 1999: Steel members under cyclic loads: A constitutive law accounting for damage. In *Proceedings of XVII Congresso CTA*, 323–334.

Zhang, R.R., Snieder, R., Gargab, L., Seibi, C. 2011: Modeling of seismic wave motion in high-rise building. *Probabilistic Engineering Mechanics*, 26, 520–527.

Chapter 6

Fire after Earthquake

6.1 INTRODUCTION

According to the EU Commission Directive issued on December 21, 1988 (1989L0106-EN-20.11.2003), the construction works must be designed and built in such a way that, in case of an outbreak of fire, the following items are gathered: the load-bearing capacity of the construction can be assured for a specific period of time; the generation and spread of fire and smoke within the building and to neighboring constructions are limited; the occupants can leave the affected place by themselves or be rescued by other means; and the safety of rescue teams is considered. Evidently, the behavior in fire of structures that have been damaged by earthquakes represents an important issue (EQE, 1995), since the earthquake-induced damage makes the structure more vulnerable to fire effects than the undamaged one (Faggiano and Mazzolani, 2011a). Frequently, fires break out after a seismic event, giving rise to a real catastrophe (Figure 6.1). In fact, the negative effects of fires on structures and human lives may be comparable to and even more important than those of the earthquake itself (Scawthorn et al., 2005). Moreover, even in case no fire develops immediately after an earthquake, the possibility of delayed fires affecting the structure must be adequately taken into account. This is because the possible consequence of a seism on a structural system is the failure of the fire protection systems and worsening of the structural fire performances, due to the possible earthquake-induced damage. In several cases, the occurrence of a fire following earthquake (FFE) in urban areas has generated catastrophic scenarios. In fact, particular building characteristics and density, electric or gas wires, fuel tanks, and meteorological conditions can combine to create a situation in which FFE is a concurrent agent of damage (Williamson and Groner, 2000).

The seriousness of the problem is also due to the probable multiple simultaneous ignitions. Such emergency is worsened by earthquake-induced impairments to communications, water supply, and transportation, leading to structural collapses, hazardous material releases, and emergency medical aid (Faggiano et al., 2008; Faggiano and Mazzolani,

Figure 6.1 Fires after earthquakes: (a) Loma Prieta 1989 (nytimes.com); (b) Northridge 1994 (guerrillalaw.com); (c) Chile 2010 (chicagotribune.com); (d) Japan 2011 (piersystem.com).

2011a). In Figure 6.2, typical pipe damages produced during different earthquakes are illustrated. Earthquake-induced fire is therefore a critical earthquake-related hazard. Large fires following an earthquake have often been of catastrophic proportions in urban areas. In the following paragraph, before examining in depth the behavior of steel structures

Figure 6.2 Pipes damaged during different earthquakes: (a) 1971 California; (b) 1994 Kobe; (c) 1995 Northridge; (d) 2009 L'Aquila.

damaged by earthquake and after being subjected to subsequent fire, for the sake of completeness, it seems to be useful to provide some information on the behavior of structural materials under fire (Section 6.2), on the related historical events (Section 6.3), and on the main issues of risk management (Section 6.4).

6.2 STRUCTURAL BEHAVIOR UNDER THE EFFECT OF FIRE

Structural materials exposed to fire are subjected to high thermal stresses, which, in many cases, compromise their mechanical resistance (Faggiano et al., 2010). Hereafter, the fire effects on the main structural materials are described in brief (Bianco, 2006).

Timber is a combustible material with low thermal conductivity. When on fire, it is not subjected to deformations and its fire resistance increases as far as the charred layer thickness enlarges, despite the resistant cross-section reduction. For the sake of reference, the dimensions of the timber elements under the effect of fire generally reduce at a rate of about 1 mm/min. So a timber element with adequate section could offer a fire resistance for a time of the order of hours.

Historically, masonry structures have demonstrated excellent fire resistance, provided that the foundations and more generally the supporting structures can keep the wall in place during the fire. In addition, the bricks, which constitute a masonry, present good behavior to high temperature. For example, tile bricks can resist temperatures up to 1000–1100°C and, in general, they suffer the fire after about 1 h if they are protected with a plaster layer 2–3 cm thick. However, thermal bowing of very tall, unreinforced cantilever masonry walls, due to a severe fire on one side of the wall, can lead to collapse.

The concrete has low thermal conductivity, which produces a slow transmission of heat into the mass. Therefore, in a concrete structure exposed to fire, the reinforcement rods are usually protected by the concrete cover. Under the effect of fires, as the temperature increases, the volume of concrete increases initially and afterward it contracts because of evaporation of the water mixture. If the cover molders, the concrete–steel solidarity is compromised and the bars are exposed to fire with strong negative consequences. In particular, for the concrete, the critical temperature, which produces significant resistance reduction of the material, is equal to 300°C.

For the steel structures, the temperature increment causes important thermal dilatations and rapid reduction of its mechanical properties. When the temperature reaches 500°C, the steel ultimate strength reduces to 50% and it is practically zero at 600°C. In particular, a steel structure, exposed to fire without protection, can collapse after about 10–20 min.

In case of fire occurring after an earthquake, the structure has been previously damaged by the seism, and therefore the structural behavior under fire is affected by the irreversible deformation due to the earthquake. In particular, the fire finds the structure in a deformed configuration and in some zones, where permanent deformation is concentrated, the full strength is already exploited. This means that on one side the collapse shapes under fire can be different with respect to the case of a structure struck by fire in its initial undeformed configuration, and on the other side that the fire resistance is varied.

6.3 HISTORICAL EVENTS TO DATE

Catastrophic fires have often followed earthquakes in urban areas (Faggiano and Mazzolani, 2011b). The most significant examples are mentioned below, starting from the historical case of Lisbon in the 18th century, which was the oldest witness of a strong fire after an earthquake.

The Great Lisbon Earthquake (Lisbon, Portugal, 1755) was one of the most destructive and deadly earthquakes in history. A total of 85% of Lisbon's buildings were destroyed. Among the Lisbon population of 275,000, up to 90,000 were killed. Another 10,000 were killed in Morocco.

The quake was followed by a tsunami and fire, resulting in the near-total destruction of Lisbon. A wide documentation testified:

> Soon after the earthquake, several fires broke out, mostly started by cooking fires and candles...Narrow streets full of fallen debris prevented access to the fire sites...the fire reached catastrophic proportions...The flames raged for five days...the downtown area...was partially burned. Several buildings which had suffered little damage due to the earthquake were destroyed by the fire [Figure 6.3; http://nisee.berkeley.edu].

The "Great Fire" of San Francisco (California, USA) followed the earthquake of 1906. The epicenter was located on the San Andreas fault at the boundary of San Francisco and San Mateo counties. The rocks snapped 6–9 miles below the surface of the earth. The magnitude 7.8 earthquake, lasting 40–65 s, was not the strongest ever experienced in California or, for that matter, in the United States. But it was the closest to a major population center. It scorched 508 city blocks, or 4.7 square miles. It is estimated that the loss due to fire after the 1906 San Francisco earthquake was 10 times larger than the one due to the ground motion.

> Firestorms fed by fierce winds raged for three days in San Francisco and caused the majority of the damage...the earthquake and fires combined to cause the nations' greatest urban tragedy... [Figure 6.4; http://bancroft.berkeley.edu; http://contentdm.marinlibrary.org].

Figure 6.3 The Great Lisbon earthquake (Portugal, 1755).

During the *Great Kanto Earthquake* (Japan, 1923), the consequent fire destroyed Tokyo and Yokohama. Deaths were estimated at nearly 100,000, with an additional 40,000 missing. Nearly all buildings were destroyed; the famous undamaged one was the Imperial Hotel, designed by Frank Lloyd Wright.

> Hundreds of thousands were left homeless in the resulting fires...A massive firestorm swept through Tokyo, destroying two thirds of the city's remaining buildings and burning thousand… [Figure 6.5; http://www.eas.slu.edu; http://www.pdc.org].

The shock of the Gediz earthquake (Turkey, 1970) reduced at least 33 towns of Turkey's high Anatolian Plateau to rubble, killed more than 1000 persons, and left thousands of refugees in their own land.

Figure 6.4 The Great Fire of the San Francisco earthquake (California, USA, 1906): (a) market street on fire after the earthquake of 1906 in San Francisco; (b) map of San Francisco showing the burned district.

> ...The terror of the night was heightened by fire, caused by broken powerlines and overturned stoves and oil lamps...Some officials have reported that fire in Gediz killed more people than the quake itself... [Figure 6.6; http://earthquake.usgs.gov].

During the Loma Prieta earthquake (San Francisco, USA, 1989), from 5:04 pm October 17 to midnight October 19, 36 fires involving structures were reported to the San Francisco Fire Department

Figure 6.5 The Great Kanto earthquake (Japan, 1923).

Figure 6.6 The Gediz earthquake (Turkey, 1970).

Figure 6.7 The Loma Prieta earthquake (USA, 1989).

> ...Fire losses from causes due to earthquake are in excess of $10 million... [Figure 6.7; www.slackermama.com; www.sfmuseum.org].

During the Northridge earthquake (California, 1994), fires caused additional damage in the San Fernando Valley and in Malibu and Venice (Figure 6.8; www.americanprogres.org; www.lafire.com).

During the Kobe earthquake (Japan, 1995), the immediate effects of the earthquake, known as the primary effects, included the collapse of buildings,

Figure 6.8 The Northridge earthquake (USA, 1994).

bridges, and roads resulting from the seismic waves shaking the crust. During the 20 s earthquake, the ground moved up to 50 cm horizontally and up to 1 m vertically. Some of the deaths were caused by these primary effects.

The secondary effects included the fires that broke out all over the city of Kobe, the congestion and chaos on the roads, the closure of businesses, and the problem of homelessness. Many more people died in the fires that followed the earthquake. Problems were made worse by the large number of aftershocks (over 1300) (http://www.georesources.co.uk/kobehigh.htm). Numerous fires, gas and water main breaks, and power outages occurred in the epicentral area... (http://earthquake.usgs.gov). More than 100,000 buildings were severely damaged or destroyed by the quake and the fires it caused. A total of 148 separate fires destroyed 6513 buildings and an area of 624,671 square miles.

> ...This earthquake resulted in nearly 143,000 deaths, primarily due to fire (1995 National Fire Protection Association, Quincy, MA) [Figure 6.9; www.cnn.com; www.asiaedu.britannica.com; www.ontario-geo-fish.blogspot.com].

During the Chile earthquake (2010), a fire was reported in a chemical plant on the outskirts of Santiago and caused the evacuation of the neighborhood. Damaged buildings and fires were reported in Concepción (Figure 6.10; http://en.wikipedia.org), where the steel structure of the building of the Chemistry Institute of the University of Chile resisted the

Figure 6.9 The Kobe earthquake (Japan, 1995).

Figure 6.10 The Chilean earthquake (2010).

Figure 6.11 The steel building of the University of Chile damaged by fire after an earthquake. (Photo by Mazzolani.)

earthquake very well, but it was seriously damaged by the subsequent fire (Figure 6.11).

This building represented a clear example demonstrating that the fire acting on a structure, even with small seismic damage, can be responsible for the out-of-service conditions for a well-designed structure.

The magnitude 9.0 Tohoku earthquake on March 11, 2011, which occurred near the northeast coast of Honshu, Japan, resulted from thrust faulting on or near the subduction zone plate boundary between the Pacific and North America plates. Some images of fires following the earthquake are shown in Figure 6.12 (*Bestpicblog.com; Stayinthenews; eurweb.com; spu.esu; asientour.info; trendsupdates.com*). Several fires occurred in Chiba and Miyagi. Tremendous induced damages and casualties were produced.

6.4 POSTEARTHQUAKE FIRE AND RISK MANAGEMENT

6.4.1 General

Despite the awareness about the hazard gained through the past evidence, large fires following earthquakes remain a serious problem. Building characteristics, meteorological conditions, and other factors can contribute

Figure 6.12 The 2011 earthquake in Japan.

to create a situation in which FFE is the predominant agent of damage. Records from historical earthquakes show that often the damage caused by the subsequent fire can be more severe than the damage caused by the ground motion itself, this being true for both single buildings and whole regions (Scawthorn et al., 2005).

In this depiction, the postearthquake fire appears in all its complexity, as it involves many sequential and situational components. Several subjects are directly or indirectly implicated in the related risk management activity, which evidently requires a multidisciplinary approach. It is arguable that, in the first instance, the leading role within the emergency is by fire service, local authorities, other utility organizations, as well as research and hazard informative services. Other interested stakeholders may be the insurance industry, building owners and/or managers, and local communities (Wellington Lifelines Group, 2002).

The general approach followed by the hazard and disaster management community in the past years consisted in operating almost solely on the relief program. So, after the occurrence of a strong earthquake followed by a large fire, in the perspective of a response-based approach to disaster management, specially trained disaster managers coordinate the relief efforts of both the affected community and the wider aid benefactors. In recent years, however, considering the high catastrophic consequences of a fire following an earthquake, relief measures have become increasingly inadequate to protect personal or community assets as well as to safeguard social and economic investments. The need for a new approach is felt, which should be aimed at identifying problems before they happen, by a systematic process of risk analysis and decisions about its acceptability. Such a general decision-making process is commonly called risk management.

Risks sources are, on the one hand, the damage, in particular, to pipelines, electric wiring, active and passive fire protection systems, and building structures; on the other hand, they are the operating difficulties for firemen, like the increase in the time needed to firemen to reach the place of the fire (due to traffic congestion, collapsed constructions, rubble in the streets, multiple fires), the possible difficulties in water supply, and the decrease in the collapse time of the structure (Faggiano et al., 2008).

6.4.2 Methodology

The risk management of a postearthquake fire requires an approach at two different scales (Chen et al., 2004): a local scale, referring to single buildings (building scale), and a global scale, referring to a whole region (regional scale).

6.4.2.1 Building scale

During the lifetime of a building, a variety of hazards may occur simultaneously or consecutively, including earthquake, wind, fire, blast, and other natural or man-made hazards. For buildings in the seismic zone, both fire and earthquake are critical design issues, although they are commonly considered as independent hazards. Actually, the case of fire after earthquake is necessary to be dealt with, because fire is more likely to be ignited after an earthquake, causing a very severe damage. At today's level of knowledge, the most correct design philosophy for integrating fire safety into the design process for structures appears to be the performance-based design (Johann et al., 2006; Bennetts and Thomas, 2002). Such a design approach has already been adopted by International Codes (the United States, Australia, the United Kingdom, New Zealand, Sweden, Eurocode system) in the field of structural fire safety, by changing the standard of care from meeting the code prescriptive requirements (height and area limits, fire resistance ratings, egress, separations, etc.) to demonstrating safe performance through design and analysis. As a general rule, the latter could be achieved by a multidisciplinary approach including fire science, structural engineering, and fire safety design.

6.4.2.2 Regional scale

Time is a key factor for the management of the earthquake-induced fire emergency at the regional scale. The occurrence of an earthquake presumably causes damage to both buildings and contents. The sources of ignitions due to earthquakes can range from overturned heat sources, such as candles or lamps, to abraded and shorted electrical wiring, to spilled chemicals having exothermic reactions, to friction of items rubbing together. At some point, if the fire is not self-extinguished, in the confusion following an earthquake, the discovery may take a very long time. If it is not possible for onlookers to immediately extinguish the fire, the intervention of the fire department is required to face the help request, then to extinguish the fire. If the firemen are successful, they move on to the next fire source; if not, they continue to attempt to control the fire, until the fuel is exhausted or the fire comes to a firebreak. Definitely the steps of the whole process are as follows: earthquake, ignition, discovery, request, response, and suppression. From a long-term perspective, the emergency management spans from routine periods for facilitating sustainable hazard management practices, to emergency periods for coordinating response and recovery requirements (Britton and Clark, 1999).

At the regional level, geographic information systems (GIS)-based approaches for earthquake hazard mitigation may be used. Such tools provide a decision support for optimizing the assignment and routing of the emergency vehicles after an earthquake, considering the geographic distribution of ignited fires and injuries, locations of emergency response facilities (including emergency operation centers, healthcare facilities, fire stations, police station, etc.), earthquake damage to the facilities, and the transportation system.

6.4.3 Building-scale issues related to postearthquake fire

Structural fire engineering, based on the theoretical knowledge, empirical information, analytical capability, and technology developed so far, should be integrated with earthquake engineering, in order to optimize structural design under combined seismic and fire actions.

Several categories of structural modifications related to the behavior under fire can occur due to an earthquake. They are either damage to the possible passive protection of structural members (like insulation, coatings, barriers) or changes in the structural configuration due to permanent damage to structural members. They should all be considered in the fire design process, in order to analyze the structure "as it is," when the fire occurs after the earthquake.

The evaluation of the effects of earthquake-induced damage on fire resistance and collapse modes is evidently a key issue. In fact, the larger the structural behavior is degraded after a seism, the shorter is the time up to collapse due to fire. Moreover, the collapse modes under fire can be different from those in the preearthquake conditions. From this perspective, a very important role is played by the modeling of earthquake-induced damage, of material behavior at elevated temperature, as well as by the modeling of fire. The analysis should reproduce the actual phases of the phenomena, from the application of vertical service loads and earthquake-induced damage up to the exposure of the structure to fire. First of all, the seismic damage state should be identified, according to pre-fixed performance levels, by means of nonlinear pushover analyses or by nonlinear time–history incremental dynamic analyses. Therefore, structures already damaged by earthquake, starting from each previously defined seismic performance level, should be analyzed under fire. Correlation between the seismic performance levels and the behavior of corresponding damaged structures under fire in terms of fire resistance and collapse mode should be determined (Della Corte et al., 2003; Faggiano et al., 2010).

The analytical tools for the evaluation of the structural response should be able to model and catch the nonlinear behavior, related either to the possible seismic inelastic damage and large displacements or to the high temperature. In the latter case, numerical programs can allow to perform only the analysis under fire conditions, resulting in the evaluation of the fields of temperature within the structural members. This requires the use of other computational codes for structural analyses, as well as for the evaluation of both the stress and strain states, taking into account the temperature variation. Besides, some programs are available that are able to carry out the fully coupled temperature and displacement analyses in a unique structural model (Faggiano et al., 2008).

The acquired knowledge on the investigated categories of structural types can lead to the definition of integrated seismic and fire design criteria. Evidently, the final goal is the development of a quantitative proposal for both fire safety and seismic design codes, aiming at guaranteeing the fire safety of buildings exposed to postearthquake fire risk in earthquake-prone countries, by fitting fire resistance according to pre-fixed performance levels.

For this purpose, it is worth considering that the exposure of a building to extreme wind, earthquake, or gravity loads represents a threat to the building structure; hence, the safety of occupants is strictly related to the safety of the structure itself. In contrast, in case of fire, the occupants may be directly threatened by smoke and flames before being hit by the indirect effect of the behavior of the building structure, due to the fire-induced weakening of the structure; in fact, the structure starts undergoing damage at temperatures far beyond human resistance. This point involves the consideration of more stringent design objectives for any structures that can be endangered by the combined hazard of earthquake and subsequent fire, aiming at guaranteeing the safety in terms of both the structural behavior in fire and the direct effects of fires on people inside the building. Therefore, the design objectives for FFE scenarios may include (Bennetts and Thomas, 2002) (1) life safety of occupants; (2) noninjury of occupants; (3) life safety of firefighters; (4) noninjury of firefighters; (5) prevention of damage to contents; (6) avoidance of damage to process; (7) prevention of damage to building; and (8) prevention of collapse of building.

6.4.4 Regional-scale issues related to postearthquake fire

Much of the early significant studies (Scawthorn, 1986, 1987; Scawthorn et al., 1991) developed models for postearthquake fire hazard in urban regions that are applicable to specific earthquakes and for determining annual expected losses on a probabilistic basis. Factors included in the models are building density, wind velocity, deterioration of firefighting response, and seismic intensity. A comprehensive overview on FFE in urban regions, covering the history of past fires, models for fire spread in urban environment, and cost-effectiveness of various mitigation strategies, is given in Scawthorn et al. (2005).

An efficient tool for a hazard assessment modeling is the HAZUS code (FEMA, 1999), which allows estimation of the damage to buildings and facilities in a GIS environment, through input data concerning both the earthquake vulnerability and, in case of fire, the number of ignitions that have the potential to strike one or more buildings, the burned area (which depends on both the fire spread rates and suppression efforts), as well as the population and building exposures affected by the fires.

The life safety risk for postearthquake fires from the perspective of gas and electricity distribution systems, which are the main risk sources, was a subject of investigation by the Pacific Earthquake Engineering Research Centre (Chen et al., 2004). In relation to the lifelines services, performance-based building codes should contain appropriate provisions to prevent undue reliance of sprinkler and other life safety systems on seismically vulnerable water and electrical services.

New Zealand is one of the countries most concerned about the risk of postearthquake fires (Cousins et al., 1991, 2002; Botting, 1998). Recommendations on fire brigade response, urban water supplies, and urban macroscale fire protection are provided on the basis of significant research carried out through the analysis of major earthquakes. In particular, Australia and New Zealand were in 1995 among the first countries in the world to formally develop and adopt a general standard on risk management. The Australian and New Zealand Risk Management Standard (1999) provides a formalized, systematic decision-making process for identifying solutions concerning the vulnerability to natural hazards. The risk management process is organized in the following steps: (1) establishment of the strategy; (2) operational and risk management context; (3) risk identification; (4) risk analysis; (5) risk evaluation; (6) risk treatment; (7) monitor and review; and (8) communication and consulting.

6.5 COMPUTATIONAL ASPECTS

6.5.1 General

In seismic zones, fire and earthquake are commonly considered as independent hazards. However, in case of a fire occurring after an earthquake, if the structure has been damaged by the seism, the fire finds the structure in a deformed configuration and, in some zones, where permanent deformation is concentrated, the full strength is already exploited and possible passive fire protection of structural members like insulation, coatings, and barriers can be broken. This means that on the one side the fire resistance is worsened; on the other side, the collapse shape under fire can be different with respect to the case of a structure failing by fire in its initial nondeformed configuration.

The evaluation of the effects of earthquake-induced damage on fire resistance and collapse modes is evidently a key issue. The analytical tools for the evaluation of the structural response should be able to both model and catch the nonlinear behavior related to possible seismic inelastic damage and large displacements and to high temperature exposure, at the level of either individual structural members and connections, or assemblies of members, or entire structural frames, or interactions between components.

In this context, a very important role is played by the modeling of earthquake-induced damage, of material behavior at elevated temperature, as well as the modeling of fire (Faggiano and Mazzolani, 2011a; CEN, EN 1993-1-2, 2005). Common numerical programs allow the heat transfer analysis to be performed as preliminary step, in order to evaluate the temperature—time law within the structural elements; subsequently, the structural analysis under design loads is carried out by imposing on the member the temperature variation obtained in the first step: the heat transfer analysis and the structural analysis are performed separately (uncoupled analyses). Besides, some programs are available that are able to carry out the fully coupled temperature and displacement analyses in a unique structural model. In this case, the mechanical and thermal aspects of the problem can be treated simultaneously and the mutual interactions caught (Faggiano et al., 2007a).

6.5.2 Structural analyses

A key aspect of the fire after analysis is the interpretation of the earthquake-induced damage. It can be done by means of a simple model that schematizes the structural damage as a combination of two damage types: a *geometrical damage*, which consists of the residual deformation of the structure, and a *mechanical damage*, which consists of the reduction of the main mechanical properties of the structural components (stiffness and strength degradation). This schematization allows for a rational evaluation of the mechanical state of the structure after the earthquake and of its mechanical behavior under external actions succeeding the earthquake. In addition, this modeling is a very useful approach for parametrical analyses.

Some examples are given in the following Sessions 6.5 and 6.6, where the results of a parametrical analysis for the evaluation of the fire resistance rating reduction are given for both single-story and multistory moment-resisting frames extracted from civil ordinary buildings.

The analysis of the portal frames should be considered as a pilot study. It allows the parameters potentially affecting the problem to be focused on and the range of residual interstory drift angles, for which the problem is significant, to be determined. The main outcome of the study is that the effect of the geometrical type of damage is quite insensitive to the level of vertical loads, when the latter corresponds to normal values.

Interesting conclusions can also be drawn from the investigation on the multistory frames. As expected, the seismic design strategy is shown to significantly affect the frame postearthquake fire performance. In fact, for the frame designed considering only the ultimate limit state (ULS) requirement, the fire resistance reduction has been found to be important, so that a high fire risk derives from earthquake-induced structural damage. In contrast, in case of structures designed considering the most severe serviceability seismic design requirement (SLS), the fire resistance reduction is relatively smaller. For such a frame, at the design seismic intensity level, the fire resistance reduction is usually less than 10% of the initial value. However, the fire resistance rating reduction usually becomes nonnegligible for very rare earthquakes, that is, earthquakes having a mean return period larger than 475 years.

In view of the development of a comprehensive methodology of performance-based design of buildings, the fire resistance performance should be taken into account by also considering the earthquake-induced damage for those buildings located in seismic areas. This consideration leads to the conclusion that the fire safety codes should distinguish between structures located in seismic and nonseismic areas, by requiring more stringent fire resistance provisions for those buildings potentially subjected to seismic actions. Clearly, a quantitative proposal for such a distinction requires the development of more comprehensive numerical simulations than that for the existing ones. Moreover, some conceptual and numerical limitations of the simplified models must be overcome and need more refined modeling for a more significant number of actual cases. In particular, a more accurate representation of seismic damage within the structure can be obtained by using the finite-element program ABAQUS (2004a,b).

6.6 ANALYSIS ASSUMPTIONS

6.6.1 Current codification approach

As far as the earthquake is concerned, the performance-based design has been generally acquired by modern technical codes such as the seismic design methodology. It allows the design structures so that, in case of the design earthquake, it behaves according to predetermined performance levels, related to both structural and nonstructural components, which are commonly the ones defined by the FEMA 356 Guidelines (2000); they are

- *Operational* (O). Very light overall damage: the postearthquake damage state in which the structural and nonstructural components are able to support the preearthquake functions present in the building.
- *Immediate Occupancy* (IO). Light overall damage: the postearthquake damage state guaranteeing that the structure remains safe for occupation and essentially retains the preearthquake design strength and stiffness.
- *Life Safety* (LS). Moderate overall damage: the postearthquake damage state related to structural components, guaranteeing that the structure retains a safety margin against the onset of partial or total collapse.
- *Collapse Prevention* (CP). Severe overall damage: the postearthquake damage state related to structural components, guaranteeing that the structure continues to support gravity loads but retains no safety margin against collapse.

Today, the most modern approach related to the design of the structures under effect of fire is *Fire Safety Engineering* (FSE), given for the first time in the ISO/TR 13387 document (1999). It is a procedure of performance type, based on simulations able to define different functions of state for the fire phenomenon. They can also be very complex and require the evaluation of several input data, like geometry of the calculation domain, conditions of ventilation, type and amount of fuel, and time–temperature curves. The fire simulations, associated to finite-element structural analyses, are used to investigate and reduce the loss of life and damage to property, quantifying the risks and hazards involved and providing an optimal solution to the application of preventive or protective measures (Purkiss, 2007).

By contrast, in the traditional approach, which is of deterministic type, the structural design is conducted through conventional evaluation adopted for all the building classes, without particular distinction concerning the real contents. In fact, it is based on prescribed requirements, consisting in rigid rules, with reference to the fire action, which is modeled through the temperature–time nominal curves of Figure 6.13, defined in Eurocode 1, Part 2-2 (CEN, 2005).

The two illustrated approaches are very different. The former, even if more complex, has good flexibility, while the second, which is more simple and characterized by homogeneity of results, is excessively rigid, neglecting the specific requirements concerning, for example, the strategic or historic–artistic importance of the building.

6.6.2 Structural and damage modeling

Modeling the behavior of buildings subject to fires following earthquakes is a challenging but very difficult task for a structural engineer (Della Corte and Mazzolani, 2002). In fact, it requires not only knowledge about the

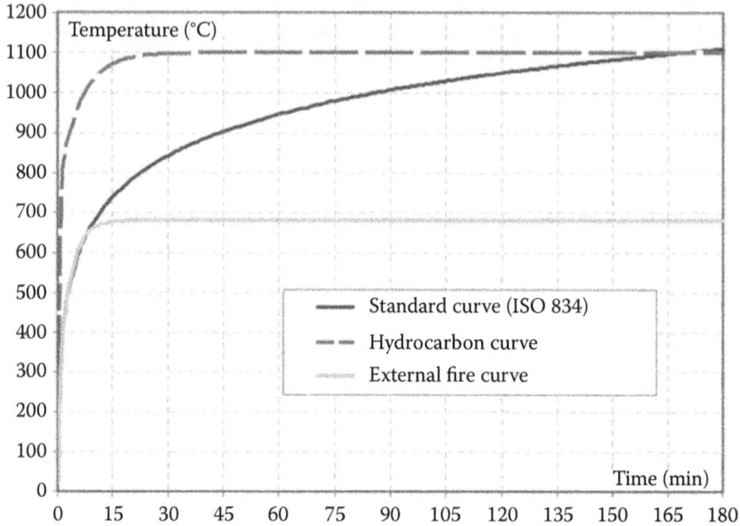

Figure 6.13 Temperature–time nominal curves.

mechanical response of the structure to the external action, but also good dominance of some interdisciplinary fields, like the seismic and the fire ones. Grossly, the following general modeling aspects could be identified: (a) modeling of the seismic action; (b) modeling of the structural response during the earthquake; (c) modeling of the fire action; (d) modeling of the thermomechanical behavior of the structure subject to fire (Della Corte et al., 2003).

The problem cannot be solved simply by facing each aspect separately, because of possible interactions of the different modeling issues. For example, the fire response of the building must be modeled taking into account the effects of the previous seismic action, which could have produced some plastic deformations in the structure, thus modifying the "initial conditions" for the subsequent fire action.

Actually, there is a strict interrelation between the geometrical and mechanical damage: in fact, as far as the strength degradation is strong, the geometrical damage is expected to be large. However, this interrelation is complex and then hard to be predicted *a priori*, because, for a given structure, it largely depends on the type of earthquake ground motion (Fajfar, 2002). Furthermore another source of complexity is due to the difficulty in predicting both strength and stiffness degradation for plastic hinges in MR steel frames: existing models are empirical ones, whose applicability out of the range of model calibration is often unreliable. However, in case of well-engineered steel structures, the assumption of nondegrading structural components is realistic in the range of plastic deformation induced by earthquakes at the design performance level (FEMA, 2001; Krawinkler, 2000; Hamburger et al. 2000). In particular, for MR steel frames designed according to Eurocode 8, strength degradation becomes significant only for very large values of the peak ground acceleration. Therefore, in real design cases, the advantage of the proposed subdivision between geometrical and mechanical damages is apparent.

A recent attempt to face the problem in a rational, but simple, way is given in Della Corte et al. (2005), where the following idealization of the earthquake-induced structural damage is proposed (Faggiano et al., 2005):

1. *Geometrical damage*, which is represented by the residual deformations of the structure after the earthquake (Della Corte and Landolfo, 2001).
2. *Mechanical damage*, which is represented by the strength and stiffness reduction produced in those parts of the structure engaged in the plastic range of deformation during the earthquake.

Figure 6.14 efficiently synthesizes the simplified scheme, evidencing that the structure after the earthquake could be subjected to significant residual P-Delta effects, which, together with the reduced lateral strength of the frame, could induce an important reduction of its fire resistance.

The level of geometrical damage is strictly dependent on the level of the mechanical damage: it seems reasonable that the stronger the strength degradation is, the larger the geometrical damage is expected to be. However, this interrelation is complex and hard to predict *a priori*: in case of both strength and stiffness degradation, the interrelation may be largely dependent on the type of earthquake ground motion for a given structure. At present, the prediction of both strength and stiffness degradation for plastic hinges in MR steel frames is not yet a simple task. The existing models are empirical ones, whose applicability out of the range of model calibration is often unreliable.

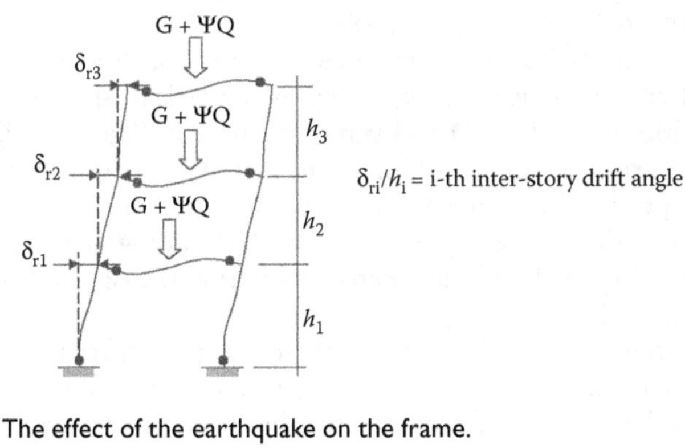

Figure 6.14 The effect of the earthquake on the frame.

The advantage of the proposed subdivision between geometrical and mechanical damages lies in the fact that the effects of strength and stiffness degradation can be considered to be negligible in some cases. For example, in case of well-engineered steel structures, the assumption of nondegrading structural components is realistic in the range of plastic deformation induced by earthquakes at the design performance level (i.e., in case of earthquakes having a 10% probability of being exceeded in 50 years, according to modern design codes). However, for structures not adequately designed against earthquakes and/or in case of rare earthquakes, that is, earthquakes having a 2% probability of being exceeded in 50 years, according to current research trends (FEMA, 2001), the effect of strength and stiffness degradation should also be taken into account (Krawinkler, 2000). Obviously, if the structure is sufficiently strong, plastic deformation demand could also be relatively small for the maximum credible earthquake the structure would be subjected to. As will be shown later, for MR steel frames designed according to Eurocode 8, strength degradation often becomes significant only for very large values of the elastic spectral acceleration, which permits to be neglected it in a wide range of practically useful seismic intensities.

The geometrical type of damage has been measured using the maximum along the building height of the residual interstory drift angles, defined as shown in Figure 6.14.

The software platform OpenSees (Mazzoni et al., 2003) has been used for modeling the steel frames under earthquake ground motions. For this purpose, the fiber beam–column element has been used, allowing for considering P-Delta effects. According to previous discussion, the effect of mechanical damage has been neglected, assuming the elastic–perfectly plastic hysteresis model.

6.6.3 Fire modeling

The more significant effect of a fire in the structural field is the increase of temperature, hence the usual representation of a fire event is the time–temperature curve (*t–T*) (Anderberg, 1988).

Different methods for fire event modeling exist, such as the method of the nominal curves, which are expressed by analytical, simple and identical relationships (e.g., the ISO 834 standard curve in Figure 6.27, which is provided by Eurocode 1, EN 1997-2-2); the method of parametrical curves, which are expressed by more complex relationships, taking account of the principle factors that influence a fire, such as the building division into compartments, the fire load and the rooms' ventilation (Feasey and Buchanan, 2002; Buchanan, 2001).

The fire scenario for the earthquake-damaged structures has been assumed to be the most dangerous, which results by applying the fire action at the first level of frames. The temperature has been assumed to be uniform in the compartment and varies according to the ISO 834 time–temperature curve, which is also provided by Eurocode 1 (CEN, EN 1991-2-2). The thermal and mechanical properties of steel have been modeled according to Eurocode 3 (CEN, EN 1993-1-2). The numerical code SAFIR (Franssen et al., 2002) has been used for computing the fire resistance ratings of structures. Within the assumptions of this finite-elements program, the temperature field is preliminarily determined on the cross-section of each beam–column element, neglecting heat transfer along the element length. Then, the mechanical analysis phase is started considering (large) displacements, nonlinearly increasing with temperature (i.e., time) under constant external gravity-related loads. The structure collapse time has been defined as the time when an instability phenomenon occurs. In this computation of the fire resistance rating, the effect of creep deformations of steel has been considered indirectly only through the use of the conventional stress–strain–temperature relationships suggested by EN 1993-1-2. The thermal elongation of fibers has been taken into account. In contrast, the effects of both residual stresses and initial geometrical imperfections have been neglected.

6.7 STRUCTURAL BEHAVIOR

6.7.1 Single-story moment-resisting frame

The influence of the geometrical type of seismic damage on the fire resistance ratings of simple portal frames is investigated by varying a number of parameters, which are chosen in such a way as to fully characterize the mechanical behavior of the structural system. They are

- L/H, which is the ratio between the span of the beam and the height of the story.
- I_b/I_c, which is the ratio between the beam and column moment of inertia.
- M_{pb}/M_{pc}, which is the ratio between the beam and column section flexural plastic strength.
- N/N_{cr}, which is the ratio between the vertical load and its elastic critical value.
- δ/H, which is the level of geometrical damage, measured by the story drift angle.

The structural scheme of the examined frames and the numerical range of the considered variable parameters are given in Figure 6.15. The results of the analyses are illustrated in Figure 6.16, by means of the nondimensional curves time (fire resistance reduction) versus damage (permanent geometrical damage), where the symbols have the following meaning:

Structural scheme	L/H	I_b/I_c	M_{pb}/M_{pc}	N/N_{cr}	δ/H
(a) 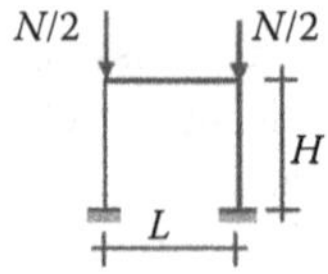	1	0.63	0.54	0.05–0.30	0.0–0.385
		4.50	2.22	0.05–0.30	0.0–0.355
(b) 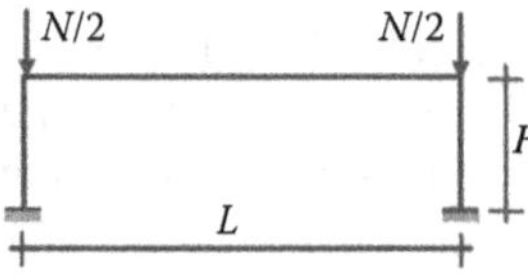	3	0.53	0.47	0.05–0.30	0.0–0.365
		4.72	2.27	0.025–0.30	0.0–0.620

Figure 6.15 Geometrical and mechanical data of portal frames for parametrical analyses: (a) $L/H = 1$; (b) $L/H = 3$.

- $t_f/t_{f,o}$ is the ratio between the fire resistance rating of the structure after the earthquake t_f (i.e., the damaged structure for a given level of seismic intensity) and the original fire resistance rating of the structure before the earthquake $t_{f,o}$ (i.e., the undamaged structure).
- $(\delta/H)_{max}$ is the maximum value of (δ/H) at room temperature, that is, the residual drift angle for which the structure is unstable under the given amount of vertical loads, at room temperature.

Figure 6.16a through d shows the $t_f/t_{f,o}$ ratio plotted versus the $(\delta/H)/(\delta/H)_{max}$ ratio. As can be expected, by increasing the level of geometrical damage, that is, the ratio $(\delta/H)/(\delta/H)_{max}$, the reduction of fire resistance rating increases, that is, the ratio $t_f/t_{f,o}$ decreases. The curves highlight that the residual story-drift angle is very well correlated to the fire resistance rating reduction. Once the fire resistance of the initial structure ($t_{f,o}$) and the level of residual drift angle are known, such type of curves in Figure 6.16 could allow a ready estimation of the amount of the postearthquake fire resistance.

It is worth noting one main difference of behavior between the frame in Figure 6.16a and the frame in Figure 6.16d. The collapse vertical load of the frame in Figure 6.16a, at room temperature and without seismic damage, is very close to the Euler lateral-buckling critical load ($(N_u/N_{cr})_o = 0.99$). In

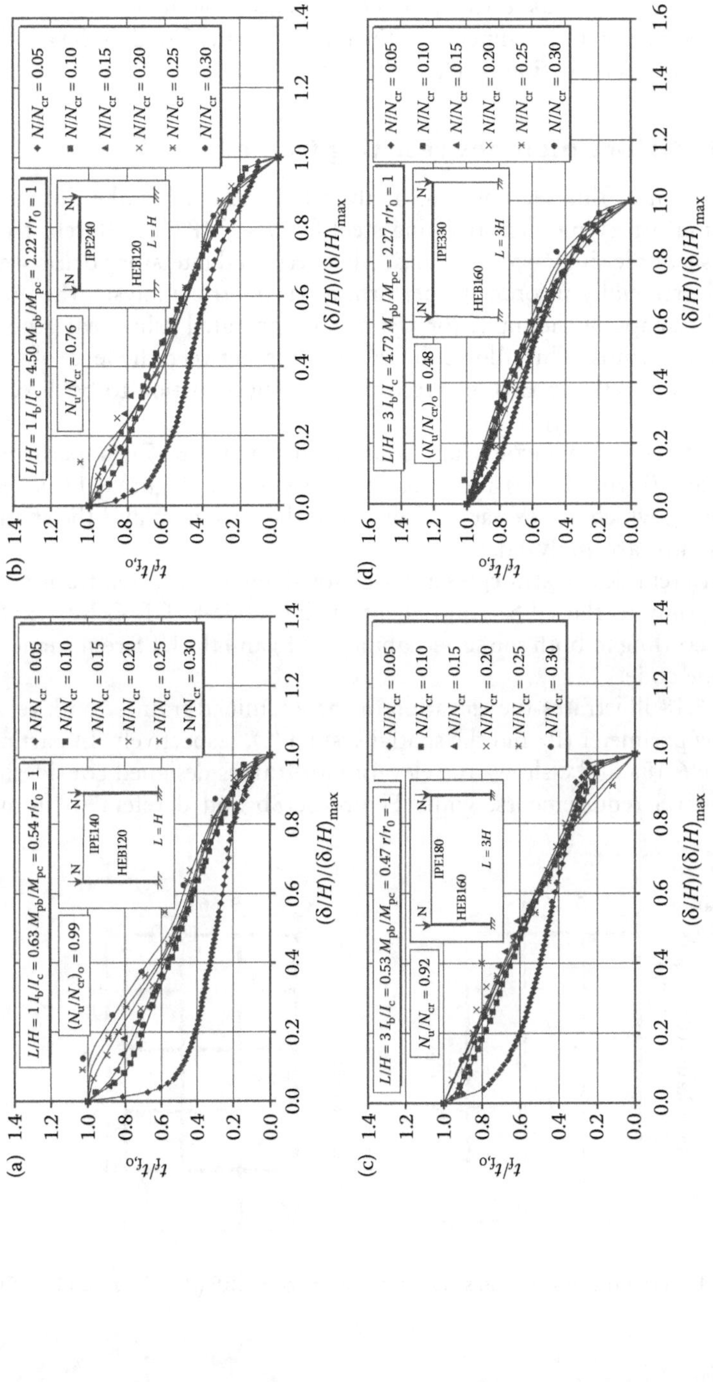

Figure 6.16 Fire resistance ratings reduction for steel portal frames subject to geometrical damage. : (a) $L = H$, IPE140; (b) $L = H$, IPE240; (c) $L = 3H$, IPE180; (d) $L = 3H$, IPE330.

contrast, the collapse vertical load of the frame shown in Figure 6.16d is appreciably lower than the lateral-buckling critical load ($(N_u/N_{cr})_o = 0.48$). In fact, the frame in Figure 6.16a is significantly more slender than the frame in Figure 6.16d, as testified by the beam and column sizes. It is possible to conclude that the influence of the level of vertical loading on the fire resistance ratings reduction is relatively small.

6.7.2 Multistory moment-resisting frame

The main aim of this section is to analyze the postearthquake fire resistance of multistory moment-resisting steel frames (MRF). Different from the analysis of the single-story frames, where residual interstory drifts were generated artificially in order to perform a parametric analysis, the post-earthquake shape of the multistory frames is generated using real ground acceleration records. This allows the relationship between the level of seismic intensity and the level of fire resistance rating reduction to be directly assessed.

Two different plan layouts are considered (Figure 6.17): a perimeter frame system (Figure 6.17a) and a spatial frame system (Figure 6.17b). Ten-story frames, which are extracted from both the perimeter and the spatial frame systems, are analyzed.

Two different design strategies are used for sizing the selected frames: (1) according only to the ULS requirement of Eurocode 8 (CEN, EN 1998); and (2) according to both the serviceability (SLS) and the ULS requirements of the same code.

Figure 6.18 illustrates the elevation of the examined frames, with reference to the perimeter (P) and the spatial system (S), respectively. In particular, Figure 6.18a and c shows the elevation of frames designed considering only the ULS requirements, while Figure 6.18b and d refers to frames

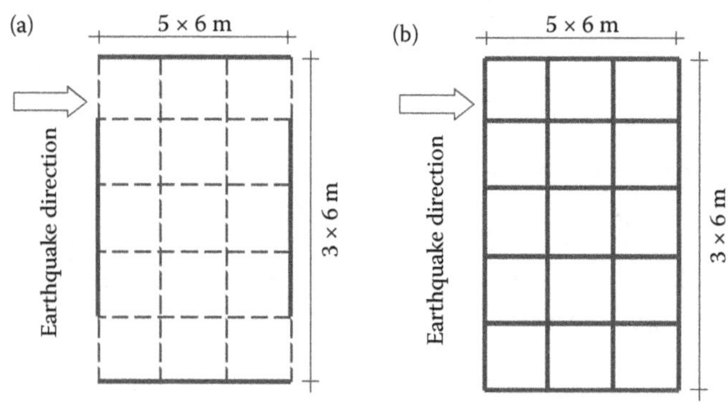

Figure 6.17 Plan layouts of the case studies: (a) perimetral MRF (P); (b) space MRF (S).

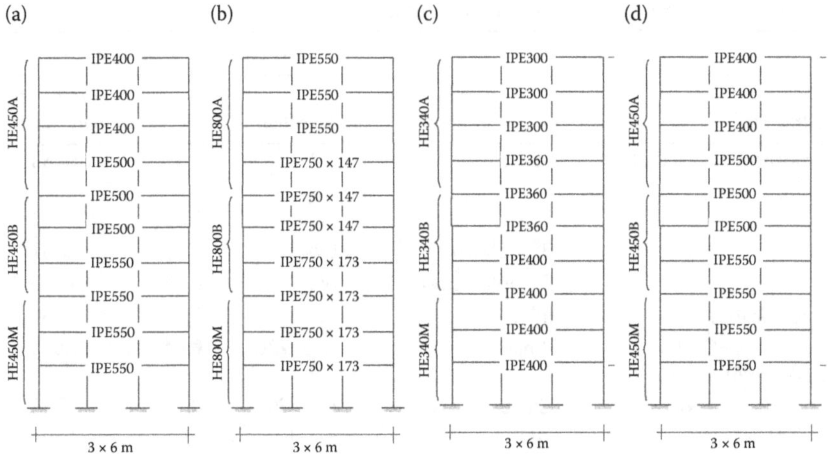

Figure 6.18 Analyzed frames. (a) P-ULS frame; (b) P-SLS frame; (c) S-ULS frame; (d) S-SLS frame.

designed considering also the serviceability requirements. As can be seen in the above figures, the following acronyms are adopted: P-ULS and P-SLS indicate either the perimeter (P) frames designed considering only the ULS or adding also the satisfaction of the SLS, respectively; S-ULS and S-SLS indicate the frames extracted from the spatial (S) system with meanings analogous to the previous ones for the remaining part of the acronyms.

Seismic analysis based on selected earthquake ground motions is scaled up to increasing values of the peak ground acceleration (PGA) in order to increase damage to the frame by increasing the first-mode 5%-damped elastic spectral acceleration ($S_{a,e}$).

Spectral accelerations are normalized by means of their design values ($S_{a,e}/S_{a,e,d}$) for the selected ground motions, where $S_{a,e,d}$ is the design value of $S_{a,e}$, corresponding to a 475-year return period and computed according to Eurocode 8. The postearthquake fire resistance ratings are normalized by means of the preearthquake values ($t_f/t_{f,0}$, see Section 6.6.1).

A summary of the main characteristics of the earthquake ground motions used in the analyses is given in Table 6.1. The selected ground motions of historical records are scaled up to increasing values of the PGA. The damage to the frame is obtained by increasing the first-mode 5%-damped elastic spectral acceleration ($S_{a,e}$), by following a well-established methodology (FEMA, 2001). After reaching the residual values of interstory drift angles, the structure is subjected to the standard ISO 834 fire (see Section 6.5.3) for computing the fire resistance rating.

Figure 6.19a through d illustrates the normalized fire resistance rating reduction obtained for the four examined cases of MR steel frames, as

Table 6.1 Selected Earthquake Ground Motions

Earthquake	*Station*	*Ground acceleration record*		
		Direction	*PGA (g)*	*Trifunac duration (s)*
Campano-Lucano (Italy, 1980)	Sturno	EW	0.32	38.53
	Sturno	NS	0.22	40.02
	Bagnoli-Irpino	EW	0.18	31.81
	Bagnoli-Irpino	NS	0.14	41.08
Kobe (Japan, 1995)	JMA	NS	0.83	8.360
Northridge (California, USA, 1994)	Rinaldi	rina_228	0.84	7.045
Michoacan (Mexico, 1985)	Caleta de Campos	cale_90	0.14	57.45
Chile (1985)	Llolleo	llol_10	0.71	35.91

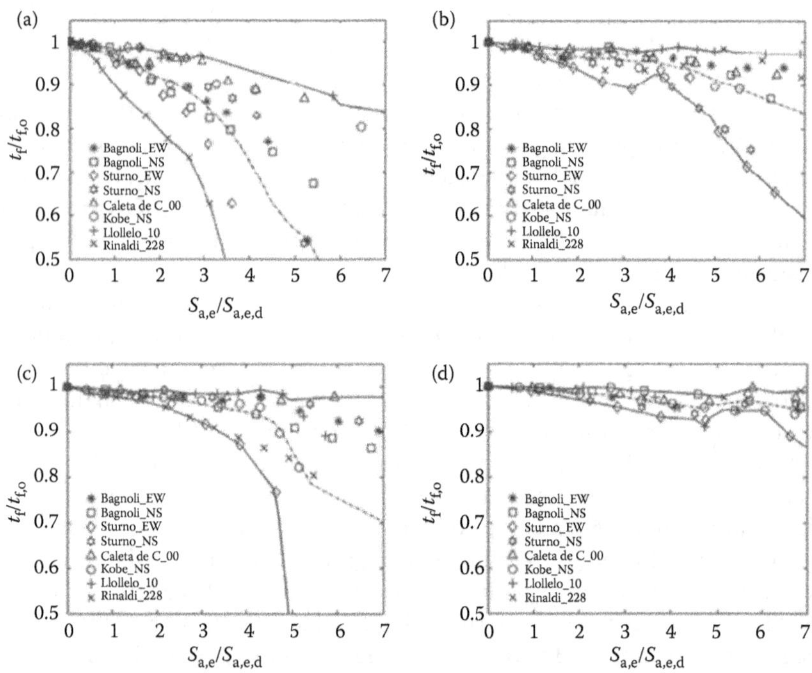

Figure 6.19 Fire resistance rating reductions for steel MR frames subjected to earthquakes. (a) P-ULS frame; (b) P-SLS frame; (c) S-ULS frame; (d) S-SLS frame.

a function of the normalized spectral acceleration, for the acceleration records given in Table 6.1.

As expected, the fire resistance reduction is larger for the perimeter systems than for the spatial ones. Moreover, the effect of the design criterion appears to be quite important, with a stronger reduction for the ULS frames than for the SLS ones. For example, at $S_{a,e}/S_{a,e,d} = 1$, the minimum values of the ratio $t_f/t_{f,0}$, deduced from Figure 6.19a through d, are 0.90, 0.97, 0.98, and 0.99 for the P-ULS, P-SLS, S-ULS, and S-SLS frames, respectively. At $S_{a,e}/S_{a,e,d} = 2$, the analogous minimum values are 0.80, 0.94, 0.96, and 0.97. Assuming that a 10% reduction of fire resistance rating is negligible, it can be concluded that in all the examined cases the effect of earthquake-induced structural damage on fire resistance ratings can be neglected under the design earthquake intensity ($S_{a,e}/S_{a,e,d} = 1$). For rare earthquakes ($S_{a,e}/S_{a,e,d} > 1$), the effect of earthquake-induced damages could be significant, depending on the design criterion and the structural system layout. In fact, the reduction still appears to be relatively small in case of spatial frame systems, but it becomes significant in case of perimeter frames designed neglecting the serviceability requirement.

As previously discussed, both stiffness and strength degradation are neglected in the structural modeling under seismic actions. Consequently, the obtained results are valid only in a limited deformation range. It is assumed that the target value of the maximum value of the transient interstory drift angle, below which degradation can be neglected, is equal to 0.03 rad. Then, the normalized spectral acceleration levels inducing a 0.03 rad maximum transient interstory drift angle are computed. Besides, the fire resistance rating reductions obtained for these limiting spectral acceleration values are computed for each of the considered acceleration records and plotted in Figure 6.20a through d, where (a), (b), (c), and (d) mean (P-ULS), (P-SLS), (S-ULS), and (S-SLS), respectively.

Under the same previous assumption of neglecting a fire resistance reduction of 10% or smaller, it can be concluded that the average fire resistance reduction is significant only for the P-ULS frame.

However, if one considers minimum values, the reduction becomes significant for the P-SLS frame, while it still remains negligible for the spatial frame systems. For values of spectral accelerations larger than those corresponding to a 0.03 rad, the maximum transient interstory drift angle, the obtained numerical results should be considered nonconservative. In fact, degradation is expected to increase the average values of the residual interstory drift angles and, then, to increase the fire resistance rating reduction.

It is interesting to notice that a significant scatter affects the fire resistance rating reduction when it is plotted versus the spectral acceleration. This is the consequence of the large differences characterizing the seismic response of a given frame subjected to various acceleration records. In contrast, the fire resistance rate reduction is very well correlated to the

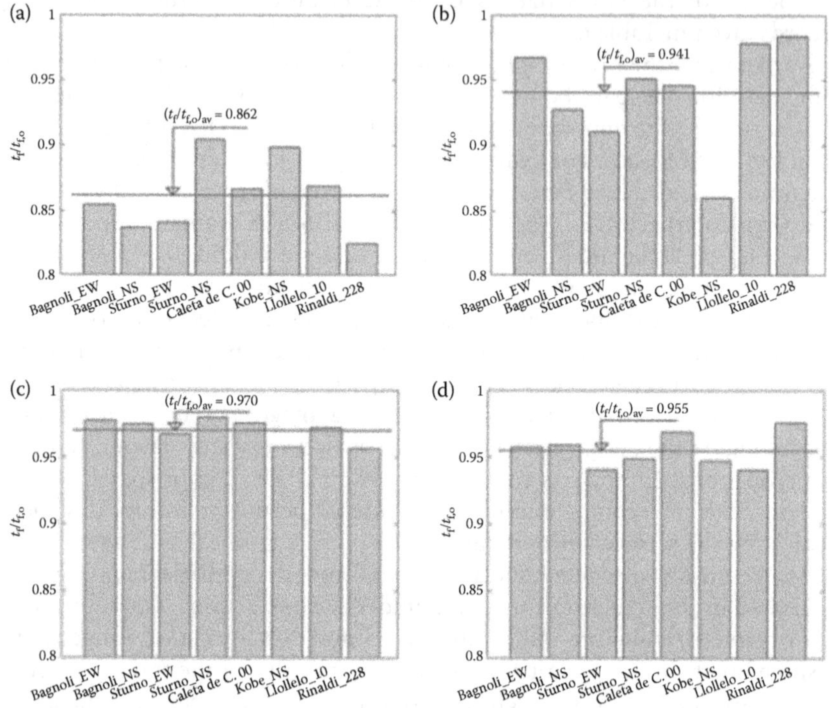

Figure 6.20 Normalized fire resistance rating computed at the spectral acceleration level inducing a 0.03 rad maximum transient interstory drift angle. (a) P-ULS frame; (b) P-SLS frame; (c) S-ULS frame; (d) S-SLS frame.

maximum value of the residual interstory drift angle. This is testified to by Figure 6.21a through d, where the (a), (b), (c), and (d) cases are related to (P-ULS), (P-SLS), (S-ULS), and (S-SLS), respectively.

6.7.3 FEM models

A more refined analysis of the influence of fire after earthquake on steel MFR can be done by means of finite-element models (Faggiano et al., 2007b). Fully coupled temperature–displacement analyses allow treatment of the mechanical and thermal aspects of the problem simultaneously by means of a finite-element model, in order to evaluate the influence of fire resistance on the following parameters:

- The span over height (L/H) ratio
- The massivity ratio (S/V), defined as the ratio between the lateral surface (S) and the volume (V), per unit length, of a structural member

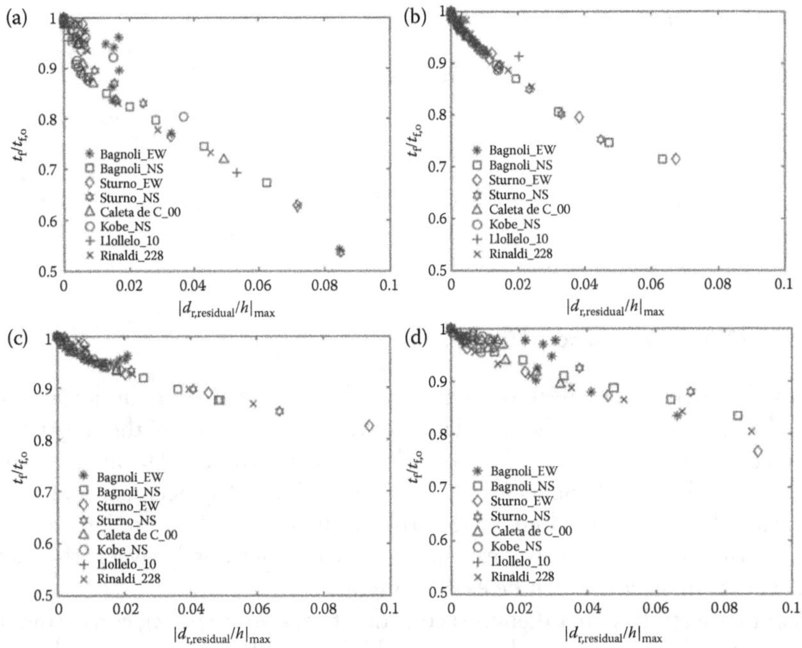

Figure 6.21 Normalized fire resistance rating versus maximum value of the residual interstory drift angle: (a) P-ULS; (b) P-SLS; (c) S-ULS; (d) S-SLS.

- The steel grade
- The exploitation ratio (σ_{Sd}/f_y), defined as the ratio between the maximum and yielding stresses at ULS, corresponding to different load combinations

The analysis is carried out by means of the general-purpose computer program ABAQUS (2004a,b), which allows coupled temperature–displacement numerical analyses to be performed. The mechanical and thermal problems are phased in a unique model, in which the actual phases of the modeled phenomenon, say, the sequential application to the structure of the design loads and, then, of the fire scenario, are reproduced in a step-by-step analysis. Such an approach differs from the usually adopted one, which consists, for the sake of simplicity, in performing the heat transfer analysis and the mechanical one separately (uncoupled analyses): the first one allows the evaluation of the temperature–time law within the structural elements exposed to fire, completely neglecting the stress–displacement aspect; the second one consists in the usual structural analysis, in which the structure is subjected to the external loads; at the end of the structural analysis, the temperature–time variation, obtained

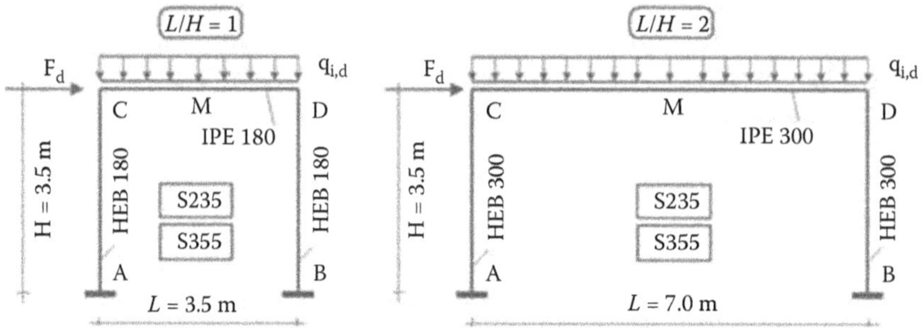

Figure 6.22 The analyzed portal frames.

from the preliminary heat transfer analysis, is imposed on the structural members, thus allowing the calculation of the fire resistance of the structure.

In contrast, in the case of fully coupled temperature–displacement analyses, the used finite elements are endowed with both displacement and temperature degrees of freedom, so that the mechanical and thermal equations are written simultaneously and the mutual interactions between the two aspects of the problem can be easily caught.

Four case studies are taken into consideration (Figure 6.22), consisting in simple steel portal frames characterized by two different span over height (L/H) ratios ($L/H = 1$ and $L/H = 2$) and two different steel grades (S235 and S355). For $L/H = 1$, the beam and column sizes are IPE 180 and HEB 180, respectively, whereas for $L/H = 2$, the beam and column sizes are IPE 300 and HEB 300, respectively. The characteristic value of the dead distributed vertical load is $G_k = 4.5$ kN/m², whereas two typical characteristic values of live vertical loads are considered, that is, $Q_{1k} = 0.5$ kN/m² and $Q_{2k} = 2.0$ kN/m². The unitary characteristic value of the horizontal concentrated force is $F_k = 10.0$ kN. The assumed interaxis is equal to 3.0 m.

Five load combinations are considered for each study case (Table 6.2 and Figure 6.23), giving rise to 20 different values of the exploitation ratio, whose importance to the fire resistance of structures is evident, if it is considered as the inverse of the overstrength ratio. In fact, the latter is a measure of the resistance reserves of the structure, and so the more the

Table 6.2 Load Combinations for the Case Studies

Number	*Load combinations*
1	$\gamma_g\, G_k + \gamma_q\, Q_{1k}$
2	$\gamma_g\, G_k + \gamma_q\, Q_{2k}$
3	$\gamma_g\, G_k + \gamma_q\, (Q_{1k} + F_k)$
4	$\gamma_g\, G_k + \gamma_q\, (Q_{2k} + F_k)$
5	$\gamma_q\, F_k$

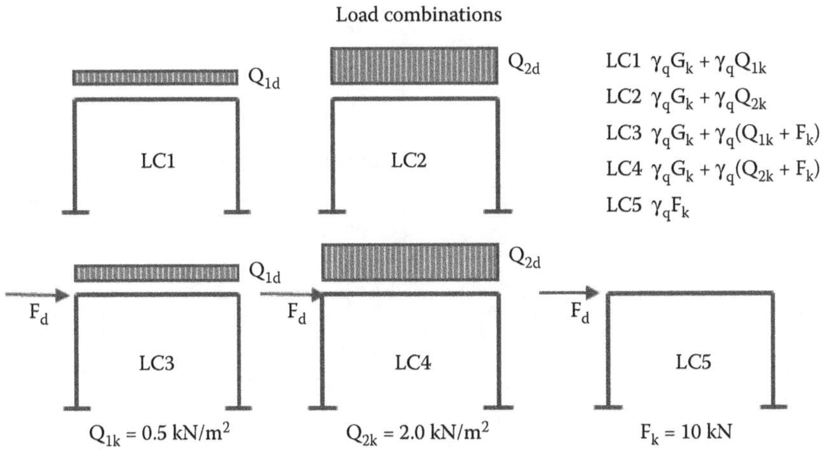

Figure 6.23 Load combinations on the analyzed portal frames.

structure is endowed with such a supply (corresponding to lower exploitation ratios), the more the resistance reduction, and then the fire duration, can be large for reaching the collapse. The partial safety factors at ULS for the load combinations are assumed as equal to $\gamma_g = 1.35$ and $\gamma_q = 1.50$, for the dead and live loads, respectively, in accordance with Eurocode 0.

The materials are modeled considering the dependence of their physical and mechanical properties on the temperature, according to Eurocode 3, part 1-2, related to the structural fire design of steel structures. Thermal elongation, specific heat, and thermal conductivity for steel are shown in Figure 6.24. A simplified assumption is made for the stress–strain law of steel, which is considered as bilinear elastic–plastic. The nondimensional yield stress ($f_{y,T}$) and Young's modulus (E_T), scaled to their values at 20°C, are plotted as a function of the temperature (T) in Figure 6.25.

The finite-element meshes of the portal frames are shown in Figure 6.26. Fixed restraints are imposed at the base of the columns; moreover, internal tie constraints, which prevent relative motions between the adjacent surfaces, are imposed at the beam-to-column interfaces and between the column surfaces and the continuity plates located in the nodal areas.

Tridimensional linear, thermally coupled, solid elements with reduced integration (C3D8RT) are used in the model, which are endowed with both translational and thermal degrees of freedom. Both the vertical distributed loads and the horizontal concentrated force (Figure 6.23) are modeled through uniform distributed pressures on the beam top flange and on the column external flange at the beam-to-column node, respectively.

The fire phenomenon is modeled by means of the ISO 834 standard curve (Figure 6.27), which represents the ambient temperature during the

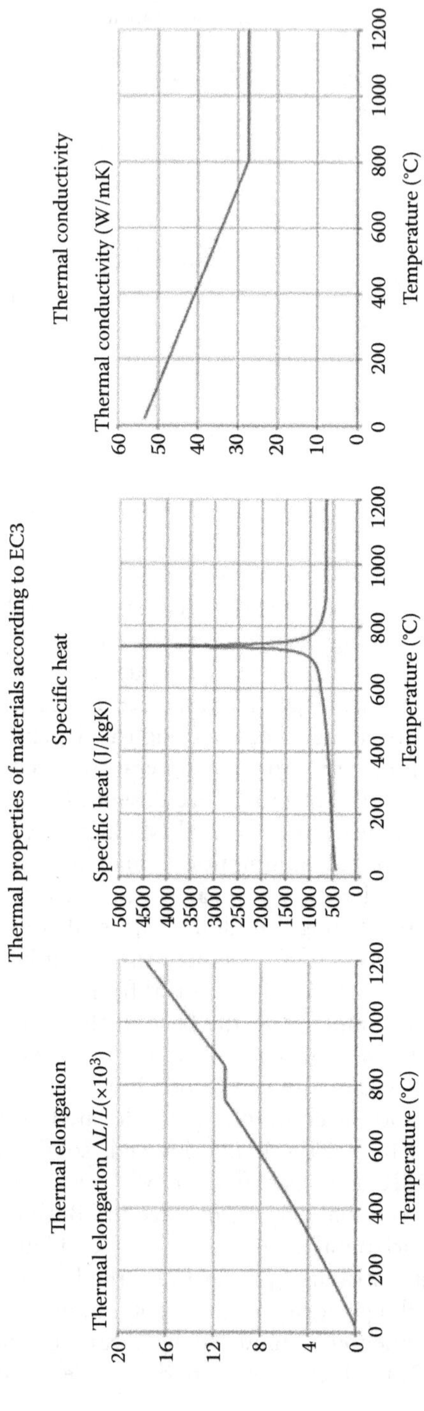

Figure 6.24 **Thermal properties for the analyzed portal frames.**

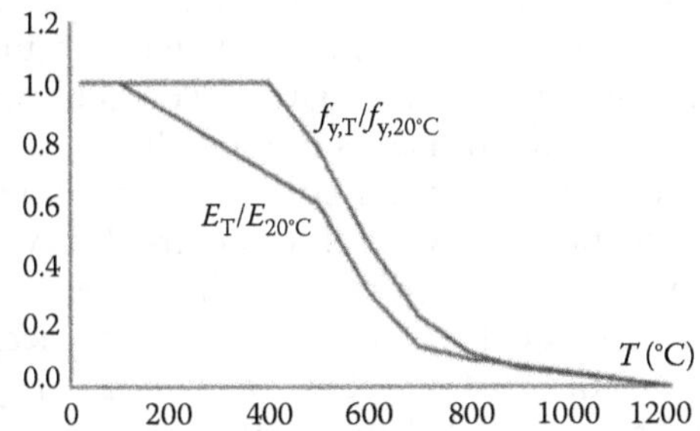

Figure 6.25 The yield point and Young's modulus variation with temperature.

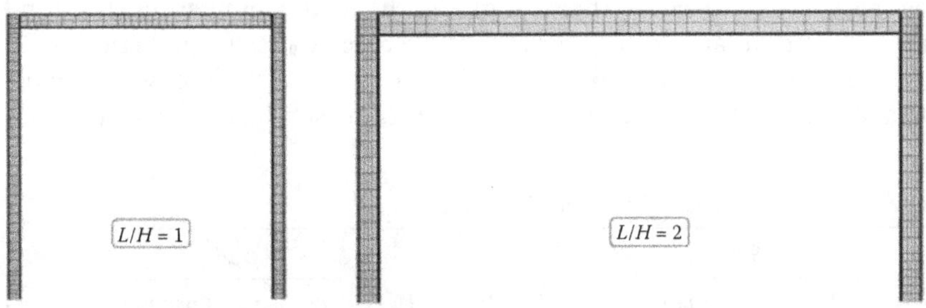

Figure 6.26 Finite-element meshes of the models.

development of the fire as a function of time. For the heat transfer phenomenon, both the radiation and the conduction mechanisms are considered. In particular, with regard to the radiation from the ambient environment, where the fire develops in the structure, all member surfaces are considered as exposed to fire. The emissivity of steel is assumed to be equal to 0.5, which is an intermediate value between the values of zinc-plated and oxidized steel. The heat transmission within the structural members is

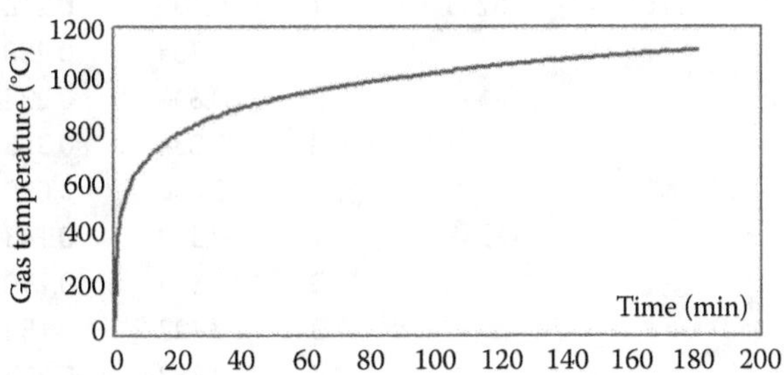

Figure 6.27 ISO 834 fire standard curve.

modeled by assigning the thermal conductivity, the mass density, and the specific heat of the material as a function of temperature, according to Eurocode 3, part 1-2 (see Figure 6.24). The initial temperature for both the ambient environment and the frames is equal to 20°C.

After being subjected to the external loads (vertical and/or horizontal ones), the analyzed portals are exposed to the fire. It is worth noting that the values of the partial safety factors used for the dead and live loads (γ_g and γ_q, respectively) are equal to 1.0, since fire is an accidental load. In both frames, a fire duration of 1 h is considered in the analyses and the fire resistance of the structures is calculated as the time necessary for reaching a 0.05 rad plastic rotation in the most engaged plastic hinge.

The results of the performed analyses are summarized in Table 6.3, where the fire resistance (R) of the portal frames is related to the L/H ratio, the massivity ratio (S/V), the steel grade, the load combination (LC), and the exploitation ratio (σ_{Sd}/f_y), for each study case. Moreover, in Figures 6.28 and 6.29, the $R - \sigma_{Sd}/f_y$ curves are plotted for each study case; the most engaged plastic hinge is evidenced on the frame schemes. The average fire

Table 6.3 Fire Resistance of the Portal Frames

	S/V				σ_{Sd}/f_y		
L/H	*Beam*	*Column*	*Steel grade*	*LC*	*Beam*	*Column*	*R (min)*
1	291	159	S235	1	0.519	0.202	15′23″
				2	0.690	0.268	13′48″
				3	0.776	0.293	15′00″
				4	0.947	0.360	13′42″
				5	0.251	0.177	27′29″
			S355	1	0.343	0.133	19′42″
				2	0.457	0.177	16′37″
				3	0.514	0.194	18′41″
				4	0.627	0.238	16′18″
				5	0.170	0.117	41′27″
2	215	116	S235	1	0.590	0.216	23′15″
				2	0.784	0.288	19′43″
				3	0.640	0.233	22′58″
				4	0.834	0.305	19′37″
				5	0.050	0.050	>60′
			S355	1	0.390	0.143	32′51″
				2	0.519	0.190	24′58″
				3	0.423	0.154	31′51″
				4	0.552	0.202	24′17″
				5	0.033	0.033	>60′

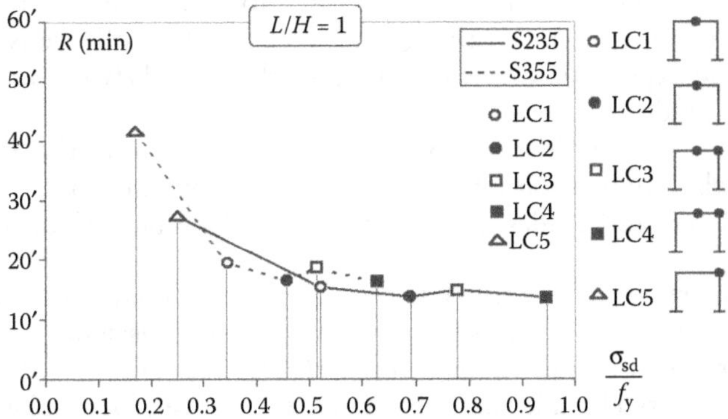

Figure 6.28 Fire resistance of the portal frames with *L*/*H* = 1.

resistance of the frames increases as the *L*/*H* ratio increases, which corresponds to the reduction of the massivity ratios in the beams and columns.

In fact, neglecting the load combination 5, for the steel grade S235, the average fire resistance ranges from 14′ 28″ (*L*/*H* = 1) to 21′ 23″ (*L*/*H* = 2), whereas for the steel grade S355, it ranges from 17′ 49″ (*L*/*H* = 1) to 28′ 29″ (*L*/*H* = 2). This was expected, since the larger the S/V ratio is, the faster the temperature increase in the members, with a consequent faster reduction of the steel mechanical properties. With regard to the steel grade influence on the fire resistance, it can be noted that, for *L*/*H* = 1, the average fire resistance ranges from 14′ 28″ (S235) to 17′ 49″ (S355), with an increase of 23%, whereas for *L*/*H* = 2, it ranges from 21′ 23″ (S235) to 28′ 29″ (S355), with an increase of 33%.

Such a result is consistent with the fact that a lower steel grade corresponds to a higher exploitation ratio, with a small amount of resistance

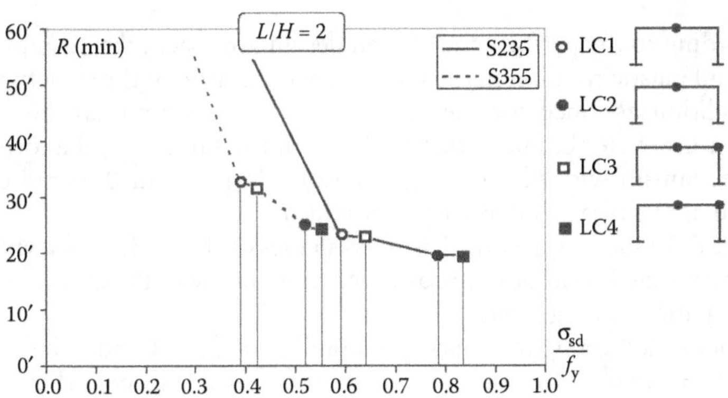

Figure 6.29 Fire resistance of the portal frames with *L*/*H* = 2.

reserves. Consequently, the time necessary for reducing the mechanical properties of the steel up to the time of reaching the collapse is lower in case of large exploitation ratios, that is, for the same *L*/*H* ratio, in case of lower steel grade.

From the considered load combinations, it can be observed that they influence both the exploitation ratio and the collapse mechanism. For different *L*/*H* ratios and steel grades, the following considerations can be made:

- An increase in the vertical live loads (from $Q_{1k} = 0.5$ kN/m^2 to $Q_{2k} = 2.0$ kN/m^2) causes a reduction in the fire resistance of the portal frames ranging from 10% ($L/H = 1$, steel grade S235) to 24% ($L/H = 2$, steel grade S355), both in the presence of vertical loads only (load combinations LC1 vs. LC2) and in the presence of vertical and horizontal loads (load combinations LC3 vs. LC4).
- The presence of the horizontal force causes a small reduction in the fire resistance of the study portal frames (load combinations LC1 vs. LC3 and LC2 vs. LC4), ranging from 0.5% ($L/H = 2$, steel grade S355) to 5% ($L/H = 1$, steel grade S355); moreover, it modifies the collapse mechanism, since in the absence of the horizontal force, the 0.05 rad plastic hinge is attained in the beam midspan section (M in Figure 6.22), whereas in the presence of the horizontal force, such plastic hinge is attained almost contemporaneously (with some seconds of difference) in the beam midspan and end sections (M and D in Figure 6.22).
- The presence of the horizontal force only (load combination 5) causes the formation of four plastic hinges in the member ends of the portal frame with $L/H = 1$, with very large values of fire resistance; in the case of $L/H = 2$, the fire resistance is larger than 1 h, due to the extremely small values of their exploitation ratios (5% and 3% for steel grades S235 and S355, respectively).

According to the performance-based design approach, the attainment of a 0.05 rad plastic rotation in the most engaged section of the structure (say, the condition assumed for the fire resistance evaluation) can be considered as a Life Safe (LS) performance level. In the same way, the formation of a mechanism with plastic hinge rotations larger than 0.08 rad can be assumed as the Near Collapse (NC) condition.

Figure 6.30 shows typically deformed shapes of the study cases where the above-mentioned conditions are attained, together with the duration of fire necessary for reaching them.

The deformed configurations corresponding to LC = 1 and 4, being similar to LC = 2 and 3, are not represented. It is possible to note that, for the considered cases, from the LS to the NC conditions there is an increment of the fire duration of about 2–3%, in the presence of vertical loads only

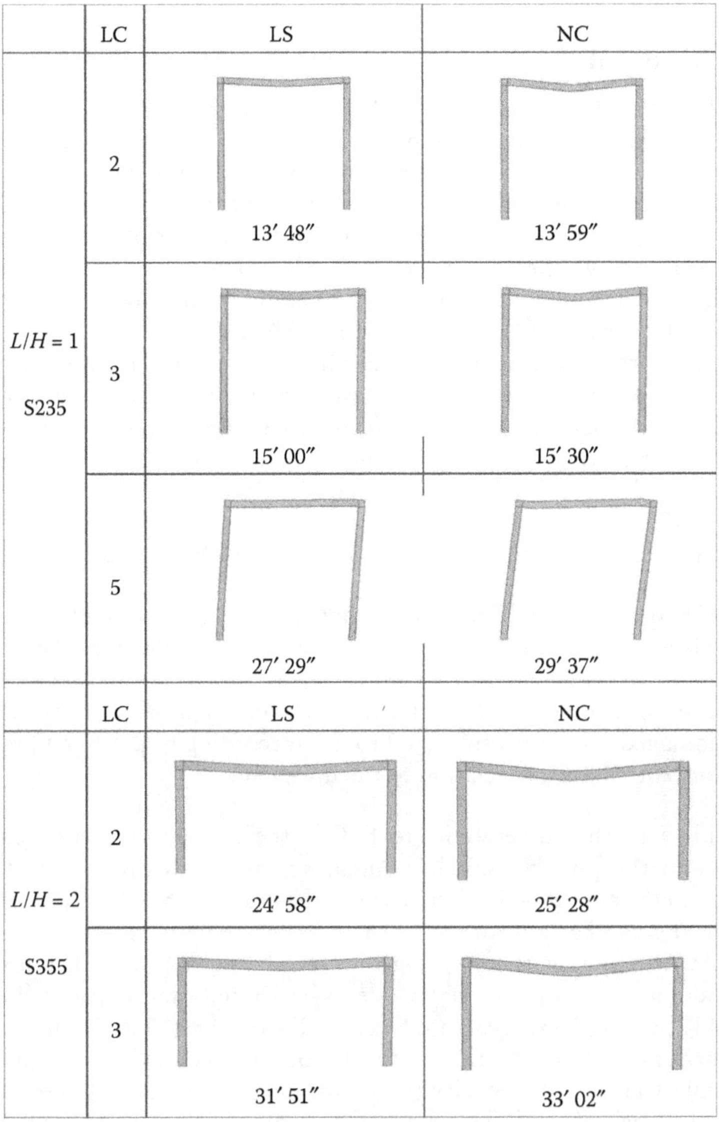

Figure 6.30 Typical deformed configurations at the LS and NC performance levels.

or vertical and horizontal ones (LC2 and LC3), whereas under horizontal loads only (LC5) such an increment is equal to 8%. Such a short time delay, from the LS to the NC conditions, confirms the severity of the fire scenario on steel structures.

6.8 METHODOLOGY FOR ASSESSING ROBUSTNESS

6.8.1 General

A methodology for the assessment of the robustness of structures subjected to FFE should be based on the application of a performance-based approach inspired from the FEMA 356 Guidelines (2000) and the philosophy of the FSE (1999). This methodology should consider each behavioral condition to be undergone by the construction, from the application of vertical service loads, through the earthquake-induced damage, up to the exposure to fire. First of all, the seismic damage state should be identified, according to pre-fixed performance levels (Faggiano and Mazzolani, 2011). Therefore, structures already damaged by an earthquake, starting from each previously defined seismic performance level, should be analyzed under fire. Correlation between the seismic performance levels and the behavior of the corresponding damaged structures under fire in terms of fire resistance and collapse mode should be determined (Della Corte et al., 2003; Faggiano et al., 2010).

The methodology should definitely consist of two main subsequent phases:

1. The identification of the seismic damage state, according to pre-fixed seismic performance levels, in relation to the intensity of the seismic event.
2. The determination of the residual bearing capacities of the seismic damaged structures subjected to fire, according to pre-fixed fire performance levels, in relation to the fire event.

In this way, the vulnerability to FFE of the existing structures can be stated and the possible suitable mitigation measures can be identified. Besides, in the case of new constructions, the design criteria and procedures against FFE can be defined.

It is worth noting that the performance-based approach has already been adopted by International Codes (the United States, Australia, the United Kingdom, New Zealand, Sweden, Eurocode system) in the field of structural fire safety. In case the former occurrence of the earthquake is a possible scenario, the novelty to be introduced is the degradation of the mechanical behavior of the structural systems due to permanent damage induced by the seism, as an initial state in the event of a fire.

First of all, the fire after an earthquake hazard should be identified, such as the probability that a fire event of a specific intensity can develop after an earthquake of a specific intensity in a built area. The probability of the occurrence of a fire after an earthquake depends on several factors, including, for example, the characteristics of the building, such as the geometry of the compartments, the conditions of ventilation, the type and amount of fuel contents, the gas and electricity wires plants, as well as the dominant meteorological conditions and, of course, the earthquake destructiveness.

Then, for every design of fire after an earthquake scenario, a structure should be designed for behaving according to the predetermined performance levels, which should integrate both seismic and fire requirements. For this purpose, it is worth considering that in case of fire, the occupants may be threatened by the effect of the behavior of the building structure, as well as by smoke and flames. This point implies considering design objectives for structures, aiming at facing the safety in terms of both the structural behavior as well as the direct effects of fires on people within the building. In this context structural systems may serve a twofold function during a fire event. On the one side, they have to continue to support loads, so that the building occupants can exit safely and the firefighters have sufficient time to react and control the fire; on the other side, they have to serve as a barrier and/or support other barriers to fire propagation (Johann et al., 2006). For example, the collapse of a floor causes the spread of a fire from one story to another; excessive deflection of a floor may contribute to the instability and failure of partition walls, with a consequent spread of fire to adjacent compartments. Accordingly, fire performance criteria for a structural system should include requirements for serviceability, for load-carrying capacity and prevention of collapse, besides time to failure requirements to allow occupant egress and suppression activities; fire containment requirements for limiting the impact of a fire at distance and prevention of room-to-room fire spread.

The fire performance levels should evidently refer to fire resistance in terms of the time up to collapse to be associated to each of the following general design objectives (Bennetts and Thomas, 2002):

1. Life safety of the occupants
2. Noninjury of occupants
3. Life safety of firefighters
4. Noninjury of firefighters
5. Prevention of damage to contents
6. Avoidance of damage to process
7. Prevention of damage to building
8. Prevention of collapse of building

Therefore, it is necessary to join the required fire resistance to the level and extent of both structural and nonstructural damage for each structural type.

Corresponding to every fire after earthquake performance levels, the optimal solution for the application of fire preventive or protective measures should be provided, such as (Purkiss, 2007)

1. Control of ignition, checking the flammability of materials within the structure
2. Control of means of escape, imposing statutory requirements on provision of suitable escape facilities
3. Detection of fire at the earliest possible stage through the installation of ad hoc systems
4. Control of the spread of fire within the building or to adjacent properties, through compartments, venting, smokescreens, or sprinklers
5. Prevention of local or global structure collapse

These issues are particularly appropriate for the analyses of constructions of particular strategic or historic–monumental importance. In the first case (hospitals, police station, government buildings), it is necessary to also guarantee performance levels able to assure the usability of all functions of the buildings under effect of a fire after seism. In the second case (churches, museums, villas, palaces, theaters), the safety measures must meet the conservation requirements relevant to the historic–artistic heritage. For both cases, the standard seismic or fire regulations cannot be applied. In this perspective, the development of the analysis must be directed to individuate and quantify both the fire scenario and the fire performance levels, which are peculiar for each particular class of buildings.

6.8.2 Case study and seismic performance characterization

For the sake of understanding (Faggiano and Mazzolani, 2011a,b), the procedure is illustrated with reference to a simple steel MRF structure made of S275 steel, whose geometrical features are shown in Figure 6.31. Both equivalent static incremental seismic analyses and fire analyses are carried out by means of the ABAQUS Ver. 6.5 software (2004a,b).

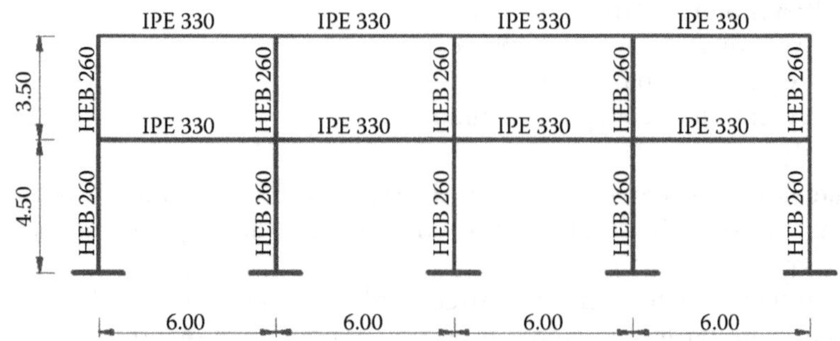

Figure 6.31 Steel MRF structure analyzed as case study [m].

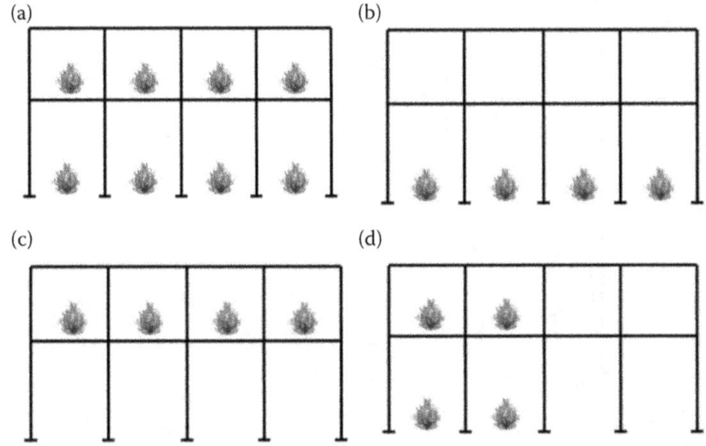

Figure 6.32 Considered fire locations: (a) fire on the whole structure; (b) fire at the ground level; (c) fire at the first level; (d) fire on the half structure.

In particular, the following four fire locations are assumed (Figure 6.32):

1. Fire on the whole structure (Figure 6.32a)
2. Fire at the ground level (Figure 6.32b)
3. Fire at the first level (Figure 6.32c)
4. Fire on the half structure (Figure 6.32d)

The seismic performance levels for a steel structure can be characterized by the extent of the interstory drift δ/h (where δ is the lateral displacement and h is the story height) and the plastic hinge rotation θ according to the FEMA 356 Guidelines. The reference values indicated in Table 6.4 are assumed. In particular, two Collapse Prevention levels are considered: (CP1) with an interstory drift equal to 5.0% and (CP2) with a plastic hinge rotation equal to 0.05 rad.

The seismic behavior of the structure is determined through a nonlinear pushover static analysis. The obtained pushover curve is shown in

Table 6.4 Reference Seismic Performance Levels of the Steel Structure

Performance level		*δ/h (%)*	*θ (rad)*
Immediate occupancy	IO	0.7	—
Life safety	LS	2.5	—
Collapse prevention 1	CP1	5.0	—
Collapse prevention 2	CP2	—	0.05

Source: Adapted from FEMA 356 2000: Guidelines for seismic rehabilitation of building.

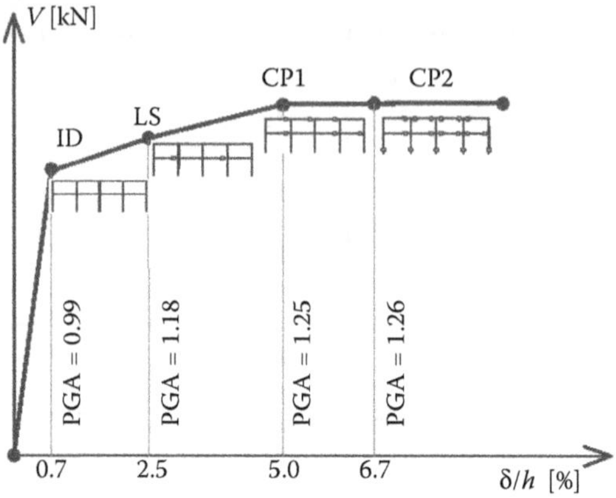

Figure 6.33 Pushover curve with the performance levels, according to FEMA 356.

Figure 6.33, where the performance levels are evidenced, in relation to the corresponding calculated PGA, together with the plastic hinge distributions.

For the sake of simplicity, the performance levels in fire are identified with reference to structural damage only according to the following definitions, which are strictly pertinent to steel structures (Figure 6.34):

- *Operational fire* (Of): Attainment of the yield stress in the most stressed section (Figure 6.34a)
- *Life Safe fire* (LSf): Formation of the first plastic hinge (Figure 6.34b)
- *Section Collapse fire* (CSf): failure of the cross-section (Figure 6.34c)
- *Local Collapse fire* (CLf): formation of the beam mechanism (Figure 6.34d)
- *Global Collapse fire* (CGf): formation of the global mechanism (Figure 6.34e)

Fire analyses are performed once the seismic analyses are already carried out and the fixed seismic performance levels reached. The fire load

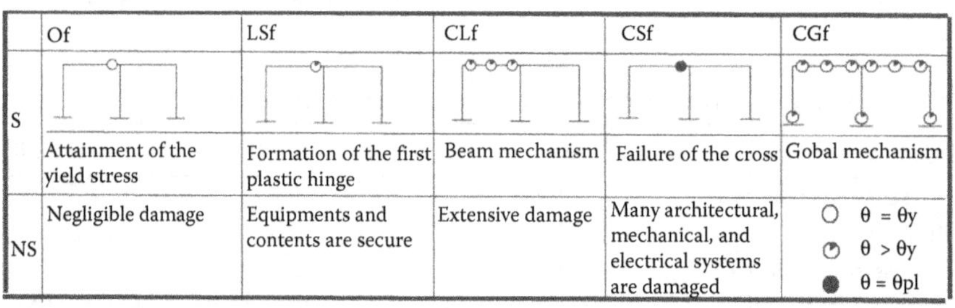

Figure 6.34 Fire after earthquake performance levels.

has been modeled through the ISO 834 time–temperature nominal curve (Figure 6.27).

With reference to the fire locations considered in Figure 6.32, the results, in terms of the time (s) necessary to reach the different performance fire levels, starting from the predetermined seismic performance levels IO, LS, CP1, and CP2 given in Table 6.5. The same results are illustrated through a three-dimensional histogram in Figure 6.35, with reference to the condition of fire on the half structure only (Figure 6.32d).

The comprehensive chart in Figure 6.35 gives an overall evidence of the FFE performances. It seems to be a very useful and powerful representation and tool both for fire after earthquake capability analysis and for fire after earthquake design. In fact for a given structural type, given a fire scenario, once fixed the seismic damage extent corresponding to the design seismic performance level, it is possible either to carry out a fire performance capability analysis with reference to the pre-fixed fire performance levels, or to design the structure in fire in order that it could reach the fire performance level required at the given acceptable time.

From Table 6.5 and Figure 6.35, it is apparent that as far as the seismic damage is large (from IO to CP2), the fixed fire performance level is reached in a shorter time.

Table 6.5 Times (s) to Reach a Fire Performance Level (Of, LSf, CLf, CSf, CGf) from a Seismic Damage State (IO, LS, CPI, CP)

Fire location	*Seismic level*	*Fire performance level*				
		Of	*LSf*	*CLf*	*CSF*	*CGf*
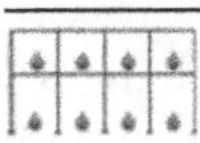	IO	387	1304	1327	1408	1568
	LS	0	0	1327	1408	1578
	CPI	0	0	1341	1407	1590
	CP2	0	0	1046	923	1292
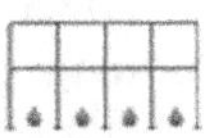	IO	310	1171	1192	1240	1490
	LS	0	0	1198	1235	1461
	CPI	0	0	1210	1159	1441
	CP2	0	0	1159	1034	1308
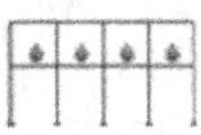	IO	361	1198	1457	1468	1510
	LS	0	0	1457	1468	1709
	CPI	0	0	1435	1438	2552
	CP2	0	0	1426	1110	1510
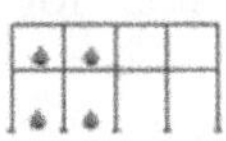	IO	387	1382	1451	1587	3076
	LS	0	0	1446	1574	3015
	CPI	0	0	1418	1544	2340
	CP2	0	0	938	1168	1980

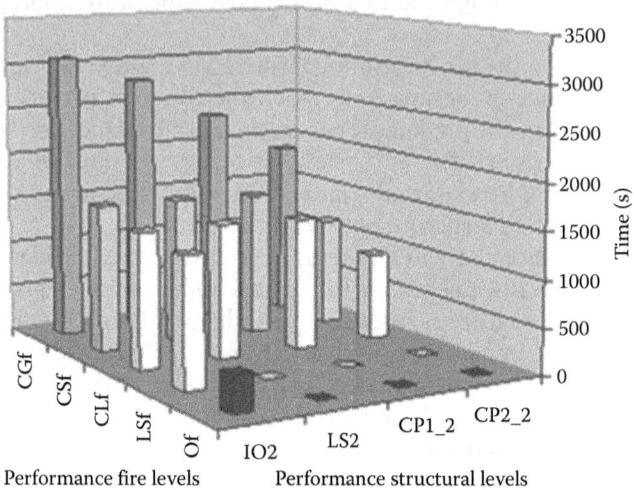

Figure 6.35 Fire after earthquake performance chart relating to the fire on the half structure.

6.9 CONCLUSIVE REMARKS

Fire after an earthquake is earthquake-related hazard causing the greatest concern. As a consequence, the analysis of the behavior of structures under this kind of scenario is a significant research field, but it is not yet fully explored. The high probability of occurrence of fires in civil and industrial buildings after being struck by a seismic shake justifies the need to investigate the response to high temperatures of the structural systems, owing to their already being in a state of irreversible deformed configuration due to the earthquake.

In order to primarily take into account in the design phases the effect of the combination of the seism and fire accidental loads, a methodology aimed at the robustness assessment under fire of structures already damaged to different extents by the earthquake, through a performance-based approach, is required. The procedure should be valuable as a design tool, as well as a vulnerability evaluation of the built heritage against FFE, in all seismic-prone areas, as it is envisaged for buildings of high strategic importance. It should distinguish between structures located in seismic and nonseismic areas, by requiring more stringent fire resistance provisions for those buildings potentially subjected to seismic actions.

The performance-based methodological approach for the robustness assessment of structures damaged by an earthquake and then subjected to fire seems to be the most appropriate analysis tool.

This methodology is particularly appropriate for the analyses of constructions of particular strategic or historic–monumental importance. In the first case (hospitals, police stations, government buildings), it is necessary to guarantee that performance levels are also able to assure the usability of all functions of the buildings under the effect of fire after earthquake. In the second case (churches, museums, villas, palaces, theaters), the safety measures must be balanced with conservation requirements of historic–artistic heritage. So for both cases, the standard seismic or fire regulations cannot be applied separately. In this perspective, for the future, the development of the research must be directed to individuate and quantify the fire performance levels peculiar to the particular classes of buildings.

However, in view of the development of a comprehensive methodology of performance-based design for buildings, the fix-safety codes should distinguish between structures located in seismic and nonseismic areas, by adopting more stringent fire resistant provisions for those building potentially subjected to seismic actions (Buchanan, 1994; Robinson, 1998; Wang and Kodur, 2000).

REFERENCES

ABAQUS. 2004a: ABAQUS Analysis User's Manual, ver. 6.5.

ABAQUS. 2004b: ABAQUS Standard User's Manual, ver. 6.5.

Anderberg, Y. 1988: Modelling steel behaviour. *Fire Safety Journal*, 13(1), 17–26.

Bennetts, I.D., Thomas, I.R. 2002: Design of steel structures under fire conditions. *Journal of Progress in Structural Engineering Materials*, 4, 6–17.

Bianco, L. 2006: Effects of the fire on structural material (in Italian). www.buildup.it.

Botting, R. 1998: The impact of post-earthquake fire on the built urban environment, Fire Engineering Research Report 98/1, University of Canterbury, New Zealand.

Britton, N.R., Clark, G.C. 1999: Emergency management and insurance: Toward a collaborate approach. In *The Changing Risk Landscape* (ed. N.R. Britton), Southwood Press, Sydney, 219–238.

Buchanan, A.H. 1994: Fire engineering for a performance based code. *Fire Safety Journal*, 23(1), 1–16.

Buchanan, A.H. 2001: *Structural Design for Fire Safety*. John Wiley & Sons, Chichester, England.

CEN, EN 1991-2-2: Eurocode 1: Basis of design and actions on structures—Part 2-2: Actions.

CEN, EN 1993-1-2: Eurocode 3: Design of steel structures—Part 1.2: General rules—Structural fire design.

CEN, EN 1998: Eurocode 8: Design provisions for earthquake resistance of structures.

Chen, S., Lee, G.C., Shinozuka, M. 2004: Hazard mitigation for earthquake and subsequent fire. *ANCER Annual Meeting*, Honolulu, Hawaii.

Cousins, W.J., Dowrick, D.J., Sritharan, S. 1991: Fire following earthquake. In *Proceedings of the Institution of Fire Engineers Conference*, New Plymouth.

Cousins, W.J., Thomas, G.C., Heron, D.W. Mazzoni, S., Lloydd, D. 2002: Modelling the spread of post-earthquake fire in Wellington city. In *Proceedings of the 2002 Technical Conference and AGM*, New Zealand Society for Earthquake Engineering, Napier.

Della Corte, G., Faggiano, B., Mazzolani, F.M. 2005: On the structural effects of the fire following earthquake. In *Proceedings of the Final Conference COST C12 "Improving Buildings' Structural Quality by New Technologies,"* Innsbruck, Austria.

Della Corte, G., Landolfo, R. 2001: Post-earthquake fire resistance of steel structures. In *Safety and Reliability, Towards a Safer World* (*Proceedings of the European Conference on Safety and Reliability—ESREL 2001*) (eds. E. Zio, M. Demichela, N. Piccinini), Politecnico di Torino, 3, 1739–1746.

Della Corte, G., Landolfo, R., Mazzolani, F.M. 2003: Post-earthquake fire resistance of moment-resisting steel frames. *Fire Safety Journal*, 38, 593–612.

Della Corte, G., Mazzolani, F.M. 2002: Seismic stability of steel frames. In *Proceedings of the International Colloquium on Stability and Ductility of Steel Structures* (ed. M. Ivanyi), September, Budapest.

EQE International 1995: The January 17, 1995 Kobe Earthquake—An EQE Summary Report. April.

Faggiano, B., De Gregorio, D., Mazzolani, F.M. 2010: Assessment of the robustness of structures subjected to fire following earthquake through a performance-based approach. In *Proceedings of the International Conference on Urban Habitat Constructions under Catastrophic Events (COST C26 Action)*, Naples, Italy, 207–212.

Faggiano, B., Della Corte, G., Mazzolani, F.M., Landolfo, R. 2005: Post-earthquake fire resistance of moment resisting steel frames. In *Proceedings of the Eurosteel Conference on Steel and Composite Structures*, Maastricht, The Netherlands.

Faggiano, B., Esposto, M., Mazzolani, F.M. 2007a: Fully coupled temperature-displacements analyses of steel portal frames under fire. In *Proceedings of the Italian Conference C.T.A.*, Catania, 1–3 October.

Faggiano, B., Esposto, M., Mazzolani, F.M. 2008: Risk assessment of steel structures under fire. In *14th World Conference on Earthquake Engineering*, Beijing, China, Paper No. S19-015.

Faggiano, B., Esposto, M., Mazzolani, F.M., Landolfo, R. 2007b: Fire analysis on steel portal frames damaged after earthquake according to performance based design. In *Proceedings Workshop Urban Habitat Constructions under Catastrophic Events (COST C26 Action)*, Prague, Czech Republic, 35–40.

Faggiano, B., Mazzolani, F.M. 2011a: Two-scale risk management in case of fire after earthquake. In *Proceedings of the 3rd International Workshop on Performance, Protection & Strengthening of Structures under Extreme Loading (PROTECT 2011)*, Lugano, Switzerland, 30 August–01 September.

Faggiano, B., Mazzolani, F.M. 2011b: Fire after earthquake robustness evaluation and design: Application to steel structures. In *Proceedings of EUROSTEEL 2011 Conference*, Budapest, August 31– September 2.

Fajfar, P. 2002: Structural analysis for earthquake engineering—A breakthrough of simplified non-linear methods. In *Proceedings of the 12th European Conference on Earthquake Engineering*, Keynote Lecture, London, Elsevier Science Ltd, Oxford, UK, CD-ROM.

Feasey, R., Buchanan, A. 2002: Post-flashover fires for structural design. *Fire Safety Journal*, 37, 83–105.

FEMA (Federal Emergency Management Agency) 1999: HAZUS Earthquake Loss Estimation Methodology, Technical Manual, USA.

FEMA 356 2000: Guidelines for seismic rehabilitation of building. Federal Emergency Management Agency, USA.

FEMA 350 2001: Seismic design criteria for new moment-resisting steel frame construction. Federal Emergency Management Agency, USA.

Franssen, J.M., Kodur, V.K.R., Mason, J. 2002: User's manual for SAFIR 2001 free—A computer program for analysis of structures submitted to the fire. University of Liege, Department: Structures du Génie Civil, Service: Ponts et Charpentes.

Hamburger, R.O., Foutch, D.A., Cornell, C.A. 2000: Performance basis of guidelines for evaluation, upgrade and design of moment-resisting steel frames. In *Proceedings of the 12th World Conference on Earthquake Engineering*, Auckland, New Zealand, Paper No. 2543, CD-ROM.

ISO/TR 13387 1999: Fire safety engineering, Technical Report, ISO TC 92/SC 4.

Johann, M.A., Albano, L.D., Fitzgerald, R.W., Meacham, B.J. 2006: Performance-based structural fire safety. *ASCE Journal of Performance of Constructed Facilities*, 20(1), 45–53.

Krawinkler, H. 2000: System performance of steel moment resisting frame structures. In *Proceedings of the 12th World Conference on Earthquake Engineering*, Auckland, New Zealand, Paper No. 2545, CD-ROM.

Mazzoni, S., McKenna, F., Scott, M., Fenves, G.L., Jeremic, B. 2003: Command Language Manual—Open System for Earthquake Engineering Simulation (OpenSees).

Purkiss, J.A. 2007: *Fire Safety Engineering. Design of Structures*, Butterworth-Heinemann, Elsevier.

Robinson, J. 1998: Fire—A technical challenge and a market opportunity. *Journal of Constructional Steel Research*, 46, 1–3, Paper No. 179.

Scawthorn, C. 1986: Simulation modelling of fire following earthquake. In *Proceedings of the Third US National Conference on Earthquake Engineering*, Charleston, August.

Scawthorn, C. 1987: Fire following earthquake: Estimates of the conflagration risk to insured property in greater Los Angeles and San Francisco. *All-Industry Research Advisory Council*. Oak Brook, Illinois.

Scawthorn, C., Eidinger, J.M., Schiff, A.J. 2005: *Fire Following Earthquake*. ASCE Publications. ISBN: 0784407398.

Scawthorn, C., Iemura, H., Yamada, Y. 1991: Model for urban post-earthquake fire hazard. *Disaster: The International Journal of Disaster Studies and Practice, London*, 5(2), 125–132.

Standards Australia and Standards New Zealand 1999: Risk Management AS/NZS4360, Sydney and Wellington.

Wang, Y.C., Kodur, V.K.R. 2000: Research toward use of unprotected steel structures. *ASCE, Journal of Structural Engineering*, 126(12), 1442–1450.

Wellington Lifelines Group 2002: Fire Following Earthquake: Identifying key issues for New Zealand.

Williamson, R.B., Groner, R. 2000: Ignition of fires following earthquakes associated with natural gas and electric distribution systems. Pacific Earthquake Engineering Research Centre, University of California.

Index

A

D

E

F

G

M

N

O

P

T

For Product Safety Concerns and Information please contact our EU
representative GPSR@taylorandfrancis.com
Taylor & Francis Verlag GmbH, Kaufingerstraße 24, 80331 München, Germany

www.ingramcontent.com/pod-product-compliance
Lightning Source LLC
LaVergne TN
LVHW011101300926
849160LV00008B/248
9781138075375